ROBUST CHAOS AND ITS APPLICATIONS

WORLD SCIENTIFIC SERIES ON NONLINEAR SCIENCE

Editor: Leon O. Chua
University of California, Berkeley

Series A.   MONOGRAPHS AND TREATISES*

Volume 63: Advanced Topics on Cellular Self-Organizing Nets and Chaotic Nonlinear Dynamics to Model and Control Complex Systems
R. Caponetto, L. Fortuna & M. Frasca

Volume 64: Control of Chaos in Nonlinear Circuits and Systems
B. W.-K. Ling, H. H.-C. Lu & H. K. Lam

Volume 65: Chua's Circuit Implementations: Yesterday, Today and Tomorrow
L. Fortuna, M. Frasca & M. G. Xibilia

Volume 66: Differential Geometry Applied to Dynamical Systems
J.-M. Ginoux

Volume 67: Determining Thresholds of Complete Synchronization, and Application
A. Stefanski

Volume 68: A Nonlinear Dynamics Perspective of Wolfram's New Kind of Science (Volume III)
L. O. Chua

Volume 69: Modeling by Nonlinear Differential Equations
P. E. Phillipson & P. Schuster

Volume 70: Bifurcations in Piecewise-Smooth Continuous Systems
D. J. Warwick Simpson

Volume 71: A Practical Guide for Studying Chua's Circuits
R. Kiliç

Volume 72: Fractional Order Systems: Modeling and Control Applications
R. Caponetto, G. Dongola, L. Fortuna & I. Petráš

Volume 73: 2-D Quadratic Maps and 3-D ODE Systems: A Rigorous Approach
E. Zeraoulia & J. C. Sprott

Volume 74: Physarum Machines: Computers from Slime Mould
A. Adamatzky

Volume 75: Discrete Systems with Memory
R. Alonso-Sanz

Volume 76: A Nonlinear Dynamics Perspective of Wolfram's New Kind of Science (Volume IV)
L. O. Chua

Volume 77: Mathematical Mechanics: From Particle to Muscle
E. D. Cooper

Volume 78: Qualitative and Asymptotic Analysis of Differential Equations with Random Perturbations
A. M. Samoilenko & O. Stanzhytskyi

Volume 79: Robust Chaos and Its Applications
Z. Elhadj & J. C. Sprott

*To view the complete list of the published volumes in the series, please visit:
http://www.worldscibooks.com/series/wssnsa_series.shtml

WORLD SCIENTIFIC SERIES ON NONLINEAR SCIENCE
Series Editor: Leon O. Chua

Series A   Vol. 79

ROBUST CHAOS AND ITS APPLICATIONS

Elhadj Zeraoulia
University of Tébessa, Algeria

Julien Clinton Sprott
University of Wisconsin-Madison, USA

World Scientific

NEW JERSEY · LONDON · SINGAPORE · BEIJING · SHANGHAI · HONG KONG · TAIPEI · CHENNAI

Published by

World Scientific Publishing Co. Pte. Ltd.
5 Toh Tuck Link, Singapore 596224
USA office: 27 Warren Street, Suite 401-402, Hackensack, NJ 07601
UK office: 57 Shelton Street, Covent Garden, London WC2H 9HE

British Library Cataloguing-in-Publication Data
A catalogue record for this book is available from the British Library.

World Scientific Series on Nonlinear Science, Series A — Vol. 79
ROBUST CHAOS AND ITS APPLICATIONS

Copyright © 2012 by World Scientific Publishing Co. Pte. Ltd.

All rights reserved. This book, or parts thereof, may not be reproduced in any form or by any means, electronic or mechanical, including photocopying, recording or any information storage and retrieval system now known or to be invented, without written permission from the Publisher.

For photocopying of material in this volume, please pay a copying fee through the Copyright Clearance Center, Inc., 222 Rosewood Drive, Danvers, MA 01923, USA. In this case permission to photocopy is not required from the publisher.

ISBN-13 978-981-4374-07-1
ISBN-10 981-4374-07-5

Printed in Singapore by B & Jo Enterprise Pte Ltd

To my family: My wife Nadjette and my father Tayeb and my brothers Zohir, Samra, Mourad, Saida, and Bachlila and to all who helped me write this book.

Elhadj

Preface

Robust chaos is defined by the absence of periodic windows and coexisting attractors in some neighborhood in the parameter space of a dynamical system. Today, there are many books devoted to nonlinear dynamics and chaos, but so far as we know, there are no books dedicated to the concept of robust chaos and its applications. This unique book explores the definition, sources, and roles of robust chaos. The book is written in a reasonably self-contained manner and aims to provide students and researchers with the necessary understanding of the subject. Most of the known results, experiments, and conjectures about chaos in general and about robust chaos in particular are collected here in a pedagogical form. Many examples of dynamical systems, ranging from purely mathematical to natural and social processes displaying robust chaos, are discussed in detail. At the end of each chapter is a set of exercises and open problems intended to reinforce the ideas and provide additional experience for both readers and researchers in nonlinear science in general and chaos theory in particular.

Robust chaos and its applications is a textbook devoted to the understanding and prediction of robust chaos in real-world dynamical systems. This book contains 260 exercises in different topics of chaos theory and dynamical systems.

Let us briefly characterize the content of each chapter:

Chapter 1. Poincaré map technique, Smale horseshoe, and symbolic dynamics. In this chapter we give an overview of the Poincaré map technique used to describe bifurcations and chaos in dynamical systems. For example, all the content of Chapter 7 is based on the construction of a Poincaré map to confirm the existence of robust chaos in such a continuous-time dynamical system. We first define the Poincaré map and the generalized Poincaré mappings for a continuous-time dynamical system with a continuous, piecewise-linear vector field. Then we discuss interval methods for calculating Poincaré maps including several methods used to prove the existence of periodic orbits using a generalized Poincaré map associated with the continuous-time flow under consideration. Secondly, we discuss the Smale horseshoe as a basic tool for studying chaos and its properties such as symbolic dynamics. In particular, the so-called *method of fixed point index* is described as a way to prove the existence of such horseshoe-type mappings in a dynamical system. This method is based on the periodic points of the so-called TS-*maps* and the existence of semi-conjugacy.

Chapter 2. Robustness of chaos. In this chapter we discuss the concept of robust chaos. First, we describe the notion of strange attractors with their topological properties, in particular the *domains of attraction* that play a crucial role in characterizing robust chaos. Secondly, we give several definitions of the word *robust* and discuss its relation to density and persistence.

Chapter 3. Statistical properties of chaotic attractors. In this chapter we discuss statistical properties of chaotic attractors with their different types of entropies along with a detailed description of the so-called *ergodic theory*. Then we present other statistical properties of chaotic attractors such as the rate of decay of correlations, the central limit theorem, and other probabilistic limit theorems. These notions are based on the correlation and autocorrelation function (ACF), which play a crucial role in characterizing hyperbolic (robust) systems.

Chapter 4. Structural stability. In this chapter we discuss the concept of structural stability as a criterion for robustness of invariant sets of dynamical systems. In particular, we define and state important properties of this notion, along with its conditions. As an example, we state and discuss in some detail a proof of Anosov's theorem on structural stability of diffeomorphisms of a compact C^∞ manifold M without boundary due to Robinson and Verjovsky, along with a weaker version called Ω-*stability*.

Chapter 5. Transversality, Invariant foliation, and the Shadowing Lemma. In this chapter we first discuss some relevant properties and the importance of transversality, invariant foliation, and the shadowing lemma in analyzing chaotic dynamical systems in general, and robust ones in particular. Indeed, transversality is the opposite of the tangency that is responsible for the coexistence of attractors. Secondly, we present and discuss the notion of the invariant foliation that is used to prove chaos and its topological properties in the singular-hyperbolic attractors. Thirdly, we present the shadowing lemma that is used to verify the existence of a true chaotic attractor. In particular, a robust chaotic attractor is more important than one obtained without verification. The possible relations between homoclinic orbits and shadowing are given along with some Shilnikov theorems concerning criteria for the existence of chaos.

Chapter 6. Chaotic attractors with hyperbolic structure. In this chapter we discuss the most interesting results concerning chaotic attractors with a hyperbolic structure. First, hyperbolic dynamics is discussed along with some concepts and definitions. Then we present Anosov diffeomorphisms and Anosov flows as typical examples of invariant sets with hyperbolic structure. The problem of classifying strange attractors of dynamical systems, which is based on rigorous mathematical analysis and is not accepted as significant from an experimental point of view, is given in some detail. Some properties of hyperbolic chaotic attractors are described with specific examples having hyperbolic structure. These include geodesic flows on compact smooth manifolds, the Smale–Williams solenoid, and Anosov diffeomorphisms on a torus, in particular Anosov automorphisms and their structure. Furthermore, expanding maps are presented because of their importance, *i.e.*, they are expanding solenoids or hyperbolic attractors. Other examples are the Blaschke product, the Arnold cat map, the Plykin attractor, and the Bernoulli map. As an example showing the required manipulations, we give a proof of the hyperbolicity of the logistic map for parameter values greater than 4. Principal properties of the so-called *generalized hyperbolic attractors* are discussed in some detail. Two methods for generating hyperbolic attractors, in particular, the transition where a periodic trajectory disappears from the Morse–Smale system to systems with hyperbolic attractors, are presented along with a discussion of the density of hyperbolicity and homoclinic bifurcations in arbitrary dimension. Finally, some hyperbolicity tests are presented and applied to the Hénon map and to the forced damped pendulum.

Chapter 7. Robust chaos in hyperbolic systems. In this chapter we give several examples of realistic models describing hyperbolic (robust) chaos. Indeed, we present a method for the construction of the Smale–Williams attractor for a three-dimensional map along with a theoretical and numerical verification of its hyperbolicity. An example of a flow system is presented with an attractor concentrated mostly at the surface of a two-dimensional torus, the dynamics of which is governed by the Arnold cat map. A model for the Bernoulli map is constructed using the Poincaré map method discussed in Chapter 1 with a hyperchaotic attractor (one with two positive Lyapunov exponents) in a perturbed heteroclinic cycle. The Poincaré map defined by a 3-D flow constructed as a bursting neuron model and exhibiting a Plykin-like attractor (a strange hyperbolic attractor) is also given and discussed.

Chapter 8. Lorenz-type systems. In this chapter, we present some recent results about Lorenz-type systems. Indeed, an overview outlining the major properties of these types of systems along with one definition resulting from observations of the dynamics of the standard Lorenz system is presented from the viewpoint of its existence and structure. Another example of Lorenz-type systems, *i.e.*, expanding and contracting Lorenz attractors is presented and discussed. Pseudo-hyperbolic theory is also presented in some detail. In particular, we discuss the essence of *wild strange attractors* along with their dynamical properties. Finally, one example of a Lorenz-type attractor (the Lozi map) realized in two-dimensional maps is presented with some properties and notes.

Chapter 9. Robust chaos in the Lorenz-type systems. In this chapter we discuss robust chaos in the Lorenz-type system and give an overview of the classical results concerning robust Lorenz-type attractors. The well-known results on structural stability and robustness of the standard Lorenz attractor are listed to clarify the similarity to the hyperbolic systems introduced in Chapter 6. Finally, robust chaos in the 2-D Lozi discrete mapping as an example of Lorenz-type attractors is discussed using the results from Chapters 3, 6, and 8.

Chapter 10. No robust chaos in quasi-attractors. In this chapter we discuss some relevant results and properties of quasi-attractors as the third type in the classification of strange attractors of dynamical systems. In particular, we give evidence for the existence of transversal homoclinic points that prove this map is a quasi-attractor, also the uniform hyperbolicity, which is discussed for both the Hénon map and other Hénon-like maps. Using the procedure described in Sec. 6.9.1, we give a proof that the Hénon attractor is a quasi-attractor. Other examples with quasi-attractors include the so-called *Strelkova–Anishchenko map*, the *Anishchenko–Astakhov oscillator*, and *Chua circuit*, which is a famous example of a system that displays quasi-attractors with homoclinic and heteroclinic orbits and double scrolls. The so-called *geometric model* of this circuit is also introduced in some detail. The distinction of this model is that it is different from the Lorenz-type presented in Chapter 8 or the quasi-attractors discussed in this chapter. This new type contains unstable points in addition to the Cantor set structure of hyperbolic points. In particular, the corresponding two-dimensional Poincaré map has a strange attractor with no stable orbits.

Chapter 11. Robust chaos in one-dimensional maps. In this chapter we give several methods to generate robust chaos in one-dimensional maps. The most mathematical tool used here is the unimodality of real functions in an interval $I \subset \mathbb{R}$ because the most important topological property of a unimodal map is that it stretches and folds the interval I into itself. The importance of S-unimodal maps and the relation between unimodality and hyperbolicity are established along with the classification of unimodal maps of the interval and their statistical properties. The concept of unimodal Collet–Eckmann maps with their properties and their relation to hyperbolicity is also presented and discussed. Some counterexamples are presented for a conjecture of Barreto–Hunt–Grebogi–Yorke claiming that robust chaos cannot appear in smooth systems. Finally, the relation between border-collision bifurcations and robust chaos is discussed in some detail.

Chapter 12. Robust chaos in 2-D piecewise-smooth maps. In this chapter we discuss robust chaos in two-dimensional piecewise-smooth maps, where we give the normal form of these maps to facilitate the study of their bifurcations and chaos. Details about bifurcations and chaos obtained using such a normal form are presented, including general classifications and different routes to chaos and regions in the parameter space with and without robust chaos. To make clear the different manipulations used to prove robust chaos in this type of map, we give three proofs (with several examples) for the unicity of the orbits for the normal form, for robustness of chaos, and for robust chaos in two-dimensional, noninvertible, piecewise-linear maps.

We would like to thank Prof. Soumitro Banerjee (Department of Electrical Engineering, IIT Kharagpur, India) for valuable discussions and for the list of open problems at the end of Chapters 11 and 12.

We would like to thank Prof. Zin Arai (Creative Research Initiative "Sousei", Hokkaido University, N21W10 Kita-ku, Sapporo, 001-0021, Japan) for offering some of his papers and figures.

We would like to thank Prof. Eyad H. Abed (V. P. for Financial Activities, IEEE Control Systems Society, Professor and Director, Institute for Systems Research, Professor, Electrical and Computer Engineering University of Maryland College Park, MD 20742 USA) for valuable discussions.

We would like to thank the following scientists for offering some of their papers concerned with the topics in this book: R. Clark Robinson

(Department of Mathematics, Northwestern University, Evanston IL 60208), Carlos Arnoldo Morales (Instituto de Matemática Pura e Aplicada, Universidade Federal do Rio de Janeiro, Brazil), Lorenzo J. Diaz Casado (Departamento de Matemática PUC-Rio, Marquês de São Vicente 225, Gávea, Rio de Janeiro 225453-900, Brazil), Guanrong Chen (Department of Electronic Engineering, City University of Hong Kong, 83 Tat Chee Avenue, Kowloon, Hong Kong SAR, P. R. China), Zbigniew Galias (Department of Electrical Engineering, AGH-University of Science and Technology, al. Mickiewicza 3030-059 Kraków, Poland), Shaobo Gan (School of Mathematical Science, Peking University, Beijing 100871, China), Juan M. Aguirregabiria (Department of Theoretical Physics, University of the Basque Country, P. O. Box 644, 48080 Bilbao, Spain), Ying-Cheng Lai (Department of Electrical Engineering, Department of Physics, Arizona State University, Tempe, Arizona 85287-5706, USA), Aubin Arroyo (Unidad Cuernavaca, Instituto de Matematicas, Universidad Nacional Autonoma de Mexico), Artur Oscar Lopes (Instituto de Matemática Universidade, Federal do Rio Grande do Sul, Av. Bento Gonçalves, 9500, 91509-900 Porto Alegre, RS - Brazil), Ulrich Hoensch (Rocky Mountain College, Mathematics and Computer Science), Mukul Majumdar (460 Uris Hall, Cornell University, Ithaca, New York 14853, USA), Yichao Chen (Department of Mechanical Engineering, University of Houston, Houston, Texas 77204-4006, USA), Mark Pollicott (Professor of Mathematics, Mathematics Institute, University of Warwick, Coventry, CV4 7AL, UK), Stefano Luzzatto (Mathematics Department, Imperial College, 180 Queen's Gate, London SW7 2BZ, UK), Warwick Tucker (Department of Mathematics, University of Bergen, 5008 Bergen, Norway), and Changming Ding (Department of Mathematics, Xiamen University, Xiamen, 361005, P. R. China).

We would also like to thank the following publishers for their kind permission to reuse some of their copyrighted materials (figures): American Physical Society, American Institute of Physics, Institute of Physics (IOP) Publishing, IEEE Publishing, Hindawi Publishing Corporation, Europhysics Letters (EPL), and Springer Science + Business Media.

Elhadj Zeraoulia
Julien Clinton Sprott
July 2011

Contents

Preface		vii
1.	Poincaré Map Technique, Smale Horseshoe, and Symbolic Dynamics	1
	1.1 Poincaré and generalized Poincaré mappings	1
	1.2 Interval methods for calculating Poincaré mappings	4
	1.2.1 Existence of periodic orbits	6
	1.2.2 Interval arithmetic	7
	1.3 Smale horseshoe	10
	1.3.1 Dynamics of the horseshoe map	12
	1.4 Symbolic dynamics	17
	1.4.1 The method of fixed point index	19
	1.5 Exercises	24
2.	Robustness of Chaos	27
	2.1 Strange attractors	27
	2.1.1 Concepts and definitions	27
	2.1.2 Robust chaos	29
	2.1.3 Domains of attraction	32
	2.2 Density and robustness of chaos	34
	2.3 Persistence and robustness of chaos	35
	2.4 Exercises	36
3.	Statistical Properties of Chaotic Attractors	39
	3.1 Entropies	39
	3.1.1 Lebesgue (volume) measure	39
	3.1.2 Physical (or Sinai–Ruelle–Bowen) measure	40

		3.1.3	Hausdorff dimension	42

		3.1.3	Hausdorff dimension	42
		3.1.4	The topological entropy	43
		3.1.5	Lyapunov exponent	47
	3.2	Ergodic theory		50
	3.3	Statistical properties of chaotic attractors		51
		3.3.1	Autocorrelation function (ACF)	51
		3.3.2	Correlations	53
	3.4	Exercises		57
4.	Structural Stability			61
	4.1	The concept of structural stability		61
		4.1.1	Conditions for structural stability	62
		4.1.2	A proof of Anosov's theorem on structural stability of diffeomorphisms	64
	4.2	Exercises		77
5.	Transversality, Invariant Foliation, and the Shadowing Lemma			80
	5.1	Transversality		80
	5.2	Invariant foliation		82
	5.3	Shadowing lemma		84
		5.3.1	Homoclinic orbits and shadowing	88
		5.3.2	Shilnikov criterion for the existence of chaos	90
	5.4	Exercises		93
6.	Chaotic Attractors with Hyperbolic Structure			95
	6.1	Hyperbolic dynamics		96
		6.1.1	Concepts and definitions	96
		6.1.2	Anosov diffeomorphisms and Anosov flows	97
	6.2	Anosov diffeomorphisms on the torus $\mathbb{T}^n$		102
		6.2.1	Anosov automorphisms	102
		6.2.2	Structure of Anosov diffeomorphisms	103
		6.2.3	Anosov torus $\mathbb{T}^n$ with a hyperbolic structure	104
		6.2.4	Expanding maps	105
		6.2.5	The Blaschke product	107
		6.2.6	The Bernoulli map	108
		6.2.7	The Arnold cat map	111
	6.3	Classification of strange attractors of dynamical systems		112
	6.4	Properties of hyperbolic chaotic attractors		114

		6.4.1	Geodesic flows on compact smooth manifolds	115
		6.4.2	The solenoid attractor	117
		6.4.3	The Smale–Williams solenoid	119
		6.4.4	Plykin attractor	120
	6.5	Proof of the hyperbolicity of the logistic map for $\mu > 4$		121
	6.6	Generalized hyperbolic attractors		126
	6.7	Generating hyperbolic attractors		134
	6.8	Density of hyperbolicity and homoclinic bifurcations in arbitrary dimension		138
	6.9	Hyperbolicity tests		139
		6.9.1	Numerical procedure	141
		6.9.2	Testing hyperbolicity of the Hénon map	144
		6.9.3	Testing hyperbolicity of the forced damped pendulum	153
	6.10	Uniform hyperbolicity test		154
	6.11	Exercises		158
7.	Robust Chaos in Hyperbolic Systems			167
	7.1	Modeling hyperbolic attractors		167
		7.1.1	Modeling the Smale–Williams attractor	168
		7.1.2	Testing hyperbolicity of system (7.1)	176
		7.1.3	Numerical verification of the hyperbolicity of system (7.1)	183
		7.1.4	Modeling the Arnold cat map	187
		7.1.5	Modeling the Bernoulli map	194
		7.1.6	Modeling Plykin's attractor	202
	7.2	Exercises		204
8.	Lorenz-Type Systems			208
	8.1	Lorenz-type attractors		208
	8.2	The Lorenz system		210
		8.2.1	Existence of Lorenz-type attractors	211
		8.2.2	Geometric models of the Lorenz equation	220
		8.2.3	Structure of the Lorenz attractor	230
	8.3	Expanding and contracting Lorenz attractors		239
	8.4	Wild strange attractors and pseudo-hyperbolicity		241
	8.5	Lorenz-type attractors realized in two-dimensional maps		253
	8.6	Exercises		255

9. Robust Chaos in the Lorenz-Type Systems — 258

- 9.1 Robust chaos in the Lorenz-type attractors 258
- 9.2 Robust chaos in Lorenz system 260
- 9.3 Robust chaos in 2-D Lorenz-type attractors 267
- 9.4 Exercises 269

10. No Robust Chaos in Quasi-Attractors — 272

- 10.1 Quasi-attractors, concepts, and properties 273
- 10.2 The Hénon map 275
  - 10.2.1 Uniform hyperbolicity of the Hénon map 276
  - 10.2.2 Hyperbolicity of Hénon-like maps 281
  - 10.2.3 Hénon attractor is a quasi-attractor 282
- 10.3 The Strelkova–Anishchenko map 286
- 10.4 The Anishchenko–Astakhov oscillator 286
- 10.5 Chua's circuit 288
  - 10.5.1 Homoclinic and heteroclinic orbits 294
  - 10.5.2 The geometric model 298
- 10.6 Exercises 299

11. Robust Chaos in One-Dimensional Maps — 303

- 11.1 Unimodal maps 303
  - 11.1.1 S-unimodal maps 305
  - 11.1.2 Relation between unimodality and hyperbolicity . 307
  - 11.1.3 Classification of unimodal maps of the interval . . 309
  - 11.1.4 Collet–Eckmann maps 312
  - 11.1.5 Statistical properties of unimodal maps 314
- 11.2 The Barreto–Hunt–Grebogi–Yorke conjecture 317
  - 11.2.1 Counter-examples to the Barreto–Hunt–Grebogi–Yorke conjecture 324
  - 11.2.2 Robust chaos without the period-n-tupling scenario 332
  - 11.2.3 The B-exponential map 335
- 11.3 Border-collision bifurcation and robust chaos 342
  - 11.3.1 Normal form for piecewise-smooth one-dimensional maps 342
  - 11.3.2 Border-collision bifurcation scenarios 343
  - 11.3.3 Robust chaos in one-dimensional singular maps . 349
- 11.4 Exercises 352

12. Robust Chaos in 2-D Piecewise-Smooth Maps			358
12.1	Robust chaos in 2-D piecewise-smooth maps		358
	12.1.1	Normal form for 2-D piecewise-smooth maps . . .	359
	12.1.2	Border-collision bifurcations and robust chaos . .	360
	12.1.3	Regions for nonrobust chaos	362
	12.1.4	Regions for robust chaos: undesirable and dangerous bifurcations	367
	12.1.5	Proof of unicity of orbits	370
	12.1.6	Proof of robust chaos	371
12.2	Robust chaos in noninvertible piecewise-linear maps . . .		378
	12.2.1	Normal forms for two-dimensional noninvertible maps	380
	12.2.2	Onset of chaos: Proof of robust chaos	382
12.3	Exercises		395

Bibliography   401

Index   451

Chapter 1

Poincaré Map Technique, Smale Horseshoe, and Symbolic Dynamics

In this chapter, we give an overview of the Poincaré map technique used to describe bifurcations and chaos in dynamical systems. For example, all the content of Chapter 7 is based on the construction of a Poincaré map to confirm the existence of robust chaos in such a continuous-time dynamical system. We first define the Poincaré map and the generalized Poincaré mappings for a continuous-time dynamical system with a continuous piecewise-linear vector field in Sec. 1.1. In Sec. 1.2, we discuss interval methods for calculating Poincaré maps, including several methods used to prove the existence of periodic orbits using a generalized Poincaré map associated with the continuous-time flow under consideration. In Sec. 1.3 we discuss the Smale horseshoe as a basic tool for studying chaos and its properties such as symbolic dynamics. In particular, the so-called *method of fixed point index* is described in Sec. 1.4.1 to prove the existence of such horseshoe-type mappings in a dynamical system. This method is based on the periodic points of the so-called *TS-maps* and the existence of semi-conjugacy. At the end of this chapter, several exercises are given to fix the ideas and provide additional experience.

1.1 Poincaré and generalized Poincaré mappings

First, we give a definition for the Poincaré map defined for a continuous-time dynamical system. Indeed, let $f : \mathbb{R} \times \Omega \longrightarrow \Omega$ be a continuous flow, where $\Omega \subset \mathbb{R}^n$ is open. Consider the n-dimensional continuous-time system given by:

$$x' = f(t, x) \tag{1.1}$$

Let φ_t denote the corresponding flow of the system (1.1).

Definition 1.1. A map P is called a C^r-diffeomorphism if both P and its inverse P^{-1} are bijective and are r-times continuously differentiable.

Let γ be a periodic orbit through a point p, and let U be an open and connected neighborhood of p. For any $x \in \Omega$, let $I(x) =]t_x^-, t_x^+[$ be an open interval in the real numbers. Then one has the following definitions:

Definition 1.2. (a) A positive semi-orbit through x is the set $\gamma_x^+ = \{f(t,x), t \in]0, t_x^+[\}$, and a negative semi-orbit through x is the set $\gamma_x^- = \{f(t,x), t \in]t_x^-, 0[\}$.

(b) A Poincaré section through a point p is a local differentiable and transversal section S of f through the point p.

Hence the Poincaré map is defined by:

Definition 1.3. A function $P : U \longrightarrow S$ is called a Poincaré map for the orbit γ on the Poincaré section S through point p if:

(1) $P(p) = p$.
(2) $P(U)$ is a neighborhood of p and $P : U \to P(U)$ is a diffeomorphism.
(3) For every point x in U, the positive semi-orbit of x intersects S for the first time at $P(x)$.

Generally, the Poincaré map P is called the *first recurrence map* because the method of analysis is to consider a periodic orbit with initial conditions on the Poincaré section S and observe the point at which this orbit first returns to the section S. The advantage of using the Poincaré map technique is that the Poincaré map is the intersection of a periodic orbit of the considered continuous-time system (1.1) with the transversal Poincaré section S [1] is one dimension smaller than the original continuous dynamical system. Note that the map P is used to analyze the original system (1.1) because it preserves many properties of periodic, quasi-periodic and chaotic orbits of the original continuous-time system. Indeed, the importance of the Poincaré map can be seen by the relation between limit sets of the Poincaré map P and limit sets of the flow of the considered system (1.1) as follows: A limit cycle of φ_t is a fixed point of P, and a period-m closed orbit of P is a subharmonic solution (relative to the considered section S) of φ_t. A chaotic orbit of P corresponds to a chaotic solution of φ_t. Furthermore, a stable periodic point of P corresponds to a stable periodic orbit of φ_t, and an unstable periodic point of P corresponds to a saddle-type periodic

---
[1] This means any periodic orbit starting on the subspace flow through S and not parallel to it.

orbit of φ_t. The Poincaré map P is not defined for points, trajectories of which never come back to the section S defining the Poincaré map. Hence the image of a given set under P can be computed rigorously only if it is continuous on the set. Conversely, if the map P is well defined, this does not imply its continuity.

A first example of this situation is a point belonging to a stable manifold of an equilibrium of (1.1) where its trajectory converges to the fixed point and hence never comes back to the Poincaré plane S. A second example is where the map P is well defined but not continuous at points for which the flow is parallel to the section S at this point or at the image $P(S)$. If the map P is not continuous at a point, then in a close neighborhood of that point, rigorous evaluation of P becomes very difficult, the sets which have to be studied become smaller, and the computation time becomes longer. All these problems can be surmounted if one knows the regions where the map P is not continuous.

Secondly, assume that the function f given in Eq. (1.1) is a continuous piecewise-linear vector field. Let $\Sigma_1, ..., \Sigma_p$ denote the hyperplanes separating the linear regions U_i of f, in which case their union is the set Ω. Let $\varphi(t, x)$ denote the trajectory of the system (1.1) starting at the point x.

Definition 1.4. The generalized Poincaré map $H : \Omega \to \Omega$ is defined by $H(x) = \varphi(\tau(x), x)$, where $\tau(x)$ is the time needed for the trajectory $\varphi(t, x)$ to reach Ω.

The generalized Poincaré map H has the same properties given above for the map P. For the evaluation of H in regions where H is continuous, one can use the analytical formulas for solutions of linear systems. To evaluate the generalized Poincaré map H on a box $X \subset \Omega$, we first find the return time for all points in X, i.e., the interval $\{\tau(x), x \in X\} \subset \tau(X)$ and then use analytic solutions to compute $\varphi(\tau(X), X)$. $H(X)$ is enclosed in the intersection of $\varphi(\tau(X), X)$ with Ω. The Jacobian of H at X can be expressed in terms of the return time $\tau(x)$, the start box X, and the image $H(X)$.

There is no general method to construct a Poincaré map. The majority of methods used in the literature are numerical and rarely analytic. In [Hénon (1982)], a method for accurately finding the intersections of a numerically integrated trajectory of a system of ordinary differential equations with a surface of section is given. A generalization of the *stopping procedure* described by Hénon is given in [Tucker (2002b)]. In [Tsuji & Ido (2002)], a computational method based on parallel computation of data tables and interpolation is given to calculate the Poincaré map. This method was successfully applied to the so-called chaotic torus magnetic field line caused by the perturbation coil. In [Fujisaka & Sato (1997)], a numerical method based on the Poincaré map, the second map constructed from the Poincaré map, and the topological degree is presented to compute the number and stability of fixed points of Poincaré maps of ordinary differential equations. The computation of the topological degree of the second map is equivalent to the calculation of the number of fixed points of the Poincaré map in a given domain of a Poincaré section. In some special cases, the Poincaré map was constructed rigorously [Chua *et al.* (1986), Chua & Tichonicky (1991), Kuznetsov & Satayev (1994), and references therein].

1.2 Interval methods for calculating Poincaré mappings

In this section, we discuss the notion of rigorous proof. This notion is very problematic because it has no *unified* definition. However, there are several opinions on how to define a *proof*, a rigorous proof, or more generally, the elements of a proof. The so-called *four-color theorem* is an example of such a situation because its proof is done by exhaustive computer testing of many individual cases that cannot be verified by hand. This type of proof is the subject of much controversy for mathematicians, some of whom think that such so-called *computer-assisted proofs* are valid since they simply replace human verifiability with machine verifiability. However, the computer-assisted proofs have some difficulties such as the inevitable numerical errors[2] due to the finite capacity of the computer. Finally, it is not possible to verify each theorem in finite computations.

Examples of computer-assisted proofs in dynamics include the works of Lanford in 1982 for proving Feigenbaum universality conjectures, Eckmann, Koch, Wittwer in 1984 for proving universality for area-preserving maps, MacKay and Percival for proving nonexistence of invariant circles for the standard map for all $k > 63/64$, Grebogi, Hammel, Yorke in 1987 for rigorous numerical shadowing of trajectories, Neumaier, Rage, Schlier in 1994 for proving chaos in the *molecular Thiele–Wilson model*, Mischaikow

---

[2]These errors are the result of the following: round-off errors, the numerical-method error, the spatial discretization error, the propagation error, and the errors involving the intersection with the section in the computation of Poincaré maps. Furthermore, it is impossible to know if there might have been a malfunction, *i.e.*, some bit was miscomputed. But a counterargument is that it is much more likely that it is a human-made error.

and Mrozek in 1995 for proving chaos in the Lorenz equations, Palmer, Coomes, Kocak, Stoffer, Kichgraber in 1996-2003 for proving chaos by shadowing for the Hénon map, Tucker in 2001 for studying the geometric model for the Lorenz attractor, Mischaikow (Rutgers), Mrozek, Zgliczynski, Wilczak, Galias, Kapela, Pilarczyk, Arioli (Milan) for proofs of chaos (semi-conjugacy with a Bernoulli shift) for the Lorenz equations, the Rössler equations, the Hénon map, Chua's circuit, homoclinic and heteroclinic orbits, the Kuramoto-Sivashinsky ODE, the Kuramoto-Sivashinsky PDE, existence of multiple steady states and their bifurcations, periodic orbits, the N-body problem, the existence of simple choreographies, *etc.*

Hardy says in [Hardy (1999), pp. 15–16] "all physicists, and a good many quite respectable mathematicians, are contemptuous about proof. I have heard Professor Eddington, for example, maintain that proof, as pure mathematicians understand it, is really quite uninteresting and unimportant, and that no one who is really certain that he has found something good should waste his time looking for proof.... [This opinion], with which I am sure that almost all physicists agree at the bottom of their hearts, is one to which a mathematician ought to have some reply."

Hardy's assertion given above was echoed by Feynman [Derbyshire (2004), p. 291] who is reported to have commented, "A great deal more is known than has been proved." However, we accept the following definitions:

Definition 1.5. (a) A proof is a rigorous mathematical argument that unequivocally demonstrates the truth of a given Proposition. (b) A proof is called rigorous if the validity of each step and the connections between the steps is explicitly made clear in such a way that the result follows with certainty.

In the interval Newton's method given in [Alefeld & Herzberger (1983), Neumaier (1990)], the existence of zeros of a function f in an n-dimensional interval X is achieved by evaluating the so-called *interval Newton operator* $N(X)$ given by:

$$N(X) = x_0 - (Df(X))^{-1} f(x_0) \qquad (1.2)$$

where $Df(X)$ is the interval matrix containing all Jacobian matrices of f for $x \in X$, and x_0 is an arbitrary point belonging to X. Hence the following theorem [Alefeld & Herzberger (1983), Neumaier (1990)] was obtained:

Theorem 1.1. *If $N(X) \subset X$, then there exists exactly one zero of f in X. If $N(X) \cap X = \emptyset$, then there are no zeros of f in X.*

Thus the interval Newton's method is used to prove the existence and uniqueness of zeros. For computation of the expression $(Df(X))^{-1} f(x_0)$, one can use for example the Gaussian algorithm, *i.e.*, Gaussian elimination (some elementary row operations) can be used to determine the solutions of linear equations, to find the rank of a matrix, and to calculate the inverse of an invertible square matrix [Lipschutz & Lipson (2001)]. In fact, Newton's method has good convergence properties if the initial point x_0 is chosen sufficiently close to the periodic point. To find all periodic orbits with a good accuracy, one can check different initial conditions located on a uniform grid in the phase space by using more sophisticated methods for locating all unstable periodic orbits of a dynamical system as developed in [Biham & Wenzel (1989), Davidchack *et al.* (2001)]. These methods do not guarantee that there are no other short periodic orbits in the specific range. However, for hyperbolic systems introduced in Chapter 6, there are many interesting results on the number and distribution of periodic orbits [Parry & Pollicott (1983)]. These results are based on the concept of the dynamical zeta function, number theory and probability theory. For piecewise-linear systems, periodic orbits can be rigorously studied by means of the interval Newton's method as in [Galias (2002a-b)]. These methods are based on the construction of a Poincaré map and the interval Newton's method to find regions where the generalized Poincaré map H is well defined and continuous and locating all low-period cycles in that region.

1.2.1 *Existence of periodic orbits*

The generalized Poincaré map H associated with a continuous-time piecewise-linear system can be used to prove the existence of periodic orbits. The method of analysis is based on the fact that to prove the existence of a period-m orbit of H, one applies the interval Newton's method to the map $G : (\mathbb{R}^n)^m \longrightarrow (\mathbb{R}^n)^m$ defined by

$$[G(z)]_k = x_{(k+1) \bmod m} - H(x_k), 0 < k < m \qquad (1.3)$$

where $z = (x_0, ... x_{m-1})$. It is easy to verify that in the subsets of Ω where the Poincaré map H can be rigorously and effectively evaluated, one has: $G(z) = 0$ if and only if x_0 is a fixed point of H^m. Thus the problem of existence of periodic orbits for the system (1.1) is translated into the problem of existence of zeros of the higher-dimensional function G.

1.2.2 Interval arithmetic

First, consider the following elements of the theory of interval arithmetic:

$$\begin{cases} X = [a,b] = \{x \in \mathbb{R} : a \leq x \leq b\} \\ V = (X_1, X_2, ..., X_n) \\ X_1 \Diamond X_2 = \{x = x_1 \Diamond x_2 \in \mathbb{R}, : x_1 \in X_1, x_2 \in X_2\} \end{cases} \quad (1.4)$$

where: X : is an interval, *i.e.*, a closed bounded set of real numbers.

V : is called an n-dimensional interval vector, and it consists of n closed intervals $X_i, i = 1,..,n$.

$\Diamond$: is any of the following usual numeric operators: $+, -, \times$ and $/$, where all operations but division are defined for arbitrary intervals. For division we assume that the interval X_2 does not contain the number 0, because a real number a can be treated as a degenerate interval $a = [a,a]$.

Generally, all the steps of *Interval arithmetic* are summarized as follows [Galias (2001)]:

(1) Reduce the continuous system (1.1) to the discrete one using the Poincaré map P.
(2) Apply the interval Newton operator G defined in (1.3).
(3) Evaluate $P(X)$ from Eq. (1.1) using direct integration by the Lohner method, which helps to reduce the wrapping effect[3] [Moore (1966a), Lohner (1992)].
(4) Find the image of $P(X)$, *i.e.*, the intersection of Ω and the trajectory computed by the rigorous integration procedure in step (3).
(5) Find $P'(X)$, the Jacobian matrix of $P(X)$ by solving the variational equation

$$\frac{dD}{dt} = \frac{\partial f}{\partial x}\left(x\left(t\right)\right) D \quad (1.5)$$

where $D\left(t, x_0\right) = \frac{\partial \varphi_t}{\partial x_0}\left(t, x_0\right)$ with the initial conditions $D\left(0, x_0\right) = I$, the unit matrix.

(6) Calculate the enclosure for the Jacobian matrix of P at $x \in X$ using the formula:

$$P'(X) = \left(I - \frac{f\left(y\right) h^T}{h^T f\left(y\right)}\right) D \quad (1.6)$$

---

[3]Wrapping occurs when a solution set of a differential equation that is not a box (a parallelepiped with edges that are parallel to the axes of an orthogonal coordinate system) is enclosed or wrapped by a box on each integration step. A result of this wrapping phenomenon is that the computed bounds become unacceptably large.

where:

D is the enclosure for the solution of the variational equation

$$\{D(t,x), x \in X, t = \tau(x)\}, \qquad (1.7)$$

h is a vector orthogonal to Ω, and

y is the enclosure for the set $\{P(x), x \in X\}$.

(7) Construct the trapping region Γ^4 for the Poincaré map P. Note that this region can be found by choosing a polygon enclosing trajectories of the Poincaré map P generated by the computer.

(8) Cover the trapping region $\tilde{\Gamma}$ by boxes of a specified size.

(9) Find all periodic orbits by applying a combination of the interval Newton's method and a generalized bisection technique. This can be done in eight steps:

(9-1) Find the graph representation of the dynamics of the system[5].

(9-2) Compute the image of each box.

(9-3) Find the set of admissible transitions (the so-called transitions represent edges of the directed graph) between boxes.

(9-4) Reduce the graph by removing vertices corresponding to boxes having no intersection with the invariant part of the trapping region. In this case a box is removed if its image has no intersection with other boxes or if it has no intersection with images of all boxes.

(9-5) Repeat this procedure until no more boxes can be removed.

(9-6) Find all period-m cycles in the resulting graph.

(9-7) Evaluate the interval operator for each period-m cycle on the corresponding interval vector z, and check what is the position of $N(Z)$ with respect to Z, i.e., if $N(Z) \subset Z$, then there exists exactly one period-m cycle of f in Z. If $N(Z) \cap Z = \emptyset$, then there are no period-m cycles of f in Z.

(9-8) If the step (9-7) does not hold, then one option is to divide the interval vector Z into smaller parts and to evaluate the interval operator on each of them.

For a good description of the dynamics of a piecewise-linear continuous-time system, it is natural to calculate the Poincaré map H and its

---

[4]A set $\tilde{\Gamma}$ is said to be a trapping region if it is positively invariant under the action of the map f, i.e., $f(x) \in \tilde{\Gamma}$ for all $x \in \tilde{\Gamma}$.

[5]For finding the graph representation of the dynamics of the system, the trapping region $\tilde{\Gamma}$ must be covered by ϵ-boxes of the form $v = [k_1\epsilon_1, (k_1+1)\epsilon_1] \times [k_2\epsilon_2, (k_2+1)\epsilon_2]$ where k_i are integer numbers, ϵ_i are fixed positive real numbers, and $\epsilon = (\epsilon_1, \epsilon_2)$. ϵ-boxes define vertices, and admissible connections between boxes define edges of the graph.

periodic orbits[6] in such a way as to reduce or eliminate some types of overestimations of the resulting solution set called *the wrapping effect* observed in the Lohner method [Galias (2002b)]. The so-called *mean value form* described below is the best method compared with the following methods: the direct evaluation of analytical formulas in interval arithmetic, bisection of the return time, generalized bisection in the Poincaré plane, Lohner method, and a combinations of these methods.

Now assume that in the linear regions Σ_k, the system (1.1) has the form:

$$x' = A_k (x - p_k). \tag{1.8}$$

Then its solution is given by

$$\varphi(t, x_0) = \exp(A_k t)(x_0 - p_k) + p_k. \tag{1.9}$$

Hence the following theorem can be used to describe the method based on the mean value form techniques:

Theorem 1.2. *Assume that $H : \mathbb{R}^n \longrightarrow \mathbb{R}^m$ is a C^1 map. Then for all $x, y \in \mathbb{R}^n$ and for all $i = 1, 2, ..., m$, there exists a point z_i in the interval $\overline{xy}$ such that*

$$H_i(x) - H_i(y) = \sum_{j=1}^{j=m} \frac{\partial H_i}{\partial x_j}(z_i)(x_j - y_j). \tag{1.10}$$

Then the mean value form technique can be done in the following steps:

(1) Take an interval vector X, and fix a point $y \in X$ as the center of X, and choose another point $x \in X$.
(2) Apply Theorem 1.2, and deduce that there exists a point $z_i \in X$ such that

$$\begin{cases} H_i(x) \in H_i(y) + \sum_{j=1}^{j=m} \frac{\partial H_i}{\partial x_j}(z_i)(x_j - y_j) \\ H(x) \in H(y) + DH(X)(x - y). \end{cases} \tag{1.11}$$

(3) Deduce that

$$\{H(x) \in X\} \subset H(y) + DH(X)(x - y). \tag{1.12}$$

---

[6]This is possible in the case where the generalized Poincaré map H is continuous in the region containing the attractor.

From the above algorithm, it is clear that narrower computational results are obtained when the Poincaré map H is computed using Eq. (1.12). An example of the application of this method to the Lorenz system (8.1) is given in Sec. 8.2. In addition to the interval Newton operator, there is also the *Krawczyk operator* and *Hansen–Sengupta operator* introduced in [Neumaier (1990), Alefeld (1994)]. Both operators provide simple computational tests for uniqueness, existence, and nonexistence of zeros of a map within a given interval vector. Generally, these two operators were used (under the assumption that the Poincaré map P is well defined and continuous in a region) for finding in a given region all low-period cycles for continuous systems. Furthermore, the method can be applied without any modifications for higher-dimensional systems, but in this case computation times may be significantly longer. The Krawczyk operator is defined as

$$K(X) = x_0 - Cf(x_0) - (Cf'(X) - I)(X - x_0) \quad (1.13)$$

where $x_0 \in X$ and C is an invertible matrix. The preconditioning matrix C is usually chosen as the inverse of $f'(x_0)$. The *Hansen–Sengupta operator* is defined as

$$H(X) = x_0 + \Gamma(Cf(X), -Cf'(x_0), Z - x_0) \quad (1.14)$$

where Γ is the *Gauss–Seidel operator*. For intervals a, b, Z, the Gauss–Seidel operator $\Gamma(a, b, Z)$ is the tightest interval enclosing the set $\{x \in Z : ax = b$ for some $a \in a, b \in b\}$, and for interval matrix A and interval vectors b, Z, the Gauss–Seidel operator $\Gamma(A, b, Z)$ is defined by

$$y_i = \Gamma(A, \mathbf{b}, Z)_i = \Gamma(A_{ii}, \left[\mathbf{b}_i - \sum_{k<i} A_{ik} y_k - \sum_{k>i} A_{ik} Z_k\right], Z_i). \quad (1.15)$$

Theorems on the existence and uniqueness of zeros for the Newton operator can be found in [Neumaier (1990)]. The basic numerical method for detection of periodic orbits described above for maps using any of the three defined operators can be applied for continuous systems using the concept of the Poincaré map, which reduces the continuous system to a discrete one [Galias (2006)].

1.3 Smale horseshoe

In this section, we give a detailed description of the Smale horseshoe as a fundamental tool for studying chaos in dynamical systems. For defining the so-called *Smale's horseshoe* map, we must define the unit square as follows:

Definition 1.6. A unit square D with side lengths 1 is the one with coordinates $(0,0), (1,0), (1,1), (0,1)$ in the real plane, or $0, 1, 1+i$ in the complex plane.

Hence the Smale horseshoe map f described in [Smale (1967), Cvitanović et al. (1988)] consists of the following sequence of operations as shown in Fig. 1.1 on the unit square D:

(a) Stretch in the y direction by more than a factor of two.
(b) Compress in the x direction by more than a factor of two.
(c) Fold the resulting rectangle and fit it back onto the square, overlapping at the top and bottom, and not quite reaching the ends to the left and right and with a gap in the middle. Hence the action of f is defined through the composition of the three geometrical transformations defined above.
(d) Repeat the above steps to generate the horseshoe attractor that has a Cantor set structure.

From Fig. 1.1, one sees that horseshoe maps cross the original square in a linear fashion, but in most applications horseshoe maps are rarely so regular, but very similar. Mathematically, the above actions (a) to (d) can be translated as follows [Smale (1967)]:

(1) Contract the square D by a factor of λ in the vertical direction, where $0 < \lambda < \frac{1}{2}$, such that D is mapped into the set $[0,1] \times [0,\lambda]$.
(2) Expand the rectangle obtained by a factor of μ in the horizontal direction, where $2+\epsilon < \mu$, and map the set $[0,1] \times [0,\lambda]$ into the $[0,\mu] \times [0,\lambda]$ (the need for this ϵ factor is explained in step 3).
(3) Steps (1) and (2) produce a rectangle $f(D)$ of dimensions $\mu \times \lambda$. This rectangle crosses the original square D in two sections after it has been bent as shown in Fig. 1.1. The ϵ in step (2) indicates the extra length needed to create this bend as well as any extra on the other side of the square.
(4) This process is then repeated, only using $f(D)$ rather than the unit square. The nth iteration of this process will be called $f^k(D)$, $k \in \mathbb{N}$.

Some higher iterations of the Smale horseshoe map f can be seen in Figs. 1.2, 1.3, and 1.4.

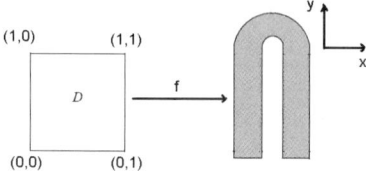

Fig. 1.1 The Smale horseshoe map f after a single iteration.

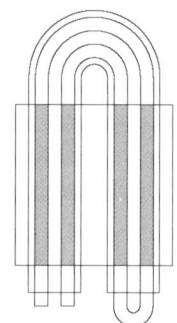

Fig. 1.2 The Smale horseshoe map f after two iterations.

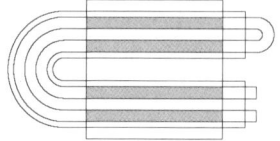

Fig. 1.3 The Smale horseshoe map f after three iterations.

1.3.1 *Dynamics of the horseshoe map*

In this section, we describe the dynamics of the horseshoe map as a tool for the reproduction of the chaotic dynamics of a flow in a small disk Δ perpendicular to a period-T orbit Γ. First, when the system evolves, some orbits diverge and the points in this disk remain close to the given periodic orbit Γ, tracing out orbits that eventually intersect the disk once again. Second, the intersection of the disk Δ and points in its neighborhood with the given period-T orbit Γ come back to themselves every period T. When this neighborhood returns, its shape is transformed. Third, inside the disk Δ are some points that will leave the disk neighborhood and others that

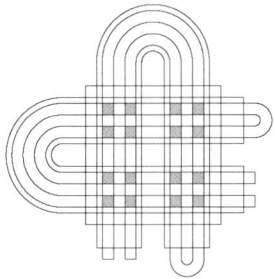

Fig. 1.4 The Smale horseshoe map f after four iterations.

will continue to return. Fourth, the set of points that never leave the neighborhood of the given period-T orbit Γ form a fractal. Hence the above four features imply that Smale's horseshoe map consists of the iterated application of both $f(D)$ and $f^{-1}(D)$.

Here are some important results that can be found in [Smale (1967)] where their proofs are carried out using the geometrical formations of horseshoes as shown in Fig. 1.1 and its repetitions.

Theorem 1.3. *(a) The set $\Pi_+ = \cap_{i=0}^{i=\infty} f^i(D)$ is an interval cross a Cantor-like set*[7]. *(b) The set $\Pi_- = \cap_{i=0}^{i=\infty} f^{-i}(D)$ is a Cantor-like set cross an interval.*

(c) The set $\Pi = \Pi_+ \cap D \cap \Pi_-$ is a Cantor-like set cross a Cantor-like set.

(d) If we consider the map $f(D)$, where f is the horseshoe map that contracts by λ and expands by μ, where $0 < \lambda < \frac{1}{2}$ and $\mu > 2 + \epsilon$, the Hausdorff dimension of Π is $\log 2 \left(\frac{1}{\log \mu} - \frac{1}{\log \lambda} \right)$.

(e) The limit set Π forms an uncountable, nowhere dense set (the interior of its closure is empty) in $\mathbb{R}^2$.

(f) $f(D)$ is equivalent to the shift map in the symbolic space.

(g) The horseshoe map f is a diffeomorphism defined from the unit square D of the plane into itself.

(h) The horseshoe map is one-to-one.

(l) The domain of f^{-1} is $f(D)$.

(m) The horseshoe map f is an axiom A diffeomorphism (see Definition 6.3(1)).

---

[7]The set Π_+ is a Cantor-like set, *i.e.*, it is a perfect, nowhere dense, and totally disconnected set.

Theorem 1.3 gives the idea that symbolic names can be given to all the orbits that remain in the neighborhood. If the initial disk Δ can be divided into a small number of regions D_i, $i = 1,..,m$ and if one knows the sequence of points $\{x_0, x_1, ...\}$ in which the orbit visits these regions D_i, then the visitation sequence $\hat{s} = \{\hat{s}_0; \hat{s}_1; ...; \hat{s}_j; ...\}$ composed of symbols $\hat{s}_j = i$ if $x_j = f^j(x_0) \in D_i$, $i = 1,..,m$ of the orbits provides a symbolic representation of the dynamics, known as symbolic dynamics introduced in Sec. 1.3. The presence of a horseshoe presented above in a dynamical system implies the following features:

(1) There are infinitely many periodic orbits, especially those with arbitrarily long period.
(2) The number of periodic orbits grows exponentially with the period.
(3) In any small neighborhood of any point of the fractal invariant set Λ, there is a point on a periodic orbit.

The proof of the above statements is based on symbolic dynamics introduced in Sec. 1.3, and a similar proof applies to those given by Cantor sets. If one considers the Smale horseshoe map as a set of topological operations, then the method of construction of an attractor of a dynamical system is the Smale horseshoe map itself. This method consists of two operations, the first is the stretching which gives sensitivity to initial conditions, and the second is the folding which gives the attraction. However, we have the following definition:

Definition 1.7. A dynamical system is chaotic in the sense of Smale if it has horseshoes of the Smale type.

Note that Definition 1.7 with the so-called *shadowing lemma* [Stoffer & Palmer (1999)] introduced in Sec. 5.3 is used to prove that the Hénon map (6.88) is chaotic in the sense of Smale. Generally, there are no rigorous methods for finding Smale horseshoes in a dynamical system with relatively small dimension [Smale (1967), Banks & Dragan (1994)], but a few works are concerned with this topic, for example [Zgliczynski (1997b)].

The most crucial role of the Smale horseshoe is its relationship to the homoclinic tangency discovered by Poincaré while investigating the three-body problem of celestial mechanics [Tufillaro et al. 1992], *i.e.*, the homoclinic tangency is the tangled intersection of such invariant manifolds with homoclinic points as shown in Fig. 1.5.

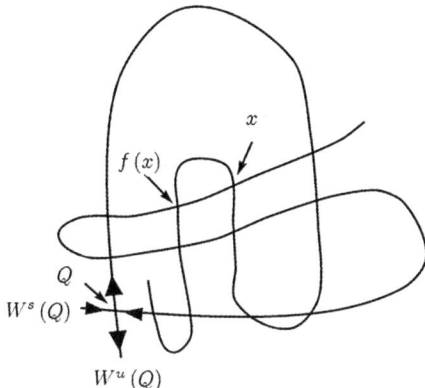

Fig. 1.5 Part of a homoclinic tangency. Q is the hyperbolic fixed point, $W^s(Q), W^u(Q)$ are the stable and unstable manifolds respectively, x is a homoclinic point, and thus $f(x)$ is also a homoclinic point.

Definition 1.8. (a) The invariant manifold of a map f is a set of points X such that $f(X) \subset X$[8].

(b) Given a fixed point Q in the invariant manifold, the manifold $W^s(Q)$ is called stable if $\forall y \in W^s(Q)$, $\lim_{n \to \infty} f^n(y) \to Q$. Similarly, a manifold is called unstable $W^u(Q)$ if $\forall y \in W^u(Q)$, $\lim_{n \to \infty} f^{-n}(y) \to Q$.

(c) A fixed point is hyperbolic if it is the intersection of one or more stable manifolds and one or more unstable manifolds.

(c) A homoclinic point is a point x different from the fixed point that lies on a stable or unstable manifold of the same fixed point Q.

Formally, the homoclinic tangency can be defined as follows:

Definition 1.9. (a) (homoclinic tangency) We say that a diffeomorphism f exhibits a homoclinic tangency if there is a periodic point Q such that there is a point $x \in W^s(Q) \cap W^u(Q)$ with $T_x W^s(Q) + T_x W^u(Q) \neq T_x M$.

(b) Given an open set V, we say that the tangency holds in V if x and Q belong to V.

Theorem 1.4. *(Poincaré, 1890) If there exists a single homoclinic point on a stable and an unstable invariant manifold corresponding to a particular hyperbolic fixed point, then there exist an infinite number of homoclinic points on the same invariant manifolds.*

---
[8]Of course every fixed point, with the exception of centers, will be an element of some invariant manifold.

Proof. The proof is done by induction on the number of homoclinic points. Assume that there exist n homoclinic points for these invariant manifolds. Let $W^s(Q)$ be the stable manifold and $W^u(Q)$ be the unstable manifold, and let x be the homoclinic point farthest from the fixed point Q along the unstable manifold $W^u(Q)$. Since $W^s(Q)$ and $W^u(Q)$ are invariant manifolds, $f(x) \in W^s(Q) \cap W^u(Q)$. Thus $f(x)$ is either a homoclinic point or the fixed point Q. Since f takes a point on the unstable manifold $W^u(Q)$ away from the fixed point Q, then $f(x)$ cannot be the fixed point. Thus it is a homoclinic point. For the same reason, it is not one of the n considered homoclinic points. Thus there exist $n+1$ homoclinic points. Therefore, the number of homoclinic points on the corresponding invariant manifolds is infinite. □

The basic theorem for the rigorous proof of chaos for a dynamical system $f : \mathbb{R}^n \to \mathbb{R}^n$ was proved by [Smale (1967)]:

Theorem 1.5. *If f is a diffeomorphism and has a transversal homoclinic point, then there exists a Cantor set $\Lambda \subset \mathbb{R}^n$ in which f^m is topologically equivalent to the shift automorphism for some m.*

The existence of the shift automorphism implies the existence of a dense set of periodic orbits and an uncountably infinite collection of asymptotically aperiodic points of the map f within the set Λ. In this case, robust chaos (see Definition 2.5) is not possible due to the existence of the transversal homoclinic point because it is possible to use some geometrical illustrations as in [Weisstein (2002)] to show that the homoclinic tangency has the same topological structure as the horseshoe map.

An example can be found in [Galias (1997b)]:

Example 1.1. Let $\tilde{N}_0 = [-1,1] \times [-1,-0.5], \tilde{N}_1 = [-1,1] \times [0.5,1], P_1 = [-1,1] \times \mathbb{R}$ be the smallest vertical stripe containing $\tilde{N}_0$ and $\tilde{N}_1, M_- = [-1,1] \times (-\infty,-1), M_0 = [-1,1] \times (-0.5, 0.5), M_+ = [-1,1] \times (1,\infty)$ M_-, M_0, and M_+ are subsets of P_1 lying below, between, and above $\tilde{N}_0$ and $\tilde{N}_1$. Let N_{0D}, N_{0U} be the lower and upper horizontal edges and N_{0L}, N_{0R} be the left and right vertical edges of N_0 and similarly $N_{1D}, N_{1U}, N_{1L}, N_{1R}$ be the lower upper left and right edges of $\tilde{N}_1$. Then the Smale horseshoe map is a map linear on $\tilde{N}_0$ and $\tilde{N}_1$ defined by

$$f(x,y) = \begin{cases} \left(\frac{1}{4}x - \frac{1}{2}, \pm 5\left(y + \frac{3}{4}\right)\right), & \text{if } (x,y) \in \tilde{N}_0 \\ \left(\frac{1}{4}x + \frac{1}{2}, \pm 5\left(y - \frac{3}{4}\right)\right), & \text{if } (x,y) \in \tilde{N}_1. \end{cases} \quad (1.16)$$

Note that the Jacobian of f^n is given by

$$Df^n(x,y) = \begin{pmatrix} (0.25)^n & 0 \\ 0 & (\pm 5)^n \end{pmatrix}. \quad (1.17)$$

The study of this example is done in Exercise 1.5.17 below.

1.4 Symbolic dynamics

In this section, we give some elements of symbolic dynamics. Indeed, let $S_m = \{0, 1, ..., m-1\}$ be the set of nonnegative successive integers from 0 to $m-1$. Let Σ_m be the collection of all bi-infinite sequences of S_m with their elements, i.e., every element s of Σ_m is $s = (..., s_{-n}, ..., s_{-1}, s_0, s_1, ..., s_n, ...)$, $s_i \in \Sigma_m$. Let $\bar{s} \in \Sigma_m$ be another sequence $\bar{s} = (..., \bar{s}_{-n}, ..., \bar{s}_0, \bar{s}_1, ...)$, where $\bar{s}_i \in \Sigma_m$. Then:

Definition 1.10. The distance between s and $\bar{s}$ is defined as

$$d(s, \bar{s}) = \sum_{-\infty}^{\infty} \frac{1}{2^{|i|}} \frac{|s_i - \bar{s}_i|}{1 + |s_i - \bar{s}_i|}, \quad (1.18)$$

Hence the set Σ_m with the distance defined as (1.18) is a metric space, and one has the following results proved in [Robinson (1995)]:

Theorem 1.6. *The space Σ_m is compact, totally disconnected, and perfect.*

A set having the three properties given in Theorem 1.4 is a Cantor set, which frequently appears in the characterization of the complex structure of invariant sets in chaotic dynamical systems. However, let an m-shift map $\sigma : \Sigma_m \Sigma_m$ be defined by:

$$\sigma(s)_i = s_{i+1}. \quad (1.19)$$

Then the map σ defined by (1.19) has the following properties proved in [Robinson (1995)]:

Theorem 1.7. *(a) $\sigma(\Sigma_m) = \Sigma_m$, and σ is continuous.*
(b) The shift map σ as a dynamical system defined on Σ_m has the following properties:
(b-1) σ has a countable infinity of periodic orbits consisting of orbits of all periods,
(b-2) σ has an uncountable infinity of nonperiodic orbits, and
(b-3) σ has a dense orbit.

Statements (b) of Theorem 1.7 imply that the dynamics generated by the shift map σ (1.19) are sensitive to initial conditions, and therefore it is chaotic in the commonly accepted sense.

Now, let X be a separable metric space and f a continuous map $f: \tilde{Q} \to X$, where $\tilde{Q} \subset X$ is locally connected and compact.

Assume that [Kennedy & Yorke (2001)]: **Assumption A**. Suppose the following assumptions hold:

A1 There exist two subsets of $\tilde{Q}$ denoted by $\tilde{Q}_1$ and $\tilde{Q}_2$, respectively. The sets $\tilde{Q}_1$ and $\tilde{Q}_2$ are disjoint and compact.
A2 Each connected component of $\tilde{Q}$ intersects both $\tilde{Q}_1$ and $\tilde{Q}_2$.
A3 The cross number m of $\tilde{Q}$ with respect to f is not less than 2,

where the cross number m is defined as follows:

Definition 1.11. (a) A connection $\tilde{\Gamma}$ for $\tilde{Q}_1$ and $\tilde{Q}_2$ is a compact subset of $\tilde{Q}$ that intersects both $\tilde{Q}_1$ and $\tilde{Q}_2$.

(b) A preconnection $\tilde{\gamma}$ is a compact connected subset of $\tilde{Q}$ for which $f(\tilde{\gamma})$ is a connection.

(c) The cross number m is the largest number such that every connection contains at least m mutually disjoint preconnections.

Definition 1.11 can be generalized to the m domain $D_1, ..., D_{m-1}$ and D_m as follows:

Definition 1.12. Let $\tilde{\gamma}$ be a compact subset of $\tilde{Q}$ such that for each $1 \leq i \leq m$; $\tilde{\gamma}_i = \tilde{\gamma} \cap D_i$ is nonempty and compact. Then $\tilde{\gamma}$ is called a connection with respect to $D_1, ..., D_{m-1}$ and D_m. Let $\tilde{F}$ be a family of connections $\tilde{\gamma}'$s with respect to $D_1, ..., D_{m-1}$ and D_m satisfying the following property:

$$\tilde{\gamma} \in \tilde{F} \implies f(\tilde{\gamma}_i) \subset \tilde{F}. \quad (1.20)$$

Then $\tilde{F}$ is said to be an f-connected family with respect to $D_1, ..., D_{m-1}$ and D_m.

Hence the following fundamental result was proved in [Kennedy & Yorke (2001)]:

Theorem 1.8. Let f be the map satisfying Assumption A. Then there exists a compact invariant set $\tilde{Q}^I \subset \tilde{Q}$ for which $f\big|_{\tilde{Q}^I}$ is semi-conjugate to an m-shift map[9].

---
[9]This holds if there exists a continuous and onto map $h : \tilde{Q}^I \to \Sigma_m$ such that $h \circ f = \sigma \circ h$.

A more applicable version of the above result was given in [Yang & Tang (2004)] for piecewise-continuous maps as follows:

Theorem 1.9. *Let $\tilde{Q}$ be a compact subset of X and $f : \tilde{Q} \to X$ be a map satisfying the following conditions:*
(a) There exist m mutually disjoint subsets $D_1, ..., $ and D_m of $\tilde{Q}$, the restriction of f to each D_i, i.e., $f|_{D_i}$ is continuous.
(b)

$$\cup_{i=1}^m D_i \subset f(D_j), j = 1, 2, ..., m, \qquad (1.21)$$

then there exists a compact invariant set $K \subset \tilde{Q}$, such that $f|_K$ is semi-conjugate to m-shift dynamics.
(c) Suppose that there exists an f-connected family $\tilde{F}$ with respect to $D_1, ..., D_{m-1}$ and D_m. Then there exists a compact invariant set $K \subset \tilde{Q}$ such that $f|_K$ is semi-conjugate to m-shift dynamics.

Theorem 1.9 is used generally to estimate the topological entropy of piecewise-linear systems in terms of half-Poincaré maps.

1.4.1 The method of fixed point index

The method of fixed point index is a purely topological one and does not require any assumptions concerning derivatives, and its assumptions can be rigorously verified by computer-assisted computations. The proof of chaos using this method can be done by introducing horseshoe-type mappings which are geometrically similar to Smale's horseshoes using the concept of semi-conjugacy introduced below to the shift map on a finite number of symbols[10]. For example, the existence of the chaotic Hénon map (6.88) was proved in [Zgliczynski (1997a)] using this method.

For a good presentation of this method we need the following definitions:
Let (X, ρ) be a metric space. Let $Z \subset X$ and $x \in X$. Let $int(Z)$, $cl(Z)$, $bd(Z)$ denote, respectively, the interior, the closure, and the boundary of the set Z. Let $f : X \to X$ be any continuous map and $N \subset X$. Let G denote the class of pairs (f, Z) such that Z is an open and bounded set and $cl(Z) \subset X$. Let $f|_N$ denote the map obtained by restricting the domain of f to the set N. $Fixf$ denotes the set of fixed points of f. Assume that f has no fixed points on $bd(X)$. The fixed point index theory introduced here using algebraic topology was developed in [Dold (1980)]. The property of

---
[10] The class of TS-maps (the topological shifts) which includes as particular cases the Smale horseshoes [Smale (1967)] that are used in this case.

the fixed point index defines the existence of a fixed point if the fixed point index is nonzero. Here we give only the axiomatic definition:

Definition 1.13. The fixed point index is an integer valued function $I : G \to \mathbb{Z}$ satisfying the following axioms: (1) If W is an open set such that $Fixf \cap Z \subset W$, then $I(f,Z) = I(f,W)$. (2) If f is constant, then $I(f,Z) = 1$ if $f(Z) \in Z$, and $I(f,Z) = 0$ if $f(Z) \notin Z$. (3) If Z is a sum of a finite number of open sets $Z_i, i = 1,...,m$ such that $Z_i \cap Z_j \cap Fixf = \emptyset$ for $i \neq j$, then $I(f,Z) = \sum_{i=1}^{m} I(f,Z_i)$. (4) If $f : X \to X, f' : X' \to X'$ are continuous maps and $(f,Z),(f',Z')$ belong to the class G, then $I((f,f'), Z \times Z') = I(f,Z)I(f',Z')$. (5) If $F_t : X \times [0,1] \to \mathbb{R}^n$ is a homotopy, $Z \subset X$, and $Fixf \cap bd(Z) = \emptyset$ for every $t \in [0,1]$ then $I(F_0, Z) = I(F_1, Z)$.

Furthermore, we need the following definition of maximal invariant part:

Definition 1.14. The maximal invariant part of N with respect to f is defined by

$$Inv(N, f) = \cap_{i \in \mathbb{Z}} f \mid_N^{-i} (Z). \tag{1.22}$$

For any set $P = \bigcup_k P_k = \cup_k [a_k, b_k] \times [c_k, d_k] \subset \mathbb{R}^2$, where P_k are disjoint rectangles, we define the sets $L(P), R(P), V(P), H(P)$ that are equal to the union of left vertical, right vertical, vertical, and horizontal edges in P, respectively:

$$\begin{cases} L(P) = \cup_k \{a_k\} \times [c_k, d_k] \\ R(P) = \cup_k \{b_k\} \times [c_k, d_k] \\ V(P) = L(P) \cup R(P) \\ H(P) = \cup_k ([a_k, b_k] \times \{c_k\} \cup [a_k, b_k] \times \{d_k\}). \end{cases} \tag{1.23}$$

Let us fix $u, d \in \mathbb{R}, u > d$ and a sequence $a_{-1} = -\infty < a_0 < a_1 < ... < a_{2K-2} < a_{2K-1} < a_{2K} = \infty$, where $a_i \in \mathbb{R}$, for $i = 0, 1, ..., 2K - 1$. Let

$$\begin{cases} N_i = [a_{2i}, a_{2i+1}] \times [d, u], \text{ for } i = 0, ..., K-1 \\ E_i = [a_{2i-1}, a_{2i}] \times [d, u], \text{ for } i = 0, ..., K \\ N = N_0 \cup N_1 \cup ... \cup N_{k-1} \\ E = E_0 \cup E_1 \cup ... \cup E_{k-1} \cup E_k. \end{cases} \tag{1.24}$$

Then one has the following result proved in [Zgliczynski (1997b)]:

Lemma 1.1. *The sets E_i, N_i are contained in the horizontal strip $(-\infty, \infty) \times [d, u]$ in the following order (if one compares x-coordinates)*

$$E_0 < N_0 < E_1 < N_1 < ... < E_{k-1} < N_{k-1} < E_k. \tag{1.25}$$

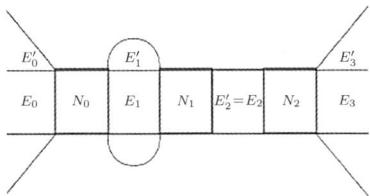

Fig. 1.6   An example of sets N_i, E_i, E'_i for $K = 3$.

For $i = 0, ..., K$, let us define the sets E'_i as follows:
$$\begin{cases} E_i = E'_i \cap (-\infty, \infty) \times [d, u] \\ cl\,(E'_i) \cap (N - V(N)) = \emptyset \\ cl\,(E'_i) \cap cl\,(E'_j) = \emptyset \text{ for } i \neq j,. \end{cases} \quad (1.26)$$
and suppose that there exist continuous homotopies $h_i : [0, 1] \times E'_i \to E'_i$ such that
$$\begin{cases} h_i(0, p) = p \text{ for } p \in E'_i \\ h_i(1, p) \in E'_i \text{ for } p \in E'_i \\ h_i(t, p) = p \text{ for } p \in E'_i \text{ and } t \in [0, 1] \end{cases} \quad (1.27)$$
Then one has the following result proved in [Zgliczynski (1997b)]:

Lemma 1.2. *(a) The set E'_i can be continuously deformed to the set of E_i without any intersection with the set N*[11]. *(b) If*
$$E' = E'_0 \cup E'_1 \cup ... \cup E'_K, \quad (1.28)$$
then
$$E'_i \cap N_j = \emptyset \text{ for } i, j = 0, 1, ..., K - 1. \quad (1.29)$$

Figure 1.6 presents a schematic drawing of the sets N_i, E_i, E'_i for $K = 3$.

Definition 1.15. Let the sets N_i, E_i, E'_i be as above. Let D be an open set such that $N \subset D$ and a map $f : D \to \mathbb{R}^2$ be continuous. We say that f is a TS-map (topological shift) (relative to the sets N, E, E') if there exist functions $l, r : \{0, 1, ..., K - 1\} \to \{0, 1, ..., K\}$ such that the following conditions hold:
$$\begin{cases} f(L(N_i)) \subset E'_{l(i)} \\ f(R(N_i)) \subset E'_{r(i)} \\ f(N) \subset E' \cup N. \end{cases} \quad (1.30)$$

---

[11]In this case, the set E_i is called the *deformation retract* of E'_i, where a deformation retraction is a map that captures the idea of continuously shrinking a space into a subspace.

1.4.1.1 Periodic points of the TS-map

In this section, we describe the concept of the periodic points of the TS-map f. These points were characterized by periodic infinite sequences $c = (c_i)_{i \in \mathbb{N}}$ of symbols $0, 1, ..., K-1$ with the property $f^i(x) \in N_{c_i}$ for $i \in \mathbb{N}$. Indeed, let

$$\begin{cases} \Sigma_K = \{0, 1, ..., K-1\}^{\mathbb{Z}} \\ \Sigma_K^+ = \{0, 1, ..., K-1\}^{\mathbb{N}}. \end{cases} \quad (1.31)$$

Then the following lemma was proved in [Zgliczynski (1997b)]:

Lemma 1.3. *(a) The sets Σ_K, Σ_K^+ are topological spaces with Tichonov topology.*
(b) On Σ_K, Σ_K^+, the shift map σ is given by

$$\sigma((c))_i = c_{i+1}. \quad (1.32)$$

Now let $A = (\alpha_{ij})$ be a $K \times K$ matrix, $\alpha_{ij} \in \mathbb{R}^+ \cup \{0\}, i, j = 0, 1, ..., K-1$. We define $\Sigma_A \subset \Sigma_K$ and $\Sigma_A^+ \subset \Sigma_K^+$ by

$$\begin{cases} \Sigma_A = \left\{c = (c_i)_{i \in \mathbb{Z}} : \alpha_{c_i c_{i+1}} > 0\right\} \\ \Sigma_A^+ = \left\{c = (c_i)_{i \in \mathbb{N}} : \alpha_{c_i c_{i+1}} > 0\right\}. \end{cases} \quad (1.33)$$

Then one has the following result proved in [Zgliczynski (1997b)]:

Lemma 1.4. *The sets Σ_A and Σ_A^+ are invariant under σ.*

Let f be a TS-map. To relate the dynamics of f on $Inv(N, f)$ with shift dynamics on Σ_K^+, we introduce the transition matrix of f denoted by $A(f)$. We define $A(f)_{ij}$, where $i, j = 0, 1, ..., K-1$ as follows:

$$A(f)_{ij} = \begin{cases} 1, & \text{if } E_{l(i)} < N_j < E_{r(i)} \text{ or } E_{l(i)} > N_j > E_{r(i)} \\ 0, & \text{otherwise}. \end{cases} \quad (1.34)$$

Then one has the following lemma proved in [Zgliczynski (1997b)]:

Lemma 1.5. *If N_j lies between the images[12] of vertical edges of N_i, then $A(f)_{ij} \neq 0$ for all $i, j = 0, 1, ..., K-1$.*

---

[12] In this case, one can deform the image by the homotopies h if necessary.

1.4.1.2 Existence of semi-conjugacy

To prove the existence of a horseshoe-type mapping in a system under consideration, the concept of semi-conjugacy is used as follows: For $i \in \mathbb{N}$ we define the map $\pi_i : Inv(N, f) \to \{0, 1, ..., K\}$ given by

$$\pi_i(x) = j \text{ if and only if } f^i(x) \in N_j. \tag{1.35}$$

Now we define the symbolic map $\pi : Inv(N, f) \to \Sigma_K^+$ by

$$\pi(x) = (\pi_i(x))_{i \in \mathbb{N}}. \tag{1.36}$$

Thus the map π given by (1.35) assigns to the point x the indices of the rectangles N_i its trajectory goes through. Then one has the following lemma proved in [Zgliczynski (1997b)]:

Lemma 1.6. *(a) We have*

$$\pi \circ f = \sigma \circ \pi. \tag{1.37}$$

(b) If f is also a homeomorphism, then the definition of π_i can be extended to all integers, and the domain of π is Σ_K.

Lemma 1.6(a) indicates the existence of a semi-conjugacy between f and σ, and this semi-conjugacy is not a sign of complicated dynamics because the set $Inv(N, f)$ is finite or empty. However, the dynamics are complicated if the set $\pi(Inv(N, f))$ is infinite as confirmed by the following theorem that gives the characterization of this set for TS-maps defined in [Zgliczynski (1997b)]:

Theorem 1.10. *Let f be a TS-map. Then $\Sigma_{A(f)}^+ \subset \pi(Inv(N, f))$. The pre-image of any periodic sequence from $\Sigma_{A(f)}^+$ contains periodic points of f. If we additionally suppose that f is a homeomorphism, then $\Sigma_{A(f)} \subset \pi(Inv(N, f))$.*

We have the following definition:

Definition 1.16. *Let $N \subset \mathbb{R}^d$ be a compact set and $f : N \to \mathbb{R}^d$ be a continuous map. Then the set N is called an isolating neighborhood if and only if*

$$Inv(N, f) \subset Int(N). \tag{1.38}$$

Then the following theorem was proved in [Zgliczynski (1996)]:

Theorem 1.11. *Let $N = \cup_{i=0}^{K-1} N_i$, where $N_i \subset \mathbb{R}^d$ are compact and disjoint. Let $F : [0, 1] \times N \to \mathbb{R}^n$ be a continuous map such that N is an* isolating neighborhood for F_λ for $\lambda \in [0,1]$. Then for every finite sequence $(\sigma_0, \sigma_1, ..., \sigma_n) \in \{0, 1, ..., K\}^{n+1}$, the fixed point index

$$I\left(F_\lambda^{n+1}, N_{\sigma_0} \cap F_\lambda^{-1}(N_{\sigma_1}) \cap ... \cap F_\lambda^{-n}(N_{\sigma_n})\right) \quad (1.39)$$

is defined and does not depend on λ.

An example of the applications of this method to the Hénon map (6.88) can be found in Sec. 10.2.

1.5 Exercises

Exercise 1. Show that a limit cycle of a flow φ_t is a fixed point of its associated Poincaré map P, and a period-m closed orbit of P is a subharmonic solution (relative to the considered section S) of φ_t, a chaotic orbit of P corresponds to a chaotic solution of φ_t, a stable periodic point of P corresponds to a stable periodic orbit of φ_t, and an unstable periodic point of P corresponds to a saddle-type periodic orbit of φ_t.

Exercise 2. Show that the Poincaré map P is not defined for points, trajectories of which never come back to the section S defining the Poincaré map.

Exercise 3. Give an example of a Poincaré map which is well defined, but not continuous.

Exercise 4. Show that the generalized Poincaré map H defined for piecewise-linear systems has the same properties as the Poincaré map P.

Exercise 5. Consider the following system [Zeraoulia (2007)]
$$\begin{cases} x' = a(y-x) + yz \\ y' = (c-a)x + cy - xz \\ z' = xy \quad bz. \end{cases} \quad (1.40)$$

(a) Show that system (1.40) is symmetric under the coordinate transformation $(x, y, z) \longrightarrow (-x, -y, z)$.

(b) Calculate the divergence of the flow of (1.40).

(c) Deduce that the system (1.40) is dissipative when $a - c + b > 0$ and that it converges exponentially to a set of measure zero.

(d) Show that system (1.40) has the following equilibria: $P_1 = (0, 0, 0)$, $P_2 = (x_2, y_2, z_2)$, $P_3 = (x_3, y_3, z_3)$, where
$$\begin{cases} y_2 = \sqrt{\frac{1}{c}\left(\frac{3}{2}abc + \frac{1}{2}\sqrt{4a^3b^2c + a^2b^2c^2}\right)}, \\ y_3 = -\sqrt{\frac{1}{c}\left(\frac{3}{2}abc + \frac{1}{2}\sqrt{4a^3b^2c + a^2b^2c^2}\right)}, \end{cases} \quad (1.41)$$

and if $c > \frac{2a}{5}$, then there are two additional fixed points $P_4 = (x_4, y_4, z_4)$, $P_5 = (x_5, y_5, z_5)$, where

$$\begin{cases} y_4 = \sqrt{\frac{1}{c}\left(\frac{3}{2}abc - \frac{1}{2}\sqrt{4a^3b^2c + a^2b^2c^2}\right)} \\ y_5 = -\sqrt{\frac{1}{c}\left(\frac{3}{2}abc - \frac{1}{2}\sqrt{4a^3b^2c + a^2b^2c^2}\right)}, \end{cases} \quad (1.42)$$

and determine their stability.

(e) Draw in the xz-plane the chaotic attractors obtained from system (1.40) for $b = 3$, $c = 28$, and (i) $a = 35$, (ii) $a = 36$.

(f) Plot the variation of the largest Lyapunov exponent of system (1.40) versus the parameter (i) $a \in [32, 48]$, with $b = 3, c = 28$, (ii) $b \in [0.4, 5]$, with $a = 36, c = 28$, (iii) $c \in [20, 30]$, with $a = 36, b = 3$ for several values of the initial conditions.

(g) Plot the bifurcation diagram using an appropriate Poincaré section Σ defined by:

$$\Sigma = \left\{(y, z) \in \mathbb{R}^2 \ / \ x = 0\right\}, \quad (1.43)$$

for (i) $a \in [32, 48]$, with $b = 3, c = 28$, (ii) $b \in [0.4, 5]$, with $a = 36, c = 28$, (iii) $c \in [20, 30]$, with $a = 36, b = 3$ for several values of the initial conditions.

(h) Deduce that there are some regions in the abc-space that support robust chaotic (see Definition 2.5) attractors (only single attractors) for system (1.40).

Exercise 6. Show the relation (1.9) according to (1.8).

Exercise 7. Show the mean value form of Theorem 1.2.

Exercise 8. Show that $G(z) = 0$ if and only if x_0 is a fixed point of H^m, where G is given by (1.3).

Exercise 9. Prove Theorem 1.1.

Exercise 10. Compare the effectiveness of the *interval Newton operator*, the Krawczyk operator, and the Hansen–Sengupta operator given in Sec. 1.2.2.

Exercise 11. Show that the image of the vertical edges defined in (1.23) does not intersect the set N given in (1.24) and the image of N is contained in a set that can be continuously deformed to the horizontal strip without any intersection with the horizontal edges of N.

Exercise 12. Draw the Smale horseshoe map f after four and five iterations.

Exercise 13. Prove that the presence of a horseshoe in a dynamical system implies the following features:
   (1) There are infinitely many periodic orbits, especially ones with arbitrarily long periods.
   (2) The number of periodic orbits grows exponentially with the period.
   (3) In any small neighborhood of any point of the fractal invariant set Λ, there is a point of a periodic orbit.

Exercise 14. Find two examples of a flow and a map that are chaotic in the sense of Smale.

Exercise 15. Show Theorem 1.3 using the geometrical formation of the Smale horseshoe map f.

Exercise 16. Prove Theorem 1.5 of Smale.

Exercise 17. (a) Draw all the sets included in the definition of the map (1.16).
   (b) Find the images of the sets $\tilde{N}_0$ and $\tilde{N}_1$.
   (c) Find the fixed points of the map (1.16).
   (d) Using the linearity of the horseshoe map, show that for every $n \in \mathbb{N}$, there is still a period-n point (u,v) of the map (1.16).
   (e) Show that the Jacobian of f^n is given by (1.17).
   (f) Show that there exists a neighborhood U of the point (u,v) which does not contain other fixed points of f^n.
   (g) Calculate the fixed point index of the pair (f^n, U), and deduce that there are 2^n fixed points of the map f^n.

Exercise 18. Prove that $d(s,\bar{s})$ defined in (1.18) is an instance of the set Σ_m in a metric space.

Exercise 19. Prove that the space Σ_m is compact, totally disconnected, and perfect (Theorem 1.6).

Exercise 20. Prove that Theorem 1.7 and the statements (b) of Theorem 1.7 imply that the dynamics generated by the shift map σ (1.19) is sensitive to initial conditions.

Exercise 21. Prove Theorem 1.9 about the semi-conjugacy to a m-shift map of a piecewise-linear map.

Chapter 2

Robustness of Chaos

In this chapter, we discuss the concept of robust chaos. First, in Sec. 2.1, we describe the notion of strange attractors with their topological properties, in particular the *domains of attraction* which play a crucial role in characterizing robust chaos as discussed in Sec. 2.1.3. Section 2.1.2 gives several definitions of the word *robust*, in particular, robust chaos is defined, and its relations with density and persistence is discussed in Sec. 2.2 and Sec. 2.3, respectively. At the end of this chapter, several exercises and one open problem are given to fix the ideas and experiences.

2.1 Strange attractors

2.1.1 *Concepts and definitions*

Chaos is the idea that a system will produce very different long-term behaviors when the initial conditions are perturbed only slightly. Several researchers have defined and studied strange attractors. The first was Lorenz [Lorenz (1963)] in 1963, where he proposed a simple mathematical model of a weather system. This system displays a complex behavior in spite of the simple from of the equations. This behavior has a sensitive dependence on the initial conditions of the model. Therefore, the prediction of a future state of the system is impossible. The first fractal shape of the Lorenz system was identified by Ruelle in 1979, and it took the form of a *butterfly* (*a butterfly in the Amazon might, in principle, ultimately alter the weather in Kansas*).

Many simple examples are now known of mathematical models that exhibit chaos, and they have been extensively studied. One of the oldest and simplest such system is the logistic map (6.25), which was used to model biological population dynamics but has now found application in many other fields. In some ways, it is the prototypical example of chaos in discrete-time systems (iterated maps). Chaotic systems are usually characterized by one or more parameters that control the behavior of the system and give rise to bifurcations and other changes in the dynamics.

Characterizing chaos is a large subject, and the most useful tests for this type of behavior are the following:

(1) Chaotic motions are more complicated than stationary, periodic, or quasi-periodic, and they have very complicated shapes, called *strange attractors*.
(2) There is a sensitive dependence on initial conditions, *i.e.*, nearby solutions diverge exponentially fast.
(3) There is a coexistence phenomenon, *i.e.*, chaotic orbits coexist with a (countable) infinity of unstable periodic orbits. See Chapter 10.

Recently, chaos has been shown to be very useful and to have great potential in many technological disciplines such as in information and computer sciences, power system protection, biomedical systems analysis, flow dynamics, liquid mixing, encryption, communications, and so on [Chen & Dong (1998), Chen (1999), Yu et al. (2007)]. A new application of chaos theory given in [Yu et al. (2007)] is the adaptive synchronization with unknown parameters as discussed for a unified chaotic system by using the Lyapunov method and the adaptive control approach. Some communication schemes, including chaotic masking, chaotic modulation, and chaotic shift key strategies, are then proposed based on the modified adaptive method. In these schemes, the transmitted signal is masked by a chaotic signal or modulated into the system, which effectively blurs the constructed return map and can resist this return map attack. The driving system with unknown parameters and functions is almost completely unknown to the attackers, and so it is more secure to apply this method to communications. More detailed analysis of some strange attractors and their major properties can be found in several places in this book. First, we give a definition of an attractor [Milnor (1985a-b)]:

Definition 2.1. (a) An attractor of a dynamical system is a subset of the state space to which orbits originating from typical initial conditions tend as time increases.

For some problems related to uniform hyperbolic dynamical systems (see Chapter 6) (convergence only for points in some *large* subset of a neighborhood of the attractor), we need the following definition:

Definition 2.2. An attractor is a compact invariant subset A such that the trajectories of all points in a neighborhood U converge to A as time goes to infinity, and A is dynamically indecomposable (or transitive); there is some trajectory dense in A.

Formally, Definition 2.2 can be reformulated as follows:

Definition 2.3. A closed subset A of the nonwandering set[1] $\Omega(f)$ is called an attractor when (1) A is f-invariant, i.e., $f(A) = A$, (2) there exists a neighborhood L of A in $\mathbb{R}^n$: $\bigcap_{n \geq 0} f^n(L) = A$, and (3) f is transitive on A.

A definition of a *strange* attractor is given by the following:

Definition 2.4. (b) (Strange attractor) Suppose $A \subset \mathbb{R}^n$ is an attractor. Then A is called a strange attractor if it is chaotic.

Generally, to prove that a dynamical system has a strange attractor, we might proceed as follows. **Step 1**: Find in the phase space of the system a *trapping region* M. **Step 2**: Show that M contains a chaotic invariant set Λ. This is done by several methods, one of which is to show that there exists a homoclinic orbit or heteroclinic cycle inside M. **Step 3**: Construct the attractor A as follows:

$$\begin{cases} \cap_{t>0} f_t(M) = A, & \text{if } f \text{ is flow} \\ \cap_{n>0} f^n(M) = A, & \text{if } f \text{ is a map}. \end{cases} \quad (2.1)$$

So $\Lambda \subset A$, and there is sensitive dependence on initial conditions. Finally, the set A is a strange attractor if

(1) (a) The sensitive dependence on initial conditions on Λ extends to A and (b) the set A is topologically transitive.

2.1.2 Robust chaos

In the dictionary, the word "*robust*" has the following meanings:

---
[1] A point x in a manifold M is said to be nonwandering if, for every open neighborhood U of x, it is true that $f^n(U) \cap U \neq \emptyset$ for a map f for some $n > 0$. In other words, every point close to x has some iterate under f which is also close to x. The set of all nonwandering points is known as the nonwandering set of f.

(1) strong and healthy; hardy; vigorous: a robust young man, a robust faith, a robust mind.
(2) strongly or stoutly built: his robust frame.
(3) suited to or requiring bodily strength or endurance: robust exercise.
(4) rough, rude, or boisterous: robust drinkers and dancers.
(5) rich and full-bodied: the robust flavor of freshly brewed coffee.

Generally, a system is said to be robust if it is capable of coping well with variations in its operating environment with minimal loss of functionality. In other words, robustness is the ability of a system to continue to operate correctly across a wide range of operating conditions. In biology, especially in genetics, mutational robustness describes the extent to which an organism's phenotype remains constant in spite of mutations. In the field of morphology, the word robust is used to describe species with a morphology based on strength and heavy build. In computer science, the word robust is used to describe some properties of an algorithm, *i.e.*, if it continues to operate despite abnormalities in input, calculations, *etc.* The so-called *"fuzz testing"*[2] is used to prove robustness since this type of testing involves unexpected inputs. In statistics, the word robust is used to describe some properties of statistical techniques: A robust statistical technique is a technique that performs well even if its assumptions are violated. In economics, the word robust defines the ability and effectiveness of a financial trading system under different markets and market conditions. In decision making, the word robust is used to describe the quality of a decision, *i.e.*, a decision is robust if it is possible, immune to uncertainty, and is good in some sense. Chaotic dynamical systems display two kinds of chaotic attractors: One type has fragile chaos (the attractors disappear with perturbations of a parameter or coexist with other attractors), and the other type has robust chaos defined as follows [Banerjee *et al.* (1998)]:

Definition 2.5. Robust chaos is defined by the absence of periodic windows and coexisting attractors in some neighborhood in the bifurcation parameter space.

The existence of these windows in some chaotic regions means that small changes of the parameters would destroy the chaos, implying the fragility

---

[2]Fuzz testing is a software testing technique that provides random data to the inputs of a program. If the program fails, the defects can be noted. This test is extremely simple and free of preconceptions about system behavior.

of this type of chaos. Contrary to this situation, there are many practical applications, such as in communications and spreading the spectrum of switch-mode power supplies to avoid electromagnetic interference [Ottino (1989), Ottino et al. (1992)] where it is necessary to obtain reliable operation in the chaotic mode, and thus robust chaos is required. Another practical example is in electrical engineering where robust chaos is demonstrated in [Banerjee et al. (1998)] and described in Sec. 12.1.6. The occurrence of robust chaos in a smooth system is proved and discussed in [Andrecut & Ali (2001a)], which includes a general theorem and a practical procedure for constructing S-unimodal maps (see Sec. 11.1.1) that generate robust chaos. This result contradicts the conjecture that robust chaos cannot exist in smooth systems [Banerjee et al. (1998)] (see Sec. 11.2). On the other hand, many methods are used to search for a smooth and robust chaotic map, for example in [Andrecut & Ali (2001a-b-c), Gabriel (2004)] where a one-dimensional smooth map that generates robust chaos in a large domain of the parameter space is presented. In [Jafarizadeh & Behnia (2002)], simple polynomial unimodal maps that show robust chaos are constructed (see Secs. 11.2.1 and 11.2,3). Other methods and algorithms are given in the discussion below. In [Avrutin & Schanz (2008)], it is shown by an elementary example that the term *robust chaos* must be used carefully since chaotic attractors which are robust in the sense of [Banerjee et al. (1998)] are not necessarily robust in the sense of [Milnor (1985)] (see Definition 2.1). This confirmation was done by the study and description of a novel bifurcation phenomenon called the *bandcount adding scenario* occurring in the two-dimensional parameter space of the one-dimensional piecewise-linear map given by:

$$x_{n+1} = \begin{cases} f_l(x_n) = ax_n + \mu + 1, & \text{if } x_n < 0 \\ f_r(x_n) = ax_n + \mu - 1, & \text{if } x_n > 0, \end{cases} \quad (2.2)$$

i.e., an infinite number of interior crises bounding the regions of multi-band attractors were detected in the region of chaotic behavior. This phenomenon leads to a self-similar structure of the chaotic region in parameter space. The past decade has seen heightened interest in the exploitation of robust chaos for applications to engineering systems. Since there are many areas for applications of robust chaos, we give here two examples of applications of robust chaos in the real world along with the other examples in this book. The first is given in [Sun et al. (2007)], which suggests a new approach (with experiments, statistical analysis, and key space analysis) for image encryption based on a robust high-dimensional chaotic map.

The new scheme employs the so-called *cat map* (6.18) to shuffle the positions and then confuses the relationship between the cipher-image and the plain-image by using the high-dimensional preprocessed Lorenz chaotic map. This work shows that the proposed image encryption scheme provides an efficient and secure means for real-time image encryption and transmission. The second example is given in [Nithin *et al.* (2006)], which uses the notion of robust chaos in another new encryption scheme. A recent bibliography on the applications of robust chaos in the real world is collected in these papers [Ottino (1989), Ottino *et al.* (1992), Majumdar & Mitra (1994), Mukul & Mitra (1994), Dogaru *et al.* (1996), Ashwin & Rucklidge (1998), Banerjee *et al.* (1998), Kei *et al.* (1998), Ruxton & Rohani (1998), Tapan & Gerhard (1999), Banerjee *et al.* (1999), Potapov & Ali (2000), Dafilis *et al.* (2001), Robert & Robert (2002), Yulmetyev *et al.* (2002), Adrangi & Chatrath (2003), Shanmugam & Leung (2003), Vikas & Kumar (2004), Vijayaraghavan & Leung (2004), Kuznetsov & Seleznev(2005), Yau & Yan (2004-2006), Vano *et al.* (2006), Drutarovsky & Galajda (2007), Sun *et al.* (2007), Henk *et al.* (2008)]. As we see throughout this book, robust chaos occurs in several types of dynamical systems: discrete, continuous-time, autonomous, nonautonomous, smooth, and nonsmooth, with different topological dimensions, and it has been confirmed either analytically, numerically, or experimentally, or in a combination of all these methods. On the other hand, there is no robust chaos in the quasi-attractor type because it is structurally unstable as confirmed in Chapter 10. There are several methods for proving the robustness of chaos in dynamical systems in the sense that there are no coexisting attractors and no periodic windows in some neighborhood of the parameter space. Some of these methods can be found individually in several places in this book, in particular, in Chapters 3, 4, 6, 7, 8, 9, 11, and 12 using the following methods: Normal form analysis, unimodality, metric entropy, construction of dynamical systems using basis of the robustness or the nonrobustness, geometric methods, detecting unstable periodic solutions, ergodic theory, weight-space exploration, numerical methods, and a combination of these methods.

2.1.3 *Domains of attraction*

Attractors of a dynamical system can correspond to stationary, periodic, quasi-periodic, or chaotic behaviors of different types, *i.e.*, stable, unstable, saddle, *etc.* Chaotic dynamical systems can have more than one attractor depending on the choice of initial conditions, and thus the notion of basin of attraction comes from the following property:

Definition 2.6. The basin of attraction $B(A)$ for an attractor A is the set of initial conditions leading to long-time behavior that approaches that attractor.

The following are some notes and remarks about the basin of attraction:

(1) The qualitative behavior of a dynamical system depends fundamentally on which basin of attraction the initial condition resides because the basic topological structure of such regions can vary greatly from system to system.
(2) The basin boundary can be a smooth or fractal; the fractacality is a result of chaotic motion of orbits on the boundary [McDonald et al. (1985)].
(3) Basin boundaries can have qualitatively different types. For example, the nature of a basin can change from a simple smooth curve to a fractal. This phenomenon is called *metamorphosis* [Grebogi et al. (1987c)].
(4) (Riddled basins) There is several types of basin topology that occur in dynamical systems such as a system with a symmetry or some other constraint or one having a smooth invariant manifold, *i.e.*, a smooth surface or hypersurface such that any initial condition in the surface generates an orbit that remains in the surface. In this case, the system under consideration has a *bizarre* type of basin structure called a *riddled basin of attraction* [Alexander et al. (1992), Ott et al. (1994)]. This type of basin can exist in a dynamical system if, for example, the following conditions exist: (1) An invariant submanifold with chaotic dynamics exists. (2) The Lyapunov exponent normal to the invariant submanifold is negative, *i.e.*, the submanifold is an attractor. (3) There exist other attractors. An example of such a situation is the 2-D map given in [Alexander et al. (1992)] by

$$z_{n+1} = z_n^2 - (1 + \lambda i)\bar{z}_n, z = x + iy, \bar{z} = x - iy, i^2 = -1 \quad (2.3)$$

or equivalently in $\mathbb{R}^2$ the map (2.3) given by:

$$\begin{cases} x_{n+1} = x_n^2 - y_n^2 - x_n - \lambda y_n \\ y_{n+1} = 2x_n y_n - \lambda x_n + y_n. \end{cases} \quad (2.4)$$

For some values of λ, the map (2.4) has three attractors, and each internal dynamic is chaotic as in the logistic map (6.25). In this case, the basins of the three attractors are *intermingled* very complexly. Other examples of systems with riddled basins can be found in [Jing-ling Shen et al. (1996), Maistrenko et al. (1998), Cao (2004)].

Formally, it is possible to define a riddled basin as follows:

Definition 2.7. (a) A closed invariant set A is said to be a weak attractor in the Milnor sense if its basin $B(A)$, *i.e.*, the set of points whose ω-limit sets of x belongs to A has positive Lebesgue measure. (b) The basin of attraction $B(A)$ of an attractor A is riddled if its complement intersects every disk in a set of positive measure.

In this case, a riddled basin corresponds to an extreme form of uncertainty because it does not include any open subset and is full of holes. A set of basins which are dense in each other is called an *intermingled set*, *i.e.*,

Definition 2.8. The basins of attraction $B(A)$ and $B(B)$ of the attractors A and B are intermingled if each disk which intersects one of the basins in a set of positive measure also intersects the other basin in a set of positive measure.

Some illustrated examples can be found in [Djellit & Boukemara (2007)] where it is shown that a three-parameter family of piecewise-linear maps of the plane display intermingled basins which explains the coexistence of several attractors. See also [Gumowski & Mira (1980), Alexander *et al.* (1992), Ding & Yang (1996), Aharonov *et al.* (1997), Bischi & Gardini (1998), Paar & Pavin (1998), Ding (2004), Kapitaniak *et al.* (2003), Yakubu (2003), Yang (2003a-b), Taixiang & Hongjian (2007), Giesl (2007), Giesl & Heiko (2007), Morales (2007), Han (2007)]. In all these examples, the resulting chaos is not robust due to the coexistence of other orbits. For quasi-attractors introduced in Chapter 10, coexistence of many attractors implies that the basins of attraction cannot be regular because it is well known that basin boundaries arise in dissipative dynamical systems when two or more attractors are present. In such situations each attractor has a basin of initial conditions that lead asymptotically to that attractor. The sets that separate different basins are called *basin boundaries*. In some cases, the basin boundaries can have very complicated fractal structure and hence pose an additional impediment to predicting long-term behavior.

2.2 Density and robustness of chaos

In this section, we discuss the possible relations between robustness and hyperbolicity, studied in detail in Chapters 6 and 7. Indeed, for linear systems, one of the most interesting features of hyperbolicity (see Chapter 6) is robustness, *i.e.*, any matrix that is close to a hyperbolic one in the sense that corresponding coefficients are close, is also hyperbolic. In this case, the stable and unstable subspaces are not necessarily identical, but the dimensions remain the same. If hyperbolicity is dense, then any matrix is close to a hyperbolic one up to arbitrarily small modifications of the coefficients. For nonlinear systems, the same properties hold by replacing the word *matrix* by the word *dynamical system*. In fact, for one-dimensional hyperbolic attractors of diffeomorphisms of surfaces, it was shown in [Zhirov et al. (2005), Zhirov (2006)] that the topological conjugacy problem is a common property for this situation, and an algorithmic solution based on the combinatorial method for describing hyperbolic attractors of surface diffeomorphisms is given in the following problem:

Problem 2.1. *(The problem of enumeration of attractors) Let Λ_f and Λ_g be one-dimensional hyperbolic attractors of diffeomorphisms $f : M \to M$ and $g : N \to N$, where M and N are closed surfaces, either orientable or not. Does there exist a homeomorphism $h : U(\Lambda_f) \to V(\Lambda_g)$ of certain neighborhoods of attractors such that $f \circ h = h \circ g$? (The topological conjugacy problem) Given $h > 0$, find a representative of each class of topological conjugacy of attractors with a given structure of accessible boundary (boundary type) for which the topological entropy is no greater than h.*

2.3 Persistence and robustness of chaos

In this section, we discuss the relation between persistence and robustness of chaos. In fact, dynamical persistence means that a behavior type, *i.e.*, equilibrium, oscillation, or chaos does not change with functional perturbation or parameter variation. Mathematically, *persistent chaos* (p-chaos) of degree-p for a dynamical system can be defined as follows:

Definition 2.9. Assume a map $f_\xi : X \to X$ ($X \subset R^d$) depends on a parameter $\xi \in R^k$. The map f_ξ has chaos of degree-p on an open set $\mathcal{O} \subset X$ that is persistent for $\xi \in \mathcal{U} \subset R^k$ if there is a neighborhood $\mathcal{N}$ of $\mathcal{U}$ such that $\forall\ \xi \in \mathcal{N}$, the map f_ξ retains at least $p \geq 1$ positive Lyapunov characteristic exponents (LCEs) Lebesgue almost every X in $\mathcal{O}$.

The choice of p is arbitrary. For example, the condition where p equals the number of positive LCEs is a very strict constraint; specifying a minimum p or ratio of p to the maximum number of positive exponents are weaker. Flexibility allows one to analyze (say) systems with 10^6 unstable directions in which a change in 1% of the geometry is undetectable, but a 50% change is. The notion of persistent chaos differs from that of a robust chaotic attractor in several ways, especially in that the uniqueness of the attractor is not required on the set U since there is little physical evidence indicating that such strict forms of uniqueness are present in many complex physical systems. For a low-dimensional system, uniqueness is markedly more difficult to establish. A recent result about persistence of chaos in high dimensions can be found in [Albers et al. (2006)] where it is shown that as the dimension of a typical dissipative dynamical system is increased, the number of positive Lyapunov exponents increases monotonically and the number of parameter windows with periodic behavior decreases. The method of analysis is an extensive statistical survey of *universal approximators*[3] given by single-layer recurrent neural networks of the form

$$x'_t = \beta_0 + \sum_{i=1}^{n} \beta_i \tanh s \left(\omega_{i0} + \sum_{j=1}^{d} \omega_{ij} x_{t-j} \right) \qquad (2.5)$$

which are maps from $\mathbb{R}^d$ to $\mathbb{R}$ and denoted by $f_{s,\beta,\omega}$. Here n is the number of neurons, d is the number of time lags which determines the system's input embedding dimension, and s is a scaling factor for the connection weights ω_{ij}. The initial condition is $(x_1, x_2, ..., x_d)$, and the state at time t is $(x_t, x_{t+1}, ..., x_{t+d-1})$. For an open problem about the occurrence of robust chaos in system (2.5), see Exercise 36. More details on robust chaos in neural networks can be found in [Sompolinsky et al. (1988), Tirozzi & Tsokysk (1991), Cessac et al. (1994), Molgedey et al. (1991), Laughton & Coolen (1994), Amit (1989), Eckhorn et al. (1988), Grey et al. (1989), Steriade et al. (1993), Dogaru et al. (1996), Potapov & Ali (2000)] and references therein.

2.4 Exercises

Exercise 22. Give five different definitions of the concept of an attractor and compare them.

---

[3]The approximation theorems of [Hornik et al. (1990)] and time-series embedding of [Sauer et al. (1991)] establish an equivalence between these neural networks and general dynamical systems [Albers & Sprott (2006)].

Exercise 23. Give five different definitions of the concept of a chaotic attractor and compare them.

Exercise 24. Describe the role of sensitive dependence on initial conditions in the formation of a chaotic attractor, and explain the sentence "*A butterfly in the Amazon might, in principle, ultimately alter the weather in Kansas.*"

Exercise 25. Describe a method that permits visualizing a chaotic attractor.

Exercise 26. Give five examples of simple dynamical systems displaying robust chaos in the sense of Definition 2.5.

Exercise 27. Give an example of a situation where a real phenomenon was modelled by the logistic map.

Exercise 28. Describe some routes to chaos in dynamical systems.

Exercise 29. Give a criterion that measures the complexity of a robust chaotic attractor.

Exercise 30. Describe a method that permits visualizing the coexistence phenomenon in dynamical systems.

Exercise 31. Describe two real applications of chaos other than those given in this chapter.

Exercise 32. Describe a method that permits visualizing the *trapping region* of an attractor.

Exercise 33. Justify the construction of the attractor in equation (2.1).

Exercise 34. Give three examples of fragile chaos.

Exercise 35. Explain the mechanism of the destruction of chaos.

Exercise 36. Review all the methods used in this book to prove the robustness of chaos.

Exercise 37. Visualize the *bandcount adding scenario bifurcation* displayed by the map (2.2).

Exercise 38. Explain the reasons that prevent robust chaos for occurring in the quasi-attractors discussed in Chapter 10.

Exercise 39. Give three examples of dynamical systems with smooth and three examples with fractal basin boundaries.

Exercise 40. Explain the *metamorphosis* phenomenon.

Exercise 41. Give three examples of dynamical systems with riddled basins.

Exercise 42. Prove that systems with a symmetry or other constraint have a smooth invariant manifold.

Exercise 43. Make a dynamical study (bifurcations and chaos) for the map (2.4).

Exercise 44. Explain the fact that riddled basins correspond to an extreme form of uncertainty.

Exercise 45. Explain the density of hyperbolic systems.

Exercise 46. Describe the possible relations between persistence and robustness of chaos.

Exercise 47. (Open problem) Determine the occurrence of robust chaos in system (2.5). See [Albers *et al.* (2006)].

Chapter 3

Statistical Properties of Chaotic Attractors

In this chapter, we discuss statistical properties of chaotic attractors. In Sec. 3.1, different types of entropies are described in some detail. In Sec. 3.2, the so-called *ergodic theory* is presented, and in Sec. 3.3 we present statistical properties of chaotic attractors, especially the rate of decay of correlations, the central limit theorem, and other probabilistic limit theorems. These notions are based on the correlation and autocorrelation function (ACF) introduced in Sec. 3.3.1 and Sec. 3.3.2, respectively, which play a crucial role in characterizing hyperbolic (robust) systems. The end of this chapter includes exercises and open problems.

3.1 Entropies

In this section, we give the definitions and some related properties of the following notions: the physical (or Sinai–Ruelle–Bowen) measure, Hausdorff dimension, topological entropy, Lebesgue (volume) measure, and Lyapunov exponent.

3.1.1 Lebesgue (volume) measure

Generally, the Lebesgue measure is a way of assigning a length, area, or volume to subsets of Euclidean space. These sets are called *Lebesgue measurable*. The following definition gives the mathematical formulation of the Lebesgue measure of a set $A \subset \mathbb{R}^n$:

Definition 3.1. (a) A box in $\mathbb{R}^n$ is a set of the form $B = \prod_{i=1}^{n}[a_i, b_i]$, where $b_i \geq a_i$. (b) The volume of the box B is $vol\,(B) = \prod_{i=1}^{n}(b_i - a_i)$. (c) The outer measure $m^*(A)$ for a subset $A \subset \mathbb{R}^n$ is defined by

$$m^*(A) = \inf \left\{ \sum_{j \in J} vol(B_j) : \text{with the property B} \right\}. \quad (3.1)$$

Property B: the set $\{B_j : j \in J\}$ is a countable collection of boxes whose union covers A. (d) The set $A \subset \mathbb{R}^n$ is Lebesgue measurable if

$$m^*(S) = m^*(S \cap A) + m^*(S - A) \text{ for all sets } S \subset \mathbb{R}^n. \quad (3.2)$$

(e) The Lebesgue measure is defined by $m(A) = m^*(A)$ for any Lebesgue measurable set A.

Generally, the invariant measure for a dynamical system describes the distribution of the sequence formed by the solution for any typical initial state.

3.1.2 Physical (or Sinai–Ruelle–Bowen) measure

Roughly speaking, a measure that is supported on an attractor describes for almost every initial condition in its corresponding basin of attraction the statistical properties of the long-time behavior of the orbits of the system under consideration. The notion of physical (or Sinai–Ruelle–Bowen) (SRB) measure was introduced and proved for Anosov diffeomorphisms introduced in Sec. 6.1.2), and for hyperbolic diffeomorphisms and flows [Sinai (1972), Ruelle (1976), Ruelle & Bowen (1975)] described in Chapter 6 of this book. The common definition of the SRB (Sinai–Ruelle–Bowen) measure is given here for both continuous-time discrete-time systems by the following:

Definition 3.2. (a) Let A be an attractor for a map f, and let μ be an f-invariant probability measure on A. Then μ is called an SRB (Sinai–Ruelle–Bowen) measure for (f, A) if one has for any continuous map g, that

$$\lim_{n \to \infty} \frac{1}{n} \sum_i g(f^i(x)) = \int g d\mu \quad (3.3)$$

where $x \in E \subset B(A)$ is in the basin of attraction for A with $m(E) > 0$, where m is the Lebesgue measure. (b) An invariant probability μ is a physical measure for the flow $f_t, t \in \mathbb{R}$ if the set $B(\mu)$ (the basin of μ) of points $z \in M$ satisfying

$$\lim_{T \to \infty} \frac{1}{T} \int_0^T \varphi(f_t(z)) \, dt = \int \varphi d\mu \text{ for all continuous } \varphi : M \to \mathbb{R} \quad (3.4)$$

has positive Lebesgue measure: $m(B(\mu)) > 0$.

The most useful notion in the theory of hyperbolic dynamical systems introduced in Chapter 6 is the invariant measure, which means that the "*events*" $x \in A$ and $f(x) \in A$ are equally probable.

Definition 3.3. (a) A probability measure μ in the ambient space M is invariant under a transformation f if $\mu(f^{-1}(A)) = \mu(A)$ for all measurable subsets A. (b) The measure μ is invariant under a flow if it is invariant under f_t for all t. (c) An invariant probability measure μ is ergodic if every invariant set A has either zero or full measure, or equivalently, μ is ergodic if it cannot be decomposed as a convex combination of invariant probability measures. That is, one cannot have $\mu = a\mu_1 + (1-a)\mu_2$ with $0 < a < 1$ and μ_1, μ_2 invariant.

A practical definition of the concept of SRB (Sinai–Ruelle–Bowen) measure is given by the following:

Definition 3.4. (a) An f-invariant probability measure μ is called an SRB measure if and only if (i) μ is ergodic, (ii) μ has a compact support, (iii) μ has absolutely continuous conditional measures on its unstable manifolds. (b) The set Λ carries a probability measure μ if and only if $\mu(\Lambda) = 1$.

This definition was used in [Kiriki et al. (2008)] to investigate the coexistence of homoclinic sets with and without SRB measures in Hénon maps (6.88). The existence of an invariant measure was proved theoretically in [Sinai (1970-1979), Ruelle (1976-1978), Bunimovich & Sinai (1980), Eckmann & Ruelle (1985)] for hyperbolic (see Chapter 6) and nearly hyperbolic systems (see Chapters 8 and 9). Some recent advances on the topic of the Sinai–Ruelle–Bowen measure, in particular its existence and smoothness can be found in [Jarvenpaa & Jarvenpaa (2001), Sanchez-Salas (2001), Jiang (2003), Bonetto et al. (2004), Gallavotti et al. (2004), Bonetto et al. (2005), Amaricci et al. (2007)]. For the nonexistence of such a measure, one can see, for example, [Hu & Young (1995)]. Because the main study of robust chaos is related directly to the notion of hyperbolicity of the system under consideration, we give here some important relations between the SRB measure and hyperbolicity. Indeed, if Λ is a hyperbolic attractor, then there exists a measure μ on Λ such that for any x in the basin of attraction of Λ and for any observable g, one has the relation (3.3) for discrete-time systems and the relation (3.4) for continuous-time systems. In [Katok & Hasselblatt (1995)], the following conjecture was given:

Conjecture 3.1. *Let Λ be a hyperbolic set and V an open neighborhood of Λ. Does there exist a locally maximal hyperbolic set $\tilde{\Lambda}$ such that $\Lambda \subset \tilde{\Lambda} \subset V$? Related questions: 1. Can this set be robust? 2. Can this happen on other manifolds, in lower dimension, on all manifolds? In [Crovisier (2001)], Crovisier answers no for a specific example on a four-torus.*

In fact, on any compact manifold M, where $\dim(M) \geq 2$, there exists a C^1 open set of diffeomorphisms, U, such that any $f \in U$ has a hyperbolic set that is not contained in a locally maximal hyperbolic set. If Λ is a hyperbolic set and V is a neighborhood of Λ, then there exists a hyperbolic set $\tilde{\Lambda}$ with a Markov partition such that $\Lambda \subset \tilde{\Lambda} \subset V$. In [Wolf (2006)], the notion of *generalized physical and SRB measures* was introduced, and it is naturally the generalization of classical physical and SRB measures defined in this section for attractor measures which are supported on invariant sets that are not necessarily attractors. An example of these measures was defined and studied in detail for certain hyperbolic Hénon maps.

3.1.3 Hausdorff dimension

Let A be an attractor for a map or flow f. If $S \subset A$ and $d \in [0, +\infty)$, the d-dimensional Hausdorff content of S is defined by:

$$C_H^d(S) = \inf \left\{ \sum_i r_i^d : \text{P is verified} \right\}. \qquad (3.5)$$

There is a cover of S with balls of radii $r_i > 0$. The Hausdorff dimension of A is defined by

$$D_H(A) = \inf \left\{ d \geq 0, C_H^d(S) = 0 \right\}. \qquad (3.6)$$

Using this definition, one has that the Hausdorff dimension of the Euclidean space $\mathbb{R}^n$ is n, for a circle $\mathbb{S}^1$ it is 1, and for a countable set it is 0. The Hausdorff dimension of a triadic Cantor set is $\frac{\ln 2}{\ln 3} = 0.63093$, and for the triadic Sierpinski triangle it is $\frac{\ln 3}{\ln 2} = 1.5850$. Some properties of the Hausdorff dimension can be summarized as follows:

(1) If the Hausdorff dimension $D_H(A)$ strictly exceeds the topological dimension of A, then the set A is a fractal.
(2) $D_H(A) \geq D_{ind}(A)$, where $D_{ind}(A)$ is the inductive dimension for A which is defined recursively as an integer or $+\infty$, $D_{ind}(A) = \inf_Y D_H(Y)$, where Y ranges over metric spaces homeomorphic to A [Szpilrajn (1937)].
(3) If $A = \cup_{i \in I} A_i$ is a finite or countable union, then $D_H(A) = \sup_{i \in I} D_H(A_i)$, and if $A = A_1 \times A_2$ then $D_H(A) \geq D_H(A_1) + D_H(A_2)$.

3.1.4 The topological entropy

The topological entropy of a dynamical system is a nonnegative real number that measures the complexity of the system. This entropy can be defined in various equivalent ways. Let A be a compact Hausdorff topological space. For any finite cover C of A, let the real number $H(C)$ be given by

$$H(C) = \log_2 \inf_j \{j = \text{The number of elements of } C \text{ that cover } A\}. \quad (3.7)$$

For two covers C and D, let $C \vee D$ be their (minimal) common refinement, which consists of all the nonempty intersections of a set from C with a set from D, and similarly for multiple covers. For any continuous map $f: A \to A$, the following limit exists:

$$H(C, f) = \lim_{n \to +\infty} \frac{1}{n} H\left(C \vee f^{-1} \vee ... \vee f^{-n+1} C\right). \quad (3.8)$$

Then one has the following definition:

Definition 3.5. The topological entropy $h(f)$ of f is the supremum of $H(C, f)$ over all possible finite covers C, i.e.,

$$h(f) = \sup_C \{H(C, f), C \text{ is a finite cover of } A\}. \quad (3.9)$$

A useful definition of the topological dimension of a space E in the case of singular-hyperbolic attractors was given in [Hurewicz & Wallman (1984)] as follows:

Definition 3.6. The topological dimension of a space E is either -1 (if $E = \emptyset$) or the last integer k for which every point has arbitrarily small neighborhoods whose boundaries have dimension less than k.

For practical applications to dynamical systems, especially maps and flows, we give a more appropriate definition for the topological entropy using the notion of (n, ϵ)-separated sets and interval arithmetic introduced in Sec. 1.2, namely the method of calculation using the number of periodic orbits of the system under consideration in a specific range. Let $f : X \to X$ be a map. Then one has the following definition:

Definition 3.7. A set $E \subset X$ is called (n, ϵ)-separated if for every two different points $x, y \in E$, there exists $0 \leq j < n$ such that the distance between $f^j(x)$ and $f^j(y)$ is greater than ϵ. Let us define the number $s_n(\epsilon)$ as the cardinality of a maximum (n, ϵ)-separated set:

$$s_n(\epsilon) = \max\{card\ E : E \text{ is } (n, \epsilon)\text{-separated}\} \tag{3.10}$$

The number

$$H(f) = \lim_{\epsilon \to 0} \limsup_{n \to \infty} \frac{1}{n} \log s_n(\epsilon) \tag{3.11}$$

is called the topological entropy of f.

Under certain assumptions [Bowen (1971)], the topological entropy can be expressed in terms of the number of periodic orbits as follows:

Definition 3.8. If f is an axiom A diffeomorphism (recall Definition 6.3(1)), then

$$H(f) = \limsup_{n \to \infty} \frac{\log C(f^n)}{n}, \tag{3.12}$$

where $C(f^n)$ is the number of fixed points of f^n.

Formula (3.12) can be used as the lower bound for the topological entropy $H(f)$ when the distance between periodic orbits of length n is uniformly separated from zero, i.e., the formula

$$H(f) = \frac{\log C(f^n)}{n} \tag{3.13}$$

for sufficiently large n. In a topological sense, a dynamical system is called *chaotic* if its topological entropy is positive. For example, in the Hénon map (6.88), the upper and lower bounds for the topological entropy were estimated based on the method given in [Misiurewicz & Szewc (1980), Zgliczynski (1997a), Galias (1998b), Stoffer & Palmer (1999)] using the interval technique introduced in Sec. 1.2 and the following result:

Theorem 3.1. *The topological entropy $H(h)$ of the Hénon map (6.88) is located in the interval*

$$0.3381 < H(h) \leq \log 2 < 0.6932. \tag{3.14}$$

More generally, let f be a continuous two-dimensional map. Let $N_1, N_2, ..., N_p$, with p chosen as pairwise disjoint quadrangles. For each N_i, we choose two opposite horizontal edges, and the two others are called vertical [Galias (2001)] as follows:

Definition 3.9. We say that N_i f-covers N_j, and we use the notation $N_i \stackrel{f}{\Longrightarrow} N_j$ if (i) The image of N_i under f has an empty intersection with the horizontal edges of N_j, (ii) The images of the vertical edges of N_i have an empty intersection with N_j and they are located geometrically on the opposite sides of N_j.

Thus the following theorem was proved in [Galias (2001)]:

Theorem 3.2. Let $N_1, N_2, ..., N_p$ be pairwise disjoint quadrangles. Let $A = (a_{ij})_{i,j=1}^p$ be a square matrix, where

$$a_{ij} = \begin{cases} 1, & \text{if } N_i \stackrel{f}{\Longrightarrow} N_j \\ 0, & \text{otherwise.} \end{cases} \quad (3.15)$$

Then f is semi-conjugate with the subshift on p symbols, with the transition matrix A.

Theorem 3.3. The topological entropy of the map f is not smaller than the logarithm of the dominant eigenvalue of the matrix A defined by Eq. (3.15):

$$H(f) \leq \log \lambda_1. \quad (3.16)$$

In [Galias (2001)], the same method of interval arithmetic was used for determining all the periodic orbits of period $n \leq 30$ for the Hénon map (6.88). Furthermore, the diameter of intervals for which existence and uniqueness of periodic orbits was proved is given in Table 3 of [Galias (2001)]. Finally, the following results were proved:

Lemma 3.1. *(a) There are no period-three and period-five orbits for the Hénon map within the trapping region for the values $a = 1.4, b = 0.3$. (b) There are exactly 109033 periodic orbits with period $n \leq 30$, and there are 3065317 points belonging to these orbits for the Hénon map with the values $a = 1.4, b = 0.3$. (c) The topological entropy of the Hénon map is close to 0.465.*

The topological entropy of the flow is defined by [Fomin et al. (1980)] as follows:

Definition 3.10. Topological entropy of a flow $f : X \to X$ is a topological entropy of the map g_t defined by $g_t : X \to X, g_t(x) = f(x,t)$, i.e.,

$$h_t(f) = H(g_t). \tag{3.17}$$

The formula (3.17) is natural since $h_t(f) = |t|.H(g_t)$ [Cornfeld et al. (1982)]. For axiom A flows (recall Definition 6.3(2)), the topological entropy can be obtained using the formula:

$$h_t(f) = \frac{\log P_t}{t} \tag{3.18}$$

where P_t is the number of periodic orbits with period smaller than or equal to t. In this case, the convergence is slow. To avoid this problem (to obtain a better convergence), Parry and Pollicott in [Parry & Pollicott (1983)] use the *prime number theorem* for flows to obtain the formula

$$P_t \sim \frac{\exp h_t}{h_t} \tag{3.19}$$

where $f(t) \sim g(t)$ means that $\frac{f(t)}{g(t)} \to 1$ as $t \to \infty$. In [Yang & Li (2004)], a formula for the topological entropy of Chua's circuit (10.15)–(10.16) in terms of the Poincaré map was presented for $\alpha = 10$, $\beta = 14.87$, $m_0 = 1.27$, and $m_1 = -1.68$ using the results from symbolic dynamics introduced in Sec. 1.3 and the following result proved in [Robinson (1995)]:

Lemma 3.2. *Let X be a compact metric space. If a map $f : X \to X$ is semi-conjugate to an m-shift σ, then*

$$H(f) \geq H(\sigma) = \log m. \tag{3.20}$$

The rigorous method of analysis was developed for nonhyperbolic continuous systems by Galias [Galias (2006)] in terms of short periodic orbits. This method is a combination of the Poincaré map technique, interval methods introduced in Chapter 1, and a *graph representation* of the system. The six steps of this method are technically similar to those for maps introduced in Sec. 1.2. These steps can be summarized in the following to find all period-n cycles enclosed in a trapping region of the flow under consideration: Step 1: Find a trapping region for the Poincaré map. Step 2: Cover the trapping region with boxes. Step 3: Find the image of each box. Step 4: Construct the set of admissible connections using the image of each box. Step 5: Construct a set of period-n cycles in the graph. Step 6: Evaluate for each cycle the interval operator, *i.e.*, if $X \cap K(X) = \emptyset$, then there are no period-n orbits in X. If $K(X) \subset intX$, then there exists exactly one period-n orbit inside X [1]. If neither of the two conditions is satisfied, the method fails and a repetition is needed using smaller boxes. At the end of this section, we note that there are some cases where the topological entropy cannot be calculated, such as the cases of iterated piecewise-affine maps, saturated linear functions in unbounded dimension, and cellular automata. Indeed, it was shown in [Koiran (2001)] that it is impossible to compute or to approximate the topological entropy for these types of functions.

3.1.5 Lyapunov exponent

The Lyapunov exponent of a dynamical system with evolution equation f_t in a n-dimensional phase space is a quantity that characterizes the rate of separation of infinitesimally close trajectories. If δx_0 is an initial separation of two trajectories in phase space, then the Lyapunov exponent λ is defined by:

$$\lambda = \lim_{t \to +\infty} \frac{1}{t} \ln \frac{|\delta x(t)|}{|\delta x_0|}. \tag{3.21}$$

In general, the spectrum of Lyapunov exponents $\{\lambda_1, \lambda_2, ..., \lambda_n\}$ depends on the starting point x_0. The Lyapunov exponents can be defined from the Jacobian matrix $J^t(x_0) = \frac{df_t}{dx}\big|_{x=x_0}$, which describes how a small change at the point x_n propagates to the final point $f_t(x_0)$.

Definition 3.11. (Lyapunov exponent for a continuous-time dynamical system) The matrix $L(x_0)$ (when the limit exists[2]) defined by

$$L(x_0) = \lim_{t \to +\infty} \left(J^t(x_0) \left(J^t(x_0) \right)^T \right)^{\frac{1}{2t}} \tag{3.22}$$

has n eigenvalues $\Lambda_i(x_0), i = 1, 2, ..., n$, and the Lyapunov exponents $\lambda_i, i = 1, 2, ..., n$ are defined by

$$\lambda_i(x_0) = \log \Lambda_i(x_0), i = 1, 2, ..., n. \tag{3.23}$$

For the zero vector, we define $\lambda_i(x_0) = -\infty$.

In the following, we present the standard definition of the Lyapunov exponents for a discrete n-dimensional mapping.

Definition 3.12. (Lyapunov exponent for a discrete dynamical system): Consider the following n-dimensional discrete dynamical system:

$$x_{k+1} = f(x_k), x_k \in \mathbb{R}^n, k = 0, 1, 2, ... \tag{3.24}$$

---
[1] In this case, the integer n is not necessarily the basic period of the orbit.
[2] The conditions for the existence of the limit are given by the Oseledec theorem.

where $f : \mathbb{R}^n \longrightarrow \mathbb{R}^n$ is the vector field associated with system (3.24). Let $J(x)$ be its Jacobian evaluated at x, and consider the matrix

$$T_r(x_0) = J(x_{r-1}) J(x_{r-2}) ... J(x_1) J(x_0). \quad (3.25)$$

Moreover, let $J_i(x_0, l)$ be the module of the ith eigenvalue of the lth matrix $T_r(x_0)$, where $i = 1, 2, ..., n$ and $r = 0, 1, 2, ...$ Now, the Lyapunov exponents of an n-dimensional discrete-time systems are defined by:

$$\lambda_i(x_0) = \ln \left(\lim_{r \longrightarrow +\infty} J_i(x_0, r)^{\frac{1}{r}} \right), i = 1, 2, ..., n. \quad (3.26)$$

Some characteristics of the Lyapunov exponents can be summarized as follows:

(1) The Lyapunov exponents are asymptotic quantities, and hence they need some conditions for existence of the solution under consideration, *i.e.*, that the phase space is a compact, boundaryless manifold and that the initial condition lies in a positively invariant region.
(2) In general, Lyapunov exponents are not continuous functions of the orbits.
(3) In general, the spectrum of Lyapunov exponents $\{\lambda_1, \lambda_2, ..., \lambda_n\}$ depends on the starting point x_0. However, for an ergodic component of the dynamical system, these exponents are the same for almost all starting points x_0.
(4) If the system is conservative, a volume element of the phase space will remain constant along a trajectory. Thus $\sum_{i=1}^{n} \lambda_i = 0$. If the system is dissipative, then $\sum_{i=1}^{n} \lambda_i < 0$.
(5) If the dynamical system is a smooth flow, then one exponent is always zero (Haken's theorem). For nonsmooth systems, this condition is not typical.
(6) For any continuous-time vector fields $f, g \in \mathbb{R}^n$, and nonzero constant $c \in \mathbb{R}$, one has

$$\begin{cases} \lambda_i(f+g) \leq \max\{\lambda_i(f), \lambda_i(g)\} \\ \lambda_i(cg) = \lambda_i(g) \end{cases}, i = 1, 2, ..., n. \quad (3.27)$$

(7) The Lyapunov exponent spectrum can be used to estimate some topological dimensions such as the so-called *Kaplan–Yorke dimension* D_{KY} that is defined as follows:

$$D_{KY} = k + \sum_{i=1}^{k} \frac{\lambda_i}{|\lambda_{k+1}|} \quad (3.28)$$

where k is the maximum integer such that the sum of the k largest exponents is still nonnegative, and in this case D_{KY} is an upper bound for the information dimension of the system [Kaplan & Yorke (1987)].

(8) Using Pesin's theorem [Pesin (1977a)], the sum of all the positive Lyapunov exponents gives an estimate of the *Kolmogorov–Sinai entropy*.

(9) The inverse of the largest Lyapunov exponent is called the *Lyapunov time* which is finite for chaotic orbits and infinite for nonchaotic orbits.

(10) The analytic calculation of Lyapunov exponents using the matrix $L(x_0)$ is not possible. In this case, there are many methods and algorithms that deal with the numerical issues. These procedures estimate the $L(x_0)$ matrix based on averaging several finite-time approximations of the limit defining $L(x_0)$. For a smooth dynamical system, the periodic Gram–Schmidt orthonormalization technique of the Lyapunov vectors was used in [Benettin *et al.* (1980), Shimada & Nagashima (1979)], and this method is the most used and effective one to calculate the Lyapunov spectrum. For nonsmooth dynamical systems, various methods have been proposed, for example the one given in [Sprott (2003)]. These types of algorithms give spurious results for piecewise-linear discontinuous systems, but they seem to work for some cases since the Lyapunov exponent plots and the bifurcation diagrams agree. The method essentially takes a numerical derivative and gives the correct result provided care is taken to ensure that the perturbed and unperturbed orbits lie on the same side of the discontinuity. This may require an occasional small perturbation into a region that is not strictly accessible to the orbit. Still a question arises about the exact value of the positive Lyapunov exponents. For all these methods, it is necessary to confirm that the calculated orbit is bounded and has adequately sampled the attractor.

In some situations, it is important to determine the upper and lower bounds for all the Lyapunov exponents of a given n-dimensional system. A recent result was given in [Li & Chen (2004)] for discrete mappings as follows:

Theorem 3.4. *If a system* $x_{k+1} = f(x_k), x_k \in \Omega \subset \mathbb{R}^n$ *that satisfies*

$$\|Df(x)\| = \|J\| = \sqrt{\lambda_{\max}(J^T J)} \leq N < +\infty, \tag{3.29}$$

with a smallest eigenvalue of $J^T J$ that satisfies

$$\lambda_{\min}(J^T J) \geq \theta > 0, \tag{3.30}$$

where $N^2 \geq \theta$, then for any $x_0 \in \Omega$, all the Lyapunov exponents at x_0 are located inside $\left[\frac{\ln \theta}{2}, \ln N\right]$. That is,

$$\frac{\ln \theta}{2} \leq l_i(x_0) \leq \ln N, i = 1, 2, ..., n, \tag{3.31}$$

where $l_i(x_0)$ are the Lyapunov exponents for the map f.

Similar results for continuous-time systems can be found in [Li & Xia (2004)].

3.2 Ergodic theory

The ergodic theory was motivated by problems of statistical physics. Its central aspect is the behavior of a dynamical system when it is allowed to run for a long time. The most important results in this field are the ergodic theorems of Birkhoff and von Neumann. In particular, the importance of ergodic theory can be seen from the following two items:

(1) For ergodic systems, the time average is the same for almost all initial conditions.
(2) Ergodic systems have stronger properties such as mixing and equidistribution.

Some examples of applications of ergodic theory are those related to the study of the problem of metric classification of systems, stochastic processes that use various notions of entropy for dynamical systems, study of the geodesic flow introduced in Sec. 6.4.1 on Riemannian manifolds, Markov chains, harmonic analysis, Lie theory (representation theory, lattices in algebraic groups), and number theory (the theory of diophantine approximations, L-functions), *etc.* For details, see [Birkhoff (1931), von Neumann (1932a-b), Hopf (1939), Birkhoff (1942), Fomin & Gelfand (1952), Mautner (1957), Moore (1966b)]. The modern general ergodic theory of smooth dynamical systems has been developed essentially by Pesin, Katok, Ledrappier, and others, but concrete examples were lacking [Arnol'd & Avez (1968), Walters (1982), Petersen (1990), Bedford (1991), Breiman (1992), Rosenblatt & Weirdl (1993), Anosov (2001)]. For axiom A attractors (recall Definition 6.3), this theory was developed by Sinai, Ruelle, and Bowen in the 1970's by constructing the SRB measures introduced in Sec. 3.1.2, with absolutely continuous conditional measures on unstable manifolds. For the one-dimensional case, the theory of chaotic maps was started in 1980 by

M. Misiurewicz where he proved the existence of absolutely continuous invariant measures for multimodal maps of the interval with some properties, *i.e.*, the critical set or set of turning points, C, of these maps has the property that for all $z_0 \in C$ and all $j \geq 1$, $dist(f_{j_{z_0}}, C) \geq \delta > 0$. In [Collet & Eckmann (1980a)], the *abundance phenomenon*, *i.e.*, positive Lebesgue measure of aperiodic behavior for a family of unimodal maps of the interval (see Sec. 11.1) was proved. In [Jakobson (1981), Benedicks & Carleson (1985)], two proofs for the abundance of existence of absolutely continuous invariant measures for the quadratic family were provided. Similar methods were later used in [Benedicks & Carleson (1991)] to prove aperiodic, chaotic behavior for a class of Hénon maps. The method of [Benedicks & Carleson (1991)] was used for several other dynamical systems [Young (1998a), Viana (1998)].

3.3 Statistical properties of chaotic attractors

Generally, statistical properties of a dynamical system are the rate of decay of correlations, the central limit theorem, and other probabilistic limit theorems. The main effective method of proving these properties is based on Markov approximations (which is an alternative to the conventional Frobenius–Perron operator techniques) to dynamical systems. This method was systematically developed by Yakov Sinai and his school.

3.3.1 *Autocorrelation function (ACF)*

Let $x(t)$ be a process, in particular $x(t)$ can be a solution of a dynamical system. Let $x_i = x(t_i)$ be the values of the process $x(t)$ at time $t_1, t_2, ..., t_N$, where N is the number of realizations of the $x(t)$ process.

The autocorrelation function (ACF) introduced in [Box & Jenkins (1976)] is used to detect nonrandomness in data and to identify an appropriate time series model if the data are not random. In this case, the time variable t is not used because the observations are assumed to be equispaced, *i.e.*, the correlation is between two values of the same variable at times t_i and t_{i+k}. If the goal is the determination of nonrandomness, then it suffices to consider only the first (lag 1) autocorrelation. But if the purpose is the identification of an appropriate time series model, then the autocorrelations are usually plotted for many lags. In the case where constant location and scale, randomness, and fixed distribution A_0 are reasonable, the univariate process Y_i can be modeled as:

$$Y_i = A_0 + E_i \qquad (3.32)$$

where E_i is an error term. However, (3.32) is not a valid model if the randomness assumption is violated. Typically, the necessary model is either a time series model or a time-independent variable nonlinear model. Autocorrelation plots given in [Box & Jenkins (1976), pp. 28–32] were used as a tool for checking randomness in a data set by computing autocorrelations for data values at varying time lags. Thus the model is random if these autocorrelations are near zero for any and all time-lag separations. Additionally, the model is nonrandom if one or more of the autocorrelations is significantly nonzero. A practical algorithm for the evaluation of the ACF for a chaotic system (more generally, a chaotic process $x(t)$) can be found in [Anishchenko et al. (2004)] as follows:

(1) Choose N, a sufficiently large number of realizations, and calculate the time-average value $\bar{x}$ using the formula

$$\bar{x} = \frac{1}{N}\sum_{i=1}^{N} x(t_i). \qquad (3.33)$$

(2) Calculate the mean product $x(t)x(t+\tau)$ by averaging over time:

$$K_l(\tau) = \frac{1}{p}\sum_{i=1}^{p} x(t_i)\,x(t_i + k\Delta t), \tau = k\Delta t_i, k = 0,1,...,n-p \qquad (3.34)$$

where $l = 1,...,N$ is the number of realizations. In this case, the limitation of a number of $x(t_i)$ values, $i = 1,2,...,N$ implies that the time-averaging converged if the number of averages p is sufficiently large, which gives the largest value of p and the smallest value of the time $\tau_{\max} = (n-p)\Delta t$ for the ACF estimation. Thus the ACF must be computed on a very large time interval since the rate of correlation splitting is not high in the regime being considered. Hence the value of p must not be too large. Finally, to attain a high precision of the ACF calculation, the obtained data must be further averaged over N realizations, i.e.,

$$\psi(\tau) = \frac{1}{N}\sum_{l=1}^{N} K_l(\tau) - \bar{x}^2. \qquad (3.35)$$

(3) Normalize the ACF to its maximal value at $\tau = 0$, that is,

$$\Psi(\tau) = \frac{\psi(\tau)}{\psi(0)}. \qquad (3.36)$$

(4) Plot $\ln \Psi(\tau)$ versus τ.

Some examples can be found Sec. 9.3 for the Lozi map (8.88), in Sec. 10.4 for the Anishchenko–Astakhov oscillator (10.13), and in Sec. 10.2.5 for the Hénon map (6.88).

3.3.2 Correlations

In this section, we give some relations between the concept of correlations defined for a dynamical system and the following notions: strong law of large numbers (SLLN), the correlation decay, and the central limit theorem. To define correlations, we need a discrete map $f : X \to X$ as a deterministic dynamical system *preserving probability measure* m on X, *i.e.*, for any measurable subset $A \subset X$, one has $m(A) = m(f^{-1}(A))$, where $f^{-1}(A)$ is the set of points mapped into A. Due to some error effects, the states $x_n \in X$ are not observable. A real-valued function F on X called an *observable* was used to generate a sequence $F_n = F(x_n)$. The function F on X is seen to be a random variable, *i.e.*, the function $F_n = F \circ f^n$ is a random variable for each n, and the sequence $\{F_n\}$ is a stationary stochastic process because the measure m is invariant. If the observable F is square integrable, *i.e.*,

$$m(F^2) < \infty, \qquad (3.37)$$

then the random variables F_n have finite mean value μ_F and variance σ_F^2 given by

$$\begin{cases} \mu_F = m(F) = \int_X F dm \\ \sigma_F^2 = m(F^2) - (m(F))^2. \end{cases} \qquad (3.38)$$

Note that the quantity σ_F^2 characterizes only one random variable F.

Now, to define the so-called *strong law of large numbers* (SLLN) related to the convergence property, we use the classical Birkhoff ergodic theorem that claims that for m-almost every initial state $x_0 \in X$, the time averages converge to the space average, *i.e.*,

$$\frac{F_0 + F_1 + \ldots + F_{n-1}}{n} \to \mu_F = m(F). \qquad (3.39)$$

Birkhoff's ergodic theorem can be stated using the partial sums of the observed sequence F_n given by $S_n = F_0 + F_1 + \ldots + F_{n-1}$ as

$$\frac{S_n - n\mu_F}{n} \to 0, \ i.e., \ S_n = n\mu_F + O(\sqrt{n}). \qquad (3.40)$$

To define the concept of correlations, we consider the covariances $C_F(n)$ (called *autocorrelations*) given by

$$C_F(n) = m(F_0 F_n) - \mu_F^2 = m(F_k F_{n+k}) - \mu_F^2, \text{ for any } k. \tag{3.41}$$

More generally, for any two square-integrable observables F and G, the correlations are defined by

$$C_{F,G}(n) = m(F_0 G_n) - \mu_F \mu_G = m(F_k G_{n+k}) - \mu_F \mu_G, \text{ for any } k. \tag{3.42}$$

The following remarks about autocorrelations should be noted:

(1) If $\sigma_F^2 = 1$ (the F_n's were normalized), then $C_F(n)$ is the *correlation coefficient* between random variables F_k and F_{n+k}, *i.e.*, between values observed at times that are n (time units) apart.
(2) If the system is chaotic, then for large n, the values of $C_F(n)$ are nearly independent in the sense that as n grows, the correlations decrease (*decay*).
(3) Generally, the quantities $C_F(n)$ are called correlations even without the normalization assumption, *i.e.*, $\sigma_F^2 = 1$.

The relation between correlations and mixing attractors can be seen as follows:

Definition 3.13. The transformation f is said to be mixing if, for any two measurable sets $A, B \subset X$, one has

$$m(A \cap f^{-n}(B)) \to m(A) m(B) \text{ as } n \to \infty. \tag{3.43}$$

The mixing property of an attractor is related to correlations, *i.e.*, the map f is mixing if and only if correlations decay[3], *i.e.*,

$$C_{F,G}(n) \to 0 \text{ as } n \to \infty. \tag{3.44}$$

The decay of correlations is essential in the study of relaxation to equilibrium in nonequilibrium statistical mechanics where the explicit formula for the autocorrelation function introduced is involved in the formulas for transport coefficients such as heat conductivity, electrical resistance, viscosity, the diffusion coefficient, *etc*. Note that the rate of decay of correlations (3.44) depends on two factors: the first is the strength of chaos in the dynamics of f, and the second is the regularity of the observables F and G. Hence the correlations decay rapidly if the system is strongly (robustly) chaotic and the observables are sufficiently regular.

---
[3]For every pair of square integrable functions F and G, the speed of the decay of correlations is crucial when one deals with particular observables.

The relation between correlations and the SLLN property can be seen as follows: Since the convergence of the time average $\frac{S_n}{n}$ to the space average μ_F is guaranteed by the Birkhoff ergodic theorem, then it is possible to show that typical values of $S_n - n\mu_F$ are of order $\sqrt{n}$, i.e.,

$$S_n = n\mu_F + O\left(\sqrt{n}\right). \tag{3.45}$$

Indeed, the root-mean-square value is given by

$$m\left[(S_n - n\mu_F)\right]^2 = nC_F(0) + 2\sum_{k=1}^{n-1}(n-k)C_F(k). \tag{3.46}$$

If the correlations decay fast enough, i.e.,

$$\sum_{n=0}^{\infty}|C_F(n)| < \infty, \tag{3.47}$$

then the following sum denoted by σ^2

$$\sigma^2 = \sum_{n=-\infty}^{\infty} C_F(n) = C_F(0) + 2\sum_{n=1}^{\infty} C_F(n) \tag{3.48}$$

is always nonnegative, and for generic observables F, it is positive. Here the quantity σ^2 characterizes the entire process $\{F_n\}$. Thus using (3.46), one has that the mean square of $S_n - n\mu_F$ grows as

$$m\left[(S_n - n\mu_F)\right]^2 = n\sigma^2 + O(n), \tag{3.49}$$

and typical values of $S_n - n\mu_F$ are of order $\sqrt{n}$ and on average grow as $\sigma\sqrt{n}$. Furthermore, the correlations can be used in the so-called *central limit theorem* (CLT) of chaotic dynamical systems as follows: The observable F satisfies the CLT if the sequence $\frac{S_n - n\mu_F}{\sqrt{n}}$ converges in distribution according to the normal law $N(0, \sigma^2)$, i.e., for every real $z \in (-\infty, \infty)$, one has

$$m\left(\frac{S_n - n\mu_F}{\sqrt{n}} < z\right) \to \frac{1}{\sqrt{2\pi\sigma^2}}\int_{-\infty}^{z}\exp\left(-\frac{t^2}{2\sigma^2}\right)dt, \text{ as } n \to \infty. \tag{3.50}$$

As an important remark, the central limit theorem holds if the correlations $C_F(n)$ decay fast enough, i.e., the asymptotics

$$|C_F(n)| = O\left(n^{-(2+\varepsilon)}\right), \ \varepsilon > 0. \tag{3.51}$$

Standard examples of strongly (robustly) chaotic systems are the angle-doubling map $f(x) = 2x \pmod{1}$ of a circle and the Arnold cat map $(x, y) \to (2x + y, x + y) \pmod{1}$ of the unit torus introduced in Chapter 6. For these maps, correlations decay exponentially fast, i.e., $|C_{F,G}(n)| =$

$O\left(n^{-na}\right)$ for some $a > 0$, and central limit theorem holds if the observables F and G are Hölder-continuous (recall Definition 6.14). However, for less regular, i.e., continuous observables, correlations decay arbitrarily slowly, and the central limit theorem fails. If the chaos is fragile (nonrobust), correlations decay more slowly[4], namely, polynomially, i.e., $|C_{F,G}(n)| = O\left(n^{-b}\right)$ for some $b > 0$. Furthermore, in Sec. 6.6, correlations decay exponentially fast, and the central limit theorem holds for some examples of two-dimensional maps called *generalized hyperbolic systems*. For systems with continuous time, the above theory can be extended as follows: Let $\phi^t : X \to X$ be a one-parameter family (a flow) that preserves a probability measure m. Let F again denote an observable. Then $F_t = F \circ \phi^t$ is a stationary stochastic process with continuous time t. In this case, the partial sums S_n are replaced by time integrals

$$S_T = \int_0^T F_t dt = \int_0^T \left(F \circ \phi^t\right) dt, \qquad (3.52)$$

and the correlation function is defined by

$$C_{F,G}(t) = m(F_0 G_t) - \mu_F \mu_G = m(F_s G_{s+t}) - \mu_F \mu_G, \text{ for any } s. \quad (3.53)$$

Hence the Birkhoff ergodic theorem claims that for almost every initial state, one has $\frac{S_T}{T} \to \mu_F$ as $T \to \infty$. The flow ϕ^t is mixing if and only if correlations decay, i.e.,

$$C_{F,G}(t) \to 0 \text{ as } t \to \infty \qquad (3.54)$$

for every pair of square integrable function F and G. If the correlations decay fast enough, then the integral

$$\sigma^2 = \int_{-\infty}^{\infty} C_{F,F}(t) dt \qquad (3.55)$$

converges absolutely. The observable F satisfies the central limit theorem for flows if

$$\frac{S_T - T\mu_F}{\sqrt{T}} \to 0 \text{ as } T \to \infty \qquad (3.56)$$

converges in distribution to a normal law $N(0, \sigma^2)$. Note that in [Anishchenko et al. (2003a-b-c)], the correlation and spectral properties are studied for three-dimensional autonomous chaotic systems such as the Rössler oscillator [Rössler (1976)], the Lorenz system (8.1) introduced in

---
[4]Sub-exponentially.

Sec. 8.2, and the Anishchenko–Astakhov oscillator introduced in Sec. 10.4, which represents a mathematical model of a real radio-technical device [Anishchenko (1990-1995)]. More details about the subject of statistical properties of chaotic attractors, in particular the rate of the decay of correlations, the central limit theorem, and other probabilistic limit theorems can be found in [Ruelle (1968-1976), Sinai (1972), Bowen (1975), Hofbauer & Keller (1982), Rychlik (1983), Denker (1989), Keller & Nowicki (1992), Liverani (1995), Chernov (1999), Young (1998-1999), Benedicks & Young (2000), Szasz (2000), Markarian (2004), Balint & Gouezel (2006)].

3.4 Exercises

Exercise 48. Explain the importance of entropies defined for a dynamical system.

Exercise 49. Describe the so-called *generalized physical and SRB measures* given in [Wolf (2006)].

Exercise 50. (Open problem) Prove Conjecture 3.1.

Exercise 51. Show that the Hausdorff dimension of the Euclidean space $\mathbb{R}^n$ is n, for a circle $\mathbb{S}^1$ is 1, for a countable set is 0, for the triadic Cantor set is $\frac{\ln 2}{\ln 3} = 0.63093$, and for the triadic Sierpinski triangle is $\frac{\ln 3}{\ln 2} = 1.5850$.

Exercise 52. Show that the Hausdorff dimension $D_H(A)$ strictly exceeds the topological dimension of A.

Exercise 53. Show that $D_H(A) \geq D_{ind}(A)$, where $D_{ind}(A)$ is the inductive dimension for A, which is defined recursively as an integer or $+\infty$, $D_{ind}(A) = \inf_Y D_H(Y)$, where Y ranges over metric spaces homeomorphic to A [Szpilrajn (1937)].

Exercise 54. Show that if $A = \cup_{i \in I} A_i$ is a finite or countable union, then $D_H(A) = \sup_{i \in I} D_H(A_i)$, and if $A = A_1 \times A_2$, then $D_H(A) \geq D_H(A_1) + D_H(A_2)$.

Exercise 55. Let $f : X \to X$ be a nilpotent map, *i.e.*, there exists an integer m such that $f^m(X) = \{0\}$, where 0 is a distinguished point of X. Show that the topological entropy of f is $h(f) = 0$.

Exercise 56. Show that the topological entropy of the tent map is $h(f) = \ln 2$.

Exercise 57. Show that if f is an axiom A diffeomorphism, then (3.12) holds, where $C(f^n)$ is the number of fixed points of f^n. See [Bowen (1971)].

Exercise 58. Show that the topological entropy $H(h)$ of the Hénon map (6.88) is located in the interval given by (3.14). See [Misiurewicz & Szewc (1980), Zgliczynski (1997a), Galias (1998b), Stoffer & Palmer (1999)].

Exercise 59. Prove Theorem 3.2. See [Galias (2001)].

Exercise 60. Prove Theorem 3.3. See [Galias (2001)].

Exercise 61. Prove Lemma 3.1. See [Galias (2001)].

Exercise 62. Using the *prime number theorem* for flows, show that P_t can be given by (3.19). See [Parry & Pollicott (1983)].

Exercise 63. Prove Lemma 3.2. See [Robinson (1995)].

Exercise 64. Give an example of a dynamical system in which the Lyapunov exponents can be found analytically.

Exercise 65. Show that if the phase space of a dynamical system is a compact, boundaryless manifold or if the initial condition lies in a positively invariant region, then it is possible to define the Lyapunov exponents of the system.

Exercise 66. Explain why Lyapunov exponents are not continuous functions of the orbits.

Exercise 67. Show that for an ergodic component of a dynamical system, the spectrum of Lyapunov exponents is the same for almost all starting points x_0.

Exercise 68. Show that if a system is conservative, then $\sum_{i=1}^{n} \lambda_i = 0$, and if a system is dissipative, then $\sum_{i=1}^{n} \lambda_i < 0$.

Exercise 69. (a) Show that if the dynamical system is a smooth flow, then one exponent is always zero (Haken's theorem).

(b) Give an example to show that this case is not typical for nonsmooth systems.

Exercise 70. Show that for any continuous-time vector fields $f, g \in \mathbb{R}^n$ and nonzero constant $c \in \mathbb{R}$, one has relations (3.27).

Exercise 71. (a) Calculate the Kaplan–Yorke dimension D_{KY} using formula (3.28) for an equilibrium and for a periodic orbit.
(b) Give an example of a dynamical system in which the Kaplan–Yorke dimension D_{KY} can be calculated rigorously.

Exercise 72. Show that the Kaplan–Yorke dimension D_{KY} is an upper bound for the information dimension of the system. See [Kaplan & Yorke (1987)].

Exercise 73. Estimate the Kolmogorov–Sinai entropy for the Hénon map (6.88).

Exercise 74. Explain why the *Lyapunov time* is finite for chaotic orbits and infinite for nonchaotic orbits.

Exercise 75. Explain a method for calculating Lyapunov exponents for nonsmooth systems.

Exercise 76. Give a proof of Theorem 3.4. See [Li & Chen (2004)].

Exercise 77. State a similar result of Theorem 3.4 for continuous-time systems. See [Li & Xiab (2004)].

Exercise 78. Find an example showing that Lyapunov exponents are not continuous functions of the orbit.

Exercise 79. Evaluate the ACF for the logistic map (6.25) using the algorithm given in Sec. 3.3.1. See [Anishchenko *et al.* 2004].

Exercise 80. Show that, if a system is chaotic, then for large n, the values of $C_F(n)$ defined in (3.41) are nearly independent in the sense that as n grows, the correlations decrease (decay).

Exercise 81. Show that a map f is mixing if and only if correlations decay, *i.e.*, that relation (3.44) holds.

Exercise 82. Explain formally the relation between correlations and the SLLN property, *central limit theorem* (CLT) of chaotic dynamical systems described in Sec. 3.3.2.

Exercise 83. Show that if the observables F and G are Holder-continuous, then the correlations decay exponentially fast and the central limit theorem holds for the angle-doubling map $f(x) = 2x \pmod 1$ of a circle and for the Arnold cat map $(x, y) \to (2x + y, x + y) \pmod 1$ of the unit torus, i.e., $|C_{F,G}(n)| = O(n^{-na})$ for some $a > 0$.

Exercise 84. Show that if the observables F and G are continuous, then correlations decay arbitrarily slowly and the central limit theorem fails.

Exercise 85. Show that for fragile chaos, correlations decay more slowly, i.e., $|C_{F,G}(n)| = O(n^{-b})$ for some $b > 0$.

Exercise 86. Show that the flow ϕ^t is mixing if and only if correlations decay for every pair of square integrable functions F and G, i.e., that relation (3.54) holds.

Exercise 87. Show that if the correlations decay fast enough, then the integral defined in (3.55) converges absolutely.

Exercise 88. Show that if the observable F satisfies the central limit theorem for flows, then (3.56) converges in distribution to a normal law $N(0, \sigma^2)$.

Chapter 4

Structural Stability

In this chapter, we discuss the concept of structural stability as a criterion for robustness of invariant sets of dynamical systems. In Sec. 4.1, we define and state important properties of this notion, along with its conditions as given in Sec. 4.1.1. An example of a proof of Anosov's theorem on structural stability of diffeomorphisms of a compact C^∞ manifold M without boundary due to Robinson and Verjovsky [Robinson & Verjovsky (1971)] is given and discussed in Sec. 4.1.2. At the end of this chapter, we present a weaker version called Ω-*stability* and then give a set of exercises and open problems concerning structural stability along with some suggested references.

4.1 The concept of structural stability

The concept of structural stability was introduced by Andronov and Pontryagin in 1937, and it plays an important role in the development of the theory of dynamical systems. Conditions for structural stability of high-dimensional systems were formulated by Smale in [Smale (1967)]. These conditions are the following: A system must satisfy both axiom A (recall Definition 6.3) and the strong transversality condition. Mathematically, let $C^r(\mathbb{R}^n, \mathbb{R}^n)$ denote the space of C^r vector fields of $\mathbb{R}^n$ into $\mathbb{R}^n$. Let $Diff^r(\mathbb{R}^n, \mathbb{R}^n)$ be the subset of $C^r(\mathbb{R}^n, \mathbb{R}^n)$ consisting of the C^r diffeomorphisms.

Definition 4.1. (a) Two elements of $C^r(\mathbb{R}^n, \mathbb{R}^n)$ are C^r ε-close ($k \leq r$), or just C^k close, if they, along with their first k derivatives, are within ε as measured in some norm. (b) A dynamical system (vector field or map) is structurally stable if nearby systems have the same qualitative dynamics.

The words *"nearby systems"* in Definition 4.1 can be translated in terms of C^k conjugate for maps and C^k equivalence for vector fields. To avoid the problem arising from the fact that $\mathbb{R}^n$ is unbounded, assume that the maps under consideration act on compact, boundaryless, n-dimensional, differentiable manifolds M, rather than on all of $\mathbb{R}^n$. This assumption induces the so-called C^k *topology* given in [Palis & de Melo (1982)] or in [Hirsch (1976)]:

Definition 4.2. The C^k topology is the topology induced on $C^r(M, M)$ by the measure of distance between two elements of $C^r(M, M)$.

Now it is possible to define formally the notion of structural stability as follows:

Definition 4.3. (Structural stability) Consider a map $f \in Diff^r(M, M)$ (resp. a C^r vector field in $C^r(M, M)$); then f is structurally stable if there exists a neighborhood N of f in the C^k topology such that f is C^0 conjugate (resp. C^0 equivalent) to every map (resp. vector field) in N.

As a result, structural stability implies a common and typical or *generic* property of a dynamical system to a dense set of dynamical systems in $C^r(M, M)$. However, this case is still not true in the case of the set of rationals, which are dense in $\mathbb{R}$ and also its complement. Topologically, this common property is true when some conditions about residual sets hold. Indeed, let X be a topological space, and let U be a subset of X.

Definition 4.4. (Residual set) (a) The subset U is called a residual set if it contains the intersection of a countable number of sets, each of which are open and dense in X. (b) If every residual set in X is itself dense in X, then X is called a Baire space.

Now we give the definition of a generic property using the notion of a residual set given in Definition 4.4 as follows:

Definition 4.5. (Generic property) A property of a map (resp. vector field) is said to be C^k generic if the set of maps (resp. vector fields) possessing that property contains a residual subset in the C^k topology.

4.1.1 *Conditions for structural stability*

First, it was proved in [Palis & de Melo (1992)] that the hyperbolic fixed points, periodic orbits, and the transversal intersection of the stable and unstable manifolds of hyperbolic fixed points and periodic orbits are structurally stable and generic. Furthermore, it was shown that structurally stable systems are generic. Indeed, for two-dimensional vector fields on compact manifolds, one has the following result due to Peixoto [Peixoto (1962)]:

Theorem 4.1. *(Peixoto's Theorem) A C^r vector field on a compact, boundaryless, two-dimensional manifold M is structurally stable if and only if: (i) The number of fixed points and periodic orbits is finite and each is hyperbolic. (ii) There are no orbits connecting saddle points. (iii) The nonwandering set consists of fixed points and periodic orbits. Moreover, if M is orientable, then the set of such vector fields is open and dense in $C^r(M,M)$ (note: this condition is stronger than generic).*

Theorem 4.1 is useful in practice because it gives precise conditions under which the dynamics of a vector field on a compact, boundaryless, two-manifold are structurally stable. Due to the presence of Smale horseshoes in higher dimensions, it was shown in [Smale (1966)] that Peixoto's Theorem 4.1 cannot be generalized for n-dimensional diffeomorphisms ($n \geq 2$) or n-dimensional vector fields ($n \geq 3$). Generally, for periodic orbits and fixed points, structural stability can be tested in terms of the eigenvalues of the linearized system. But this approach fails for homoclinic and quasi-periodic orbits since the nearby orbit structure may be exceedingly complicated and defy any local description. Definition 4.1 can be reformulated in terms of C^k topology as follows:

Definition 4.6. (a) A diffeomorphism f is C^r structurally stable if, for any C^r small perturbation g of f, there is a homeomorphism h of the phase space such that $h \circ f(x) = g \circ h(x)$ for all points x in the phase space. (b) A flow f is C^r structurally stable if there exists a homeomorphism h sending trajectories of the initial flow f to the trajectories of any small C^r perturbation g.

Peixoto's theorem for diffeomorphisms on the circle $\mathbb{S}^1$ can be stated as follows:

Theorem 4.2. *(Peixoto) A diffeomorphism $f \in Diff^1\left(\mathbb{S}^1\right)$ is structurally stable if and only if its nonwandering set $\Omega(f)$ consists of finitely many fixed points or periodic orbits.*

Generally, it was conjectured that hyperbolicity (see Chapter 6) is equivalent to structural stability in the C^k topology. Indeed, it was proved that stability holds for hyperbolic systems in [Anosov (1967), Palis & Smale (1968), Palis (1969), Robbin (1971), de Melo (1973), Robinson (1974)]. The converse was completed in [Mañé (1988)] for diffeomorphisms and in [Hayashi (1997)] for flows, using the results given in Liao [Mañé (1982), Liao (1980-1983), Sannami (1983)].

Conjecture 4.1. *(Stability conjecture), A system is C^r structurally stable if and only if it is hyperbolic and all the stable and unstable manifolds associated with the orbits in the limit set are transversal.*

In this section, we provide a proof given in [Robinson & Verjovsky (1971)] of Anosov's theorem [Moser (1969)] on structural stability of diffeomorphisms of a compact C^∞ manifold M without boundary given by Theorem 4.5 below. The proof of Anosov's Theorem 4.4 below uses the shadowing results described in Sec. 5.3. This concept is narrow because the existence of an Anosov diffeomorphism imposes severe topological restrictions on the underlying manifold M. For the three-dimensional case, the basic sets of C^1 structurally stable flows are isolated fixed points, isolated closed orbits, and suspensions of nontrivial irreducible shifts of finite type. These sets have infinitely many periodic orbits but rational zeta functions and two-dimensional attractors or repellers just like the suspension of Plykin's attractor introduced in Sec. 6.4.4.

4.1.2 A proof of Anosov's theorem on structural stability of diffeomorphisms

In this section, we state a proof of Anosov's theorem on structural stability of diffeomorphisms of a compact C^∞ manifold M without boundary due to Robinson and Verjovsky [Robinson & Verjovsky (1971)] using the results in [Moser (1969)], Mather's appendix in [Smale (1967)] and the *Hirsch–Pugh lemma* given in [Hirsch & Pugh (1970)] where they show that the Anosov diffeomorphisms form an open (maybe empty according to M) set in the set of C^r diffeomorphisms of M with the C^r topology, $r \geq 1$ denoted by D. The objective of this proof is to give the reader some tools and techniques used in situations like this. This proof can be decomposed into eight lemmas, in particular, the Hirsch–Pugh lemma stated and proved in this section. A generalization of the definition of Anosov diffeomorphisms ($f \in D$) (see Definition 6.3) to an arbitrary C^∞ manifold M, is the following:

Definition 4.7. The map $f \in D$ is said to be an Anosov diffeomorphism if and only if: (a) The tangent bundle of M splits into a Whitney direct sum of continuous subbundles $TM = E^s \oplus E^u$, where E^s and E^u are Df-invariant, (b) There exist constants $c, c' > 0$ and $0 < \lambda < 1$ such that

$$\begin{cases} |Df_x^n v| < c\lambda^n |v| \\ |Df_x^{-n} w| < c'\lambda^n |w| \end{cases} \tag{4.1}$$

for all $x \in M$, $v \in E_x^s$, $w \in E_x^u$, and $n > 0$.

As a remark, the manifold M is compact, Definition 4.7 is independent of the Riemannian metric $<,>$, and the subbundles E^s and E^u are uniquely determined according to the above conditions. The proof of Anosov's theorem on structural stability of diffeomorphisms needs the following definitions of the concepts: normed vector bundle, vertical map of class C^r ($r \geq 0$), and vertical map of class C^r along fibers.

Definition 4.8. (a) (Normed vector bundle) A vector bundle $\pi : E \to M$ of class C^r is said to be normed if there is a C^s ($0 \leq s \leq r$) real function $F : E \to \mathbb{R}$ such that $F|_{\pi^{-1}(x)}$ defines a norm on $\pi^{-1}(x)$ for every $x \in M$. (b) (Vertical map of class C^r). Let K_1 and K_2 be compact metric spaces, U an open subset of a Banach space F_1, and V an open subset of a Banach space F_2. Suppose that we have $f : K_1 \to K_2$ and $\overline{f} : K_1 \times U \to K_2 \times V$ continuous such that the following diagram is commutative:

$$\begin{array}{ccccc} K_1 \times U & \xrightarrow{\overline{f}} & K_2 \times V & \xrightarrow{p_2} & V \\ \pi \downarrow & & \downarrow p_1 & & \\ K_1 & \xrightarrow{f} & K_2 & & K_2 \end{array} \tag{4.2}$$

where π, p_1, and p_2 are projections. We say that $\overline{f}$ is vertically of class C^r ($r \geq 0$) if $p_2 \circ \overline{f}$ has r partial derivatives with respect to the variable in U and the partials are continuous mappings

$$D_2^k(p_2 \circ \overline{f}) : K_1 \times U \to L_s^k(F_1, F_2), k = 0, ..., r. \tag{4.3}$$

Here $L_s^k(F_1, F_2)$ are symmetric k-multilinear mappings from F_1 to F_2. In particular, for each fixed $x \in K_1$, $p_2 \circ \overline{f}(x,.) : U \to V$ is of class C^r. (c) (Vertical map of class C^r along fibers) Let $\pi_1 : E_1 \to M$ and $\pi_2 : E_2 \to N$ be two Riemannian vector bundles of class C^0 over compact metric spaces M and N. Let $\overline{f} : E_1 \to E_2$ be a continuous map that preserves fibers, i.e., there exists a map $f : M \to N$ such that $f \circ \pi_1 = \pi_2 \circ \overline{f}$. We say that $\overline{f}$ is vertically of class C^r or $\overline{f}$ is of class C^r along the fibers, $0 \leq r \leq \infty$ if the local representatives of $\overline{f}$ in local vector bundle charts are vertically of class C^r (using Definition 4.8(b)).

To prove the Hirsch–Pugh lemma 4.1 below, one needs a Banach space E and two closed subspaces E_1 and E_2 such that E is the direct sum of E_1, E_2, i.e., $E = E_1 \oplus E_2$. Given $0 < \tau < 1$, let $\mathfrak{L}_\tau$ be the hyperbolic isomorphisms L of E leaving E_1, E_2 invariant such that

$$\begin{cases} \|L|_{E_1}\| < \tau \\ \|L^{-1}|_{E_2}\| < \tau. \end{cases} \quad (4.4)$$

Hence the Hirsch–Pugh lemma given in [Hirsch & Pugh (1970)] was proved by J. Palis.

Lemma 4.1. *Given $\tau, 0 < \tau < 1$, there exists an $\epsilon > 0$ such that if the isomorphism $T : E \to E$ with respect to the splitting $E = E_1 \oplus E_2$ has the form $\begin{pmatrix} A & B \\ C & D \end{pmatrix}$ with $L = \begin{pmatrix} A & 0 \\ 0 & D \end{pmatrix} \in \mathfrak{L}_\tau$ and $\|B\| < \epsilon, \|B\| < \epsilon$, then T is hyperbolic.*

Proof. According to Palis [Palis (1968)], it was proved that if there exists an $\epsilon > 0$ (which depends only on τ) such that if $\|B\| < \epsilon, \|C\| < \epsilon$, then T is locally conjugate to L and there is a globally uniform continuous conjugacy h between T and L, i.e., $\tilde{T} \circ h = h \circ L$, where $\tilde{T} = T$ near the origin. In this case, the local images of E_1 and E_2, $h(E_1)$ and $h(E_2)$ generate closed linear subspaces $\tilde{E}_1$ and $\tilde{E}_2$, invariant by T and $\tilde{E}_1 \cap \tilde{E}_2 = 0$, and they are characterized by the fact that $T^n v \to 0$ and $T^{-n} w \to 0$ as $n \to \infty$ for any $v \in \tilde{E}_1$ and $w \in \tilde{E}_2$. Furthermore, $\|T^n|_{\tilde{E}_1}\| < 1$ and $\|T^{-n}|_{\tilde{E}_2}\| < 1$ for some integer n. Thus the spectral radii of $T|_{\tilde{E}_1}$ and of $T^{-1}|_{\tilde{E}_2}$ are less than one. To prove that $E = \tilde{E}_1 \oplus \tilde{E}_2$, it suffices to show that $h(v+w) - h(v) \in \tilde{E}_2$ for small $v \in \tilde{E}_1$ and $w \in \tilde{E}_2$. In fact, the quantities

$$\begin{cases} \|L^{-n}(v+w) - L^{-n}v\| = \|L^{-n}w\| < \lambda^n \|w\| \\ h(L^{-n}(v+w)) - h(L^{-n}v) = T^{-n}(h(v+w) - h(v)) \end{cases} \quad (4.5)$$

converge to the origin as $n \to \infty$ for h uniformly continuous, which implies that $E = \tilde{E}_1 \oplus \tilde{E}_2$, and since the spectral radii of $T|_{\tilde{E}_1}$ and of $T^{-1}|_{\tilde{E}_2}$ are less than one, T is hyperbolic. $\square$

Another obvious lemma was also in the proof:

Lemma 4.2. *Let $\pi : E \to N$ be a Riemannian vector bundle of class C^0. Let M and N be compact metric spaces. Let $f : M \to N$ be a continuous function. Let $A_f : \Gamma(E) \to \Gamma(f_* E)$ be defined by $\gamma \to \gamma \circ f$. Then for fixed f, A_f is a continuous linear function in γ and hence C^∞.*

In what follows, we present in brief the proof of Robinson and Verjovsky. Let $<,>$ be a C^∞ Riemannian metric on M and $|.|$ its induced norm on T_xM for each $x \in M$. Let D be the space of diffeomorphisms on M with the C^1 topology and H be the space of homeomorphisms on M with the C^0 topology. Let C be the space of continuous functions from M to M. We give M a C^∞ Riemannian metric. The topology of C is given by the metric $\bar{d}$:

$$\bar{d}(f,g) = \sup\{d(f(x),g(x)) : x \in M\}, \tag{4.6}$$

where d is the distance between points of M induced by the Riemannian structure on M. In Theorem 4.5, one has

$$H = \{h \in C : h \text{ is a homeomorphism}\}. \tag{4.7}$$

The proof of Theorem 4.5 requires only the use of the local coordinate chart at the identity given by the following lemma:

Lemma 4.3. *C admits the structure of a C^∞ manifold modeled on a Banach space.*

Proof. Let $\mathcal{U}$ be an open cover of M by convex neighborhoods[1]. Let $\delta > 0$ be a Lebesgue number associated with the open cover, *i.e.*, given a ball B of radius less than or equal to δ, there exists a $U \in \mathcal{U}$ such that $B \subset U$. Let $f \in C$ and let $\Gamma(f)$ denote the Banach space of continuous sections of $f_*(TM)$, $\Gamma(f_*(TM))$. Let $U(f) = U_\delta(f)$ be the open ball in $\Gamma(f)$ of radius δ centered at the zero section. Let $B(f) = B_\delta(f)$ be the open ball in C centered at f of radius δ. The open ball $B(f)$ can be parameterized by $U(f)$ as follows: Let $\phi_f : U(f) \to B(f)$ be given by

$$(\phi_f(\sigma))(x) = \exp_{f(x)}(\sigma(x)) \text{ for } \sigma \in U(f). \tag{4.8}$$

Hence

$$\bar{d}(\phi_f(\sigma_1),\phi_f(\sigma_2)) \leq \sup_{x \in M}\{|\sigma_1(x) - \sigma_2(x)|\} \leq \|\sigma_1 - \sigma_2\|. \tag{4.9}$$

Thus ϕ_f is continuous. On the other hand, ϕ_f has an inverse $\phi_f^{-1} : B(f) \to U(f)$ defined by

$$\phi_f^{-1}(g)(x) = (x, (\exp_{f(x)})^{-1}(g(x))). \tag{4.10}$$

In this case, the expression $(\exp_{f(x)})^{-1}(g(x))$ is well defined because the neighborhoods in U are convex. The uniform continuity of the exponential

---
[1] Convex with respect to the Riemannian structure.

on M implies that there is a constant e such that

$$\begin{cases} \left\|\phi_f^{-1}(g_1) - \phi_f^{-1}(g_2)\right\| = \sup_{x \in M} h(x) \\ \left\|\phi_f^{-1}(g_1) - \phi_f^{-1}(g_2)\right\| \leq e \sup_{x \in M} h_2(x) \leq e\, \bar{d}(g_1, g_2) \\ h_1(x) = \{|(exp_{f(x)})^{-1}(g_1(x)) - (exp_{f(x)})^{-1}(g_2(x))|\} \\ h_2(x) = \{\bar{d}(g_1(x), g_2(x))\}. \end{cases} \quad (4.11)$$

Thus ϕ_f^{-1} is continuous. An atlas was defined for C, whose local charts are modeled on the Banach spaces $f_*(TM)$ with $f \in C$. Finally, it suffices to show that the changes of coordinates are C^∞. Let $\phi_f : U(f) \to B(f)$ and $\phi_g : U(g) \to B(g)$ be two charts. We need to prove that $\phi_g^{-1}\phi_f : U(f) \to U(g)$ is a diffeomorphism of class C^∞ on its domain of definition. Let

$$\begin{cases} V(f) = \{v \in f_*(TM) : |v| < \delta\} \\ V(g) = \{v \in g_*(TM) : |v| < \delta\}. \end{cases} \quad (4.12)$$

Then one has

$$U(f) = \Gamma(V(f)), U(g) = \Gamma(V(g)). \quad (4.13)$$

Define the homeomorphism $G : V(f) \to V(g)$ by

$$G(x, v) = (x, (exp_{g(x)})^{-1} \circ exp_{f(x)} v).G. \quad (4.14)$$

Hence G is well defined by the convexity of the neighborhoods, and one has that

$$\phi_g^{-1}\phi_f(v) = G \circ v = \Omega G(v).G \quad (4.15)$$

preserves fibers, and G is vertically of class C^∞ because along a fixed fiber

$$G(x,.) = \left(x, \left(exp_{g(x)}\right)^{-1} \circ exp_{f(x)}\right). \quad (4.16)$$

By Lemma 4.5, $\phi_g^{-1}\phi_f$ is of class C^∞. The same method implies that $(\Omega G)^{-1} = \Omega G^{-1} = \phi_f^{-1}\phi_g$ is also of class C^∞. □

Let $\pi : E \to M$ be a normed vector bundle over M. Let $\Gamma(E)$ be the Banach space of continuous sections of E with norm

$$\|\sigma\| = \sup_{x \in M} |\sigma(x)|, \sigma \in \Gamma(E). \quad (4.17)$$

The space $\Gamma(TM)$ is denoted simply by $\Gamma(M)$. If $f \in D$, then it is possible to define a continuous operator $f_* : \Gamma(M) \to \Gamma(M)$ given by

$$f_*\sigma = Df\sigma \circ f^{-1}, \sigma \in \Gamma(M) \text{ or } f_*\sigma(x) = Df_{f^{-1}(x)}\sigma(f^{-1}(x)). \quad (4.18)$$

Hence f_* is linear and continuous due to the fact that M is compact, i.e., the operator f_* is an isomorphism, where $(f_*)^{-1} = (f^{-1})_*$. To prove that the Anosov diffeomorphisms form an open set, we need the following lemmas:

Lemma 4.4. *The map $f \in D$ is Anosov if and only if f_* is hyperbolic. Furthermore, if f is Anosov, then there is a C^∞ structure of normed vector bundles on TM for which we can take $c = c' = 1$ in Definition 4.7.*

Proof. According to Definition 4.7, if f is Anosov, then the set $\Gamma(M)$ splits into a direct sum of closed subspaces as follows:

$$\Gamma(M) = \Gamma(E^s) \oplus \Gamma(E^u) \tag{4.19}$$

where

$$\sigma \in \Gamma(E^s) \Leftrightarrow \sigma(x) \in E^s, \forall x \in M, \sigma \in \Gamma(E^u) \Leftrightarrow \sigma(x) \in E^u. \tag{4.20}$$

The sets $\Gamma(E^s)$ and $\Gamma(E^u)$ are f_*-invariant because E^s and E^u are Df-invariant. Let $f_s = f|_{\Gamma(E^s)}$ and $f_u = f|_{\Gamma(E^u)}$. Then one has $f_* = f_s \oplus f_u$, and f_s, f_u are (continuous) isomorphisms of $\Gamma(E^s), \Gamma(E^u)$, respectively. This implies that

$$Spectrum(f_*) = Spectrum(f_s) \cup Spectrum(f_u). \tag{4.21}$$

However, f being Anosov implies $\|f_s^n\| \leq c\lambda^n, \| f_u^{-n} \| \leq c'\lambda^n$. Hence the spectral radii of f_s and f_u^{-1} are not larger than $\lambda < 1$. Thus f_* is hyperbolic. Now if $f_* : \Gamma(M) \to \Gamma(M)$ is hyperbolic for $f \in D$ and $\Gamma(M)$ with the norm (4.17), then using the approach given in [Palis (1968)], the set $\Gamma(M)$ can be decomposed into a direct sum of f_*-invariant subspaces $\Gamma(M) = \Gamma^s \oplus \Gamma^u$ so that the spectral radii of $f_s = f_*|_{\Gamma^s}$ and of $f_u^{-1} = f_*^{-1}|_{\Gamma^u}$ are smaller than 1. For each $x \in M$, define

$$E_x^s = \{\sigma(x) : \sigma \in \Gamma^s\}, E_x^u = \{\sigma(x) : \sigma \in \Gamma^u\}. \tag{4.22}$$

Hence one has that the following subsets

$$\begin{cases} E^s = \cup_{x \in M} E_x^s \\ E^u = \cup_{x \in M} E_x^u \end{cases} \tag{4.23}$$

are continuous subbundles of TM, Df-invariant, and $TM = E^s \oplus E^u$. To verify that this sum is direct, let $v \in E_x^s \cap E_x^u$ for some $x \in M$. Since the spectral radii of f_s and f_u^{-1} are smaller than 1, there is an integer n_0 such that

$$\|f_s^{n_0}\| < k, \|f_u^{-n_0}\| < k, 0 < k < 1. \tag{4.24}$$

Define $\sigma^s \in \Gamma^s$ and $\sigma^u \in \Gamma^u$ such that
$$\begin{cases} \sigma^s(f^{-n_0}(x)) = Df_x^{-n_0}v, \\ \|\sigma^s\| = |Df_x^{-n_0}v|, \sigma^u(x) = v \\ \|\sigma^u\| = |v|. \end{cases} \quad (4.25)$$

From (4.25) one has
$$\begin{cases} \|f_s^{n_0}(\sigma^s)\| \leq k|Df_x^{-n_0}v| \\ \|f_u^{-n_0}(\sigma^u)\| \leq k|v|. \end{cases} \quad (4.26)$$

This means that
$$\begin{cases} |v| \leq k|Df_x^{-n_0}v| \\ |Df_x^{-n_0}v| \leq k|v|. \end{cases} \quad (4.27)$$

Thus (4.27) implies that $v = 0$ since $0 < k < 1$. Thus $TM = E^s \oplus E^u$. Now set
$$\begin{cases} \lambda = \frac{k_1}{n_0} < 1 \\ c = \sup_{0 \leq i < n_0}\{\|f_s\|^i \lambda^{-i}\} \\ c' = \sup_{0 \leq i < n_0}\{\|f_u^{-1}\|^i \lambda^i\}. \end{cases} \quad (4.28)$$

From the inequalities $\|f_s^{n_0}\| < k$ and $\|f_u^{-n_0}\| < k$, one has
$$\begin{cases} |Df_x^n v| < c\lambda^n |v| \\ |Df_x^{-n} w| < c'\lambda^n |w| \end{cases} \quad (4.29)$$

for each $x \in M$, $v \in E_x^s$, and $w \in E_x^u$. Thus the map f is Anosov. The second part of this proof is to verify that if f is Anosov, then there is a C^∞ norm on TM so that one can take $c = c' = 1$ in the above inequalities. Following the work of Palis [Palis (1968)], let ρ be such that
$$\lambda < \rho < 1, \quad (4.30)$$

and define
$$|v|_s = \sum_{n=0}^{\infty} \rho^{-n} |Df_x^n v|, |w|_u = \sum_{n=0}^{\infty} \rho^{-n} |Df_x^{-n} w|, v \in E_x^s, w \in E_x^u. \quad (4.31)$$

Hence for any element $\alpha \in T_x M$, α can be written as $\alpha = v + w$ with $v \in E_x^s$ and $w \in E_x^u$. Define
$$|\alpha|_1 = |v|_s + |w|_u. \quad (4.32)$$

Thus the norm $|.|_1{}^2$ is a norm equivalent to the original one $|.|$, and
$$\begin{cases} |Df_x v|_1 \leq \rho |v|_1 \\ |Df_x^{-1} w|_1 \leq \rho |w|_1 \end{cases} \quad (4.33)$$

for $v \in E_x^s$ and $w \in E_x^u$. $\square$

---
[2] In this case the norm $|.|_1$ is a C^0 norm.

Now the proof of the fact that Anosov diffeomorphisms form an open set in $Diff(M)$ can be obtained from Lemmas 4.1 and 4.4 using the continuity of the norm of the operators corresponding to the decomposition of g_* with respect to the splitting $\Gamma(M) = \Gamma^s \oplus \Gamma^u$ defined in Lemma 4.4 because the map $\rho: Diff(M) \to Isom(\Gamma(M))$ defined by $\rho(g) = g_*$, is not continuous.

Theorem 4.3. *The Anosov diffeomorphisms form an open set in $Diff(M)$.*

Proof. If f is an Anosov diffeomorphism, then f_* is a hyperbolic isomorphism of $\Gamma(M)$. Thus $\Gamma(M) = \Gamma^s \oplus \Gamma^u$, where the subsets Γ^s and Γ^u are given by Lemma 4.4, and they are f_*-invariant and satisfy $\|f_*|_{\Gamma^s}\| < \tau, \|f_*^{-1}|_{\Gamma_u}\| < \tau$ for some τ such that $0 < \tau < 1$. For a given $\epsilon > 0$, there is neighborhood $N(f) \subset Diff(M)$ with the property that for any $g \in N(f)$, $g_* = \begin{pmatrix} A & B \\ C & D \end{pmatrix}$ with respect to the splitting $\Gamma(M) = \Gamma^s \oplus \Gamma^u$, where $\|A\| < \tau, \|D^{-1}\| < \tau, \|B\| < \epsilon$, and $\|C\| < \epsilon$. If ϵ as in the Hirsch–Pugh Lemma 4.1, then g_* is hyperbolic, and by Lemma 4.4, g is Anosov$\square$

Now let $f: M \to N$ be a continuous function and $\pi: E \to M$ be a Riemannian vector bundle. $f_*(E)$ is the subset of $M \times E$ of pairs (x, v) such that $f(x) = \pi(v)$. Let $\pi(f)$ be the projection on the first factor of $M \times E$. $\pi(f): f_*(E) \to M$ is a vector bundle. There is a Riemannian metric induced on $f_*(E)$ by the inclusion in $M \times E$. Let $\pi_i: E_i \to M_i = 1, 2$ be two Riemannian vector bundles. Let $U \subset E_1$ be an open subset such that $\pi_1|_U : U \to M$ is a surjection. Let $\Gamma(U) \subset \Gamma(E_1)$ be the open subset of sections with images in U. We assume U is connected enough so the $\Gamma(U)$ is nonempty. Let $\overline{f}: U \to E_2$ be a continuous function that preserves fibers covering $f: M \to M$. Let $\Omega_{\overline{f}}: \Gamma(U) \to \Gamma(f_*E_2)$ be the map induced by composition on the left by $\overline{f}$, i.e., $\Omega_{\overline{f}}: \gamma \to \overline{f} \circ \gamma$. Hence one can obtain the following result:

Lemma 4.5. *If $\overline{f}$ is vertically of class C^r, $0 \leq r \leq \infty$, then $\Omega_{\overline{f}}: \Gamma(U) \to \Gamma(f_*E_2)$ is of class C^r.*

Proof. For $r = 0$, the map $\Omega_{\overline{f}}$ is the composition of continuous functions on a compact set. This implies that $\Omega_{\overline{f}}$ is continuous. Let $\gamma, \sigma \in \Gamma(U)$ be small enough in norm so that $\gamma + \sigma \in \Gamma(U)$. For each $x \in M$, Taylor's theorem applied to the function $\overline{f}x: E_x^1 \to E_{f(x)}^2$ at the point $\gamma(x)$ gives the following relation:

$$\Omega_{\overline{f}}(\gamma+\sigma)(x) = \left(\Omega_{\overline{f}}(\gamma) + \sum_{k=1}^{r} \frac{1}{k!} D^k \overline{f}_x(\gamma)(\sigma)^k \left(R(\gamma,\sigma)(\sigma)^r\right)\right)(x) \quad (4.34)$$

where

$$\begin{cases} (\sigma(x))^k = (\sigma(x),...,\sigma(x)), \\ R(x,y) \in L_x^r\left(E_x^1, E_{f(x)}^2\right).L_x^r\left(E_x^1, E_{f(x)}^2\right) \end{cases} \quad (4.35)$$

are symmetric r-multilinear functions from E_x^1 to $E_{f(x)}^2$. Formula (4.34) without evaluation at x is given by

$$\Omega_{\overline{f}}(\gamma+\sigma)(x) = \Omega_{\overline{f}}(\gamma) + \sum_{k=1}^{r} \frac{1}{k!} D^k \overline{f}_x(\gamma)(\sigma)^k + R(\gamma,\sigma)(\sigma)^r \quad (4.36)$$

if one takes only the derivative of $\overline{f}$ along the fiber and

$$R(\gamma,\sigma) \in L_x^r(\Gamma(E^1), \Gamma(f_*E^2)). \quad (4.37)$$

In this case, the function $R(.,.)$ is continuous, and $R(\gamma,0) = 0$. Using the converse of Taylor's theorem described in [Abraham & Robbin (1967), Smale (1967), Moser (1969)], it follows that $\Omega_{\overline{f}}$ is of class C^r and that

$$D^k \Omega_{\overline{f}}(\gamma)(\sigma_1,...,\sigma_k) = Dk\overline{f}_x(\gamma)(\sigma_1,...,\sigma_k) \text{ for } \sigma_1,...,\sigma_k \in \Gamma(E^1). \quad (4.38)$$

Then $D^k \Omega_{\overline{f}} : \Gamma(U) \to L_x^k(\Gamma(E^1), \Gamma(f_*E^2))$. $\square$

Finally, the Anosov theorem on structural stability is given by:

Theorem 4.4. *(Anosov) If f is an Anosov diffeomorphism, then f is structurally stable. In particular, there exists a neighborhood V of f in D, a neighborhood U of the identity $id : M \to M$ in H, and a continuous function $h : V \to U$ such that if $g \in V$, then $h = h(g)$ is the unique solution in U of the functional equation*

$$h \circ g = f \circ h. \quad (4.39)$$

Proof. The proof is based on the study of the map $D \times D \times C \to C$ defined by $(g_1, g_2, h) \to g_1 \circ h \circ g_2^{-1}$. If $g_1 \circ h \circ g_2^{-1} = h$, then $g_1 \circ h = h \circ g_2$. Thus fixed points of the map give a semi-conjugacy between g_1 and g_2[3]. Also $g \circ id \circ g^{-1} = id$. To prove the stability of this fixed point, it suffices to take local coordinates in C near id, i.e., $\phi : U \subset \Gamma(M) \to C$ with

$$\phi(\sigma)(x) = \exp_x \sigma(x). \quad (4.40)$$

---
[3]There is a conjugacy if h is a homeomorphism.

For neighborhoods V of f in D and U of 0 in $\Gamma(M)$, the operator $A : V \times V \times U \to \Gamma(M)$ is well defined by

$$A(g_1, g_2, h) = \phi^{-1}(g_1 \circ \phi(h) \circ g_2^{-1}), \tag{4.41}$$

or

$$A(g_1, g_2, h)(x) = exp_x^{-1}(g_1 \circ exp_{g_2^{-1}(x)} \circ (h \circ g_2^{-1}(x))), \tag{4.42}$$

and for $g_1, g_2 \in V$, define $G(g_1, g_2) : TM \to TM$ by

$$G(g_1, g_2)(vx) = exp_{g_2^{-1}(x)}(g_2 \circ exp_x v_x). \tag{4.43}$$

Then

$$\begin{cases} \left(\Omega_{G(g_1,g_2)} A'_{g_2^{-1}}\right)(x) = h_3(x) = A(g_1, g_2, h)(x), \\ h_3(x) = exp_x^{-1}(g_1 \circ exp_{g_2^{-1}(x)}(h \circ g_2^{-1}(x))) \end{cases} \tag{4.44}$$

where $A'_{g_2^{-1}}$ is the map given by Lemma 4.2. □

Lemma 4.6. *The operator A has a partial derivative with respect to the third variable. When $g_1 = g_2 = g$, we have that*

$$D_3 A(y, g, 0)k = Dg(g^{-1})k \circ g^{-1} = g_* k . D_3 A(g_1, g_2, h) \tag{4.45}$$

is continuous in the first and third variables and uniform in the second variable, i.e., given (g_1, h) and $\epsilon > 0$, there exist neighborhoods V' of g_1 and U' of h such that for $f_{11}, f_{12} \in V'$, $f_2 \in V$, and $h_1, h_2 \in U'$

$$\|D_3 A(f_{11}, f_2, h_1) - D_3 A(f_{12}, f_2, h_2)\| < \epsilon. \tag{4.46}$$

In particular, given $\epsilon > 0$, there exist neighborhoods V' of f and U' of 0 in $\Gamma(M)$ such that the Lipschitz constant is

$$L(A(f_{11}, f_2, .)|_{U'} - D_3 A(f_{11}, f_2, 0)|_{U'}) < \epsilon \text{ for } g_1, g_2 \in V'. \tag{4.47}$$

Proof. Using Lemmas 4.1 and 4.2, the partial derivative of A with respect to the third variable exists, and since $D(exp_x)(0_x) = id : T_x M \to T_x M$, it follows that

$$D_3 A(g, g, 0)k = Dg(g^{-1})k \circ g^{-1}. \tag{4.48}$$

Let $G_1 = G(f_{11}, f_2)$ and $G_2 = G(f_{12}, f_2)$. Then one has

$$\begin{cases} \|D_3 A(f_{11}, f_2, h_1) - D_3 A(f_{12}, f_2, h_2)\| = \sup_{k \in \Gamma(M) \text{ with} \|k\|=1} \{s\} \\ s_1 = \|DG_1(h_1 \circ f_2^{-1})k \circ f_2^{-1} - |DG_2(h_2 \circ f_2^{-1})k \circ f_2^{-1}\|. \end{cases} \tag{4.49}$$

Thus
$$\begin{cases} \|D_3 A(f_{11}, f_2, h_1) - D_3 A(f_{12}, f_2, h_2)\| \leq \sup_{x \in M}\{s_2\} \\ s_2 = \|DG_1(h_1 \circ f_2^{-1}(x)) - DG_2(h_2 \circ f_2^{-1}(x))\|. \end{cases} \quad (4.50)$$

Using the uniformity in the exponential, and letting $f_{11}, f_{12} \to g$ and $h_1, h_2 \to h$, one has that the limit

$$\|DG_1(h_1 \circ f_2^{-1}(x)) - DG_2(h_2 \circ f_2^{-1}(x))\| \to 0 \quad (4.51)$$

converges uniformly in f_2 and x because $G_i(f_2^{-1}(x), v) = exp_x^{-1}(f_{1i} \circ exp_{f_2^{-1}(x)} v), i = 1, 2$. Hence $D_3 A$ is continuous, and the Lipschitz constant follows from the above results using the mean value theorem given in [Dieudonné (1960)]. □

The operator $D_3 A(g_1, g_2, h)$ is not continuous in g_2. To prove this, one can consider the case of a map defined in the plane so that the exponentials can be ignored. Let $h = 0$, and take g_2' arbitrarily near g_2 in the C^1 topology with the condition $(g_2)^{-1}(x_0) = (g_2')^{-1}(x_0)$. Hence for each such g_2', there exists a $k \in \Gamma(M)$ such that

$$\begin{cases} \|k\|_0 = 1 \\ |k \circ (g_2)^{-1}(x_0) - k \circ (g_2')^{-1}(x_0)| = 1. \end{cases} \quad (4.52)$$

Then
$$\begin{cases} \|D_3 A(g_1, g_2, 0) - D_3 A(g_1, g_2', 0)\| \geq |s_3| \\ s_3 = D(g_1)_{(g_2)^{-1}(x_0)} k \circ g_2^{-1}(x_0) - D(g_1)_{(g_2')^{-1}(x_0)} k \circ (g_2')^{-1}(x_0). \end{cases} \quad (4.53)$$

Thus, as g_2' goes to g_2, the quantity $\|D_3 A(g_1, g_2, 0) - D_3 A(g_1, g_2', 0)\|$ remains bounded away from zero, and hence $D_3 A(g_1, g_2, h)$ is not continuous. A partial result in this direction can be obtained using the result:

Lemma 4.7. *Let* $T : D \times D \times \Gamma(M) \to \Gamma(M)$ *be defined by*

$$T(g_1, g_2, h) = D_3 A(g_1, g_2, 0)h. \quad (4.54)$$

Then T is continuous in all variables.

The following lemma was used to prove the stability of the fixed point of A, and it is based on the last paragraph of page 144 in [Hirsch & Pugh (1970)]. In this case, a standard fixed point theorem or the implicit function theorem can be used if $D_3 A : D \times D \times \Gamma(M) \to L(\Gamma(M), \Gamma(M))$ is continuous.

Lemma 4.8. *Let P be a topological space. Let $F_1 \oplus F_2$ be a Banach space with the norm equal to the maximum of the norms on the two factors. Let*

$T: P \times F_1 \oplus F_2 \to F_1 \oplus F_2$ *be a function (not necessarily continuous) such that for each* $x \in P, T(x,.) : F_1 \oplus F_2 \to F_1 \oplus F_2$ *is a continuous linear isomorphism. Assume*

$$\begin{cases} \|T_1(x,.,0)^{-1}\| \leq \tau \\ \|T_2(x,0,.)\| \leq \tau \\ \|T_1(x,0,.)\| \leq \mu \\ \|T_2(x,.,0)\| \leq \mu \end{cases} \tag{4.55}$$

where $T_i(x,.,0) : F_1 \to F_i$. *We also have* $\epsilon > 0$ *such that*

$$\tau + \mu + \epsilon < 1. \tag{4.56}$$

Let $U_1 \oplus U_2 \subset F_1 \oplus F_2$ *be a ball about the origin of radius R. Assume* $f : P \times U_1 \oplus U_2 \to F_1 \oplus F_2$ *is a function such that for all* $x \in P$, *(i) the Lipschitz constant*

$$L(f(x,.) - T(x,.)|_{U_1 \oplus U_2}) < \epsilon \tag{4.57}$$

and (ii)

$$|f(x,0,0)| \leq (1 - \tau - \mu - \epsilon) R. \tag{4.58}$$

Then there exists a function $u : P \to U_1 \oplus U_2$ *such that*

$$\begin{cases} f(x, u(x)) = u(x) \\ |u(x)| \leq \frac{|f(x,0,0)|}{1-\tau-\mu-\epsilon}. \end{cases} \tag{4.59}$$

Furthermore, if f and T are continuous, then so is u.

Proof. Define $g : P \times U_1 \oplus U_2 \to F_1 \oplus F_2$ by

$$\begin{cases} g(w_1) = (T_1(x,.,0)^{-1}(y_1 + T_1(x, y_1, 0) - f_1(w)), f_2(w)) \\ w_1 = (x, y_1, y_2). \end{cases} \tag{4.60}$$

Hence the fixed points of $g(x,.)$ are the same as those of $f(x,.)$. In fact, $g(x,.)$ is a contraction with contraction constant $\tau + \mu + \epsilon$ because if $y = (y_1, y_2)$ and $y' = (y'_1, y'_2)$, then one has

$$\begin{cases} |g_1(x,y) - g_1(x,y')| \leq w_2 \\ w_2 = \tau(|y_1 - y'_1| + L(T_1 - f_1)|y - y'| + |T_1(x, 0, y_2 - y'_2)|). \end{cases} \tag{4.61}$$

Hence

$$|g_1(x,y) - g_1(x,y')| \leq \tau(1 + \epsilon + \mu)|y - y'| \leq (\tau + \epsilon + \mu)|y - y'| \tag{4.62}$$

$$|g_2(x,y) - g_2 x, y')| \leq (\tau + \epsilon + \mu)|y - y'|. \tag{4.63}$$

Using the results in [Dieudonné (1960), 10.1.1, Hirsch & Pugh (1970), 1.1], g has a fixed point, $u(x)$, for each $x \in P$ with

$$|u(x)| \leq \frac{|g(x,0)|}{1 - \tau - \mu - \epsilon} \leq \frac{|f(x,0)|}{1 - \tau - \mu - \epsilon}. \qquad (4.64)$$

Now if f and T are continuous, then g is continuous. For $x_0 \in P$, by [Dieudonné (1960), 10.1.1, Hirsch & Pugh (1970), 1.1], one has

$$|u(x_0) - u(x)| \leq \frac{|g(x, u(x_0)) - u(x_0)|}{1 - \tau - \mu - \epsilon}. \qquad (4.65)$$

Hence u is continuous. Since the operator $T(f, f, .) = D_3 A(f, f, 0) = f_* : \Gamma(M) \to \Gamma(M)$ is hyperbolic and has a splitting $\Gamma(M) = \Gamma(E^u) \oplus \Gamma(E^s)$, it is possible to approximate E^u and E^s by smooth bundles F^u and F^s. Let $F^1 = \Gamma(F^u)$ and $F^2 = \Gamma(F^s)$ and for a small neighborhood V of f, one has that $T(g_1, g_2, h) = D_3 A(g_2, g_2, 0) h$ satisfies Lemma 4.8 due to the continuity of the norms of the coordinate functions and Lemmas 4.6 and 4.7. On $F_1 \oplus F_2$, the following norm was defined:

$$|(y_1, y_2)| = \max\{|y_1|, |y_2|\}. \qquad (4.66)$$

Thus there exist neighborhoods V_1 of f in D and U_1' of 0 in $\Gamma(M)$ such that for $g_1, g_2 \in V_1$, there exists a unique $k = u'(g_1, g_2) \in U_1'$ such that $A(g_1, g_2, k) = k$. Let $U_1 = \phi(U_1') = exp(U_1') \subset C$, $u = \phi \circ u'$, and $h = u(g_1, g_2)$. Then $g_1 \circ h = h \circ g_2$. Also u is a continuous function of g_1 and g_2. Let U_2 be a smaller neighborhood of id in C such that for all $h_1, h_2 \in U_2$, $h_1 \circ h_2 \in U_1$. This exists since composition is continuous. Continuity of u or continuity of A and the estimate

$$|u'(g_1, g_2) - id| \leq \frac{|A(g_1, g_2, id) - id|}{1 - \tau - \mu - \epsilon} \qquad (4.67)$$

implies that there exists a smaller neighborhood V_2 of f in D such that for $g_1, g_2 \in V_2, u(g_1, g_2) \in U_2$. If $g \in V_2$, let $h = u(g, f)$ and $h' = u(f, g)$. Then $g \circ h = h \circ f$ and $f \circ h' = h' \circ g$. Thus

$$h \circ h' \circ g = h \circ f \circ h' = g \circ h \circ h'. \qquad (4.68)$$

Also

$$h' \circ h \circ f = f \circ h' \circ h. h' \circ h, h \circ h' \in V_1, \qquad (4.69)$$

and so by uniqueness one has

$$h' \circ h = h \circ h' = id. \qquad (4.70)$$

Finally, h is a homeomorphism. This is the end of the proof. $\square$

As a recent result, it was shown in [Medvedev & Zhuzhoma (2004)] that there are no structurally stable diffeomorphisms of odd-dimensional manifolds with codimension-one, nonorientable, expanding attractors, namely the following result:

Theorem 4.5. *Let $f : M \to M$ be a structurally stable diffeomorphism of a closed $(2m + 1)$-manifold M, $m \geq 1$. Then the spectral decomposition of f does not contain codimension-one, nonorientable expanding attractors.*

The proof of Theorem 4.6 does not work for even-dimensional manifolds for which the existence of codimension-one, nonorientable, expanding attractors remains an open question (except for $d = 2$). At the end of this section, we present a weaker property of structural stability called "Ω-stability". This concept is defined requiring equivalence only restricted to the nonwandering set $\Omega(f)$, *i.e.*,the Ω-stability conjecture of Palis–Smale is given in [Palis & Takens (1993)] as follows:

Conjecture 4.2. *For any $r \geq 1$, Ω-stable systems should coincide with the hyperbolic systems with no cycles, that is, such that no basic pieces in the spectral decomposition are cyclically related by intersections of the corresponding stable and unstable sets.*

The Ω-stability theorem of Smale [Bonatti et al. (2005)] states that these properties are sufficient for C^r Ω-stability. Palis in [Palis & Takens (1993)] showed that the no-cycles condition is also necessary for C^r Ω-stability. In Mané [Mané (1987)], it was proved that for C^1 diffeomorphisms, hyperbolicity is necessary for Ω-stability. This was extended to C^1 flows by Hayashi in [Hayashi (1997)]. The reader interested in the implications of structural stability to real applications can see the book of Simitses and Hodges [Simitses & Hodges (2006)]. This book discusses many practical examples such as mechanical stability models, elastic buckling of columns, buckling of frames, the energy criterion and energy-based methods, columns on elastic foundations, buckling of rings and arches, buckling of shafts, lateral-torsional buckling of deep beams, instabilities associated with rotating beams, nonconservative systems, and dynamic stability and related topics.

4.2 Exercises

Exercise 89. Show that a high-dimensional system is structurally stable if it satisfies both axiom A and the strong transversality condition. See [Smale (1967)].

Exercise 90. Show that the structure induced on $C^r(M, M)$ by the measure of distance between two elements of $C^r(M, M)$ defines a topology called the C^k topology. See [Palis & de Melo (1982)] or [Hirsch (1976)].

Exercise 91. Explain the relation between structural stability and *genericity*.

Exercise 92. Show that the hyperbolic fixed points, periodic orbits, and the transversal intersection of the stable and unstable manifolds of hyperbolic fixed points and periodic orbits are structurally stable and generic. See [Palis & de Melo (1992)].

Exercise 93. Prove Peixoto's Theorem 4.1 for vector fields on a compact, boundaryless, two-dimensional manifold. See [Peixoto (1962)].

Exercise 94. Show that Peixoto's Theorem 4.1 cannot be generalized for n-dimensional diffeomorphisms ($n \geq 2$) or n-dimensional vector fields ($n \geq 3$). See [Smale (1966)].

Exercise 95. Describe a method for testing structural stability for periodic orbits and fixed points.

Exercise 96. Describe a method for testing structural stability for homoclinic and quasi-periodic orbits.

Exercise 97. Show Peixoto's Theorem 4.2 for diffeomorphisms on the circle $\mathbb{S}^1$.

Exercise 98. Show that hyperbolicity (see Chapter 6) is equivalent to structural stability in the C^k topology. See [Anosov (1967), Palis & Smale (1968), Palis (1969), Robbin (1971), de Melo (1973), Robinson (1974), Liao (1980-1983), Mañé (1982), Sannami (1983), Mañé (1988), Hayashi (1997)].

Exercise 99. (a) Show that for the three-dimensional case, the basic sets of C^1 structurally stable flows are: isolated fixed points, isolated closed orbits, and suspensions of nontrivial irreducible shifts of finite type.

(b) Show that these sets have infinitely many periodic orbits but rational zeta functions and two-dimensional attractors or repellers just like the suspension of Plykin's attractor introduced in Sec. 6.4.4.

Exercise 100. Show that Definition 4.7 is independent of the Riemannian metric $<,>$, and the subbundles E^s and E^u are uniquely determined according to the above conditions.

Exercise 101. (Open problem) Prove Stability Conjecture 4.1.

Exercise 102. (Open problem) Show the structural stability for diffeomorphisms of even-dimensional manifolds for which codimension-one, nonorientable, expanding attractors exist. See [Medvedev & Zhuzhoma (2004)].

Exercise 103. (Open problem) Prove the Ω-stability Conjecture 4.2 of Palis–Smale. See [Palis & Takens (1993)].

Chapter 5

Transversality, Invariant Foliation, and the Shadowing Lemma

In this chapter, we discuss some relevant properties and the importance of transversality, invariant foliation, and the shadowing lemma in analyzing chaotic dynamical systems in general, and robust ones in particular. Indeed, transversality introduced in Sec. 5.1 is the opposite of tangency introduced in Sec. 1.3.1 responsible for the coexistence of attractors. In Sec. 5.2, we present and discuss the notion of the invariant foliation that is used to prove chaos and its topological properties in the singular-hyperbolic attractors. In Sec. 5.3, we present the shadowing lemma that is used to verify the existence of a true chaotic attractor. In particular, a robust chaotic attractor is more important than one obtained without verification. The possible relations between homoclinic orbits and shadowing are given in Sec. 5.3.1, and some Shilnikov theorems concerning criteria for the existence of chaos are presented in Sec. 5.3.2.

5.1 Transversality

In this section, we assume that the reader is familiar with the basic notions and definitions related to abstract manifolds, especially the notions of tangency, submanifolds, tangent space, codimension, oriented manifolds, *etc*. A good reference discussing basic differential geometry, especially the notion of transversality, is the book of Hirsch [Hirsch (1976)].

Transversality is a fundamental concept in differential topology, and it is the opposite of tangency. It describes the methods of generic intersection of several spaces (generally, the transversality of a pair of submanifolds). This notion is defined by considering the linearizations of the intersecting spaces at the points of intersection. The notion can be easily extended to transversality of a submanifold M_1 and a map $f : M_1 \to M$ to the ambient manifold

M, or to a pair of maps $f : M_1 \to M$ and $g : M_2 \to M$ to the ambient manifold M. The method of analysis is based on the fact that whether the *pushforwards* (the derivative) of the tangent spaces $df_x : T_x M_1 \to T_{f(x)} M$ and $dg_x : T_x M_2 \to T_{g(x)} M$ at points of intersection of the images generate the entire tangent space of $T_x M$. Because the transversality is a stable and generic condition (recall Definition 4.5), if $f : M_1 \to M$ is transverse to M_2, then small perturbations of f are also transverse to $B \subset M$. Furthermore, any smooth map $f : M_1 \to M$ can be perturbed slightly to obtain a smooth map that is transverse to a given submanifold $B \subset M$.

Definition 5.1. (a) Two submanifolds M_1 and M_2 of the manifold M intersect transversely if at every point x of intersection $M_1 \cap M_2$, their separate tangent spaces $T_x M_1$ and $T_x M_2$ at that point, together generate the tangent space $T_x M$ of the ambient manifold M at the point x.

(b) Let M and N be manifolds, and M_1 be a submanifold of N. A map $f : M \to N$ is transverse to M_1, written $f \pitchfork M_1$, if, for every $x \in M$ such that $f(x) \in M_1$, we have $Tf(T_x M) \oplus T_{f(x)} M_1 = T_{f(x)} N$.

In this case, the subset $f^{-1}(M_1)$ is a submanifold of M, and $co\dim\left(f^{-1}(M_1)\right) = co\dim(M_1)$, namely the following result:

Theorem 5.1. *(a) (Preimage). Let $f : M \to N$ be smooth, and let M_1 be a codimension-k submanifold of N. If f is transverse to M_1, then $f^{-1}(M_1)$ is either empty or a smooth codimension-k submanifold of M.*

(b) (Elementary transversality). Let M and N be manifolds, and let M_1 be a submanifold of N. Then the smooth functions $f : M \to N$ that are transverse to M_1 form a residual subset (recall Definition 4.4) of $C^\infty_{M_1}(M, N)$. If M_1 is closed, then this subset is open and dense in $C^\infty_{M_1}(M, N)$.

A first example of transversality is arcs in a surface, where an intersection point between two arcs is transverse if and only if their lines tangent to the surface inside the tangent plane are distinct. In the case of a three-dimensional space, the following things are true: (a) Transverse curves do not intersect. (b) Curves transverse to surfaces intersect in points, and surfaces transverse to each other intersect in curves. (c) Curves that are tangent to a surface at a point do not intersect the surface transversely. The second example of such a situation is the case of a hyperbolic periodic orbit of a C^r, $r \geq 1$ vector field on $\mathbb{R}^n$ where its Poincaré map (see Sec. 1.1) (linearized about the fixed point) has $n - k - 1$ eigenvalues with modulus greater than one and k eigenvalues with modulus less than one.

Some important remarks about the notion of transversality are the following:

(1) Transversality persists under sufficiently small perturbations, *i.e.*, transversality is a robust generic property.
(2) In the case where the intersection of M_1 and M_2 is transverse, then the intersection $M_1 \cap M_2$ is a submanifold whose codimension is equal to the sums of the codimensions of the two manifolds, *i.e.*, $co\dim(M_1 \cap M_2) = co\dim(M_1) + co\dim(M_2)$. On the other hand, the absence of transversality implies that the intersection is not a submanifold having some sort of singular point.
(3) If the manifolds M_1 and M_2 are of complementary dimension, then transversality implies that the tangent space TM defined for the ambient manifold M is the direct sum of the two smaller tangent spaces, *i.e.*, $TM = TM_1 \oplus TM_2$. In this case, the submanifolds M_1 and M_2 are oriented if M is oriented, and they intersect in isolated points, *i.e.*, a 0-manifold.
(4) If the maps $f : M_1 \to M$ and $g : M_2 \to M$ are embeddings, then their submanifolds are transverse.
(5) If $M_1 = \{b\}$ is a single point, then f is transverse to M_1, where b is a regular value[1] for f.

At the end of this section, we note that transversality is an essential tool in studying hyperbolic systems (see Chapters 6 and 7) as can be seen from several notes about the proofs of theorems for these type of systems. Moreover, details about the concept can be found in the books [Hirsch (1976), Spivak (1999)].

5.2 Invariant foliation

In this section, we define the notion of the *invariant foliation* that is used to prove chaos and its topological properties in singular-hyperbolic attractors. This notion is a geometric device used to study manifolds, or it is a kind of *clothing* worn on a manifold, cut from a striped fabric that gives a local product structure of the manifold. Let M be an n-dimensional manifold. In other words, the foliation is a partition of a subset of the ambient space

---
[1]A critical point of f is a point such that its differential Df does not have the full rank. A regular value of f is a point which is not the image of any critical point of f [Spivak (1999)].

M into smooth submanifolds called *leaves of the foliation* with the same dimension and varying continuously from one point to the other.

Definition 5.2. (a) A dimension-p foliation f of M is a covering by charts U_i together with maps $\phi_i : U_i \to R^n$ such that on the overlaps $U_i \cap U_j$, the transition functions $\phi_{ij} : \mathbb{R}^n \to \mathbb{R}^n$ defined by $\phi_{ij} = \phi_j \circ \phi_i^{-1}$ take the form $\phi_{ij}(x,y) = (\phi_{ij}^1(x,y), \phi_{ij}^2(x,y))$, where x denotes the first $n-p$ coordinates and y denotes the last p coordinates. That is, $\phi_{ij}^1 : \mathbb{R}^{n-p} \to \mathbb{R}^{n-p}$ and $\phi_{ij}^2 : \mathbb{R}^p \to \mathbb{R}^p$. (b) The stripes at $x = constant$ are called *plaques of the foliation F*. (c) The leaves of the foliation f are the union of all the plaques. (d) If $y_0 \in U_{iy}$, then $\Gamma_1 = U_{ix} \times \{y_0\}$ is called a *local transversal section of the foliation F*.

In this case, one can obtain the following results proved in [Moerdijk & Mrčun (2003)]:

(1) The plaques $x = constant$ are $n-p$-dimensional submanifolds in each chart (U_i, ϕ_i).
(2) The leaves of foliation f form a maximal-connected, injectively-immersed submanifold.
(3) Each chart U_i can be written in the form $U_i = U_{ix} \times U_{iy}$, where $U_{ix} \subset R^{n-p}$ and $U_{iy} \subset \mathbb{R}^p$ and U_{iy} is isomorphic to the plaques and the points of U_{ix} that parameterize the plaques in U_i.
(4) The set Γ_1 is a submanifold of U_i that intersects every plaque $x = constant$ exactly once[2].

The first example of different foliation is the so-called *flat space* based on the fact that $\mathbb{R}^n = \mathbb{R}^{n-p} \times \mathbb{R}^p$, which can be covered with a single chart U with the leaves $\mathbb{R}^{n-p}$ being enumerated by $\mathbb{R}^p$.

The second example of a foliation is the covers defined by the following: If $M \to N$ is a covering between manifolds and f is a foliation on N, then it pulls back to a foliation on M.

The third example can be found by using a submersion of manifolds[3] $f : M^n \to N^q$ with $q \leq n$. Indeed, using the inverse function theorem, it follows that the connected components of the fibers of the submersion define a codimension-q foliation of M. An example of such a situation is the so-called *fiber bundles* that form a space (E, B, π, C), where E is the

---
[2]Due to the *monodromy effect*, there are no global transversal sections of the foliation F.
[3]A submersion is a differentiable map between differentiable manifolds whose derivative is everywhere surjective.

total space, B is the base space of the bundle, and C is the fiber. E, B, and C are topological spaces, and π (the bundle projection $E \to B$) is a continuous surjection satisfying a local triviality condition. In particular, the *Reeb foliation* of the hypersphere $\mathbb{S}^3$ is a foliation constructed as the union of two solid tori with a common boundary [Rolfsen (1976)].

The fourth example of a such situation is the trajectories of a vector field f forming a one-dimensional foliation (the leaves are curves) of the complement of the set of zeros of f.

In conclusion, note that the importance of invariant foliations can be seen in the context of hyperbolic systems (see Chapters 6 and 7) in particular. For partially hyperbolic systems, the existence of invariant foliations is a crucial geometric feature, *i.e.*, the strong-stable and strong-unstable foliations are absolutely continuous if the system under consideration is at least twice differentiable. This is a remark of Anosov and Sinai, *i.e.*, the *holonomy maps*[4] of these foliations send zero Lebesgue measure sets of one cross-section to zero Lebesgue measure sets of the other cross-section.

5.3 Shadowing lemma

In this section, we present the shadowing lemma along with its applications to determining the true chaotic attractors for both discrete and continuous-time dynamical systems. Indeed, it is well known that there are rounding errors at every step when numerically calculating the trajectory of a dynamical system. Especially if the dynamical system has chaotic attractors, then these errors will grow exponentially, and the resulting orbits will differ wildly from those in the exact system. The so-called *shadowing lemma* introduced in [Palmer (1988)] eliminates this problem by demonstrating that under rather general conditions there is a trajectory of the true system starting from a slightly perturbed initial state that shadows the computed one. An essential feature is that this technique only holds for hyperbolic systems (see Chapters 6 and 7) because the unstable periodic orbit theory for multifractal characterization of chaotic systems given in [Grebogi *et al.* (1988)] applies only for such systems.

The general shadowing lemma can be stated in an arbitrary Riemannian manifold as follows:

Theorem 5.2. *Let Ω be a Riemannian manifold, with $f : \Omega \longrightarrow \Omega$ a diffeomorphism and $\tilde{\Lambda} \subset \Omega$ a compact hyperbolic set for f. Then there is a*

---

[4]Holonomy maps are the projections along the leaves between two given cross-sections to the foliation.

neighborhood U of $\tilde{\Lambda}$ such that for every $\delta > 0$ there is an $\epsilon > 0$ such that every ϵ-orbit in U is δ-shadowed by an orbit of f.

Moreover, there is a $\delta_0 > 0$ such that if $\delta < \delta_0$ and if the pseudo-orbit is bi-infinite, then the shadowing orbit is unique, and if $\tilde{\Lambda}$ has a local product structure, then the shadowing orbit is in $\tilde{\Lambda}$.

Generally, the proof of the density of periodic points in dynamical systems theory is a standard use of the shadowing lemma. In real systems, *i.e.*, when $\Omega \subset \mathbb{R}^n$, assume that the set Ω is a Riemannian manifold. However, the following definitions help in understanding Theorem 5.2:

Definition 5.3. (a) An ϵ-*pseudo-orbit* for f is a sequence (x_n) such that $|x_n - f(x_n)| < \epsilon$, for all $n \in \mathbb{Z}$.

(b) A sequence $\{x_k, k \in \mathbb{Z}\}$ is said to be *an ϵ pseudo-periodic orbit of period N* of f if $|x_{k+1} - f(x_k)| < \epsilon$ and $x_{k+N} = x_k$ for $k \in \mathbb{Z}^5$.

(c) Let $\{x_k, k \in \mathbb{Z}\}$ and $\{y_k, k \in \mathbb{Z}\}$ be two ϵ pseudo-periodic orbits. A sequence $\{z_k, k \in \mathbb{Z}\}$ is said to be *an ϵ pseudo-connecting orbit* connecting $\{x_k, k \in \mathbb{Z}\}$ to $\{y_k, k \in \mathbb{Z}\}$ if

(i) $|z_{k+1} - f(z_k)| < \epsilon$, for $k \in \mathbb{Z}$.

(ii) $z_k = x_k$ for $k \leq p$ and $z_k = y_k$ for $k \geq q$ for some integers $p < q$.

(d) In the case $\{x_k, k \in \mathbb{Z}\} = \{y_{k+\tau}, k \in \mathbb{Z}\}$ for some τ, the ϵ pseudo-connecting orbit $\{z_k, k \in \mathbb{Z}\}$ is called *an ϵ pseudo-homoclinic orbit*.

The definition of a shadow of an ordinary differential equation system is the following:

Definition 5.4. An approximate trajectory $y = \{y_n\}_{n \in \mathbb{Z}}$ with time steps $\{h_n\}_{n \in \mathbb{Z}}$ is ϵ-shadowed by a true solution if there exists a sequence of points $x = \{x_n\}_{n \in \mathbb{Z}}$ with time steps $\{\tau_n\}_{n \in \mathbb{Z}}$ such that $x_{n+1} = \varphi_{\tau_n}(x_n)$ where $\varphi_{\Delta t}$ is the Δt-flow of the system, and $|y_n - x_n| < \epsilon$ and $|\tau_n - h_n| < \epsilon$.

Now let us define the set $l^\infty(\mathbb{Z}, \mathbb{R}^n)$ as the space of $\mathbb{R}^n$-valued bounded sequences $x = \{x_n\}_{n \in \mathbb{Z}}$ with norm $\|x\| = \sup_{n \in \mathbb{Z}} |x_n|_2$, and the set $C^{1, Lip}(\Omega, \Omega)$ as the ensemble of C^1-valued Lipschizian functions on $\Omega \subset \mathbb{R}^n$. Hence a discrete version of Theorem 5.2 was given with its detailed proof in [Stoffer & Palmer (1999)] as follows:

Theorem 5.3. Let $\Omega \subset \mathbb{R}^n$ be open, $f \in C^{1, Lip}(\Omega, \Omega)$ be injective, $y = \{y_k\}_{k \in \mathbb{Z}} \in \Omega^{\mathbb{Z}}$ be a given sequence, let $\{A_k\}_{k \in \mathbb{Z}}$ be a bounded sequence of

---
[5] More precisely, a periodic orbit $\{x_n, n \in \mathbb{Z}\}$ is a finite set of points.

$k \times k$ matrices, and let δ, δ_1, m be positive constants. Assume that the operator $L : l^\infty(\mathbb{Z}\mathbb{R}^n) \longrightarrow l^\infty(\mathbb{Z}, \mathbb{R}^n)$ defined by

$$(Lz)_k = z_{k+1} - A_k z_k \tag{5.1}$$

is invertible and that

$$\|L^{-1}\| \leq \frac{1}{\delta_1 + \sqrt{2m\delta}}. \tag{5.2}$$

Then the numbers

$$r_0 = \frac{2\delta}{\frac{1}{\|L^{-1}\|} - \delta_1 + \sqrt{\left(\frac{1}{\|L^{-1}\|} - \delta_1\right)^2 - 2m\delta}} \tag{5.3}$$

$$r_1 = \frac{\frac{1}{\|L^{-1}\|} - \delta_1 + \sqrt{\left(\frac{1}{\|L^{-1}\|} - \delta_1\right)^2 - 2m\delta}}{m} \tag{5.4}$$

satisfy $0 < r_0 \leq r_1$. Let $\rho \in [r_0, r_1]$. Moreover, assume that the set $\overline{\cup_{n \in \mathbb{Z}} B_\rho(y_n)}$ (the closure) is in Ω and that for every $n \in \mathbb{Z}$

$$|y_{k+1} - f(y_k)| < \delta \tag{5.5}$$

$$|A_k - Df(y_k)| < \delta_1 \tag{5.6}$$

$$|Df(u) - Df(v)| < m|u - v|, u, v \in B_\rho(y_k). \tag{5.7}$$

Then there is a unique r_0-shadowing orbit $x = \{x_k\}_{n \in \mathbb{Z}}$ of y. Moreover, there is no orbit $\tilde{x}$ other than x with

$$|\tilde{x} - y| < \rho. \tag{5.8}$$

There are three important things to note about the notion of the shadow of an orbit. First, the numerical solution is shadowed if it closely follows the path of a true solution. Second, the linear growth of errors for the system considered is due to a lack of hyperbolicity in the direction of the flow in phase space as explained in [Eric & Vleck (1995)]. Third, the shadowing lemma used for proving the existence of a shadow requires the computation of the Jacobian of the map or solving the variational equations of the ODE's, and then estimating its hyperbolicity (see Sec. 6.6.1), and so this method is direct [Palmer (1988), Chow & Palmer (1992)].

As an application of Theorem 5.3 to prove the existence of a true homoclinic orbit for the Hénon map (10.88), we need the following definitions for a map $h : \mathbb{R}^2 \to \mathbb{R}^2$ of class C^1.

Definition 5.5. A sequence $w = (w_n)_n \in \mathbb{Z} \in l^\infty(\mathbb{Z}, \mathbb{R}^2)$ is called a pseudo-orbit of h with error $d = (d_n)_n \in \mathbb{Z} \in l^\infty(\mathbb{Z}, \mathbb{R}^2)$ if $w_{n+1} - h(w_n) = d_n$ holds for $n \in \mathbb{Z}$.

For $n \in \mathbb{Z}$, let Ω_n be an open and convex set containing w_n.

Definition 5.6. An R-shadowing orbit z of w is a sequence $z = (z_n)_n \in \mathbb{Z} \in l^\infty(\mathbb{Z}, \mathbb{R}^2)$ with $z_n \in \Omega_n$, $z_{n+1} = h(z_n)$ and $|z_n - w_n| \le R$ for $n \in \mathbb{Z}$.

Assume that there are (constant) 2×2-matrices A_n with $\sup_{n \in \mathbb{Z}} |A_n| < \infty$ and such that $|Dh - A_n|$ is small enough[6]. Let L be a linear operator, for $z \in \Omega = \prod_{n \in \mathbb{Z}} \Omega_n$ the linear operator Δ_z in $l^\infty(\mathbb{Z}, \mathbb{R}^2)$ is defined as follows:

$$\begin{cases} L : (L\zeta)_n = \zeta_{n+1} - A_n \zeta_n \\ \Delta_z : (\Delta_z \zeta)_n = (Dh(z_n) - A_n)\zeta_n. \end{cases} \quad (5.9)$$

Hence the following theorem was proved in [Pilyugin (1999), Kirchgraber & Stoffer (2004)]:

Theorem 5.4. *(Shadowing Theorem). Let $h \in C^1(\mathbb{R}^2, \mathbb{R}^2)$ be a map of class C^1, and let δ_0 and δ_1 be positive constants. For $n \in \mathbb{Z}$ let $\Omega_n \subset \mathbb{R}^2$ be open and convex. Let $w = (w_n)_n \in \mathbb{Z} \in l^\infty(\mathbb{Z}, \mathbb{R}^2)$ with $w_n \in \Omega_n$ be a pseudo-orbit of h with error $d \in l^\infty(\mathbb{Z}, \mathbb{R}^2)$. Let the operators L and Δ_z be defined as above. Assume that L is invertible and that the following estimates are satisfied:*

$$\|L^{-1} d\| \le \delta_0 \quad (5.10)$$

$$\|L^{-1} \Delta_z\| \le \delta_1 < 1, \text{ for all } z \in \Omega = \prod_{n \in \mathbb{Z}} \Omega_n \quad (5.11)$$

$$\bar{B}_R(w_n) \subset \Omega_n, \text{ for } R = \frac{\delta_0}{1 - \delta_1}, n \in \mathbb{Z}. \quad (5.12)$$

Then there is an R-shadowing orbit $z^* \in \Omega$ of w, and z^* is the only orbit in Ω. Moreover, z^* is hyperbolic.

For recent results on the shadowing property, see those given for the Lozi map (8.88) in [Kiriki & Soma (2007)] where a study was given for a certain shadowing property for this map with the y-axis as the singularity set and strange attractors. For the Lorenz attractor L (see Chapter 8 and 9), similar work was done in [Kiriki & Soma (2005)].

Theorem 5.5. *(Parameter-shifted shadowing property for the Lorenz map (PSSP)). There exists an open set $O \subset L$ such that every Lorenz map $L \in O$ has the parameter-shifted shadowing property.*

---
[6]This means that the Jacobian Dh does not vary too much in the set Ω_n.

The two concepts, *weak and strong parameter-shifted shadowing property*, were introduced in [Kiriki & Soma (2005)] for a set C_0 of vector fields generated from the geometric Lorenz method controlled by Lorenz maps (see Sec. 8.2.2) in O as follows:

Theorem 5.6. *(Weak PSSP for the Lorenz flow). Every geometric Lorenz flow generated by a vector field in C_0 has the weak parameter-shifted shadowing property.*

Theorem 5.7. *(Strong PSSP for the Lorenz flow). Every geometric Lorenz flow has no strong parameter-shifted shadowing property.*

5.3.1 Homoclinic orbits and shadowing

In this section, we present the possible existing relations between homoclinic orbits and shadowing. Indeed, a new computer-assisted technique with two main components was presented in [Coomes et al. (2005)]. The objective of this method is the rigorous proof of the existence of a true transversal homoclinic orbit to a periodic orbit (or a fixed point) of diffeomorphisms in $\mathbb{R}^n$ using the global Newton's method introduced in Sec. 1.2 and the homoclinic shadowing theorems, namely Theorems 5.8 and 5.9 given below, where all the quantities in the hypotheses of these theorems are computable as shown in [Coomes et al. (2005)].

Theorem 5.8. *(Connecting Orbit Shadowing Theorem) Suppose $f : \mathbb{R}^n \to \mathbb{R}^n$ is a C^2 diffeomorphism, and $\{x_k, k \in \mathbb{Z}\}$ to $\{y_k, k \in \mathbb{Z}\}$ are two ϵ pseudo-periodic orbits with periods N and N' of f, respectively. Let $z = \{z_k, k \in \mathbb{Z}\}$ be an ϵ pseudo-orbit of f connecting $\{x_k, k \in \mathbb{Z}\}$ to $\{y_k, k \in \mathbb{Z}\}$. Suppose that the operator Lz defined in (5.1) is invertible, and set*

$$\epsilon_z = 2\left\|L^{-1}z\right\|\epsilon \qquad (5.13)$$

and

$$M_z = \sup\left\{\left\|D^2 f(u)\right\| : u \in \mathbb{R}^n, \|u - z_k\| \le \epsilon_z \text{ for some } k \in \mathbb{Z}\right\}. \qquad (5.14)$$

Then if

$$2M_z\left\|L^{-1}z\right\|^2 \epsilon < 1: \qquad (5.15)$$

(i) The pseudo-periodic orbits $\{x_k, k \in \mathbb{Z}\}$ to $\{y_k, k \in \mathbb{Z}\}$ are ϵ_z-shadowed by unique true hyperbolic periodic orbits $\{v_k, k \in \mathbb{Z}\}$ of period-N and $\{w_k, k \in \mathbb{Z}\}$ of period-N'.

(ii) The pseudo-periodic orbit $\{z_k, k \in \mathbb{Z}\}$ is ϵ_z-shadowed by a unique true transversal connecting orbit $\{e_k, k \in \mathbb{Z}\}$ connecting the true periodic orbit $\{v_k, k \in \mathbb{Z}\}$ to the periodic orbit $\{w_k, k \in \mathbb{Z}\}$. In fact, $\lim_{k \to -\infty} \|e_k - u_k\| = 0$ and $\lim_{k \to \infty} \|e_k - v_k\| = 0$.

Theorem 5.9. *(Homoclinic Orbit Shadowing Theorem). Suppose $f : \mathbb{R}^n \to \mathbb{R}^n$ is a C^2 diffeomorphism and $\{x_k, k \in \mathbb{Z}\}$ is an ϵ pseudo-periodic orbit of f with period-N. Let $\{z_k, k \in \mathbb{Z}\}$ be an ϵ pseudo-homoclinic orbit of f connecting $\{x_k, k \in \mathbb{Z}\}$ to $\{y_{k+\tau}, k \in \mathbb{Z}\}$, where $0 \leq \tau < N$. Suppose that the operator Lz is invertible, and set*

$$\epsilon_z = 2 \|L^{-1}z\| \epsilon \tag{5.16}$$

and

$$M_z = \sup\left\{\|D^2 f(u)\| : u \in \mathbb{R}^n, \|u - z_k\| \leq \epsilon_z \text{ for some } k \in \mathbb{Z}\right\}. \tag{5.17}$$

Then if

$$\begin{cases} 2M_z \|L^{-1}z\|^2 \epsilon < 1 \\ \|x_k - x_j\| > 2\epsilon_z, \text{ for } j \neq k, k < N : \end{cases} \tag{5.18}$$

(i) The pseudo-periodic orbit $\{x_k, k \in \mathbb{Z}\}$ is ϵ_z-shadowed by a unique true hyperbolic periodic orbit $\{v_k, k \in \mathbb{Z}\}$ of minimal period-N.

(ii) The pseudo-homoclinic orbit $\{z_k, k \in \mathbb{Z}\}$ is ϵ_z-shadowed by a unique true orbit $\{e_k, k \in \mathbb{Z}\}$. When $\tau > 0$, the point e_0 is a transversal homoclinic point to the periodic orbit $\{v_k, k \in \mathbb{Z}\}$ with phase shift τ. When $\tau = 0$, the point e_0 is a transversal homoclinic point to the periodic orbit $\{v_k, k \in \mathbb{Z}\}$ with phase shift 0 provided that $\|z_k - x_k\| > 2\epsilon_z$ for some k with $p < k < q$.

For the proof of Theorems 5.8 and 5.9, the Lemma 5.3 below is required. This lemma establishes infinite-time shadowing of a pseudo-orbit by a true hyperbolic orbit without any uniform hyperbolicity requirement for the considered diffeomorphism. The second lemma gives a hyperbolicity-type condition under which two orbits shadowing the same pseudo-orbit must be asymptotic to each other. See [Coomes et al. (2005)].

The following lemma establishes the so-called shadowing lemma that confirms that a pseudo-orbit or chain is always close to an actual orbit of the system:

Lemma 5.1. *(Shadowing lemma). If Λ is a hyperbolic set for a diffeomorphism f and $\delta > 0$, then there exists an $\epsilon > 0$ such that for each ϵ-pseudo-orbit $(x_i)_{i \in \mathbb{Z}}$ in Λ, there is a point x such that $d\left(x_i, f^i(x)\right) < \delta$ for all $i \in \mathbb{Z}$. If Λ is locally maximal, then $x \in \Lambda$.*

Theorem 5.10. *(Shadowing Theorem) A continuous family of pseudo-orbits for a perturbation of a hyperbolic diffeomorphism is shadowed by a continuous family of genuine orbits for the diffeomorphism itself.*

The relationship between the shadowing theorem and structural stability discussed in Chapter 4 can be seen from the following facts: First, the shadowing theorem is closely related to structural stability, it holds in any structurally stable system, and structural stability is equivalent to hyperbolicity, *i.e.*, the stability theorem of Smale, Robbin, Robinson, Palis, Mañé, Liao, and Hayashi. Second, the shadowing theorem is a tool for proving that hyperbolic sets are structurally stable. Indeed, given a hyperbolic set Λ_f for a diffeomorphism f, there is an $\epsilon > 0$ such that every ϵ-perturbation g in the C^1-topology of f has a hyperbolic set Λ_g in an ϵ-neighborhood of Λ_f and such that $f | \Lambda_f$ and $g | \Lambda_g$ are topologically conjugate. In this case, the conjugacy can be uniquely chosen and moves points very little; it is always Hölder continuous (recall Definition 6.14) but rarely smooth. For any $\alpha \in (0, 1)$, there is a symplectic automorphism A of a torus and a C^k-neighborhood U of A in the space of symplectic diffeomorphisms such that for an open dense subset of $f \in U$, the conjugacy to A and its inverse are α-Hölder only on a set of measure zero. The last fact is that if a system is structurally stable, then hyperbolic systems form an open subset of the space of diffeomorphisms or flows. This implies that all sufficiently small perturbations of Anosov flows are also Anosov flows. See Chapters 4, 6, and 7 for some details on these topics.

5.3.2 *Shilnikov criterion for the existence of chaos*

In this section, we present some Shilnikov theorems about the existence of chaos in dynamical systems. Indeed, homoclinic and heteroclinic orbits arise in applications to the study of bifurcation and chaotic phenomena in mechanics, biomathematics, and chemistry [Aulbach & Flockerzi (1989), Balmforth (1995), Feng (1998)]. One of the commonly agreed-upon analytic criteria for proving chaos in autonomous systems is the *Shilnikov theorem* given in [Shilnikov (1965-1970), Silva (2003)]. The resulting chaos is called the *horseshoe type* or *Shilnikov chaos*, and these horseshoes give the extremely complicated behavior typically observed in chaotic systems [Guckenheimer & Holmes (1983)].

Let $f : \mathbb{R}^n \longrightarrow \mathbb{R}^n$ be a real function defining a discrete mapping, and let $Df(x)$ be its Jacobian matrix. To simplify the notion of *homoclinic bifurcations*, we need the following definitions:

Definition 5.7. (a) A saddle point q of f is a point where $Df(q)$ has some eigenvalues λ such that $|\lambda| < 1$, and the rest satisfy $|\lambda| > 1$.

(b) A homoclinic point p of a map $f : \mathbb{R}^n \longrightarrow \mathbb{R}^n$ lies inside the intersection of its stable and unstable separatrix (invariant manifold), *i.e.*, $\lim_{n \longrightarrow +\infty} f(x) = \lim_{n \longrightarrow -\infty} f(x) = P$.

(c) The map $f : \mathbb{R}^n \longrightarrow \mathbb{R}^n$ has a hyperbolic period-k orbit γ if there exists a $q \in \mathbb{R}^n$ such that $f^k(q) = q$, and q is a saddle point with its stable and unstable manifold $W^s(q), W^u(q)$ intersecting transversely in a point p.

(d) The point p in the Definition 5.7(b) is called a *transversal homoclinic point*.

The most important results for the existence of a transversal homoclinic point are the existence of infinitely many periodic and homoclinic points in a small neighborhood of this point as shown in the *Smale–Moser Theorem* [Smale (1965)]. More precisely,

Theorem 5.11. *In the neighborhood of a transversal homoclinic point, there exists an invariant Cantor set on which the dynamics are topologically conjugate to a full shift on N symbols.*

More precisely, there is a positively and a negatively invariant Cantor set containing infinitely many saddle-type (unstable) periodic orbits of arbitrarily long periods, uncountably many bounded nonperiodic orbits, and a dense orbit. Thus the horseshoe persists under perturbations. Indeed, if q is a transversal homoclinic point to a hyperbolic fixed point p of a diffeomorphism f, then for any neighborhood U of $\{p, q\}$, there is a positive integer k such that f^k has a hyperbolic invariant set $\{p, q\} \in \Gamma \subset U$ on which f^k is topologically conjugate to the two-sided shift map on two symbols introduced in Sec. 1.3. Furthermore, in any neighborhood of a transversal homoclinic point, there are infinitely many periodic and nonperiodic points.

In this case, the definition of homoclinic bifurcations is given by the following:

Definition 5.8. A homoclinic bifurcation occurs when periodic orbits appear from homoclinic orbits to a saddle, saddle-focus, or focus-focus equilibrium.

A good reference for homoclinic bifurcations in dynamical system is [Kuznetsov (2004)].

Now, consider the third-order autonomous system
$$x' = f(x) \tag{5.19}$$
where the vector field $f: \mathbb{R}^3 \longrightarrow \mathbb{R}^3$ belongs to class $C^r (r \geq 1)$, $x \in \mathbb{R}^3$ is the state variable of the system, and $t \in \mathbb{R}$ is the time. Suppose that f has at least one equilibrium point P.

Definition 5.9. (a) The point P is called a hyperbolic saddle focus for system (5.19) if the eigenvalues of the Jacobian $A = Df(P)$ are γ and $\rho + i\omega$, where $\rho\gamma < 0$ and $\omega \neq 0$.

(b) A homoclinic orbit $\gamma(t)$ refers to a bounded trajectory of system (5.19) that is doubly asymptotic to an equilibrium point P of the system, i.e., $\lim_{t \to +\infty} \gamma(t) = \lim_{t \to -\infty} \gamma(t) = P$.

(c) A heteroclinic orbit $\delta(t)$ is similarly defined except that there are two distinct saddle foci, P_1 and P_2, being connected by the orbit, one corresponding to the forward asymptotic time, and the other to the reverse asymptotic time limit, i.e., $\lim_{t \to +\infty} \delta(t) = P_1$, and $\lim_{t \to -\infty} \gamma(t) = P_2$.

The homoclinic and heteroclinic Shilnikov methods are summarized in the following theorem [Shilnikov (1965-1970)]:

Theorem 5.12. *Assume the following:*
(i) The equilibrium point P is a saddle focus and $|\gamma| > |\rho|$.
(ii) There exists a homoclinic orbit based at P.
Then
(1) The Shilnikov map, defined in a neighborhood of the homoclinic orbit of the system, possesses a countable number of Smale horseshoes in its discrete dynamics.
(2) For any sufficiently small C^1-perturbation g of f, the perturbed system
$$x' = g(x) \tag{5.20}$$
has at least a finite number of Smale horseshoes in the discrete dynamics of the Shilnikov map defined near the homoclinic orbit.
(3) Both the original system (5.19) and the perturbed system (5.20) exhibit horseshoe chaos.

Similarly, there is also a heteroclinic Shilnikov theorem [Shilnikov (1965-1970)] given by:

Theorem 5.13. *Suppose that two distinct equilibrium points, denoted by P_1 and P_2, respectively, of system $\dot{x} = f(x)$ are saddle foci whose characteristic* values γ_k and $\rho_k + i\omega_k$ ($k = 1, 2$) satisfy the following Shilnikov inequalities: $\rho_1 \rho_2 > 0$ or $\gamma_1 \gamma_2 > 0$. Suppose also that there exists a heteroclinic orbit joining P_1 and P_2. Then the system $\dot{x} = f(x)$ has both Smale horseshoes and the horseshoe type of chaos.

5.4 Exercises

Exercise 104. Show that transversality is a stable and generic condition.

Exercise 105. Let M and N be manifolds, and M_1 a submanifold of N. Let $f : M \to N$ be a transverse map to M_1. Show that the subset $f^{-1}(M_1)$ is a submanifold of M and $co \dim (f^{-1}(M_1)) = co \dim (M_1)$.

Exercise 106. Prove Theorem 5.1.

Exercise 107. Show that an intersection point between two arcs in a surface is transverse if and only if their tangent lines inside the tangent plane to the surface are distinct.

Exercise 108. Show that in the case of three-dimensional space, the following results are true:
  (a) Transverse curves do not intersect.
  (b) Curves transverse to surfaces intersect in points, and surfaces transverse to each other intersect in curves.
  (c) Curves that are tangent to a surface at a point do not intersect the surface transversally.

Exercise 109. Show that there is transversality in the case of a hyperbolic periodic orbit of a C^r, $r \geq 1$ vector field on $\mathbb{R}^n$, where its Poincaré map linearized about the fixed point has $n - k - 1$ eigenvalues with modulus greater than one and k eigenvalues with modulus less than one.

Exercise 110. Show that transversality is a robust generic property, *i.e.*, that transversality persists under sufficiently small perturbations.

Exercise 111. (a) Show that if the intersection of M_1 and M_2 is transverse, then the intersection $M_1 \cap M_2$ is a submanifold, and $co \dim (M_1 \cap M_2) = co \dim (M_1) + co \dim (M_2)$.
  (b) Show that the absence of transversality implies that the intersection is not a submanifold having some sort of singular point.

Exercise 112. (a) Show that if the manifolds M_1 and M_2 are of complementary dimension, then transversality implies that $TM = TM_1 \oplus TM_2$.

(b) Show that the submanifolds M_1 and M_2 are oriented if M is oriented, and that they intersect in isolated points, i.e., a 0-manifold.

Exercise 113. Show that if the maps $f : M_1 \to M$ and $g : M_2 \to M$ are embeddings, then their submanifolds are transverse.

Exercise 114. Show that if $M_1 = \{b\}$ is a single point, then f is transverse to M_1 and b is a regular value for f.

Exercise 115. Using Definition 5.2, show the following:

(a) The plaques $x = constant$ are $n - p$-dimensional submanifolds in each chart (U_i, ϕ_i).

(b) The leaves of foliation f are maximal-connected, injectively-immersed submanifolds.

(c) Each chart U_i can be written in the form $U_i = U_{ix} \times U_{iy}$ where $U_{ix} \subset R^{n-p}$ and $U_{iy} \subset \mathbb{R}^p$, and U_{iy} is isomorphic to the plaques and the points of U_{ix} parameterize the plaques in U_i.

(d) The set Γ_1 is a submanifold of U_i that intersects every plaque $x = constant$ exactly once.

Exercise 116. Using Definition 5.2, show the following:

(a) The so-called *flat space* is a foliation.

(b) If $M \to N$ is a covering between manifolds and f is a foliation on N, then it pulls back to a foliation on M.

(c) The submersions of manifolds $f : M^n \to N^q$ with $q \leq n$ can define a foliation.

(d) The trajectories of a vector field f form a one-dimensional foliation (the leaves are curves) of the complement of the set of zeros of f.

Exercise 117. Explain the relationship between the shadowing theorem and structural stability.

Chapter 6

Chaotic Attractors with Hyperbolic Structure

In this chapter, we discuss the most interesting results concerned chaotic attractors with hyperbolic structure. First, the hyperbolic dynamics is discussed in Sec. 6.1 along some concepts and definitions. In Sec. 6.1.2 we present Anosov diffeomorphisms and Anosov flows as typical examples of invariant sets with hyperbolic structure. Section 6.3 deals with the problem of classification of strange attractors of dynamical systems, which is based on the rigorous mathematical analysis and is not accepted as significant from the experimental point of view. In Sec. 6.2.2 and Sec. 6.4, we describe some properties of hyperbolic chaotic attractors. Some concrete examples of chaotic attractors with hyperbolic structure are presented in Sec. 6.2.4 to Sec. 6.2.7 and Sec. 6.4.2 to Sec. 6.4.4. These include geodesic flows on compact smooth manifolds, the Smale–Williams solenoid, Anosov diffeomorphisms on a torus, in particular Anosov automorphisms, and their structure. Furthermore, expanding maps are presented due their importance, *i.e.*, they are expanding solenoids or hyperbolic attractors. Other examples are the Blaschke product, the Arnold cat map, the Plykin attractor, and the Bernoulli map.

As an example showing the required manipulations for such a proof of hyperbolicity, we give a proof of the hyperbolicity of the logistic map (6.25) below for the values of the parameter greater than 4. In Sec. 6.6, we discuss principal properties of the so-called *generalized hyperbolic attractors*. In Sec. 6.7, we describe two methods for generating hyperbolic attractors, in particular, the transition when a periodic trajectory disappears from Morse–Smale systems to systems with hyperbolic attractors is presented in some detail. The density of hyperbolicity and homoclinic bifurcations in arbitrary dimension is discussed in Sec. 6.8. In Sec. 6.9, some hyperbolicity tests are presented and applied to the Hénon map (6.88) and to the forced, damped, pendulum (6.94). At the end of this chapter, some exercises and open problems are listed on the topic of hyperbolicity and robustness.

6.1 Hyperbolic dynamics

6.1.1 Concepts and definitions

The most important map in chaos theory is the so-called *Smale horseshoe map* introduced in [Smale (1967)] and discussed in Sec. 1.1. This simple geometric model is characterized by the existence of infinitely many robust periodic orbits. There are now many models that exhibit this phenomenon. The crucial common feature of these models is hyperbolicity, *i.e.*, the tangent space at each point splits into two complementary directions such that the derivative contracts one of these directions and expands the other at uniform rates. The mathematical theory of hyperbolic systems is now well developed, and it is a main paradigm for the behavior of chaotic systems. In this book, we omit local aspects of hyperbolicity, *i.e.*, linear systems and local behavior, and concentrate our attention on the global theory, *i.e.*, hyperbolic sets in $\mathbb{R}^n$ given Sec. 6.1 and hyperbolic sets (partial hyperbolicity, uniform hyperbolicity, nonuniform hyperbolicity...) in a Riemmmanian manifold M (in particular, a closed 3-manifold), *etc.*

Let $f : \Omega \subset \mathbb{R}^n \longrightarrow \mathbb{R}^n$ be a C^r real function that defines a discrete map also called f, and Ω is a manifold. Then one has the following definitions given in [Abraham & Marsden (1978)]:

Definition 6.1. (a) A point x is a nonwandering point for the map f if for every neighborhood U of x there is a $k \geq 1$ such that $f^k(U) \cap U$ is nonempty.

(b) The set of all nonwandering points is called the nonwandering set of f.

(c) An f-invariant subset Λ of $\mathbb{R}^n$ satisfies $f(\Lambda) \subset \Lambda$.

(d) If f is a diffeomorphism defined on some compact smooth manifold $\Omega \subset \mathbb{R}^n$, an f-invariant subset Λ of $\mathbb{R}^n$ is said to be hyperbolic if there exists $0 < \lambda_1 < 1$ and $c > 0$ such that

(d-1) $T_\Lambda \Omega = E^s \oplus E^u$, where $\oplus$ mean the algebraic direct sum.

(d-2) $Df(x) E^s_x = E^s_{f(x)}$, and $Df(x) E^u_x = E^u_{f(x)}$ for each $x \in \Lambda$.

(d-3) $\left\|Df^k v\right\| \leq c\lambda_1^k \left\|v\right\|$, for each $v \in E^s$ and $k > 0$.

(d-4) $\left\|Df^{-k} v\right\| \leq c\lambda_1^k \left\|v\right\|$, for each $v \in E^u$ and $k > 0$.

where E^s and E^u are, respectively, the stable and unstable submanifolds of the map f, i.e., the two Df-invariant submanifolds, and E^s_x and E^u_x are the two $Df(x)$-invariant submanifolds. For a compact surface, a result of Plykin implies that there must be at least three holes for a hyperbolic attractor and it looks locally like a Cantor set × an interval. In this case, the interval is the unstable direction, and the Cantor set is the stable one. Furthermore, hyperbolic attractors have dense periodic points and a point with a dense orbit.

Definition 6.2. A hyperbolic set Λ is locally maximal (or isolated) if there exists an open set U such that

$$\Lambda = \bigcap_{n \in \mathbb{Z}} f^n(U). \tag{6.1}$$

Note that the Smale horseshoe discussed in Sec. 1.1 and the Plykin attractor introduced in Sec. 6.4.4 are examples of locally maximal attractors. In fact, if a Λ is a hyperbolic set with a nonempty interior for a map f, then f is Anosov if it is transitive, locally maximal, and Ω is a surface.

Locally maximal, transitive, hyperbolic sets have properties including the following: shadowing introduced in Sec. 5.3, structural stability introduced in Sec. 4.1, Markov partitions, and SRB measures (for attractors) introduced in Sec. 3.1.2.

6.1.2 Anosov diffeomorphisms and Anosov flows

When $\Lambda = \Omega$, then the diffeomorphism f is called an *Anosov diffeomorphism*[1].

Definition 6.3. (1) The map f is uniformly hyperbolic or an *axiom A* diffeomorphism if:

(a) The nonwandering set $\Omega(f)$ has a hyperbolic structure.

(b) The set of periodic points of f is dense in $\Omega(f)$, i.e., $Cl\,(Per\,(f)) = \Omega(f)$, whose closure is the nonwandering set itself.

(2) If φ^t is a flow, then $\varphi^t : \Omega \subset \mathbb{R}^n \longrightarrow \mathbb{R}^n$ is an Anosov flow if for every $x \in \Omega$ there is a splitting of the tangent space $T_x\Omega = E^s \oplus E^0 \oplus E^u$, where E^0 is the flow direction and there are constants $C > 0$ and $\lambda \in (0,1)$ such that for every $t > 0$, one has $\|D\varphi^t(v)\| \leq c\lambda^t \|v\|$ for each $v \in E^s$ and $\|D\varphi^{-t}(v)\| \leq c\lambda^t \|v\|$ for each $v \in E^u$.

---

[1] or uniformly hyperbolic.

The essential point of Definition 6.3 is that the so-called *homoclinic tangencies* (recall Definition 1.9(a)) are excluded, and the resulting attractor is robust in the sense of Definition 2.5. Note that the axiom A diffeomorphism serves as a model for the general behavior at a transversal homoclinic point where the stable and unstable manifolds of a periodic point intersect, and it plays a crucial role in the study of homoclinic bifurcations [Anosov (1969)]. More generally, if M is a C^∞ compact manifold without boundary and if $\mathcal{F}^1(M)$ denotes the set of diffeomorphisms $f \in Diff^1(M)$ having a C^1 neighborhood $\mathcal{U}$ such that all the periodic points of every $g \in \mathcal{U}$ are hyperbolic, then a proof of a conjecture of Mañé given in [Mañé (1982)] claiming that every element of $\mathcal{F}^1(M)$ satisfies axiom A was given in [Hayashi (1992)]:

Theorem 6.1. *If $f \in \mathcal{F}^1(M)$, then f satisfies axiom A.*

The proof of Theorem 6.1 was reduced to finding a *string*[2] to which Mañé's perturbation technique used in Lemma 3.1 of [Mañé (1982)] to create a homoclinic point can be applied. Due to the fact that a local perturbation makes the string part of a homoclinic orbit for a diffeomorphism C^1 near to f, this string was considered almost homoclinic since one end of it is on or near a local stable manifold of a basic set and the other end is on or near the local unstable manifold of the same basic set, and besides no other points of the string are in a domain containing a neighborhood of the basic set.

Some remarks and notes resulting from Definition 6.3 are the following:

(1) The stable and unstable subspaces E^s and E^u depend continuously on the point x, and they are invariant and interchanged when one passes from a map to its inverse.
(2) Not every manifold admits an Anosov diffeomorphism or flow. For instance, the *hairy ball theorem* shows that there is no Anosov diffeomorphism on the 2-sphere.
(3) It is unknown whether the universal cover of a manifold that admits an Anosov diffeomorphism must be $\mathbb{R}^n$ for some n.
(4) (The cone criterion, or the Alekseev cone criterion) [Anosov (1969)]: Due to the difficulties that arise when testing the conditions of Definition 6.3, an alternate robust (under perturbations) definition that can be checked with limited accuracy is the so-called *cone criterion:* At each point x, it requires the existence of two complementary closed

---
[2]A *string* is a finite backward orbit $\{x, f^{-1}(x), f^{-2}(x), ..., f^{-j}(x)\}$ of distinct points.

cones (or sectors) $C_\alpha^s(x)$ and $C_\alpha^u(x)$ in the tangent space $T_x\Omega$ that are strictly invariant in the sense that there is a $\gamma \in (0,1)$ such that

$$Df(C_\alpha^u(x)) \subset C_{\gamma\alpha}^u(f(x)) \text{ and } Df^{-1}(C_\alpha^s(x)) \subset C_{\gamma\alpha}^s(f^{-1}(x)) \tag{6.2}$$

where the cone $C_\alpha^s(x)$ is defined to be the set of vectors in the tangent space at x that make an angle less than α with E^s, and similarly for E^u. In this case, the stable and unstable subspaces are given by

$$\begin{cases} E^s = \cap_{n\in\mathbb{N}} Df^{-n}(C_\alpha^s f^n(x)) \\ E^u = \cap_{n\in\mathbb{N}} Df^n(C_\alpha^u(f^{-n}(x))). \end{cases} \tag{6.3}$$

(5) Anosov diffeomorphisms are rather rare, and every known one of them is a generalization of automorphisms of a nilmanifold up to topological conjugacy. An example of a diffeomorphisms that is hyperbolic on a proper invariant subset of interest is the one presented in Sec. 1.1, namely, the Smale horseshoe[3] extracted by Smale [Smale (1967)] from a study of relaxation oscillations by discerning a geometric picture in horseshoe shape.

(6) An example of Anosov flows is *free-particle motion*, or the mechanical system called a *geodesic flow* on a compact surface of negative curvature introduced in Sec. 6.4.1. Such surfaces locally resemble a mountain-pass landscape or the inner rim of a doughnut. In this case, opposing curvatures appear in contrast to a spheroid surface, and the effect on this system is that nearby trajectories quickly diverge from one another. The free-particle motion is an Anosov flow because at every point the phase space can be decomposed into contracting and expanding directions (their rates are related to the curvature) plus a 1-D flow direction[4].

(7) The coexistence of highly complicated long-term behavior, sensitive dependence on initial conditions, and the overall stability of the orbit structure are the most important features resulting from hyperbolicity.

(8) All Anosov systems are expansive, *i.e.*, there is a universal distance by which any two orbits will be separated at some time. Mathematically, there is a $\delta > 0$ such that for any points x and y, if $d(f^n(x), f^n(y)) \leq \delta$ for all n, then $x = y$.

---

[3]The Smale horseshoe is the invariant set obtained from an embedding.

[4]Furthermore, spaces for which all small 2-D cross sections have negative curvature as described for surfaces, *i.e.*, the higher-dimensional spaces with negative sectional curvatures are Anosov flows.

To define the Morse–Smale diffeomorphisms, we need the definition of the *Kupka–Smale diffeomorphism* given by the following:

Definition 6.4. The diffeomorphism f is Kupka–Smale if all its periodic orbits are hyperbolic and if moreover for any periodic points p and q, the unstable manifold $W^u(p)$ of p and the stable manifold $W^s(q)$ of q are in a general position (*i.e.*, at any intersection point $x \in W^u(p) \cap W^s(q)$, we have $T_x M = T_x W^u(p) + T_x W^s(q)$).

In the context of the Newhouse phenomena, Kupka [Kupka (1963)] and Smale [Smale (1963)] have shown that for any $r \geq 1$, the set of Kupka-Smale G_{KS} diffeomorphisms is a dense subset of diffeomorphisms defined in Ω and of class C^r denoted by $Diff^r(\Omega)$ and equipped with the C^r-topology (recall Definition 4.2).

A very recent result concerning some Kupka–Smale diffeomorphisms in dimension-three is given in [Pujals (2008)] where it was proved for a given topologically hyperbolic attracting set of a smooth three-dimensional Kupka–Smale diffeomorphism, that either the set is hyperbolic or the diffeomorphism is C^1-approximated by another one exhibiting either a heterodimensional cycle or a homoclinic tangency. This work has a relation to the famous Palis conjecture. To this end, we give the following definition for the Morse–Smale diffeomorphism:

Definition 6.5. The diffeomorphism f is Morse–Smale if it is a Kupka–Smale diffeomorphism whose nonwandering set $\Omega(f)$ (or equivalently whose chain-recurrent set $R(f)$) is finite.

A very recent result concerning Anosov flows, is given in [Araujo & Bessa (2008)] where it was proved that there exists an open and dense subset of the incompressible 3-flows of class C^2 such that, if a flow in this set has a positive volume regular invariant subset with dominated splitting for the linear Poincaré flow, then it must be an Anosov flow. Indeed, if $C^r(M)$ is the space of C^r vector fields, for any $r \geq 1$, then $C^r_\mu(M)$ is the subset of divergence-free vector fields defining incompressible (or conservative) flows. Then the following result was proved in [Araujo & Bessa (2008)]:

Theorem 6.2. *(a) There exists a generic subset $\mathcal{R} \subset C^1_\mu(M)$ such that for $f \in \mathcal{R}$,*
- *either f is Anosov,*
- *or else for Lebesgue almost every $p \in M$, all the Lyapunov exponents of f^t are zero.*

(b) Let $\varepsilon > 0$, an open subset U of M, and a non-Anosov vector field $f \in C^1_\mu(M)$ be given. Then there exists a $g \in C^1_\mu(M)$ such that g is C^1-ε-close to g and g^t has an elliptic closed orbit intersecting U.

(c) There exists an open and dense subset $G \subset C^2_\mu(M)$ such that for every $f \in G$ with a regular invariant set Λ (not necessarily closed) satisfying the following:

• the linear Poincaré flow over Λ has a dominated decomposition; and
• Λ has a positive volume: $\mu(\Lambda) > 0$; then f is Anosov and the closure of Λ is the whole of M.

Using the results given in Theorem 6.2, the dichotomies of Bochi–Mañé studied in [Mañé (1996), Bochi (2002), Bessa (2007)] and of Newhouse studied in [Newhouse (1977), Bessa & Duarte (2007)] for flows with singularities can be extended, *i.e.*, for a residual subset of the C^1 incompressible flows on 3-manifolds the following cases hold:

(i) Either all Lyapunov exponents are zero or the flow is Anosov.
(ii) Either the flow is Anosov or else the elliptic periodic points are dense in the manifold.

The proofs of Theorems 6.2(a) and (b) can be done using the standard arguments given in [Avila & Bochi (2006), Bessa (2007), Bessa & Duarte (2007)] assuming Theorem 6.2(c) together with the denseness of C^2 incompressible flows among C^1 incompressible ones given by Zuppa in [Zuppa (1979)].

The most important property of a hyperbolic diffeomorphism is the structure of its invariants, in particular, corresponding iterated attractors, especially, hyperbolic set attractors[5] *i.e.*, attractors with an isolating neighborhood, or Lyapunov-stable attractors. This implies that all unstable manifolds lie in the attractor. The Plykin attractor introduced in Sec. 6.4.4 and the Smale–Williams solenoid introduced in Sec. 6.4.3 are examples of such a situation. Note also that in two dimensions only the torus supports Anosov diffeomorphisms, and all are topologically conjugate to hyperbolic toral automorphisms.

Definition 6.6. (a) A hyperbolic set is defined to be a compact invariant set Λ of a diffeomorphism f such that the tangent space at every $x \in \Lambda$ admits an invariant splitting that satisfies the contraction and expansion conditions described in (d-3) and (d-4) of Definition 6.1(d).

---
[5]They are hyperbolic sets that are trapped attracting sets.

(b) A locally maximal hyperbolic set is the largest invariant set in a small neighborhood of the hyperbolic set.

The horseshoe attractor introduced in Sec. 1.1 is an example of Definition 6.6(b), where every point for which all iterates lie in the rectangle D shown in Fig. 1.1 indeed belongs to the horseshoe. Hyperbolic flows can be obtained from diffeomorphisms using the so-called *special flows* constructed as follows:

(a) Choose a hyperbolic diffeomorphism f on a space Ω.
(b) Define using the unit-speed upward motion on

$$\{(x,t) : x \in \Omega, 0 \leq t \leq \varphi(x)\}, \tag{6.4}$$

the special flow over f and under a *roof* function φ^6.

As a result, the obtained flow has expanding and contracting directions in Ω because an orbit that reaches the roof at a point $(x, \varphi(x))$ continues its upward motion from the point $(f(x), 0)$ at the bottom.

For further reading about the hyperbolicity and in particular the subject of Anosov systems, the reader is redirected to the following basic references: [Morse (1921), Smale (1965), Smale (1967-1968), Anosov (1967), Hirsch & Pugh (1968), Sinai (1968), Adler & Weiss (1970), Bowen (1970), Hirsch et al. (1970), Bowen (1971-1975), Sinai (1972)].

6.2 Anosov diffeomorphisms on the torus $\mathbb{T}^n$

In this section, we describe the major properties of the Anosov diffeomorphisms on the torus $\mathbb{T}^n$, in particular, topological properties of toral automorphisms and their relation to Anosov diffeomorphisms of $\mathbb{T}^n$. Examples of such hyperbolic and structurally stable mappings are described in some detail.

6.2.1 *Anosov automorphisms*

As a continuation of the content of Sec. 6.1.2, we give here some important results about Anosov automorphisms:

---

[6]The *suspension flow is a* special flow under the function $\varphi = 1$. This flow is never topologically mixing because at integer times the image of $\Omega \times \left(0, \frac{1}{2}\right)$ is disjoint from $\Omega \times \left(\frac{1}{2}, 1\right)$. But, for a generic roof function, the corresponding special flow over a topologically mixing Anosov flow is itself topologically mixing.

Definition 6.7. An automorphism is a hyperbolic, linear diffeomorphism of the torus $\mathbb{T}^n = \mathbb{R}^n/\mathbb{Z}^n$ into itself.

Automorphism on the torus $\mathbb{T}^n$ are called *toral automorphisms, i.e.*, any automorphism F_L induced on $\mathbb{T}^n$ by a hyperbolic linear map L of $\mathbb{R}^n$ with integer entries and determinant ± 1 is an Anosov diffeomorphism. For these mappings, there is a Euclidean norm in $\mathbb{R}^n$ that makes L contracting in $E^s(L)$ and expanding in $E^u(L)$. Thus there exists an invariant splitting into subspaces parallel to $E^s(L)$ and $E^u(L)$. In this case, toral automorphisms have the following properties (and for any transitive Anosov diffeomorphisms) that can be obtained using a symbolic model [Yoccoz (1993)]:

(a) The periodic points are dense.
(b) There exists a point whose orbit is dense.
(c) There are many ergodic invariant probability measures with full support.

For example, the map defined by the matrix $\begin{pmatrix} 1 & 1 \\ 1 & 0 \end{pmatrix}$ and its square $\begin{pmatrix} 2 & 1 \\ 1 & 1 \end{pmatrix}$ has an entropy value of $\log \frac{3+\sqrt{5}}{2}$.

From the result in [Manning (1974)], it is sufficient to consider only automorphisms on the torus $\mathbb{T}^n$ instead of nonlinear Anosov mappings in the general case.

Theorem 6.3. *Every Anosov diffeomorphism f of $\mathbb{T}^n$, such that $\Omega(f) = \mathbb{T}^n$, is topologically conjugate to some Anosov automorphism of $\mathbb{T}^n$.*

6.2.2 Structure of Anosov diffeomorphisms

Using Theorem 6.3, it was shown in [Arrowsmith & Place (1990)] that such Anosov diffeomorphisms have complicated nonwandering sets.

Proposition 6.1. *A point $\theta \in \mathbb{T}^n$ is a periodic point of the Anosov automorphism $f : \mathbb{T}^n \to \mathbb{T}^n$ if and only if $\theta = \pi(x)$ where $x \in \mathbb{R}^n$ has rational coordinates.*

Here, the map π is defined in some way, and it is related to the automorphism $f : \mathbb{T}^n \to \mathbb{T}^n$. Proposition 6.2 implies that f has infinitely many periodic points that are dense on the torus $\mathbb{T}^n$. All these points lie in the nonwandering set $\Omega(f)$ of f, and, since $\Omega(f)$ is closed, and one concludes that $\Omega(f) = \mathbb{T}^n$.

In [Mather (1968(a–b),1999)], it was shown that Anosov diffeomorphisms on $\mathbb{T}^n$, $n \geq 2$, are structurally stable diffeomorphisms on a compact manifold whose nonwandering set contains infinitely many points.

Theorem 6.4. *The Anosov diffeomorphisms on $\mathbb{T}^n$ are structurally stable in $Diff^1(\mathbb{T}^n)$.*

Hence the dynamics on $\Omega(f)$ are very complicated involving infinitely many hyperbolic periodic orbits densely distributed over the torus $\mathbb{T}^n$.

A recent result on the topic of hyperbolic automorphisms of the 2-torus is the work of Anosov himself with Klimenko and Kolutsky given in [Anosov et al. (2008)] where it was explained the existence of an isomorphism between a deterministic dynamical system and a random process for an example of the circle expanding map. A classification of hyperbolic toric automorphisms was done, and the notion of the simplest Markov partitions was discussed in a new way with their suggested classification.

6.2.3 *Anosov torus $\mathbb{T}^n$ with a hyperbolic structure*

The first example of a hyperbolic attractor [Arnold & Avez (1968), Arnold (1983)] is the Anosov torus $\mathbb{T}^n$ with a hyperbolic structure on it. Indeed, consider a solid torus $\Pi \in \mathbb{R}^n$, i.e., $\mathbb{T}^2 = \mathbb{D}^2 \times \mathbb{S}^1$ where $\mathbb{D}^2$ is a disk and $\mathbb{S}^1$ is a circumference. If one realizes the following operations:

(1) Expand $\mathbb{T}^2$ m-times (m is an integer) along the cyclic coordinate on $\mathbb{S}^1$.
(2) Shrink $\mathbb{T}^2$ q-times along the diameter of $\mathbb{D}^2$ where $q \leq \frac{1}{m}$.
(3) Embed this deformated torus Π_1 into the original one so that its intersection with $\mathbb{D}^2$ consists of m smaller disks.
(4) Repeat this routine with Π_1 and $\Sigma = \sum_{i=1}^{\infty} \Pi_i$.

Then the so-called *Witorius–Van Danzig solenoid* is obtained [Shilnikov (2002)]:

Proposition 6.2. *(a) The local structure of the Witorius–Van Danzig solenoid is represented as the direct product of an interval and a Cantor set.*

(b) The Wictorius–Van Danzig solenoids are quasi-minimal sets[7].

(c) [Smale (1967)] The Witorius–Van Danzig solenoid has hyperbolic attractors of diffeomorphisms on solid tori.

Using the so-called *surgery* operation over the automorphism of a two-dimensional torus with a hyperbolic structure, another example of a hyperbolic attractor was constructed[8] by Smale on this torus. This is the so-called DA-(*derived from Anosov*) diffeomorphism introduced in Sec. 6.4.2.

6.2.4 Expanding maps

The importance of expanding maps is that they are expanding solenoids or hyperbolic attractors [Williams (1974)] which lead to robust chaotic attractors because their method of construction is similar to that of minimal sets of limit-quasi-periodic trajectories.

Definition 6.8. A map is called expanding in length if any tangent vector field grows exponentially under the action of the differential of the map.

An example is the algebraic hyperbolic automorphisms of the torus given by

$$\bar{\theta} = A\theta \, (\mathrm{mod}\, 1), \tag{6.5}$$

such that the spectrum of the integer matrix A lies strictly outside the unit circle, and any neighboring map is also expanding. Shub [Shub (1971)] established that expanding maps are structurally stable. The study of expanding maps and their connection to smooth diffeomorphisms was continued by Williams [Williams (1970)]. It is easy to verify that the condition of hyperbolicity given in Definition 6.1(d) are still holds for the class of algebraic hyperbolic automorphisms of the torus,

$$\bar{\theta} = A\theta \, (\mathrm{mod}\, 2\pi) \tag{6.6}$$

under some conditions on the matrix A.

Proposition 6.3. *(a) Automorphisms (6.6) are conservative systems whose set Ω of nonwandering trajectories coincide with the torus $\mathbb{T}^n$ itself.*

(b) The set Ω can be represented by a finite union of nonintersecting, closed, invariant, transitive sets $\Omega_1, ...\Omega_p$ called basis sets of Smale systems.

---

[7]In the sense of the theory of sets of limit-quasi-periodic functions.

[8]This construction is designed as that of minimal sets known from the Poincaré–Donjoy theory in the case of C^1-smooth vector fields on a two-dimensional torus.

(c) In the case of cascades, any such Ω_i can be represented by a finite number of sets having these properties which are mapped to each other under the action of the diffeomorphism.

The basis sets of Smale systems may be of the following three types: attractors, repellers, and saddles.

Definition 6.9. (a) Repellers are the basis sets which become attractors in backward time.

(b) Saddle basis sets are such that they may both attract and repel outside trajectories.

Saddle basis sets are one-dimensional in the case of flows, *i.e.*, they are homeomorphic to the suspension over topological Markov chains and null-dimensional in the case of cascades, *i.e.*, they are homeomorphic to simple topological Markov chains [Bowen (1970)].

The trajectories passing sufficiently close to an attractor of a Smale system, satisfy the condition

$$dist\left(\varphi\left(t,x\right),A\right) < ke^{-\lambda t}, t \geq 0 \tag{6.7}$$

where k and λ are positive constants.

Proposition 6.4. (a) *Attractors of a Smale system are transitive, i.e., periodic, homoclinic, and heteroclinic trajectories as well as Poisson-stable ones are everywhere dense in them.*

(b) *The unstable manifolds of all points of such an attractor lie within it, i.e., $W^u(x) \in A$ where $x \in A$.*

This implies that hyperbolic attractors may be smooth or nonsmooth manifolds, have a fractal structure, not be locally homeomorphic to a direct product of a disk and a Cantor set, and so on....

The second example can be realized as a limit of the inverse spectrum of the expanding cycle map [Williams (1967)],

$$\bar{\theta} = m\theta, (\bmod 1). \tag{6.8}$$

The most interesting property of maps of the form (6.8) is that they are expanding solenoids (see Sec. 6.4.2) or hyperbolic attractors[9] [Williams (1974)]. These attractors are generalized (extended) solenoids, and their

---

[9] In this case, its dimension coincides with the dimension of the unstable manifolds of the points of the attractor. This type of attractor is called the Smale–Williams solenoids introduced in Sec. 6.4.3.

method of construction is similar to that of minimal sets of limit-quasi-periodic trajectories. More generally, an example of the Anosov diffeomorphism is a mapping of an n-dimensional torus,

$$\bar{\theta} = A\theta + f(\theta) \,(\mathrm{mod}\, 1) \tag{6.9}$$

where A is a matrix with integer entries other than 1, $\det |A| = 1$, the eigenvalues of A do not lie on the unit circle, and $f(\theta)$ is a periodic function of period-1.

6.2.5 The Blaschke product

A (finite) Blaschke product is a map of the form

$$B(z) = \theta_0 \prod_{i=1}^{n} \frac{z - a_i}{1 - z\bar{a}_i} \tag{6.10}$$

where $n \geq 2$, $a_i \in \mathbb{C}$; $|a_i| < 1, i = 1, ..., n$, and $\theta_0 \in \mathbb{C}$ with $|\theta_0| = 1$. The map (6.10) is rational in $\mathbb{C}$, and it is an analytic function in a neighborhood of the unit disk $\mathbb{D}$. The map B maps the unit circle $\mathbb{T}$ to itself. In [Pujals et al. (2006)], the family of Blaschke products given by

$$\{B_\theta\}_{\theta \in \mathbb{T}} = \{\theta B\}_{\theta \in \mathbb{T}} \tag{6.11}$$

was considered and some properties were proved, in particular, the map (6.10) is expanding or has a unique attracting or indifferent fixed point.

Theorem 6.5. *Given a family of Blaschke products $\{B_\theta\}_{\theta \in \mathbb{T}}$, one of the next two options holds for any $\theta \in \mathbb{T}$:*

(1) B_θ is an expanding map, i.e., there are $n = n(\theta)$ and $\lambda = \lambda(\theta) > 1$ such that

$$\left|(B_\theta^n)'(x)\right| > \lambda. \tag{6.12}$$

(2) B_θ has a unique attracting or indifferent fixed point in $\mathbb{T}$. Moreover, the set of $\theta \in \mathbb{T}$ satisfying the first option is a nonempty open set.

Let λ be Lebesgue measure on $\mathbb{T}$ normalized to be a probability measure, $\lambda(\mathbb{T}) = 1$. Hence some statistical behaviors of B_θ were proved also in Pujals et al. (2006)]

Theorem 6.6. *Given a family of Blaschke products $\{B_\theta\}_{\theta \in \mathbb{T}}$, it follows that for all θ, the push forwards of Lebesgue measure $B_{\theta *}^n(\lambda)$ converges to a measure μ_θ which is:*

(1) Absolutely continuous with respect to Lebesgue if B_θ satisfies condition 1 of Theorem 6.5, or

(2) A Dirac delta measure supported on an attracting or indifferent fixed point of B_θ on $\mathbb{T}$.

For more details about the topic of expanding maps, the reader can see [Ledrappier (1981), Mané (1985), Shub & Sullivan (1985), Burns et al. (2001), Dedieu & Shub (2003), Ledrappier et al. (2003)].

6.2.6 The Bernoulli map

The essential property of the Bernoulli map given by

$$\phi_{n+1} = 2\phi_n (\bmod 2\pi) \qquad (6.13)$$

is the doubling of the cyclic coordinate around the torus. Sometimes map (6.13) is called the *sawtooth map* which exhibits chaotic dynamics as shown in [Sinai (1979), Devaney (1989), Ott (1993), Katok & Hasselblatt (1999)]. In particular, Eq. (6.13) implies sensitivity to initial conditions, *i.e.*, a small deviation from the initial state doubles at each iteration step. The map (6.13) was used essentially in studying the robustness of chaos generated by some nonautonomous continuous-time systems as shown in Sec. 7.1.5.

The map (6.13) is expanding and it transforms the angle variable ϕ is a nonuniform way, but in any case it must be monotonic and possess the characteristic topological property. In particular, the map (6.13) displays *homogeneous chaotic dynamics, i.e.*, the rate of exponential divergence of two close trajectories is identical at each point of the phase space and equal to $\ln 2$.

In some applications, such as in the topic of modeling hyperbolic chaos (see Chapter 7), the form

$$\phi_{n+1} = 2\phi_n + const (\bmod 2\pi) \qquad (6.14)$$

was used to describe the dynamics of such a situation, where the constant in (6.14) depends on delay times and details of the manipulation with signals in the feedback loops of a continuous-time dynamical system. This constant has no principal significance because it can be removed from (6.14) by an appropriate selection of the offset for the variable ϕ_n.

The Bernoulli map (6.13) has the following properties:

(1) The Bernoulli map (6.13) is an exactly solvable model of deterministic chaos.

(2) The transfer operator (which encodes information about an iterated map), or *Frobenius–Perron operator*, of the Bernoulli map (6.13) is solvable [Driebe (1999)].
(3) The eigenvalues are multiples of $\frac{1}{2}$, and the eigenfunctions are the *Bernoulli polynomials* which occur in the study of many special functions such as the Riemann and Hurwitz zeta functions [Gaspard (1992)].

More details about the properties of the Bernoulli map (6.13) can be found in [Helstrom (1984), Gaspard (1992), Driebe (1999), Bertsekas & Tsitsiklis (2002)].

More generally, it was shown in [Dolgopyat & Pesin (2002)] that a smooth compact Riemannian manifold of dimension 2 admits a Bernoulli diffeomorphism with nonzero Lyapunov exponents, as in the following theorem:

Theorem 6.7. *Given a compact smooth Riemannian manifold* $\mathbb{K} \neq \mathbb{S}^1$, *there exists a* C^∞ *diffeomorphism* f *of* $\mathbb{K}$ *such that*
(1) f *preserves the Riemannian volume* m *on* $\mathbb{K}$,
(2) f *has nonzero Lyapunov exponents at* m-*almost every point* $x \in \mathbb{K}$, *and*
(3) f *is a Bernoulli diffeomorphism.*

Theorem 6.7 was proved for surface diffeomorphisms in [Katok (1979)] and for any compact smooth Riemannian manifold $\mathbb{K}$ of dimension 5 in [Brin (1981)] where a C^1 Bernoulli diffeomorphism which preserves the Riemannian volume and has all but one Lyapunov exponents nonzero was constructed. A result from [Katok (1979), Brin (1981), Brin et al. (1981)] is the fact that any manifold $\mathbb{K}$ admits a diffeomorphism with l zero exponents, where

$$l = \begin{cases} 0, & \text{if } \dim \mathbb{K} = 2 \\ 2, & \text{if } \dim \mathbb{K} = 4 \\ 1, & \text{otherwise.} \end{cases} \quad (6.15)$$

Some properties of the hyperbolic Bernoulli diffeomorphisms found in [Mañé (1984-1996), Bochi (2001)] are the following:

Definition 6.10. (1) Let f be a diffeomorphism of $\mathbb{K}$ preserving a smooth volume m and $T\mathbb{K} = E \oplus F$ the splitting of $T\mathbb{K}$ into two invariant subbundles. We say that F *dominates* E (and write $E < F$) if there exists $\theta < 1$ such that

$$\max_{v \in E, \|v\|=1} \|df(v)\| \leq \theta \min_{v \in F, \|v\|=1} \|df(v)\|. \quad (6.16)$$

Theorem 6.8. *(1) If f admits a dominating splitting, then any diffeomorphism sufficiently close to f also admits a dominating splitting. (2) If for any sufficiently small perturbation of f, the subspace E_2 does not admit further splitting, then f can be approximated by a diffeomorphism g such that all Lyapunov exponents of g along E_2 are close to each other*[10].

And in [Shub & Wilkinson (2000), Dolgopyat (2001)], we find the following:

Theorem 6.9. *If $T\mathbb{K} = E_1 \oplus E_2 \oplus E_3$ where $E_1 < E_2 < E_3$, then the function*

$$f \to \int \log \det \left(df \mid_{E_2} \right)(x) \, dm(x) \tag{6.17}$$

is not locally constant.

The results in Theorem 6.8 and Theorem 6.9 were used for constructing nonuniformly hyperbolic systems (with nonzero exponents on a set of positive measure) on manifolds carrying diffeomorphisms with dominated decomposition [Brin (1981), Brin *et al.* (1981)]. The approach used is based on transitivity of foliations to pass from local to global ergodicity [Pesin (1977), Brin & Pesin (1974)]. Examples of such constructions of open sets of hyperbolic Bernoulli diffeomorphisms on some manifolds are given in [Alves *et al.* (2000), Bonnatti & Viana (2000), Dolgopyat (2001), Shub & Wilkinson (2000)].

The same results can be obtained for flows as shown in [Hu *et al.* (2002)] where every compact smooth Riemannian manifold M of $\dim M \geq 3$ admits a volume-preserving Bernoulli flow with nonzero Lyapunov exponents except for the Lyapunov exponent along the direction of the flow. Namely, we have the following theorem:

Theorem 6.10. *Given a compact smooth Riemannian manifold M of $\dim M \geq 3$, there exists a C^∞ flow f^t such that for each $t \neq 0$,*

(1) f^t preserves the Riemannian volume on M,

(2) f^t has nonzero Lyapunov exponents (except for the exponent along the flow direction) at m-almost every point $x \in M$, and

(3) f^t is a Bernoulli diffeomorphism.

The construction of the flow f^t in Theorem 6.10 is based on the special Katok diffeomorphism defined in a two-dimensional unit disk $\mathbb{D}^2$ [Katok (1979)].

---
[10]This situation is very important where $\dim \mathbb{K} = 4$.

6.2.7 The Arnold cat map

As a 2-D example, consider the *Arnold cat map* [Anosov (1967)] given by

$$\begin{cases} x_{n+1} = x_n + y_n \,(\text{mod}\,1) \\ y_{n+1} = x_n + 2y_n \,(\text{mod}\,1). \end{cases} \quad (6.18)$$

The map (6.18) is the Anosov torus $\mathbb{T}^2$. This map is conservative, and one cannot speak of an attractor. The introduction of a small perturbation

$$\begin{cases} x_{n+1} = x_n + y_n + \delta \sin 2\pi y_n \,(\text{mod}\,1) \\ y_{n+1} = x_n + 2y_n \,(\text{mod}\,1) \end{cases} \quad (6.19)$$

to map (6.19) makes it invertible, dissipative, and a diffeomorphism of the torus for $\delta < \frac{\pi}{2}$. One of the methods used to avoid *dangerous tangencies* of manifolds in a two-dimensional map is to consider a map on a torus. In this case, no nonrobust homoclinic orbits are obtained because manifolds of a saddle bend around the torus surface intersecting transversally each time. The *modified cat map* (6.19) was considered in [Farmer et al. (1983), Sinai (1972)]. The qualitative behavior of the manifolds of a saddle is the same for both maps (6.18) and (6.19) and numerically displays a robust hyperbolic attractor for $\delta < \frac{\pi}{2}$. See Exercise 141, where it is clear that the map (6.19) repeats qualitatively the results obtained for the Lozi map (8.88). Thus it is possible to conclude from the experimental point of view that the modified cat map (6.19) has a Lorenz-type attractor as introduced in Chapter 8. Due to the importance of the Arnold cat map (6.18) in modeling robust hyperbolic chaotic attractors as seen in Sec. 7.1.4, we give here some of its important properties:

(1) It is a conservative map, *i.e.*, the *cat face* conserves under iteration the area of any domain in the xy-plane.
(2) The map (6.18) has a hyperbolic[11] chaotic attractor [Arnold (1988), Bunimovich et al. (2000)]. This implies the existence of continuous invariant measure, a possibility of description in terms of Markov partition and symbolic dynamics, positive topological and metric entropies, etc.
(3) The map (6.18) has two Lyapunov exponents expressed by eigenvalues of the matrix associated with it, *i.e.*, $\Lambda_1 = \frac{\ln(3+\sqrt{5})}{2} = 0.9624$ and $\Lambda_2 = -\frac{\ln(3+\sqrt{5})}{2} = -0.9624$, and their sum vanish.
(4) The second iteration of the Fibonacci map yields the Arnold cat map (6.18).

---
[11] *i.e.*, in the sense of the theory of Smale and Anosov.

6.3 Classification of strange attractors of dynamical systems

In this section, we give a common classification of strange attractors of dynamical systems. Generally, at the present time, strange attractors can be classified into three principal classes [Anishchenko & Strelkova (1997), Plykin (2002)]: *hyperbolic*, *Lorenz-type*[12], and *quasi-attractors*.

The hyperbolic attractors are the limit sets for which Smale's *axiom A* (recall Definition 6.3) is satisfied and are structurally stable (recall Definition 4.3). Periodic orbits and homoclinic orbits are dense and are of the same saddle type, which is to say that they have the same index[13]. This definition is a result of a rigorous axiomatic foundation that exploits the notion of hyperbolicity [Katok & Hasselblatt (1995), Ott (1993)]. Hyperbolic chaos is often called *true chaos* from a rigorous mathematical point of view and is characterized by a homogeneous and topologically stable structure as shown in [Anosov (1967), Smale (1967), Ruelle & Takens (1971), Guckenheimer & Holmes (1981)]. However, the Lorenz-type attractors presented in Chapters 8 and 9 are almost not structurally stable, although their homoclinic and heteroclinic orbits are structurally stable (hyperbolic), and no stable periodic orbits appear under small parameter variations, as for example in the Lorenz system [Lorenz (1963)].

The *quasi-attractors*[14] introduced in Chapter 10 are the limit sets enclosing periodic orbits of different topological types (for example, stable and saddle periodic orbits) and structurally unstable orbits. For example, the attractors generated by Chua's circuit [Chua *et al.* (1986)] presented in Sec. 10.5 associated with saddle-focus homoclinic loops are quasi-attractors. Note that this type is more complex than the above two attractors and thus are not suitable for some cases of potential applications of chaos such as secure communications and signal masking. For further information about these types of chaotic attractors, see Chapter 10.

The above classification is based on the rigorous mathematical analysis, and it not accepted as significant from an experimental point of view. Recent studies show that properties of hyperbolic, Lorenz-type attractors, and quasi-attractors are basically different [Anishchenko & Strelkova (1998)].

The most common property of quasi-attractors and Lorenz-type ones is that both admit homoclinic tangencies of the trajectories, and the

---

[12] Or pseudo-hyperbolic.
[13] *i.e.*, the same dimension for their stable and unstable manifolds.
[14] They are the most frequently observed limit sets in problems of nonlinear dynamics.

difference between them is that quasi-attractors contain a countable subset of stable periodic orbits. For the above types of attractors, it is possible to prove chaos by finding a positive Lyapunov exponent, a continuous frequency spectrum, fast decaying correlation functions, *etc*. Furthermore, the chaotic behavior of Lorenz-type attractors can be studied using statistical methods because they admit the introduction of reasonable invariant measures introduced in Sec. 3.1 contrary to the quasi-attractors[15]. This gives a rigorous foundation for studying characteristics of these attractors such as Lyapunov exponents. Hence hyperbolic and Lorenz attractors are called *stochastic*.

Using the so-called *autocorrelation function* (ACF) introduced in Sec. 3.3.1, it is possible to distinguish the quasi-attractors from Lorenz attractors by the presence of a periodic component in the ACF and sudden peaks at certain characteristic frequencies in the spectrum. The latter being typical features of quasi-attractors. Note that the results about homoclinic tangencies for 3-D systems and 2-D diffeomorphisms also hold for the general case where there may be some other peculiarities, as for instance the co-existence of a countable sets of saddle periodic orbits of distinct topological types [Gonchenko *et al.* (1993)]. Hence the complete theoretical analysis of the models, which admit homoclinic tangencies, including complete bifurcation diagrams and so forth, is nonrealistic.

For more details about the problem of topological classification of strange attractors of dynamical systems, the reader can consult the following references which present the most common results known in the literature: [Vietoris (1927), van Dantzig (1930), Bourbaki (1942-1947), Cassels (1961), Arov (1963), Hewitt & Ross (1963), Novikov (1965), Smale (1967), Franks (1970), Ruelle & Takens (1971), Perov & Egle (1972), Aleksandrov (1977), Plykin (1977), Rand (1978), Plykin (1980), Watkins (1982), Plykin (1984), Ustinov (1987), Aarts & Fokkink (1991), Fokkink (1991), Plykin *et al.* (1991), Pesin (1992), Plykin & Zhirov (1993), Zhirov (1994a-b–1995a-b), Gorodetski & Ilyashenko (1996), Anishchenko & Strelkova (1997), Zhirov (2000), Plykin (2002)].

For the remaining of this chapter, we assume that the reader is well familiar with the following concepts: homeomorphism, diffeomorphism, smooth flow and maps, C^k topology, foliations, attractors, limit sets, invariant measure, and the elements of local stability theory, in particular the Hartman–Grobman theorem and the stable manifold theorem.

---

[15]They hardly admit the introduction of an invariant measure.

Some of these concepts are presented and discussed in several places in this book.

6.4 Properties of hyperbolic chaotic attractors

The theory of *structural stability* introduced in Chapter 4 was initiated in the works of Anosov and Smale. The Anosov systems are a class of systems satisfying the hyperbolicity conditions given in Definition 6.1(d) in the whole phase space [Anosov (1967)]. Some examples of Anosov systems are geodesic flows on compact smooth manifolds of a negative curvature [Anosov (1967)]. These systems are presented in Sec. 6.4.1, and they are conservative with the set of nonwandering trajectories introduced in Definition 6.1(a) that coincides with the phase space.

Generally, hyperbolic attractors satisfy the following three conditions [Afraimovich (1989-1990)]:

(1) A hyperbolic attractor consists of a continuum of *unstable leaves* or curves, which are dense in the attractor and along which close trajectories exponentially diverge.
(2) A hyperbolic attractor (in the neighborhood of each point) has the same geometry defined as a product of the Cantor set on an interval.
(3) A hyperbolic attractor has a neighborhood foliated into *stable leaves* along which the close trajectories converge to the attractor.

The notion of robustness introduced in Chapter 2 means that properties (1)–(3) hold under perturbations. Some important remarks about the dynamical behavior of hyperbolic attractors are summarized in the following topics:

(1) The structure of hyperbolic attractors is homogeneous, *i.e.*, a neighborhood of any point on the attractor has the same geometry. This is a result of the fact that all trajectories of a hyperbolic attractor belong to the same saddle type.

(2) Dissipative hyperbolic systems, *i.e.*, with phase-volume contracting exhibit strange chaotic attractors with strong chaotic properties such as positivity,[16] smoothness with respect to parameters of at least one Lyapunov exponent,[17] a fractional dimension, *etc*.

(3) Hyperbolic chaotic attractors are structurally stable (robust) under small changes in their governing equations [Katok & Hasselblatt (1995),

---
[16] *i.e.*, never drops to negative values. Otherwise, the attractor is nonhyperbolic.
[17] These exponents quantify the sensitivity of chaotic dynamics to initial conditions.

Afraimovich & Hsu (2003), Kuznetsov (1998)]. These attractors can be characterized both topologically, *i.e.*, in terms of symbolic dynamics introduced in Sec. 1.4 and probabilistically, *i.e.*, in terms of the Sinai–Bowen–Ruelle measures introduced in Sec. 3.1.2. In this case, Cantor-like structure persists under these small changes without qualitative change (bifurcations). For example, it has been proved that the Smale–Williams solenoid introduced in Sec. 6.4.3 is structurally stable [Sinai (1979), Afraimovich & Hsu (2003)].

(4) The absence of homoclinic tangencies implies robust hyperbolicity.

(5) No dissipative mathematical model (governed by differential equations or maps) is known for which the existence of the robust hyperbolic attractor was strictly proved, but there are examples of conservative systems for which all trajectories of the phase space are hyperbolic. Furthermore, only geometrical constructions of the hyperbolic strange attractors were known before the work of Hunt and Kuznetsov [Hunt (2000), Kuznetsov (2001), Hunt & MacKay (2003), Belykh *et al.* (2005)]. More on the geometry of hyperbolic attractor can be found in the main works of [Plykin (1984), Shilnikov (1993a)]. In this chapter, some examples of hyperbolic attractors are discussed such as geodesic flows on compact smooth manifolds, Plykin's attractor, the Smale–Williams solenoid, the Anosov torus $\mathbb{T}^n$, the Arnold cat map, the sawtooth map, and the triple linkage system. More details on these systems can be found in [Anosov (1967), Sinai (1979), Devaney (1989), Ott (1993), Katok & Hasselblatt (1999), Hunt (2000), Kuznetsov (2001), Hunt & MacKay (2003), Belykh *et al.* (2005)]. Without concrete examples, hyperbolic strange attractors appear also in systems obtained from the perturbation of quasi-periodic motions on tori of dimension ≥ 3 as in [Newhouse *et al.* (1978)]. On the other hand, autonomous or periodically forced nonlinear oscillators, like the Rössler system and Chua's circuit are not completely hyperbolic [Ott (1993), Anishchenko *et al.* (2003d)], *i.e.*, they are quasi-attractors as introduced in Chapter 10.

6.4.1 *Geodesic flows on compact smooth manifolds*

In this section, we introduce the first important example of uniform hyperbolicity which was the geodesic[18] flow G^t on Riemannian manifolds of negative curvature[19] M [Anosov (1967)]. Indeed, let M be a compact

---

[18] A geodesic is a generalization of the notion of a *straight line* to *curved spaces*.
[19] The negativity of the curvature is an indicator of the hyperbolic behavior of the flow under consideration.

Riemannian manifold. Let v be any tangent vector and $\gamma_v : R \to TM$ be the geodesic with initial condition $v = \gamma_v(0)$. Let $\gamma'_v(t)$ be the velocity vector at time t. Because $\|\gamma'_v(t)\| = \|v\|$ for all t, it is no restriction to consider only unit vectors. Hence one has the following definition:

Definition 6.11. The geodesic flow is the flow $G^t : T^1 M \to T^1 M$ on the unit tangent bundle $T^1 M$ of the manifold defined by $G^t(v) = \gamma'_v(t)$.

If M has negative sectional curvature then the Liouville measure on the unit tangent bundle $T^1 M$ is ergodic (see Sec. 3.2) for the flow, *i.e.*, any invariant set has zero or full Liouville measure. This result is due to Anosov, and this is not the first result in this subject because the case when M is a surface had been studied before by Hedlund and Hopf. The importance of the result of Anosov can be seen in the proof that the geodesic flows $G^t(v)$ with negative sectional curvature are uniformly hyperbolic. If M is a surface, then the stable and unstable invariant subbundles are differentiable, but this is not true in higher dimensions. A weaker form of the regularity property was used by Anosov to overcome this obstacle.

A few examples of hyperbolic chaos in differential equations were discussed in a theoretical manner. It was found in [Hunt & MacKay (2003)] that the frictionless motion of a mechanical system called *triple linkage* can be described in terms of a geodesic flow on a surface with everywhere negative Gaussian curvature. In the presence of friction, the system, supplemented with an appropriate feedback control, is expected to have a hyperbolic chaotic attractor.

Some recent results on the topic of geodesic flows and their generalization can be found in [Butler & Gelfreich (2008)] where the positivity of entropy was discussed for geodesic flows on nilmanifolds, *i.e.*, the Euler flows of the standard Riemannian and sub-Riemannian structures of T_4 (the nilpotent group of real 4×4 upper-triangular matrices with 1's on the diagonal) have transversal homoclinic points on all regular coadjoint orbits. Another estimate for the entropy of Hamiltonian flows can be found in [Chittaro (2007)] where a generalization of the entropy estimate inequality for geodesic flow on a compact Riemannian manifold of nonpositive sectional curvature was proved by Ballmann and Wojtkovski in [Ballmann & Wojtkowski (1989)] for Hamiltonian flows on symplectic manifolds of dimension $2n$. In [Agrachev (2007)], a self-contained description of the so-called *curvature-type invariants* of Hamiltonian systems, which is a generalization of sectional curvatures of Riemannian manifolds, was given. Related constructions and facts lead to a natural extension of the classical results about Riemannian geodesic flows and indicate some new phenomena. The uniqueness of central foliations[20] of geodesic flows for compact surfaces was discussed in [Gomes & Ruggiero (2007)] where it was shown that the central stable and the central unstable foliations are the only continuous, codimension-one foliations which are invariant under the action of the geodesic flow of a compact surface without conjugate points.

6.4.2 *The solenoid attractor*

The concept of a *solenoid* has an interesting history in geometry and dynamics [Takens (2005)]. This concept was first introduced by Vietoris in [Vietoris (1927)] as part of the construction of an example of a continuum, *i.e.*, based on a special type of homology, cohomology, and fundamental groups for compact metric spaces. In dynamics, the concept of a solenoid was first used by Nemytskii and Stepanov [Nemytskii & Stepanov (1960)] as an example of the closure of an almost periodic motion which is not quasi-periodic. Later the solenoid was used by Williams in [Williams (1955)] as a first example of a space on which an *unstable* homeomorphism (expansive homeomorphism) could be defined, and later in [Williams (1967)] as one of the first examples of a hyperbolic, chaotic, fractal attractor. This example is given as follows:

Consider a solid torus $\mathbb{T} = \mathbb{S}^1 \times \mathbb{D}^2$ in dimension 3 and a diffeomorphism of $\mathbb{T}$ to a subset of $\mathbb{T}$ such that in the direction of $\mathbb{S}^1$ one has an expanding map (see Definition 6.13) of degree 2 and in the direction of $\mathbb{D}^2$ one has a (strong) contraction. Let $s \in \mathbb{R} \pmod 1$ be a coordinate on $\mathbb{S}^1$ and x, y, with $x^2 + y^2 < 1$, as coordinates on $\mathbb{D}^2$. An example of such a map is given by

$$\begin{cases} \varphi(s, x, y) = (2s \pmod 1, c_1 \cos(2\pi s) + c_2 x, c_1 \sin(2\pi s) + c_2 y), \\ 0 < c_2 < c_1, c_1 + c_2 < 1. \end{cases} \quad (6.20)$$

Thus the solenoid is defined as

$$\mathbb{S} = \bigcap_i \varphi^i(\mathbb{T}), \quad (6.21)$$

and the dynamics on $\mathbb{S}$ is given by $\varphi|_S$ which has the following characteristics:

(1) The diffeomorphism $\varphi|_S$ is structurally stable because φ is hyperbolic.

---

[20]This uniqueness is typical of Anosov systems and expansive geodesic flows in compact manifolds without conjugate points.

(2) The attractor $\mathbb{S}$ is chaotic and has sensitive dependence on initial conditions in the $\mathbb{S}^1$ direction under repeated application of the map φ.

(3) The attractor $\mathbb{S}$ is not a manifold, and it is a fractal set because for each $s \in \mathbb{R} \pmod 1$, $\mathbb{S} \cap (\{s\} \times \mathbb{D}^2)$ is a Cantor set.

Note that in the theory of dynamical systems, the concept of a solenoid is a general set that includes all the attractors which are locally homeomorphic to the product of a Cantor set and an interval just like the Plykin attractor [Plykin (1974)] introduced in Sec. 6.4.4.

An elementary example of the above situation is the so-called *Smale attractor* [Smale (1967), Katok and Hasselblatt (1995)] on the solid torus $\mathbb{T} = \mathbb{S}^1 \times \mathbb{D}^2$ given by $f : \mathbb{T} \to \mathbb{T}$ where

$$f(\varphi, x, y) = \left(2\varphi, \frac{1}{10}x + \frac{1}{2}\cos\varphi, \frac{1}{10}y + \frac{1}{2}\sin\varphi\right). \quad (6.22)$$

Smale called this map the *DE-map*, for *derived from expanding*, and its attractor is the natural extension of the double self-covering $x \to 2x$ of the circle. Hence f is expanding, and the set $\Lambda = \bigcap_{l \in \mathbb{N}} f^l(\mathbb{T})$ is an attractor of f of the local form (Cantor set $\times$ interval), but it is connected. In this case, the stable manifolds are the sections $C = \{\theta\} \times \mathbb{D}^2$, and the unstable manifold of each point is entirely contained in the attractor.

The above construction of a solenoid attractor can be generalized to a k-dimensional map φ_k (so $\varphi = \varphi_2$) taking in account the following tasks:

(a) The set $\mathbb{S}^1$ is taken as an expanding map of degree $k > 2$ (instead of degree 2).

(b) The assumption "$2s \mod 1$" must be replaced by "$ks \mod 1$" in the definition of φ.

(c) A further restriction on the parameter c_2 must be added to keep this map injective.

In this case, the attractor is defined as

$$\mathbb{S}_k = \bigcap_i \varphi_k^i(\mathbb{T}), \quad (6.23)$$

and for the sake of completeness, the orientation reversing attractors can be obtained by taking negative values for k.

The *solenoid attractor* $\mathbb{S}$ given by (6.21) can be reconsidered in the set of complex numbers as follows: Consider the solid torus $J = \mathbb{S}^1 \times \mathbb{D}$, where $\mathbb{S}^1 = \mathbb{R}/\mathbb{Z}$ is the circle and $\mathbb{D} = \{z \in \mathbb{C} : |z| < 1\}$ is the unit disk in the complex plane. Consider the map $f : J \to J$ given by

$$f(\theta, z) = (2\theta, az + \beta e^{i\theta/2}) \quad (6.24)$$

where $\theta \in \mathbb{S}^1$ and $\alpha, \beta \in \mathbb{R}$ with $\alpha + \beta < 1$ which implies that $f(J) \subset J$.

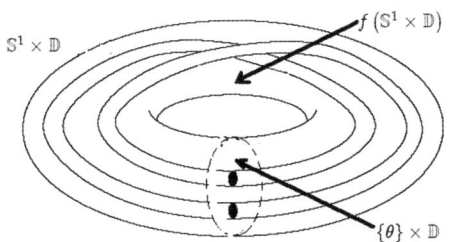

Fig. 6.1 The solenoid attractor $\Lambda = \cap_{n \geq 0} f^n(J)$.

Hence the so-called *solenoid attractor* is constructed as follows:

(1) The image $f(J)$ is a long thin domain going around the solid torus twice, as described in Fig. 6.1.
(2) For any $n \geq 1$, the corresponding iterate $f^n(J)$ is an increasingly thinner and longer domain that winds $2k$ times around J. In this case, the maximal invariant set $\Lambda = \cap_{n \geq 0} f^n(J)$ is called the *solenoid attractor*.

The solenoid attractor has the following properties:

(a) The forward orbit under f of every point in J accumulates on Λ.
(b) The restriction of f to the attractor Λ is transitive.
(c) The set of periodic points of f is dense in Λ.
(d) The set Λ has a dense subset of periodic orbits and also a dense orbit. In this case, every point in a neighborhood of Λ converges to Λ.

The reader is asked to show these properties in Exercise 142.

6.4.3 *The Smale–Williams solenoid*

First, note that the *Smale–Williams attractor* appears in a mapping of a toroidal domain of dimension 3 or more into itself. This mapping cannot be produced by a continuous flow in the three-dimensional phase space, and the minimal dimension for such a situation is four where the Smale–Williams attractor can occur in the three-dimensional Poincaré map. One iteration of this map leads to a longitudinal stretch of the torus with contraction in the transversal directions, and insertion of the doubly folded tube into the original domain. Furthermore, this type of attractor (without explicit examples of differential equations) is a result of a codimension-one bifurcation as shown in [Shilnikov & Turaev (1997)]. Fig. 6.2 shows the method of

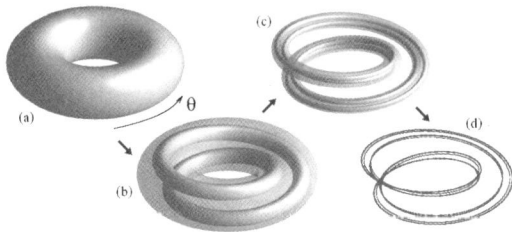

Fig. 6.2 Generation of an attractor by mapping a 3-D phase space into itself.

generating the Smale–Williams solenoid by a dissipative dynamical system that maps a three-dimensional phase space into itself iteratively as follows:

(1) The 3-D torus shown in Fig. 6.2(a) is stretched to twice its original length, folded in half, and squeezed into its original volume as in Fig. 6.2(b).
(2) Repeating this procedure gives a sequence of nested disks whose number doubles at each iteration step as each disk is cropped into two smaller ones that are contained in it as shown in Fig. 6.2(c) and Fig. 6.2(d) for two iteration steps.

In this case, the Smale–Williams solenoid has a fractal cross section similar to the Cantor set, generated as the process is continued *ad infinitum*. Note that the Smale–Williams solenoid was used in Chapter 7 to generate robust chaos in the sense of this book.

6.4.4 Plykin attractor

The Plykin attractor is an example of a hyperbolic attractor of a diffeomorphism on a two-dimensional sphere, which was built by Plykin in [Plykin *et al.* (1991)]. This attractor occurs in a special delicately organized map f in a domain V on a plane with three holes as shown in Fig. 6.3 and Fig. 6.4. In fact, the map f is a diffeomorphism of a two-dimensional torus projected onto a two-dimensional sphere having four repelling fixed points in its simplest case. In [Hunt (2000), Belykh *et al.* (2005)], some attempts to compose a three-dimensional set of ordinary differential equations with an attractor of Plykin type were given. The method of analysis is based on the study of the associated Poincaré map. Hence this construction can be implemented as a physical system. For some details see Sec. 7.1.6.

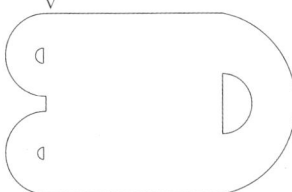

Fig. 6.3 The domain V on a plane with three holes for the definition of the Plykin attractor.

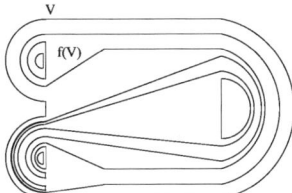

Fig. 6.4 Iteration of the Plykin attractor.

More details on the topic of the Plykin attractors and their applications can be found in [Sinai (1991), Pollicott (1993), Anosov *et al.* (1995), Cook *et al.* (1995), Palis & Takens (1995), Katok & Hasselblatt (1996), Berger (2001), Nikolaev (2001), Devaney & Devaney (2003), Bonatti *et al.* (2004), Walczak *et al.* (2005), Bertau *et al.* (2007), Forni *et al.* (2007), Hasselblatt (2007), Pinto *et al.* (2008)].

In the following section, we give a proof of the hyperbolicity of the logistic map for $\mu > 4$ to give the reader an idea about the nature of such a proof.

6.5 Proof of the hyperbolicity of the logistic map for $\mu > 4$

First, note that the logistic map was originally a demographical model and is now known as a population model of species including the effects of reproduction and starvation. The parameter μ represents a combined rate for reproduction and starvation. The logistic map is the subject of many works in topics such as dynamics, bifurcation, and chaos.

Using only elementary real analysis, with a good estimate of the expansion rate on the invariant set, Glendinning gives in [Glendinning (2001)] a new and elementary proof of the classic results due to Guckenheimer and

Misiurewicz [Guckenheimer (1979), Misiurewicz (1981)] which imply that the invariant set of the logistic map given by

$$F_\mu(x) = \mu x(1-x) \tag{6.25}$$

with $\mu \in (4, 2+\sqrt{5}]$ is hyperbolic, where they use relatively sophisticated ideas from complex variable theory. If $\mu \in (0, 4]$, then the interval $[0,1]$ is invariant under the map (6.25), and many interesting dynamical phenomena can be observed as is μ varied. If $\mu > 4$ then there is no invariant interval, but if we define

$$\Omega_n = \left\{ x \in [0,1] : F_\mu^r(x) \in [0,1] \right\}, r = 0, ..., n-1, \tag{6.26}$$

then the set defined as the intersection of all the sets Ω_n

$$\Omega = \bigcap_{n=1}^{\infty} \Omega_n \tag{6.27}$$

is a closed invariant set, and its dynamics is topologically conjugate to a one-sided shift on two symbols (see Sec. 1.4).

Since the map (6.25) is a one-dimensional system, then the following was proved in [Devaney (1989)]:

Lemma 6.1. *An unstable set of (6.25) is hyperbolic if there exists a $\lambda > 1$ (the expansion rate) such that for all $x \in \Omega$ and all $n \in \mathbb{Z}^+$*

$$\left| DF_\mu^n(x) \right| \geq c\lambda^n, \tag{6.28}$$

where DF_μ^n denotes the derivative of F^n (the nth iterate of F) and c is a constant.

To proves that the map (6.25) is hyperbolic for $\mu > 2+\sqrt{5}$, it suffices to conclude from direct calculation that $\left| DF_\mu^n(x) \right| > 1$ for all $x \in \Omega_2$ if $\mu > 2+\sqrt{5}$.

The set Ω is hyperbolic for $4 < \mu \leq 2+\sqrt{5}$. This statement was proved by several authors [van Strien (1981), de Melo & van Strien (1993), Robinson (1995), Katok & Hasselblatt (1997)]. Robinson in [Robinson (1995)]) uses the techniques developed by Guckenheimer [Guckenheimer (1979)] and Misiurewicz [Misiurewicz (1981)]. The proof of Robinson is shorter and uses some ideas from complex variable theory.

The new result of Glendinning [Glendinning (2001)] is given by the following:

Theorem 6.11. *If $\mu > 4$, then the set Λ is hyperbolic.*

The proof of Theorem 6.11 uses some ideas (namely, the standard topological conjugacy between the logistic map with $\mu = 4$ and the tent map with slope two) employed in [de Melo & van Strien (1993), Luzzatto (2000)] to study expansion in the logistic map (6.25) with $\mu < 4$. Indeed, Glendinning shows that this conjugacy for a related quadratic map can be used to conjugate the quadratic map to a map which has a slope with modulus greater than one for an appropriate range of μ values. This technique permits one to obtain estimates of the derivative of the original map, a good estimate of the expansion rate λ, and hence to prove hyperbolicity.

The proof can be done in five steps as follows:

Step 1: Determination of the one-dimensional quadratic map which is equivalent to the logistic map (6.25).

Step 2: Introduction of a change of variables in order study families of rescaled versions of the conjugating function of the logistic map.

Step 3: Estimation bound of the first term in a resulting formula from the change of variables introduced in Step 2.

Step 4: Estimation bound of the second term in a resulting formula from the change of variables introduced in Step 2.

Step 5: Achievement of the proof by substituting the bounds obtained in Steps 3 and 4.

Step 1. Quadratic map

It is well known that the logistic map (6.25) is topologically conjugate to the quadratic map family given by

$$f_r(x) = 1 - rx^2, r > 2 \qquad (6.29)$$

for all $r > 0$ with μ and r related by the formula

$$r = \frac{1}{4}\mu(\mu - 2). \qquad (6.30)$$

In particular, $\mu = 4$ gives $r = 2$, and $\mu > 4$ corresponds to $r > 2$.

Thus the map (6.29) is used instead of the standard logistic map (6.25). Hence the map f_r has the following properties:

(1) The most interesting dynamics of (f_r) lies in an interval $I_r = [-a, a]$ where $x = -a$ is the fixed point of f_r with

$$a = \frac{1 + \sqrt{1 + 4r}}{2r}. \qquad (6.31)$$

(2) If $r < 2$, then one has $I_r \subset [-1, 1]$.

(3) The conjugating function p, satisfying $F_\mu = p^{-1} \circ f_r \circ p$ is affine of the form $p(x) = Ax + B$, where the precise values of A and B are not important.

(4) The conjugating function, p maps the interval $[0,1]$ onto I_r, and maps Ω and Ω_n onto the corresponding sets for f_r denoted Λ and Λ_n.

Hence for the map (6.29), the following result was obtained:

Theorem 6.12. *If $r > 2$, then for all $x \in \Lambda$ and $n \in \mathbb{Z}^+$, one has*

$$|Df_r^n(x)| \geq C_r \lambda_r^n \qquad (6.32)$$

where $r = \sqrt{2r}$ and $C_r > 0$ with

$$C_r = \frac{2r^2 - 2r - 1 - \sqrt{1+4r}}{2r^2 - 2r + 1 + \sqrt{1+4r}}. \qquad (6.33)$$

Note that as $r \downarrow 2$, $C_r \downarrow 0$, and as $r \to \infty$, $C_r \uparrow 1$.

Step 2. A change of variable

It is easy to prove that if $r = 2$ then the change of variable $h : [-1,1] \to [-1,1]$ defined by

$$h(\theta) = \cos \frac{\pi(\theta-1)}{2}, -1 \leq \theta \leq 1 \qquad (6.34)$$

conjugates the quadratic map f_2 to the tent map $g_2 : [-1,1] \to [-1,1]$

$$g_2(\theta) = \begin{cases} 1 + 2\theta & \text{if } \theta \in [-1,0] \\ 1 - 2\theta & \text{if } \theta \in [0,1], \end{cases} \qquad (6.35)$$

i.e.,

$$g_2(\theta) = h^{-1} \circ f_2 \circ h(\theta). \qquad (6.36)$$

The proof of Theorem 6.12 uses the function h defined by (6.34) to conjugate f_r ($r > 2$) given by (6.29) to a new map g_r given by (6.37) which has $|Dg_r(\theta)| > 1$ for all $\theta \in h^{-1}(\Lambda)$.

Thus for $r > 2$, define a new map g_r by

$$g_r(\theta) = h^{-1} \circ f_r \circ h(\theta). \qquad (6.37)$$

The most important feature of the map g_r is when $\theta \in h^{-1}(\Lambda)$ because the invariant set Λ of f_r is contained in the set $\Lambda_2 = I_r/\Delta$ where $\Delta = (-b, b)$ with

$$b^2 = \frac{2r - 1 - \sqrt{1+4r}}{2r^2} \qquad (6.38)$$

where the value of b is found by computing the set of points near $x = 0$ for which $f_r(x) = a$. Differentiating (6.37) and rearranging a little gives

$$|Df_r^n(x)| = \left|\frac{Dh(g_r^n(y))}{Dh(y)}\right| \cdot |Dg_r^n(y)|, \; y = h^{-1}(x). \qquad (6.39)$$

Note that Eq. (6.39) explains implicitly the choice of the form of the function f_r instead of the logistic map F_μ. This is a result of the fact that if $r > 2$, then the invariant set of f_r is contained inside the interval $[-1, 1]$ on which the conjugacy (6.34) was defined for f_2, and h is a diffeomorphism on any closed interval contained in $(-1, 1)$.

Step 3. Bounding the first term of (6.39) for $y \in h^{-1}(\Lambda)$

Using the result that if $r > 2$ then the invariant set of f_r is contained in $\Lambda_2 = [-a, b] \cup [b, a]$, and one has the following inequality for all $y \in h^{-1}(\Lambda_2)$:

$$\left|\frac{Dh(g_r^n(y))}{Dh(y)}\right| \geq C_r \qquad (6.40)$$

where

$$C_r = \min_{0, 0' \in h^{-1}(\Lambda_2)} \left|\frac{Dh(\theta)}{Dh(\theta')}\right| = \min_{\theta, \theta' \in h^{-1}(\Lambda_2)} \left|\frac{\sin\frac{\pi}{2}(\theta - 1)}{\sin\frac{\pi}{2}(\theta' - 1)}\right|. \qquad (6.41)$$

Now, since $\Lambda_2 = [-a, b] \cup [b, a]$ with $0 < b < a < 1$, $h^{-1}(\Lambda_2)$ is also a union of two intervals, $[-\alpha, -\beta] \cup [\beta, \alpha]$ with $0 < \beta < \alpha < 1$, $\cos\frac{\pi}{2}(\alpha - 1) = a$ and $\cos\frac{\pi}{2}(\beta - 1) = b$.

Hence

$$C_r = \left|\frac{\sin\frac{\pi}{2}(\alpha - 1)}{\sin\frac{\pi}{2}(\beta - 1)}\right|. \qquad (6.42)$$

Using relations (6.31) and (6.38), one has

$$C_r = \frac{1 - a^2}{1 - b^2} = \frac{2r^2 - 2r - 1 - \sqrt{1 + 4r}}{2r^2 - 2r + 1 + \sqrt{1 + 4r}}. \qquad (6.43)$$

Step 4. Bounding the second term of (6.39) for $y \in h^{-1}(\Lambda)$.

The map g_r can be written explicitly as

$$g_r(\theta) = 1 + \frac{2}{\pi}\cos^{-1}\left(1 - r\cos^2\frac{1}{2}\pi(\theta - 1)\right). \qquad (6.44)$$

This is a result of (6.34) and (6.36). By the chain rule, one has

$$Dg_r(\theta) = \sqrt{2r}\frac{\sin\frac{1}{2}\pi(\theta - 1)}{\left(1 - \frac{1}{2}r\cos^2\frac{1}{2}\pi(\theta - 1)\right)^{\frac{1}{2}}}. \qquad (6.45)$$

Set $\omega = \left|\cos^2 \frac{1}{2}\pi(\theta-1)\right| = \left|h(\theta)\right|^2$, so if $\theta \in h^{-1}(\Lambda)$, then $b^2 \leq \omega \leq a^2$ and

$$|Dg_r(\theta)| = \sqrt{2r}\left(\frac{1-\omega}{1-\frac{1}{2}r\omega}\right)^{\frac{1}{2}}. \qquad (6.46)$$

If $r > 2$, then provided $1 - \frac{1}{2}r\omega > 0$, this implies that

$$|Dg_r(\theta)| \geq \sqrt{2r}. \qquad (6.47)$$

For ω in the interval $[b^2, a^2]$, $1 - \frac{1}{2}r\omega \geq 1 - \frac{1}{2}ra^2$. The derivative of the expression $1 - \frac{1}{2}ra^2$ with respect to r shows that if $r > 2$, then $1 - \frac{1}{2}ra^2 > 0$. Hence $1 - \frac{1}{2}r\omega > 0$ if $r > 2$, and so (6.47) does indeed hold for all $r > 2$.

Step 5. Completing the proof. The proof of Theorem 6.12 is completed if the bounds obtained in Steps 3 and 4 were substituted in the previous two sections into (6.39) to obtain

$$|Df_r^n(x)| \geq C_r |2r|^{\frac{n}{2}} \text{ for all } n > 0 \text{ and } x \in \Lambda. \qquad (6.48)$$

Taking the minimization of the right-hand side of (6.46) for $r > 2$, then it is possible to obtain the slightly better estimate given by

$$|Df_r^n(x)| \geq C_r \lambda_r^n \qquad (6.49)$$

with

$$\lambda_r = 2\left(\frac{2r^2 - 2r + 1 + \sqrt{1+4r}}{2r + 1 + \sqrt{1+4r}}\right)^{\frac{1}{2}}. \qquad (6.50)$$

Hence the proof is completed.

6.6 Generalized hyperbolic attractors

In this section, we discuss some statistical properties of a class of hyperbolic attractors introduced in [Pesin (1992)] called *generalized hyperbolic attractors*. This class include the 2-D hyperbolic attractors of Belykh and generalized Lozi introduced in Sec. 8.5 and the Lorenz attractor given by (8.1). The Smale spectral decomposition was also established and yields countably many components, each ergodic with respect to any Gibbs u-measure called SRB-measures and introduced in Sec. 3.1.2. This decomposition is characterized by the fact that every ergodic component is decomposed into finitely many subsets which are cyclically permuted, and on each of those subsets the corresponding iteration of the map is of the mixing and Bernoulli type. In [Sataev (1992)], it was shown that the number of ergodic components is finite and the Belykh, generalized Lozi, and Lorenz attractors satisfy the assumption required by Pesin in [Pesin (1992)] and Sataev in [Sataev (1992)]. Hence these attractors are ergodic-even-mixing for some value of their parameters as in [Bunimovich & Sinai (1979)] for the Lorenz attractors, and in [Misiurewicz (1983)] for the Lozi map (8.88).

The class of these 2-D generalized hyperbolic attractors can be defined as follows:

Let M be a smooth two-dimensional manifold equipped by a Riemannian metric ρ and Riemannian volume $vol(.)$. Let $U \subset M$ be an open connected subset with compact closure, and $\Gamma \subset U$, a closed subset. Assume that the set $S^+ = \Gamma \cup \partial U$ consists of a finite number of compact smooth curves. The set $U \setminus \Gamma$ consists of a finite number of open connected components. Let $f : U \setminus \Gamma \to f(U \setminus \Gamma) \subset U$ be a C^2-diffeomorphism. Assume that f is twice differentiable up to the boundary $\partial(U \setminus \Gamma) = S^+$ which is the singularity set for the map f. Hence the boundary $\partial(f(U \setminus \Gamma)) = S^-$ (which is the set of singularities for f^{-1}) is then a finite union of compact smooth curves. In this case, the inverse map f^{-1} is twice differentiable up to S^-, and the first and the second partial derivatives of both f and f^{-1} are uniformly bounded because $cl(U)$ is a compact set in M_1. Let S_m^+ be the union of the curves on which the map f^m is singular. Let Γ_ε be the ε-neighborhood of the set Γ and ν the Lebesgue measure on M. Hence a smooth curve γ in U is unstable (stable) if its tangent line belongs in the cone $C^u(z)$ (resp. $C^s(z)$) defined by (6.53) below at any $z \in \gamma$.

Let

$$\begin{cases} U^+ = \{x \in U : f^n(x) \notin S^+, n = 0, 1, 2, ...\} \\ D = \cap_{n \geq 0} f^n(U^+). \end{cases} \quad (6.51)$$

In this case, the set D is invariant under both f and f^{-1} and its $\Lambda = cl(D)$ is called the attractor for f for systems with $Vol(\Lambda) = 0$ [21]. This last property can be assured if $cl(f(U \setminus \Gamma)) \subset U$ as shown in [Sataev (1992)].

To define a hyperbolic structure for the map f, let $z \in U$ be any point and P any line lying in the tangent plane T_zM and any real number $\alpha > 0$. Let $C(z, \alpha, P)$ be the *cone set* defined by

$$C(z, \alpha, P) = \{v \in T_zM : \angle(v, P) \leq \alpha\}. \quad (6.52)$$

Assume that for each point $z \in U \setminus S^+$ there are two cones

$$\begin{cases} C^u(z) = C(z, \alpha^u(z), P^u(z)) \\ C^s(z) = C(z, \alpha^s(z), P^s(z)) \end{cases} \quad (6.53)$$

---

[21] This case includes also piecewise-linear toral automorphism attractors, *i.e.*, with $Vol(\Lambda) > 0$.

having the following three properties:

(1) The angle between $C^u(z)$ and $C^s(z)$ is uniformly bounded away from zero.
(2) The inclusions are
$$\begin{cases} df\left(C^u(z)\right) \subset C^u(fz) & \text{for any } z \in (U\backslash S^+) \\ df^{-1}\left(C^s(z)\right) \subset C^s\left(f^{-1}z\right) & \text{for any } z \in f(U\backslash S^+). \end{cases} \quad (6.54)$$
(3) There exist constants $C > 0$ and $\lambda \in (0,1)$ such that for any integer $n > 0$,
    (a) if $z \in U^+$ and if $v \in C^u(z)$, then
$$\|df^n v\| \geq C\lambda^{-n} \|v\|. \quad (6.55)$$
    (b) if $z \in f^n(U^+)$ and if $v \in C^s(z)$, then
$$\|df^{-n} v\| \geq C\lambda^{-n} \|v\|. \quad (6.56)$$

The class of mappings satisfying the above properties (1)–(3) is not empty because the Lozi and the Belykh attractors are examples of such a situation. Furthermore, it is possible to assume that $C = 1$ because it was shown in [Sataev (1992)] that there exists an integer $m \geq 1$ so that f^m enjoys the properties (1)–(3). Thus the following definition is given in [Pesin (1992)]:

Definition 6.12. An attractor Λ is called a *generalized hyperbolic attractor* if the two families of cones given by (6.53) exist.

Now if $z \in D$, then properties (1)–(3) yield families of invariant subspaces E_z^u and E_z^s in T_zM with the following two properties:
$$\begin{cases} E_z^u \subset C^u(z) \text{ and } E_z^s \subset C^s(z) \\ df E_z^{u,s} = E_{f(z)}^{u,s}. \end{cases} \quad (6.57)$$

To guarantee a uniform hyperbolic structure for the map f, it suffices to assume the following [Pesin (1992)]:

(4) The two cones $C^u(z)$ and $C^s(z)$ depend continuously on $z \in U^+$.

In this case, for any $z \in \Gamma$, the two limit cones $C^{u,s}(z) = \lim_{z' \to z} C^{u,s}(z')$ exist on both sides of Γ, and the angle between the tangent line Γ at z and the unstable limit cone $C^u(z)$ is uniformly bounded away from zero.

To guarantee that expansion and contraction prevail over discontinuities (that the singularities of f and f^{-1} are *mild* and they are concentrated on a finite union of smooth compact curves, and the first and second derivatives of f and f^{-1} have one-sided limits) it suffices to assume the following conditions given in [Afraimovich et al. (1995)]:

Condition A1: There exists an integer $\tau \geq 1$ such that $f^{-k}(\Gamma) \cap \Gamma = \emptyset$ for $k = 1, 2, ..., \tau$ and $\lambda^{-\tau} > 2$, where $\lambda^{-1} > 1$ is, as before, the minimal factor of expansion of vectors in the unstable cones $C^u(z)$ at all points $z \in U \setminus S^+$. Moreover, there is a neighborhood of the attractor Λ in which the smooth components of Γ do not intersect one another.

Condition A2: There exist constants $C_0 > 0$ and $K_0 < \lambda^{-1}$ such that for any integer $m \geq 1$, no more than $C_0 K_0^m$ smooth components of the union $\cup_{l=0}^{m} = S_l^+$ can meet at any point $z \in U$.

Condition A3: There exist constants $B > 0, \beta > 0$, and $\varepsilon_0 > 0$ such that for any integer $n \geq 1$ and any $\varepsilon \in (0, \varepsilon_0)$, one has $\nu(f^{-n}\Gamma_\varepsilon) < B\varepsilon^\beta$.

Condition A4: There is a constant $\varepsilon_0 > 0$ such that for any unstable curve W^u, there exists an integer $n_0 = n_0(W^u)$ and a constant $B_0 = B_0(W^u)$ such that for any $\varepsilon \in (0, \varepsilon_0)$ one has (a) $\nu^u(W^u \cap f^{-n}\Gamma_\varepsilon) < \varepsilon^\beta \nu^u(W^u)$ for all integers $n > n_0$, and (b) $\nu^u(W^u \cap f^{-n}\Gamma_\varepsilon) < B_0 \varepsilon^\beta \nu^u(W^u)$ for all integers $n \geq 1$. Some important remarks about these conditions are the following:

(a) Some (but not all) Lorenz, Lozi, and Belykh attractors satisfy Condition A1, but in some cases a weaker assumption than Condition A1 is sufficient [Pesin (1992), Afraimovich et al. (1995)].
(b) Condition A1 implies Condition A2 with $K_0 = 2^{\frac{1}{\tau}}$. When $K_0 = 1$, Condition A2 in a more stringent form was used in [Bunimovich et al. (1990), Chernov et al. (1993)]. In this case, Condition A2 holds for a single, sufficiently large value of $m \geq 1$. Two more conditions were adduced in [Sataev (1992)].
(c) Condition A3 and a weaker version of Condition A4 were also assumed in [Pesin (1992)]. Furthermore, it is very hard to check Conditions A3 and A4 for particular examples. In fact, Condition A2 implies Conditions A3 and A4 [Afraimovich et al. (1995)] as shown in Proposition 6.6 below.

The following proposition was proved in [Afraimovich et al. (1995)]:

Proposition 6.5. *If a generalized hyperbolic attractor Λ satisfies Condition A2, then it satisfies Conditions A3 and A4.*

Now if $\varepsilon > 0$ and $l = 1, 2, ...$, then it is possible to define the following subsets [Pesin (1992)]:

$$\begin{cases} \hat{D}^+_{\varepsilon,l} = \left\{ z \in U^+ : \rho\left(f^n\left(z\right), S^+\right) \geq l^{-1}e^{-\varepsilon n}, n = 0, 1, ... \right\} \\ D^-_{\varepsilon,l} = \left\{ z \in D : \rho\left(f^{-n}\left(z\right), S^-\right) \geq l^{-1}e^{-\varepsilon n}, n = 0, 1, ... \right\} \\ D^+_{\varepsilon,l} = \hat{D}^+_{\varepsilon,l} \cap \Lambda \\ D^0_{\varepsilon,l} = D^+_{\varepsilon,l} \cap D^-_{\varepsilon,l} \\ D^{\pm}_{\varepsilon} = \cup_{l \geq 1} D^{\pm}_{\varepsilon,l} \\ D^0_{\varepsilon} = \cup_{l \geq 1} D^0_{\varepsilon,l}. \end{cases} \quad (6.58)$$

The subset $D^+_{\varepsilon,l} \left(D^-_{\varepsilon,l} \right)$ consists of points that do not approach the singularity set too rapidly in the future (respectively, in the past). In this case, the sets $\hat{D}^+_{\varepsilon,l}, D^{\pm}_{\varepsilon,l}$, and D^0_{ε} are closed, $D^0_{\varepsilon} = D^+_{\varepsilon} \cap D^-_{\varepsilon}$, the set D^+_{ε} is f-invariant, D^-_{ε} is f^{-1}-invariant, and D^0_{ε} is both f and f^{-1} invariant with $D^0_{\varepsilon} \subset D$ for any $\varepsilon > 0$. See Exercise 158. In fact, it was proved in [Pesin (1992)] that the attractor Λ is regular[22] under weaker assumptions than Condition A3 and Condition A4, so under Condition A2, the attractor Λ is regular.

Proposition 6.6. *There exists an $\varepsilon > 0$ such that for any point $z \in D^+_{\varepsilon,l} \left(z \in D^-_{\varepsilon,l} \right)$, there is a local stable fiber, LSF, denoted by $V^s(z)$ (resp. a local unstable fiber, LUF, denoted by $V^u(z)$). An LSF (LUF)[23] is a C^1-curve in M. It is tangent to the line E^s_z (resp. to E^u_z) at z. The ρ-distance of the point z from the endpoints of that fiber is at least $\delta_l = \frac{1}{l}$, a quantity determined by l and independent of z.*

In this case, the maximal smooth local stable and unstable fiber passing through z were denoted by $V^{s,u}(z)$, and one has $V^u(z) \subset D^-_{\varepsilon}$ for any $z \in D^-_{\varepsilon}$. In this case, the Gibbs u-measures on Λ are defined as follows: Let $J^u(z)$ denote a one-step expansion factor in E^u_z, i.e., the Jacobian of the map $df|_{E^u_z}$ at z. Then for any $z \in D^-_{s,l}$ such that $V^u(z)$ exists, define for all $y \in V^u(z)$

$$k(z, y) = \lim_{\pi \to \infty} \Pi^n_{j=1} \left[J^u \left(f^{-j}(z) \right) \right] \cdot \left[J^u \left(f^{-j}(y) \right) \right]^{-1}. \quad (6.59)$$

Limit (6.59) exists, is positive, continuous on $D^-_{s,l}$ and uniformly bounded away from zero and infinity on D^-_{ε} because f is smooth up to S^+.

---

[22]The attractor Λ is said to be regular if $D^0_{\varepsilon} \neq \emptyset$ for all sufficiently small $\varepsilon > 0$ [Pesin (1992)].

[23]In some cases, such as the linear toral automorphism, the LUF's and LSF's have infinite length, and hence the LUF's and LSF's were redefined for a large $L > 0$ and denote by $V^{s,u}(x)$ a *segment* of the LUF (LSF) at the point x that has length $L > 0$ and is centered at x.

Definition 6.13. A measure μ on Λ is called a Gibbs u-measure or a Sinai–Ruelle–Bowen measure (SRB measure) if (a) it is f-invariant; (b) $\mu\left(D_\varepsilon^0\right) = \mu\left(\Lambda\right) = 1$ for some $\varepsilon > 0$; (c) the conditional measure on the LUF's $V^u\left(z\right)$ induced by μ has a density with respect to the Lebesgue measure on $V^u\left(z\right)$ proportional to $k\left(z, y\right)$.

Hence the Gibbs u-measures are constructed as follows:

(a) Consider a point $z \in D_\varepsilon^-$, and take the normalized Lebesgue measure ν^u on $V^u\left(z\right)$.
(b) Pull the normalized Lebesgue measure forward under f_*: $\nu_k = f_*^k \nu^u$, i.e., for any Borel set $A \subset U$, one takes $\nu_k\left(A\right) = \nu^u \left(f^{-k} A \cap V^u\left(z\right)\right)$.
(c) The sequence of measures

$$\mu'_n = \frac{1}{n} \sum_{k=0}^{n-1} \nu_k \qquad (6.60)$$

has a limit point (a measure) in the weak topology and defines a Gibbs u-measure[24] for any $z \in D_\varepsilon^-$.

If $Vol(\Lambda) = 0$, then any Gibbs u-measure is singular with respect to the Lebesgue measure on U and with respect to the Lebesgue measure on any LSF. Thus a Gibbs measure μ has no atoms, and any particular LUF or LSF has μ measure zero. In fact, it was shown in [Sataev (1992)] that there is always one satisfying the following three additional properties:

(d) For every $l > 0$, the sets $\Lambda_i^j \cap D^-_{\varepsilon, l}$ are closed.
(e) For every $l > 0$, every $i = 1, ..., r$, and every open subset $Q \subset U$ such that $Q \cap \Lambda_i^j \cap D^-_{\varepsilon, l} \neq \emptyset$, we have $\mu_i \left(Q \cap \Lambda_i^j \cap D^-_{\varepsilon, l}\right) > 0$.
(f) If $z \in \Lambda_i^j$, then $V^u(z) \subset \Lambda_i^j$.

The following proposition called *Smale spectral decomposition*, was proved in [Pesin (1992)] for generalized hyperbolic attractors:

Proposition 6.7. *There are subsets Λ_i ($i = 0, 1, ..$) and Gibbs u-measures μ_i ($i \geq 1$) such that (a) $\Lambda = \cup_{i \geq 0} \Lambda_i$ and $\Lambda_i \cap \Lambda_j = \emptyset$ for $i \neq j$; (b) for $i \geq 1$: $\Lambda_i \subset D, f\left(\Lambda_i\right) = \Lambda_i, \mu\left(\Lambda_i\right) = 1$ and $f|_{\Lambda_i}$ is ergodic with respect to μ_i; (c) for $i \geq 1$: there exists a finite decomposition $\Lambda_i = \cup_{j=1}^{r_i} \Lambda_i^j$, where $\Lambda_i^j \cap \Lambda_i^{j'} = \emptyset$ for $j \neq j'$, $f\left(\Lambda_i^j\right) = \Lambda_i^{j+1}$ and $f\left(\Lambda_i^{r_i}\right) = \Lambda_i^1$, and $f^{r_i}|_{\Lambda_i^1}$ is*

---
[24]One can take any measure equivalent to the Lebesgue measure on LUF. Thus a Gibbs u-measure is not unique.

a Bernoulli automorphism; (d) any Gibbs u-measure μ is a weighted sum $\mu = \sum_{i \geq 1} \alpha_i \mu_i$ with some $\alpha_i \geq 0$ and $\sum \alpha_i = 1$. In particular, $\mu(\Lambda_0) = 0$.

As a continuation of the work in [Pesin (1992)], Afraimovich et al. give a more detailed study of this class in [Afraimovich et al. (1995)] where some statistical properties of 2-D generalized hyperbolic attractors were studied, in particular, the stretched exponential bound on the decay of correlations and the central limit theorem. The method of analysis is based on the Markov approximation[25] to hyperbolic dynamical systems developed for hyperbolic billiards and similar models[26] in [Bunimovich et al. (1990), Chernov (1994)]. Now let H denote the class of Hölder continuous (HC) functions on the attractor, i.e.,

Definition 6.14. A function $F(x)$ is said to be Hölder continuous if

$$|F(x) - F(y)| \leq C(F) [\rho(x,y)]^\beta \qquad (6.61)$$

where $\beta > 0$ is called the *Hölder exponent*. More generally, let ξ be a partition of U into a finite number of domains separated by a finite number of compact smooth curves. Then let $H_\beta(\xi)$ be the class of functions that are Hölder continuous (with the exponent β) within each of those domains. We say that such functions are piecewise Hölder continuous (PHC) (with respect to the given partition ξ).

To estimate the decay of correlations (see Sec. 3.3.2), it suffices to take an arbitrary subcomponent $\Lambda_* = \Lambda_i^j$ of any ergodic component Λ_i of the attractor and define $f_* = f^{r_i}|_{\Lambda_*}$, μ_* as the normalized measure $\mu_i|_{\Lambda_*}$, $r_* = r_i$ and denote by $\langle . \rangle$ the expectation with respect to μ_*. Hence Proposition 6.7 implies that the triple (Λ_*, f_*, μ_*) is a Bernoulli dynamical system. In particular, it is mixing. Thus the following results about decay of correlations, the central limit theorem, and relaxation to an equilibrium distribution for generalized hyperbolic attractors were proved in [Afraimovich et al. (1995)] using a Markov approximation to the dynamical system (Λ_*, f_*, μ_*) for Theorem 6.6(a) and Theorem 6.6(b) below, in particular, the so-called *Markov sieves* developed in [Bunimovich et al. (1990-1991), Chernov (1992-1994), Chernov et al. (1993)].

---

[25] Note that the techniques of Markov approximations can work in a multidimensional as shown in the work of Chernov [Chernov (1994)] and in a different form, the Markov approximation techniques were used to establish good statistical properties for the Lorenz attractor as in [Bunimovich (1983)].

[26] For example, the Lorentz gas with an external field studied in [Chernov et al. (1993)] which is a hyperbolic attractor of special kind.

Theorem 6.13. *(a) (Decay of correlations)* Let $F(x)$ and $G(x)$ be two HC or PHC functions on M. Then for any integer N

$$\left|\left\langle \left(F \circ f_*^N\right).G\right\rangle - \langle F\rangle \langle G\rangle\right| \leq c(F,G)\,\alpha^{\sqrt{|N|}} \tag{6.62}$$

where $c(F,G) > 0$ depends on F and G and $\alpha < 1$ is determined by the subcomponent $\Lambda_* = \Lambda_i^j$ and the class of HC or PHC functions under consideration. *(b) (Central limit theorem)* Again, let $F(x)$ be an HC or a PHC function. Assume that $\langle F\rangle = 0$. Then the quantity

$$\sigma_F^2 = \sum_{n=-\infty}^{\infty} \left\langle \left(F \circ f_*^N\right).F\right\rangle \tag{6.63}$$

is finite and nonnegative. If $\sigma_F \neq 0$, then the sequence

$$\frac{F(x) + F(f_*x) + \ldots + F\left(f_*^{N-1}x\right)}{\sqrt{\sigma_F^2 N}} \tag{6.64}$$

converges in distribution to the standard normal law as $N \to \infty$. *(c) (Relaxation to equilibrium distribution)* For any integers $k \geq l > 1$ and $1 \leq i_1 < i_2 < \ldots < i_k \leq N$, there is a subset $R_* = R_*(i_1, \ldots, i_k) - \mathfrak{L}^{k-l+1}$ of $(k-l+1)$-tuples of indices such that (i) if $(j_l, \ldots, j_k) \in R_*$, then

$$\begin{cases} \displaystyle\sum_{j_1,\ldots,j_{l-1}=0}^{I} |\mu_*(A_1) - \mu_*(A_2)| \leq \Delta. \\ A_1 = f_*^{i_1}A_{j_1} \cap \ldots \cap f_*^{i_{l-1}}A_{j_{l-1}}/f_*^{i_l}A_{j_l} \cap \ldots \cap f_*^{i_k}A_{j_k} \\ A_2 = f_*^{i_1}A_{j_1} \cap \ldots \cap f_*^{i_{l-1}}A_{j_{l-1}}, \end{cases} \tag{6.65}$$

(ii) and one has

$$\sum_{(j_l,\ldots,j_k) \in R_*}^{I} \left|\mu_*\left(f_*^{i_l}A_{j_l} \cap \ldots \cap f_*^{i_k}A_{j_k}\right)\right| \geq 1 - \Delta \tag{6.66}$$

where

$$\Delta = \max\left\{c_4\alpha_4^n, \left(1 - \frac{g_1}{2}\right)^{\left[\frac{L}{2}\right]}\right\} \text{ with } L = \left[\frac{i_l - i_{l-1}}{g_0 n}\right]. \tag{6.67}$$

Finally, note that Theorem 6.6(c) is still true for *reverse time*, i.e., if $1 \geq i_1 > i_2 > \ldots > i_k \geq N$, the conditional distributions relax to equilibrium exponentially fast in the parameter $|i_l - i_{l-1}|$, which represents the *interval* between the *future* and the *past*, at least as long as that interval is less than $const.n^2$.

6.7 Generating hyperbolic attractors

There are several ways to prove hyperbolicity of a dynamical system. One method is to prove the existence of a homeomorphism between two systems, one of which is hyperbolic. An example is given in [Tresser (1983)] in the context of discussing some theorems of L. P. Shilnikov on $C^{1,1}$ where it was proved that there exists a map h_m which is a homeomorphism of $\Sigma^m = \{1,...,m\}^{\mathbb{Z}}$ onto $O_m = h_m(\Sigma^m)$ and the set O_m is hyperbolic. The construction of the map h_m is based on a very complicated geometrical method.

Theorem 6.14. *Consider the system*

$$\begin{cases} x' = \rho x - \omega y + P(x,y,z) \\ y' = \omega x + \rho y + Q(x,y,z) \\ z' = \lambda z + R(x,y,z) \end{cases} \quad (6.68)$$

where P, Q, R are $C^{1,1}$ functions which vanish as well their first derivatives at O. Suppose that there exists a homoclinic orbit Γ_0, biasymptotic to O, which remains a finite distance from any other singularity. Suppose at least that

$$\lambda > -\rho > 0. \quad (6.69)$$

Then if θ is a first return map correctly defined on a well-chosen piece of surface π_0, one gets the following conclusion: (a) For each positive integer m, there exists a map

$$h_m : \Sigma^m = \{1,...,m\}^{\mathbb{Z}} \to \pi_0 \quad (6.70)$$

which is a homeomorphism of Σ^m onto $O_m = h_m(\Sigma^m)$ such that

$$\theta|_{O_m} = h_m \circ \sigma \circ h_m^{-1}. \quad (6.71)$$

(b) The set O_m is hyperbolic. (c) For each real α with $1 \leq \alpha \leq \frac{\lambda}{\rho}$, there exists a map

$$h_{*,\alpha} : \Sigma^{*,\alpha} = \left\{ s = \{s_i\}_{i=-\infty}^{i=\infty} : s_i \in \mathbb{Z} - \{0\} : |s_{i+1}| \geq \frac{|s_i|}{\alpha} \right\} \to \pi_0 \quad (6.72)$$

which is a homeomorphism of $\Sigma^{*,\alpha}$ onto $O_{*,\alpha} = h_{*,\alpha}(\Sigma^{*,\alpha})$ such that

$$\theta|_{O_{*,\alpha}} = h_{*,\alpha} \circ \sigma \circ h_{*,\alpha}^{-1}. \quad (6.73)$$

It is known that hyperbolic attractors may be generated from Morse–Smale systems, for example through Ω-explosion, period doubling cascade, *etc.* But these bifurcations do not lead explicitly to the appearance of hyperbolic strange attractors. In this section, we describe only the transition when a periodic trajectory disappears from a Morse–Smale system to systems with hyperbolic attractors [Turaev & Shilnikov (1995)]. Indeed, consider a C^r-smooth, one-parameter family of dynamical systems X_μ in $\mathbb{R}^n$. Assume that X_μ has a periodic orbit L_0 of the saddle-node type at $\mu = 0$. Let U_0 be a neighborhood of L_0 which is a solid torus partitioned by the $(n-1)$-dimensional strongly stable manifold $W^{ss}_{L_0}$ into two regions, the node region U^+ where all trajectories tend to L_0 as $t \to +\infty$, and the saddle region U^- where the two-dimensional unstable manifold $W^u_{L_0}$ bounded by L_0 lies. Suppose that all the trajectories of $W^u_{L_0}$ return to L_0 from the node region U^+ as $t \to +\infty$ and do not lie in W^{ss}. In this case, $cl\left(W^u_{L_0}\right)$ is compact since any trajectory of W^u is bi-asymptotic to L_0. Observe that systems close to X_0 and having a simple saddle-node periodic trajectory close to L_0 form a surface B of codimension-one in the space of dynamical systems. We assume also that the family X_μ is transverse to B. Thus when $\mu < 0$, the orbit L_0 is split into two periodic orbits, namely, L^-_μ of the saddle type and a stable L^+_μ. When $\mu > 0$, L_0 disappears. It is clear that X_μ is a Morse–Smale system in a small neighborhood U of the set W^u for all small $\mu < 0$. The nonwandering set here consists of the two periodic orbits L^+_μ and L^-_μ. All trajectories of $U \backslash W^s_{L^-_\mu}$ tend to L_μ as $t \to +\infty$. At $\mu = 0$ all trajectories on U tend to L_0. The situation is more complicated when $\mu > 0$. In this case the Poincaré map is given by

$$\begin{cases} \bar{x} = f(x, \theta, \mu), \\ \bar{\theta} = m\theta + g(\theta) + \omega + h(x, \theta, \mu), (\text{mod } 1), \end{cases} \quad (6.74)$$

where f, g, and h are periodic functions of θ. Moreover, $\| f \|_{C_1} \to 0$ and $\| h \|_{C_1} \to 0$ as $\mu \to 0$, where m is an integer and $\omega \in [0, 1)$. Diffeomorphism (6.74) is defined in a solid-torus $\mathbb{D}^{n-2} \times \mathbb{S}^1$, where $\mathbb{D}^{n-2} = \{\|x\| < r\}, r > 0$. Then one has the following theorem for the so-called *blue sky catastrophe* proved in [Turaev & Shilnikov (1995)]:

Theorem 6.15. *If $m = 0$ and if $\max \|g'(\theta)\| < 1$, then for sufficiently small $\mu > 0$, the original flow has a periodic orbit whose length and period both tend to infinity as $\mu \to 0$.*

Proof. The proof can be done with the following remarks: (a) The map (6.74) is a strong contraction along x. (b) The map (6.74) is close to the degenerate map

$$\begin{cases} \bar{x} = 0, \\ \bar{\theta} = m\theta + g(\theta) + \omega \pmod{1}, \end{cases} \quad (6.75)$$

and its dynamics is determined by the circle map

$$\bar{\theta} = m\theta + g(\theta) + \omega \pmod{1} \quad (6.76)$$

where $0 \le \omega < 1$. If $n = 3$, then the integer m may assume the values $0, 1$.

Depending on m, the nature of the closure $cl\left(W_{L_0}^u\right)$ can be found as follows: If $m = 1$, then the closure $cl\left(W_{L_0}^u\right)$ is a two-dimensional torus, and it is smooth if (6.74) is a diffeomorphism. If $m = -1$, then $cl\left(W_{L_0}^u\right)$ is a *Klein bottle*, also smooth if (6.74) is a diffeomorphism. In Theorem 6.15, the set $cl\left(W_{L_0}^u\right)$ is not a manifold. In the case of $\mathbb{R}^n$ ($n \ge 4$), the constant m may be any integer. □

The following result was proved in [Turaev & Shilnikov (1995)]:

Theorem 6.16. *Let $|m| \ge 2$, and let $|m + g'(\theta)| > 1$. Then for all $\mu > 0$ sufficiently small, the Poincaré map (6.74) has a hyperbolic attractor homeomorphic to the Smale–Williams solenoid (see Sec. 6.4.3), while the original family has a hyperbolic attractor homeomorphic to a suspension over the Smale–Williams solenoid.*

Consider now a one-parameter family of smooth dynamical systems

$$x' = X(x, \mu) \quad (6.77)$$

which possesses an invariant m-dimensional torus $\mathbb{T}^m$ with a quasi-periodic trajectory at $\mu = 0$. Assume that the vector field has the form

$$\begin{cases} y' = C(\mu)y \\ z' = \mu + z^2 \\ \theta' = \Omega(\mu) \end{cases} \quad (6.78)$$

in a neighborhood of $\mathbb{T}^m$. Here, $(z, y, \theta) \in \left(\mathbb{R}^1, \mathbb{R}^{n-m-1}, \mathbb{T}^m\right)$ and $\Omega(0) = (\Omega_1, ..., \Omega_m)$. Assume that:

(1) The matrix $C(\mu)$ is stable[27].
(2) At $\mu = 0$, the equation of the torus is $y = 0$.
(3) The equation of the unstable manifold W^u is $y = 0, z > 0$.
(4) The strongly unstable manifold W^{ss} partitioning the neighborhood of $\mathbb{T}^m$ into a node and a saddle region, is $z = 0$.

---
[27]Its eigenvalues lie to the left of the imaginary axis in the complex plane.

(5) All the trajectories of the unstable manifold W^u of the torus come back to it as $t \to +\infty$. Hence they do not lie in W^{ss}.

Therefore, we arrive at the following statement proved in [Turaev & Shilnikov (1995)]:

Proposition 6.8. *If the shortened map is an Anosov map for all small ω, then for all $\mu > 0$ sufficiently small, the original flow possesses a hyperbolic attractor which is topologically conjugate to the suspension over the Anosov diffeomorphism.*

Proof. With the assumptions listed above, and on a cross-section transverse to $z = 0$, the associated Poincaré map may be written in the form

$$\begin{cases} \bar{y} = f(y, \theta, \mu) \\ \bar{\theta} = A\theta + g(\theta) + \omega + h(x, \theta, \mu), (\text{mod } 1), \end{cases} \quad (6.79)$$

where A is an integer matrix, and f, g, and h are 1-periodic functions of θ. Moreover, $\|f\|_{C_1} \to 0$ and $\|h\|_{C_1} \to 0$ as $\mu \to 0$, $\omega = (\omega_1, ..., \omega_m)$ where $0 \leq \omega_k < 1$. Thus it is easy to observe that the restriction of the Poincaré map on the invariant torus is close to the shortened map

$$\bar{\theta} = A\theta + g(\theta) + \omega, (\text{mod } 1). \quad (6.80)$$

This implies that if (6.80) is an Anosov map for all ω[28], then the restriction of the Poincaré map is also an Anosov map for all $\mu > 0$. □

Proposition 6.10 holds true if the shortened map is expanding. If $\|(G'(\theta))^{-1}\| < 1$, where $G = A + g(\theta)$, it follows then that the shortened map

$$\bar{\theta} = \omega + A\theta + g(\theta) \,(\text{mod } 1) \quad (6.81)$$

is an expansion for all $\mu > 0$. Using the result of [Williams (1970)], one comes to the following result proved in [Turaev & Shilnikov (1995)] which is analogous to Theorem 6.16:

Proposition 6.9. *If $\|(G'(\theta))^{-1}\| < 1$, then for all small $\mu > 0$, the Poincaré map possesses a hyperbolic attractor locally homeomorphic to a direct product of $\mathbb{R}^{m+1}$ and a Cantor set.*

---

[28] For example, when the eigenvalues of the matrix A do not lie on the unit circle of the complex plane and $g(\theta)$ is small.

Definition 6.15. An endomorphism of a torus is called an *Anosov covering* if there exists a continuous decomposition of the tangent space into the direct sum of stable and unstable submanifolds[29].

In this case, one has [Turaev & Shilnikov (1995)] the following:

Lemma 6.2. *The map (6.80) is an Anosov covering if $|\det \Lambda| > 1$ and if $g(\theta)$ is sufficiently small.*

Thus the following result which is similar to the previous proposition was also proved in [Turaev & Shilnikov (1995)]:

Proposition 6.10. *If the shortened map (6.80) is an Anosov covering for all ω, then for all small $\mu > 0$ the original Poincaré map possesses a hyperbolic attractor locally homeomorphic to a direct product of $\mathbb{R}^{m+1}$ and a Cantor set.*

For more details on generating hyperbolic attractors, see [Williams (1970), Tresser (1983), Turaev & Shilnikov (1995)] and references therein.

6.8 Density of hyperbolicity and homoclinic bifurcations in arbitrary dimension

In [Pujals (2006)], it was proved that given a maximal invariant attracting homoclinic class for a smooth three-dimensional Kupka-Smale diffeomorphism (recall Definition 6.4), either the diffeomorphism is C^1 approximated by another one exhibiting a homoclinic tangency or a heterodimensional cycle, or it follows that the homoclinic class is conjugate to a hyperbolic set, *i.e.*, the homoclinic class is *topologically hyperbolic*, in this case, a characterization of the dynamics of a topologically hyperbolic homoclinic class, in arbitrary dimension. In particular, the continuation of the homoclinic class for a perturbation of the initial system was described and a proof that under some topological conditions, the homoclinic class is contained in a two-dimensional manifold and it is hyperbolic. For attracting topologically hyperbolic sets of a three-dimensional Kupka-Smale diffeomorphism, the same property was also proved in [Pujals (2008)].

---

[29]This is just like the case of the Anosov map, but the Anosov covering is not a diffeomorphism because it is not a one-to-one map.

6.9 Hyperbolicity tests

In this section, we present some results about detecting hyperbolic and non-hyperbolic behaviors for chaotic systems. A common definition of chaotic saddles can be found in [Grebogi et al. (1983)] as follows:

Definition 6.16. Chaotic saddles (in the plane) are closed, bounded, nonattracting, chaotic invariant sets with a dense orbit such that each point of the set has a stable direction and an unstable direction.

First, detecting nonhyperbolic behavior is a negative criterion for the nonexistence of hyperbolic (robust) chaotic attractors, *i.e.*, proving the absence of robust chaos in such types of chaotic systems. Indeed, one extreme form of nonhyperbolicity[30] for chaotic attractors having periodic orbits with a different number of unstable directions is the unstable dimension variability (UDV) [Abraham & Smale (1970)].

The UDV is characterized by the following:

(1) The finite-time Lyapunov exponent nearest to zero fluctuates about zero. This is a result of the visits of the periodic trajectory to regions of the attractor with a varying number of stable and unstable directions [Dawson et al. (1994)].
(2) An analytical estimate for the pointwise dimension agrees with the scaling of the probability distribution of the attractor near each periodic point as shown in [Alligood et al. (2006)].
(3) There is no general method for verification of UDV, and the known methods in the literature are still restricted to a few dynamical systems in certain intervals of their parameters [Ott et al. (1992)].

The failure of the transversality conditions is a common source of nonhyperbolic behavior, *i.e.*, the existence of an infinite number of homoclinic and heteroclinic tangencies between unstable and stable subspaces or the unstable dimension variability [Abraham & Smale (1970), Grebogi et al. (1987-1988)]. In [Kubo et al. (2008)], a new mechanism for the onset of unstable dimension variability for invertible discrete-time maps and continuous-time flows[31] was proposed. The method of analysis is based on an interior crisis or a collision between a chaotic attractor and an unstable periodic orbit. This method requires that the dynamical systems under consideration have

---
[30] *i.e.*, there is no continuous splitting between stable and unstable manifolds along the chaotic invariant set.
[31] Using Poincaré sections.

an invariant subspace with a chaotic set to guarantee the existence of a tunable bifurcation parameter. The two mechanisms known in the literature for this onset of UDV are the *bubbling transition* or *riddling bifurcation* at the parameter value where a low-period orbit embedded in the chaotic attractor loses transversal stability as shown in [Lai et al. (1996), Viana et al. (2004)]. The second mechanism requires an unstable periodic orbit with the property that its period goes to infinity as one approaches the onset of UDV [Pereira et al. (2007)]. The third mechanism was discussed in [Kubo et al. (2008)], and it is qualitatively different from the previous ones because it does not require the existence of an invariant subspace or a collision between a chaotic attractor and an unstable periodic orbit. An example of such a situation induced UDV is the mechanical system called a *kicked double rotor* studied in [Ott et al. (1992)].

The presence of nonhyperbolic parameter values in a dynamical system implies that nonhyperbolic chaotic saddles are common in chaotic systems and the so-called *Newhouse intervals* can be quite large in the parameter space. These results can be seen in the work of [Lai et al. (1993)] where numerical investigation of the fraction of nonhyperbolic parameter values in chaotic dynamical systems with tangencies between some stable and unstable manifolds. In particular, the fraction of nonhyperbolic parameter values was computed for the Hénon map (10.1) in the parameter range where there exist only chaotic saddles, *i.e.*, nonattracting invariant chaotic sets. The importance of these nonhyperbolic parameter values is that near each one, there is another parameter value at which there are infinitely many coexisting attractors, *i.e.*, the dynamics are *pathological*. Newhouse and Robinson proved in [Newhouse (1979), Robinson (1983)] that the existence of one nonhyperbolic parameter value typically implies the existence of an interval called the *Newhouse interval* of nonhyperbolic parameter values.

Second, there are many known results that give sufficient conditions for hyperbolicity, for example see [Hirsch & Pugh (1970), Newhouse & Palis (1971), Moser (1973), Palis & Takens (1993), Diamond et al. (1995), Robinson (1999), Anishchenko et al. (2000), Arai (2007a), Mazure (2008), Mazur & Tabor (2008), Mazur et al. (2008)]. In [Newhouse (2004)], some conditions for dominated and hyperbolic splittings on compact invariant sets for a diffeomorphism in terms of its induced action on a cone field and its complement was given in Sec. 10.2.1. These results were applied to prove the well-known theorems for hyperbolicity of the set of bounded orbits in real and complex Hénon mappings given in Sec. 10.2.2 using complex methods [Bedford & Smillie (2006c)]. Furthermore, Hruska gives in [Hruska (2006a-b)] a rigorous numerical method for proving hyperbolicity of complex Hénon maps. In [Kuznetsov & Sataev (2007)], a numerical verification of the hyperbolic nature (robustness) of a chaotic attractor in a system of two coupled nonautonomous van der Pol oscillators was described in detail (see Sec. 7.1.2 and 7.1.3).

6.9.1 *Numerical procedure*

The most direct method for numerical analysis, whether a trajectory on the chaotic attractor is hyperbolic or not, is the exploration of the behavior of the angle between stable and unstable manifolds of the chaotic trajectory when moving on the attractor. The algorithm for such an investigation was proposed in [Lai et al. (1993)] where it was used in general to analyze hyperbolicity of chaotic saddles. This procedure consists of the forward and backward transformation of an arbitrary vector by a linearized evolution operator along the trajectory considered. It allows one to find the angle between the directions of stability and instability for various points of the trajectory on the attractor. Naturally, such calculations can be more easily carried out for two-dimensional invertible maps.

Generally, the angle between the stable and unstable directions of a map T is defined as follows:

Definition 6.17. (a) We say s is the stable direction for a map T at x if there is a constant $K < 1$ such that

$$\|DT^n(x)(s)\| \leq K^n \|s\| \text{ as } n \to \infty. \tag{6.82}$$

(b) We say u is the unstable direction at x if

$$\|DT^{-n}(x)(u)\| \leq K^n \|u\| \text{ as } n \to \infty. \tag{6.83}$$

Let $\theta(x, a, b) \in [0, 2\pi]$ be the angle between s and u at x in the invariant set. In some cases, $\theta(x, a, b) = 0$ if s and u are identical. Let x_0 be an initial condition and

$$\theta_{\inf}(x_0, a, b) = \inf_{i=0,1,2,\ldots,\infty} \theta\left(T^i(x_0), a, b\right) \tag{6.84}$$

be the lower bound of the angle of a trajectory $x_{n+1} = T(x_n, a, b)$. Due to the limitation of the number of iterates, one must replace (6.84) by

$$\theta_m(x_0, a, b, N) = \min_{i=0,1,2,\ldots,N} \theta\left(T^i(x_0), a, b\right) \tag{6.85}$$

where $\theta\left(T^{i}\left(x_{0}\right),a,b\right)$ is the angle at a point and $\theta_{m}\left(x_{0},a,b,N\right)$ is the minimum of that angle for $(N+1)$ points on the trajectory. If the dependence on initial data was ignored, then one can write $\theta_{\inf}\left(a,b\right)$ and $\theta_{m}\left(a,b,N\right)$ instead of $\theta_{\inf}\left(x_{0},a,b\right)$ and $\theta_{m}\left(x_{0},a,b,N\right)$, i.e.,

$$\theta_{\inf}(a,b) = \lim_{N\to\infty} \theta_m(a,b,N). \tag{6.86}$$

The numerical procedure given in [Lai et al. (1993)] for calculating the angles is as follows:

Let $\theta(x) = \theta(x, a, b)$ be the angles between the stable and unstable directions of a map T for points x at fixed parameter values along a trajectory on the chaotic set.

Step 1: Calculate a single orbit on the chaotic set using direct iterations of the map T or by using the PIM-triple method given in Appendix A.

Step 2: Calculate the stable and unstable directions for each point x on the chaotic set.

(1) (a) Calculate the stable direction at point x. (a-1) Iterate the point x forward under the map T, N times (It is possible to use $N = 100$, but $N = 20$ is quite adequate) to get a trajectory $T^1(x), T^2(x), ..., T^N(x)$, as shown schematically in Fig. 6.6(a). (a-2) Consider a circle of radius $\epsilon > 0$ centered on the point $T^N(x)$, and iterate this circle backward once. This gives an ellipse at the point $T^{N-1}(x)$ with the major axis along the stable direction of the point $T^{N-1}(x)$. (a-3) Iterate this ellipse backwards N times and keep the ellipse's major axis of order ϵ by some necessary normalizations. In this case, all the way back to the point x, the ellipse becomes very *thin* with its major axis along the stable direction at point x. In practice, a unit vector at the point $T^N(x)$ is used instead of a small circle (for sufficiently large N, the unit vector at point x is a good approximation to the stable direction at x). Hence all the steps in this case are the following: (a-3.1) Iterate the unit vector backward to point x by multiplying by the Jacobian matrix DT^{-1} of the inverse map T^{-1} at each point on the already existing orbit. The Jacobian matrix DT^{-1} was calculated by storing the inverse Jacobian matrix at every point of the orbit $T^i(x)$ ($i = 1, ..., N$). (a-3.2) Normalize the vector after each multiplication to the unit length.

(b) Calculate the unstable direction at point x. To find the unstable direction, use a method similar to (a) as shown schematically in Fig. 6.6(b). This is done by backward iteration of x under the inverse map N times to get a backward orbit $T^{-j}(x)$ with $j = N, ..., 1$, and

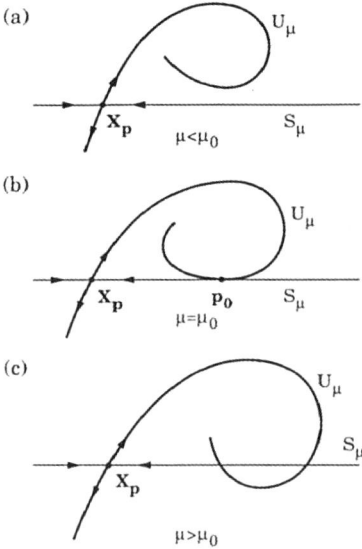

Fig. 6.5 The stable and unstable manifolds of a saddle point x_p for (a) $\mu < \mu_0$, (b) $\mu = \mu_0$, and (e) $\mu > \mu_0$. Reused with permission from Lai, Y. C., Grebogi, C., Yorke, J. A., and Kan, I., Nonlinearity (1993). Copyright 1993, IOP Publishing Ltd.

hence iterate forward a chosen unit vector at point $T^{-N}(x)$ to the point x by multiplying by the Jacobian matrix of the map N times, applying a normalization of this vector to unit length at each step. The final vector is a good approximation to the unstable direction at x.

Step 3: Choose $0 \leq \theta(x) \leq 2\pi$ to be the smaller of the two angles defined by the two straight lines along the stable and unstable directions at x. The convergence rate (the average rate at which the error decreases as N increases) of this method can be estimated as follows: Let $\Delta \theta_N^{s,u} = (\Delta \theta_N^{s,u}(x))$ be the error between the calculated unit vectors at point x and the local stable or unstable direction at x. Since this average is over a large number of x values chosen with respect to the natural measure on the chaotic set, the error is given by:

$$\Delta \theta_N^{s,u} = \Delta \theta_0 \exp\left(-\delta_{s,u} N\right) \qquad (6.87)$$

where δ_s and δ_u are the convergence rates for the stable and unstable directions, respectively, and $\Delta \theta_0$ is the error in the initial angle.

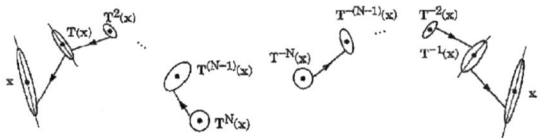

Fig. 6.6 Schematic illustration of the numerical method used to calculate (a) the stable direction and (b) the unstable direction for a point x on the chaotic set. Reused with permission from Lai, Y. C., Grebogi, C., Yorke, J. A., and Kan, I., Nonlinearity (1993). Copyright 1993, IOP Publishing Ltd.

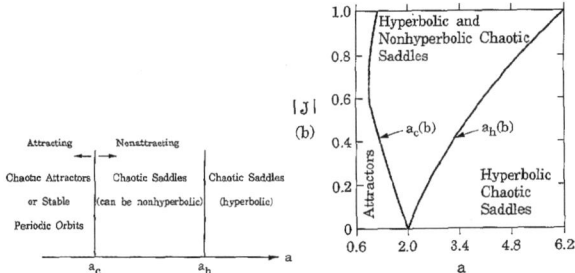

Fig. 6.7 Parameter regime for the Hénon map (6.88) (a) For a given b value, the nature of various chaotic invariant sets in different ranges of a. (b) The parameter boundaries of interest [the $a_c(b)$ and $a_h(b)$ curves]. For $a < a_c(b)$, there are chaotic attractors, while for $a > a_h(b)$ the chaotic saddles are hyperbolic. Reused with permission from Lai, Y. C., Grebogi, C., Yorke, J. A., and Kan, I., Nonlinearity (1993). Copyright 1993, IOP Publishing Ltd.

6.9.2 Testing hyperbolicity of the Hénon map

For the Hénon map

$$H_{a,b}(x,y) = (a - x^2 + by, x), \quad (6.88)$$

let $b \in (0, 1)$, and let $a_c(b)$ be the crisis value of a, i.e., the value at which the main attractor is destroyed or the attractor collides with the basin boundary [Grebogi et al. (1983)], and let $a_h(b)$ be the *last tangency value* so that for $a > a_h(b)$, the chaotic saddle is hyperbolic. For different ranges of parameter a, for a given b value, Fig. 6.7(a) shows the nature of various chaotic invariant sets, and Fig. 6.7(b) shows the boundary of the parameters $a_c(b)$ and $a_h(b)$. For $a \in [a_c(b), a_h(b)]$ and for most values of b, the fraction of nonhyperbolic parameter values is larger than 0.2, and the minimum length of the Newhouse interval is the order of 10^{-1}, i.e., the nonhyperbolic fraction of the interval $[a_c(b), a_h(b)]$ tends to 1 as $b \to 0$. Figure 6.8, shows numerically calculated nonhyperbolic parameter values for $b \in (0, 1)$ and

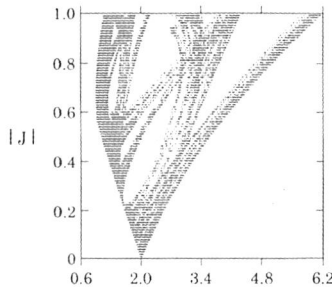

Fig. 6.8 Hénon map (6.88) set of parameter values with nonhyperbolic chaotic saddles. An average of 700 values of a were tested for each of 100 values of b. About 33% of the nonhyperbolic parameter points are missed, and about 11% are incorrectly plotted. These errors should have little effect on the overall configuration. The left and right edges of this set coincide with the curves shown in Fig. 6.7(b). Reused with permission from Lai, Y. C., Grebogi, C., Yorke, J. A., and Kan, I., Nonlinearity (1993). Copyright 1993, IOP Publishing Ltd.

$a \in [a_c(b), a_h(b)]$ where 70000 parameter pairs (a, b) have been tested in the region shown in Fig. 6.7(b) between the curves $a_c(b)$ and $a_h(b)$ in which about 0.9% of them give attractors. For the Hénon map (6.88), the above method was applied for $a > 0$ and $b = 0.3$. For $a = 1.4$, the map has a chaotic attractor with stable and unstable directions at 2000 points on the chaotic attractor as shown in Figs. 6.11(a) and 6.11(b), respectively. Figure 6.11(c) shows the histogram of angles for 20000 points on the attractor. From Fig. 6.11(c), it is clear that the standard Hénon chaotic attractor is nonhyperbolic because there exist points on the attractor at which the angles between stable and unstable directions are arbitrarily close to zero. In this case, complicated structure holds for chaotic sets of map (6.88) without much regularity near $\theta = 0$. This irregularity persists even with much longer trajectories, and it is impossible to extrapolate reliably to very small angles.

The Hénon chaotic attractor is destroyed when the boundary crisis occurs at $a_c \simeq 1.426$ and it becomes an invariant chaotic saddle for $a > a_c$ as shown in [Grebogi et al. (1983)]. Figure 6.9(a) shows the chaotic saddle at $a = 1.6$, calculated using the PIM-triple algorithm described in Appendix A. Figure 6.9(b) shows a histogram of 20000 points on the chaotic saddle for $a = 1.6$ and suggests the nonhyperbolicity of the attractor due to the existence of tangency points of stable and unstable manifolds. For this case, Figs. 6.12(a) and 6.12(b) show the dependence of $\Delta \theta_N^s$ and $\Delta \theta_N^u$ on N on a semi-logarithmic scale, where the average is over 1000 points. With

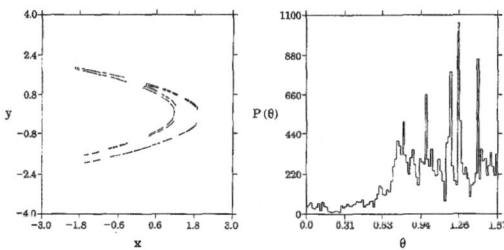

Fig. 6.9 For the Hénon map (6.88) with $a = 1.6$ and $b = 0.3$, (a) the nonhyperbolic chaotic saddle, and (b) a histogram of 20000 angles on the chaotic saddle. Reused with permission from Lai, Y. C., Grebogi, C., Yorke, J. A., and Kan, I., Nonlinearity (1993). Copyright 1993, IOP Publishing Ltd.

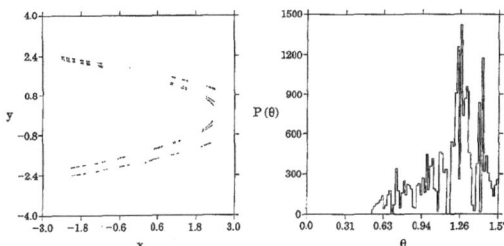

Fig. 6.10 For the Hénon map (6.88) with $a = 3.0$ and $b = 0.3$, (a) the hyperbolic chaotic saddle, and (b) a histogram of 20000 angles on the chaotic saddle. Note that the angles are bounded away from 0. Reused with permission from Lai, Y. C., Grebogi, C., Yorke, J. A., and Kan, I., Nonlinearity (1993). Copyright 1993, IOP Publishing Ltd.

$N = 10$, the error in the stable direction is about $e^{-10} \simeq 10^{-5}$, and with $N = 20$, the error is about 10^{-8}. The convergence rates (least squares fit) are $\delta_s \simeq 2.13 \pm 0.02$ and $\delta_u \simeq 2.07 \pm 0.02$. For a point x near a tangency in which the angle θ between the stable and unstable manifolds is less than 1.91×10^{-5}, one has $\delta_s = 2.54 \pm 0.01$ and $\delta_u = 2.15 \pm 0.02$, which implies that the method for calculating the angle converges quickly for any point x either on the chaotic set or near tangencies of stable and unstable manifolds. Figure 6.13 shows the minimum angle $\theta_m(a, b, N)$ versus a for $b = 0.3$ with $N = 10^4$ iterates for each of the 10000 tested values of a. In this case, parameter values at which there exist attractors were denoted by diamonds of height 0.5, and for $a > a_c$, these attractors only occur in small parameter intervals with small basins of attraction [Tedeschini-Lalli & Yorke (1985)].

For $a = 3.0$, the Hénon map (6.88) has a hyperbolic chaotic saddle as shown in Fig. 6.10(a) where it is clear from Fig. 6.10(b) that the angle

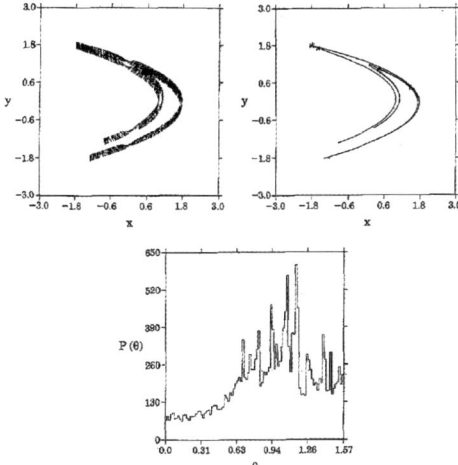

Fig. 6.11 For the Hénon map (6.88) with $a = 1.4$ and $b = 0.3$, (a) the stable directions for 2000 points on the attractor, (b) the unstable directions also for 2000 points on the attractor, and (c) the histogram of 20000 angles on the attractor. Reused with permission from Lai, Y. C., Grebogi, C., Yorke, J. A., and Kan, I., Nonlinearity (1993). Copyright 1993, IOP Publishing Ltd.

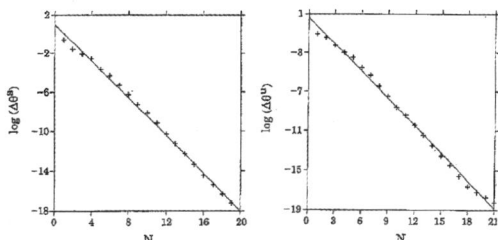

Fig. 6.12 For the Hénon map (6.88) with $a = 1.6$ and $b = 0.3$, the average error over 1000 points on the chaotic saddle versus the number of iterates in calculating (a) the stable direction, and (b) the unstable direction. Reused with permission from Lai, Y. C., Grebogi, C., Yorke, J. A., and Kan, I., Nonlinearity (1993). Copyright 1993, IOP Publishing Ltd.

distribution calculated for 20000 points of all the points considered are bounded away from zero with a minimum angle of approximately 0.57.

6.9.2.1 *Detect whether the minimum angle tends to zero*

To detect whether the angle $\theta_m(a, b, N)$ tends to zero as $N \to \infty$, one can calculate $\theta_m(a, b, N)$ for $N = 10^4$. If this minimum angle is small

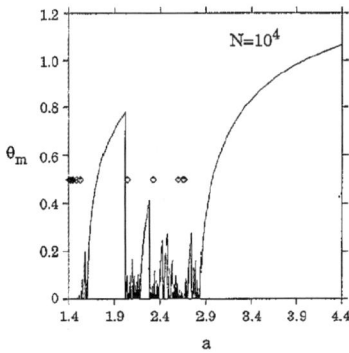

Fig. 6.13 Hénon map (6.88) with $b = 0.3$, a plot of the minimum angle $\theta_m(a,b,N)$ versus a, where we use $N = 10^4$ iterates for each of the 10000 tested values of a. The diamonds plotted with a height 0.5 denote parameter values at which there exist attractors. Of these 10000 tested a values, about 1% result in attractors. Reused with permission from Lai, Y. C., Grebogi, C., Yorke, J. A., and Kan, I., Nonlinearity (1993). Copyright 1993, IOP Publishing Ltd.

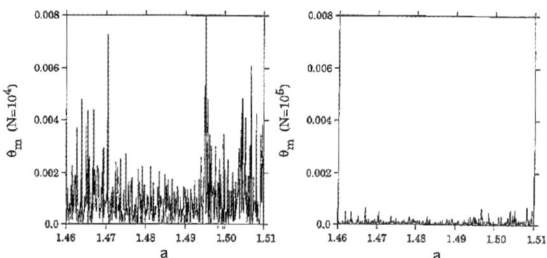

Fig. 6.14 The effect of increasing the number of iterates of the Hénon map (6.88) on the minimum angle. For $b = 0.3$ and $a \in [1.46, 1.51]$, $\theta_m(a,b,N)$ versus a for (a) $N = 10^4$ and (b) $N = 10^5$. Reused with permission from Lai, Y. C., Grebogi, C., Yorke, J. A., and Kan, I., Nonlinearity (1993). Copyright 1993, IOP Publishing Ltd.

enough, i.e., if $\theta_m(a,b,N)$ go to zero as $N \to \infty$, then the chaotic saddle is nonhyperbolic. While if $\theta_m(a,b,N)$ is not small, then the chaotic saddle is hyperbolic, and $\theta_m(a,b,N)$ will not decrease even if one increases N. For $b = 0.3$, $a \in [1.46.1.51]$, $N = 10^4$, and 10^5, Fig. 6.14(a) and Fig. 6.14(b) show the result of increasing the number of iterates N.

The criterion mentioned for detecting whether the minimum angle is zero is as follows:

Let θ_{crit} be such a value, and count the pair as nonhyperbolic if $\theta_m(a,b,N) \leq \theta_{crit}$. As shown in Fig. 6.15, the interval $(10^{-5}, 10^{-2})$ was

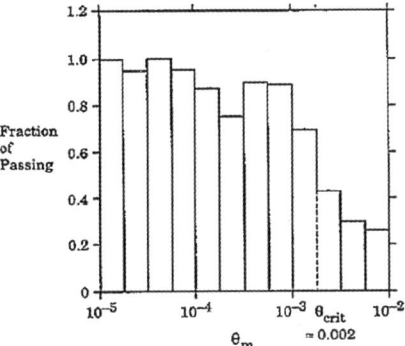

Fig. 6.15 The fraction of parameter values for the Hénon map (6.88) that pass the *ten-times-improvement test* versus the minimum angle θ_m $(a, b, 10^4)$. Reused with permission from Lai, Y. C., Grebogi, C., Yorke, J. A., and Kan, I., Nonlinearity (1993). Copyright 1993, IOP Publishing Ltd.

partitioned into twelve subintervals, and for each subinterval, fifty pairs (a, b) were randomly selected having θ_m $(a, b, 10^4)$ in that subinterval. If

$$\theta_m\left(a, b, 10^7\right) < \frac{1}{10}\theta_m\left(a, b, 10^4\right), \tag{6.89}$$

then the pair (a, b) passes a *ten-times-improvement test* and gives nonhyperbolic invariant sets, while almost all those failing are presumably hyperbolic. Figure 6.15 shows that the smaller θ_m $(a, b, 10^4)$ is, the more likely (a, b) is pass this test. Thus the following criterion was used to detect hyperbolicity:

Criterion: Given a parameter pair (a, b) of the Hénon map (6.88), if the minimum angle between the stable and unstable manifold chosen among a trajectory of 10^4 iterates is less than $\theta_{crit} = 0.002$, then the chaotic saddle is counted as nonhyperbolic.

In this case, 33% of the parameter pairs are nonhyperbolic, and 11% are hyperbolic, *i.e.*, the fraction of nonhyperbolic parameter values for $a_c < a < a_h$ varies from 0.2 (for $b > 0.25$) to 1 (for b approaching 0).

6.9.2.2 *Theoretical model for the fraction of nonhyperbolic parameter values*

This last result was proved theoretically, *i.e.*, there is a theoretical model for the fraction of nonhyperbolic parameter values for small b. This model equates the probability that a particular parameter is nonhyperbolic to the probability of unbounded growth in a stochastic *birth–death* process.

Integration of this probability over the parameter range gives the fraction of nonhyperbolic parameter values denoted by f. Let q and p be the two fixed points of the Hénon map (6.88) as shown in Fig. 6.16(a). Let $U_a(x)$ be the unstable manifold of $x = p, q$. In this case, for $a \leq a_c$, every point sufficiently close to $U_a(p)$ is attracted to $U_a(p)$, and for $a \geq a_c$, a subset of $U_a(p)$ limits on the left branch of $U_a(q)$ goes to infinity under iteration.

Definition 6.18. *Bends* are the points at which maximum bending of the curves of the closure of $U_a(p)$ takes place under further iteration, and they also called *critical points C*.

These bends [Benedicks & Carleson (1991)] are the analogue of critical points in one-dimensional maps. Hence a parameter value is hyperbolic if all the *bends* in the curves comprising the closure of $U_a(p)$ go to infinity under iteration. If not, then the parameter is nonhyperbolic, or there are small bounded attractors [Tedeschini-Lalli & Yorke (1985)]. Geometrically, the set of critical points C is (at least when the chaotic saddle is hyperbolic) the intersection of the curve l (containing the critical points) with the closure of $U_a(p)$ as shown in Fig. 6.16. Thus C is a Cantor set with λ_1^n pieces of size λ_2^n, where λ_1^n and λ_2^n are the two Lyapunov numbers for the chaotic saddle. For small positive b, and $a_c \leq a \leq a_h$, the two Lyapunov numbers for the chaotic saddle are approximately $\lambda_1 \simeq 2$ and $\lambda_2 \simeq \frac{b}{2}$. Thus on average, the critical points locally separate at an exponential rate close to $\ln(\lambda_1)$. After $\frac{n(\ln 2 - \ln b)}{\ln 2}$ iterations, the size of the piece of size $\left(\frac{b}{2}\right)^n$ has lengthened to of order 1, and this confirms that the number of effective critical points grows at an exponential rate of $\frac{(\ln 2)^2}{\ln 2 - \ln b}$. The above analysis describes the behavior of the critical points under iteration. Indeed, on some iterations, the critical points go to infinity because they lie outside the U-shaped region bounded by the part of the stable manifold $S_a(q)$ depicted in Fig. 6.16(a).

The *birth–death* process was used to estimate the probability that the parameter a is nonhyperbolic. Assume the following:

(1) On any given iteration, each effective critical point has a probability of order $\frac{(\ln 2)^2}{\ln 2 - \ln b}$ of stretching into two effective critical points (birth).
(2) Each effective critical point that has not yet fallen outside of the U-shaped region will fall on each iteration along a curve of $U_a(p)$ in a random place inside the union of the U-shaped region and the small box.
(3) Each effective critical point behaves independently of all other points on each iteration, and thus this is a classic birth–death process.

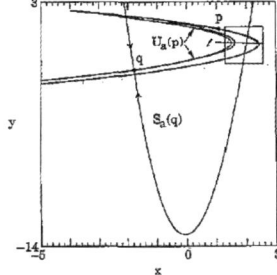
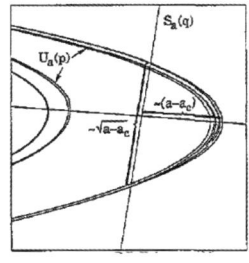

Fig. 6.16 (a) The fixed points p and q, the curve l, the unstable manifold of p, and the U-shaped portion of the stable manifold of q are shown for the Hénon map with $a = 2$ and $b = 0.3$. (b) An enlargement of the small box. Reused with permission from Lai, Y. C., Grebogi, C., Yorke, J. A., and Kan, I., Nonlinearity (1993). Copyright 1993, IOP Publishing Ltd.

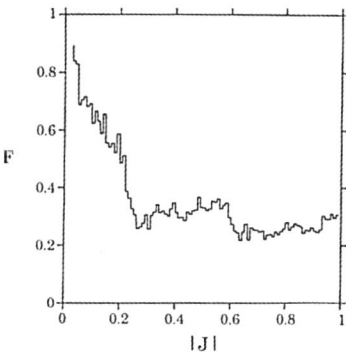

Fig. 6.17 Fraction of nonhyperbolic parameter values for the Hénon map (6.88) versus $|J|$ $(= b)$ for $a_c(b) < a < a_h(b)$. Note that for all the $|J|$ values considered, the fraction of nonhyperbolic parameter values is larger than 0.2. We have not plotted the fraction when $|J| < 0.03$. See the next figure for the range $0 < |J| < 0.03$. Reused with permission from Lai, Y. C., Grebogi, C., Yorke, J. A., and Kan, I., Nonlinearity (1993). Copyright 1993, IOP Publishing Ltd.

Hence on any iteration, each effective critical point has probability of order $\sqrt{a - a_c}$ of falling outside the U-shaped region (death). This implies that for each effective critical point, there is a probability of one minus a term of order $\sqrt{a - a_c} + \frac{(\ln 2)^2}{\ln 2 - \ln b}$, which means that each of these points neither stretches into two effective critical points nor falls outside the U-shaped region.

Let G be the probability that the effective critical points all eventually iterate to infinity (the probability that the parameter is hyperbolic). Thus

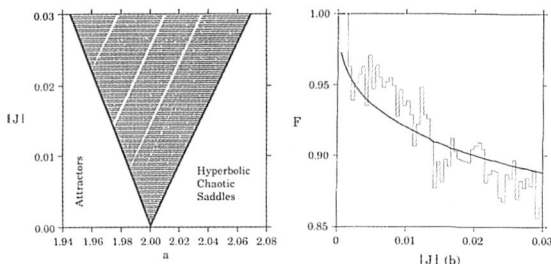

Fig. 6.18 (a) Nonhyperbolic parameter values for the Hénon map (6.88) for $0 < |J| < 0.03$ and $a_c(b) < a < a_h(b)$. (b) The fraction of nonhyperbolic parameter values for small Jacobian ($0 < |J| < 0.03$). Note that the fractions are close to one. The continuous curve in the figure derives from a heuristic argument in Sec. 6.9.2 for small Jacobian. Reused with permission from Lai, Y. C., Grebogi, C., Yorke, J. A., and Kan, I., Nonlinearity (1993). Copyright 1993, IOP Publishing Ltd.

the probability that the effective critical points all eventually iterate to infinity, given that we start with exactly two effective critical points, is G^2, and the probability that a given effective critical point stretches into two before it falls outside the U-shaped region is of order

$$H = \frac{(\ln 2)^2}{(\ln 2)^2 + (\ln 2 - \ln b)\sqrt{a - a_c}}. \tag{6.90}$$

Thus one has using conditional probabilities that

$$G = (1 - H) + HG^2. \tag{6.91}$$

Thus G is given by

$$G = \frac{1 - H}{H} = \frac{(\ln 2 - \ln b)}{(\ln 2)^2}\sqrt{a - a_c}. \tag{6.92}$$

The fraction of hyperbolic parameter values $a_c \lesssim a \lesssim a_h$ is of order $-\sqrt{b}\ln\frac{b}{2}$ because $a_h - a_c$ is of order b and

$$\lim_{b \to 0} \frac{1}{b} \int_0^b \frac{(\ln 2 - \ln b)}{(\ln 2)^2}\sqrt{x}dx = \lim_{b \to 0} \frac{2(\ln 2 - \ln b)}{3(\ln 2)^2}\sqrt{b} = 0. \tag{6.93}$$

Conjecture 6.1. *The fraction of nonhyperbolic parameter values tends to 1 as $|J| \to 0$.*

Indeed, the nonhyperbolic parameter values (a, b) for $b \in (0, 1)$ ($|J| = b$) and $a \in [a_c(b), a_h(b)]$ of the Hénon map (6.88) are shown in Fig. 6.8, where

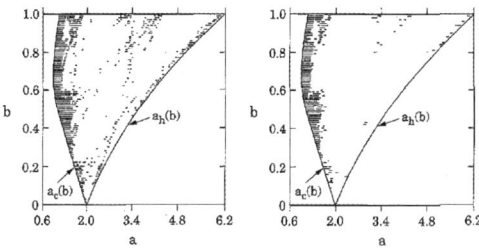

Fig. 6.19 Computed Newhouse intervals for the Hénon map (6.88) with (a) length > 0.02, and (b) length > 0.04. Reused with permission from Lai, Y. C., Grebogi, C., Yorke, J. A., and Kan, I., Nonlinearity (1993). Copyright 1993, IOP Publishing Ltd.

the dots denote the nonhyperbolic parameter values. Figure 6.17 shows that for small values of the Jacobian, the fraction of nonhyperbolic parameter values is close to one, *i.e.*, at least 20% of the parameter pairs tested are nonhyperbolic for $a \in [a_c(b), a_h(b)]$. Figure 6.18(a) shows the nonhyperbolic parameter values, and Fig. 6.18(b) shows the fraction of nonhyperbolic parameter values and supports the theoretical result for $a \in [a_c(b), a_h(b)]$ and $0 < |J| < 0.03$. Figures 6.19(a) and 6.19(b) show the distribution of calculated Newhouse intervals with lengths larger than 0.02 and 0.04, respectively. Furthermore, for all the b values considered, the maximum Newhouse interval is large (about 0.12), and this extends the results in [Kan *et al.* (1995)]. (Newhouse intervals are typically small and of order 10^{-6}.) The method of computation is as follows: For $b = 0.3$ and $b = 0.7$ and for a given a value θ, one can choose a grid of 700 values of a in (a_c, a_h), and for each a one compute the angle $\theta\left(a, N = 10^4\right)$ and then choose the largest interval called the *apparent Newhouse interval* for which all the tested a values are nonhyperbolic according to the above criterion.

6.9.3 Testing hyperbolicity of the forced damped pendulum

The forced damped pendulum presented in [Grebogi *et al.* (1987c)] is given by

$$\begin{cases} x' = y \\ y' = -0.2y - \sin x + 2\cos z \\ z' = 1. \end{cases} \quad (6.94)$$

For system (6.94), the time-2π map has two stable fixed points denoted by A and B and several invariant chaotic saddles [Grebogi *et al.* (1987c)], where two of which are in the basins of A and B as shown in Fig. 6.20(a).

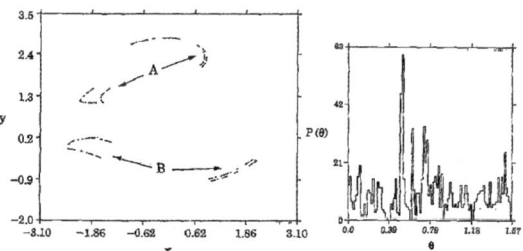

Fig. 6.20 For the forced damped pendulum of equation (6.94). (a) The nonhyperbolic chaotic saddles A and B, and (b) the histogram of 1000 angles on the chaotic saddle A, that is, the angle between stable and unstable manifolds from a numerical trajectory of 1000 points lying on A. The minimum angle observed here is about 0.01. Reused with permission from Lai, Y. C., Grebogi, C., Yorke, J. A., and Kan, I., Nonlinearity (1993). Copyright 1993, IOP Publishing Ltd.

The Poincaré surface of section was taken at $t = 2n\pi$, $n = 0, 1, ...$, to obtain a discrete mapping. The evolution of an infinitesimal tangent vector of a set of N first-order differential equations:

$$\frac{dx_i}{dt} = F_i(x_1, x_2, ..., x_N), i = 1, 2, ..., N \quad (6.95)$$

is given by

$$\frac{d\delta x}{dt} = DF(x).\delta x \quad (6.96)$$

where $DF(x)$ is the $N \times N$ Jacobian matrix with matrix elements

$$DF_{ij}(x) = \frac{dF_i}{dx_j}. \quad (6.97)$$

Equations (6.95) and (6.96) were solved using the fourth-order Runge–Kutta algorithm, and the local stable and unstable directions were calculated for 1000 points on each chaotic saddle A and B in Fig. 6.20(a). The resulting histogram of these 1000 angles on the chaotic saddle A is shown in Fig. 6.20(b), and it is clear that both chaotic saddles A and B are nonhyperbolic because there exist angles approaching zero, or if one chaotic saddle is nonhyperbolic, the other must also be nonhyperbolic due to the symmetry $(t \to t + \pi, z \to z + \pi, x \to -x)$.

6.10 Uniform hyperbolicity test

In this section, we present a method for testing uniform hyperbolicity due to Arai and given in [Arai (2007a)]. The main difference between the method of Arai and the one of Hruska [Hruska (2006a-b)] is that the method of Arai does not prove hyperbolicity directly. It proves quasi-hyperbolicity, which is equivalent to uniform hyperbolicity under the assumption of chain recurrence. Furthermore, the method of Arai uses the subdivision algorithm given in [Dellnitz & Junge (2002)] which implies the effectiveness of this method for inductive search of hyperbolic parameters. Hence this method avoids the computationally intensive procedure of constructing a metric adapted to the hyperbolic splitting. Furthermore, the method of Arai is applicable to higher-dimensional dynamical systems, especially the complex Hénon map [Bedford & Smillie (2006c), Arai (2007b)].

To present the Arai's algorithm, assume that $M = \mathbb{R}^n$, and consider a family of diffeomorphisms $f_a : \mathbb{R}^n \to \mathbb{R}^n$ that depend on r-tuples of real parameters $a = (a_1, \ldots, a_r) \in \mathbb{R}^r$. Define $F : \mathbb{R}^n \times \mathbb{R}^r \to \mathbb{R}^n$ and $DF : T\mathbb{R}^n \times \mathbb{R}^r \to T\mathbb{R}^n$ by

$$F(x, a) = f_a(x) \quad \text{and} \quad DF(x, v, a) = Df_a(x, v) \qquad (6.98)$$

where $x \in \mathbb{R}^n$ and $v \in T_x\mathbb{R}^n$. Let $\mathbb{F}$ be the set of *floating point numbers*, or the set of numbers the computer can handle. Let $\mathbb{IF}$ be the set of intervals whose end-points are in $\mathbb{F}$, namely,

$$\begin{cases} \mathbb{IF} = \{I = [a, b] \subset \mathbb{R} \mid a, b \in \mathbb{F}\} \\ \mathbb{IF}^n = \{I_1 \times \cdots \times I_n \subset \mathbb{R}^n \mid I_i \in \mathbb{IF}\} \end{cases} \qquad (6.99)$$

where $\mathbb{IF}^n$ is a set of n-dimensional cubes. Let $X, F \in \mathbb{IF}^n$ and $A \in \mathbb{IF}^r$ where X is the tangent space of M and F is the parameter space. Note the images $F(X \times A)$ and $DF(X \times V \times A)$ are not objects of $\mathbb{IF}^n$ nor $\mathbb{IF}^{2n}$ in general. Due to rounding errors, these images are not determined exactly by computer. To solve this problem, one requires that the computer enclose these images using elements of $\mathbb{IF}^n$ and $\mathbb{IF}^{2n}$.

Condition: There exists a computational method such that for any $X, V \in \mathbb{IF}^n$ and $A \in \mathbb{IF}^r$, one can compute $Y \in \mathbb{IF}^n$ and $W \in \mathbb{IF}^{2n}$ such that $F(X \times A) \subset \text{int} Y$ and $DF(X \times V \times A) \subset \text{int} W$ hold rigorously.

In this case, rigorous interval arithmetic (see Sec. 1.2) can be used to satisfy the condition and gives a good outer approximations if Y and W are too large.

Now, let $K \subset \mathbb{R}^n$ be a compact set that contains Λ, which is a compact invariant set of f and $L = \subset T\mathbb{R}^n$, the product of K and $[-1, 1]^n$. Assume that K and the fiber $[-1, 1]^n \subset T_x\mathbb{R}^n$ are decomposed into a finite union of elements of $\mathbb{IF}^n$, i.e.,

$$\begin{cases} K = \bigcup_{i=1}^k K_i & \text{where} \quad K_i \in \mathbb{IF}^n \\ L = \bigcup_{j=1}^\ell L_j & \text{where} \quad L_j \in \mathbb{IF}^{2n}. \end{cases} \qquad (6.100)$$

The condition above permits one to compute $Y_i \in \mathbb{IF}^n$ and $W_j \in \mathbb{IF}^{2n}$ as follows:
$$\begin{cases} F(K_i \times A) \subset Int(Y_i) \\ DF(L_j \times A) \subset Int(W_j), \end{cases}, 1 \leq i \leq k \text{ and } 1 \leq j \leq \ell \quad (6.101)$$
and directed graphs $\mathcal{G}(F, K, A)$ and $\mathcal{G}(DF, L, A)$ can be constructed to enclose the chain recurrent set of f_a and the maximal invariant set of N.

- $\mathcal{G}(F, K, A)$ has k vertices: $\{v_1, v_2, \ldots, v_k\}$.
- There exists an edge from v_p to v_q if and only if $Y_p \cap K_q \neq \emptyset$.

And similarly,

- $\mathcal{G}(DF, L, A)$ has ℓ vertices: $\{w_1, w_2, \ldots, w_\ell\}$.
- There exists an edge from w_p to w_q if and only if $W_p \cap N_q \neq \emptyset$.

For the set $\mathcal{G}(F, K, A)$, if there exists $x \in K_p$ that is mapped into K_q by f_a for some $a \in A$, then there must be an edge of $\mathcal{G}(F, K, A)$ from v_p to v_q. This property also holds for $\mathcal{G}(DF, L, A)$. Let G be a directed graph and G' be a subgraphs of $\mathcal{G}(F, K, A)$ and G'' of $\mathcal{G}(DF, L, A)$.

Definition 6.19. (a) The vertices of $\operatorname{Inv} G$, the *invariant set* of G is defined by
$$\{v \in G \mid \exists \text{ bi-infinitely long path through } v\}. \quad (6.102)$$
(b) The vertices of $\operatorname{Scc} G$, the set of *strongly connected components* of G is
$$\{v \in G \mid \exists \text{ path from } v \text{ to itself}\}. \quad (6.103)$$
The edges of these graphs are defined to be the restriction of that of G.

(c) The *geometric representations* $|G| \subset \mathbb{R}^n$ and $|G'| \subset \mathbb{R}^{2n}$ are defined by
$$|G'| = \bigcup_{v_i \in G} K_i, \quad \text{or} \quad |G''| = \bigcup_{w_j \in G'} L_j. \quad (6.104)$$

Note that $\operatorname{Scc} G$ is a subgraph of $\operatorname{Inv} G$ and $|\mathcal{G}(F, K, A)| = K$ and $|\mathcal{G}(DF, L, A)| = L$. For the computation of $\operatorname{Inv} G$, $\operatorname{Scc} G$, the algorithms of [Szymczak (1997), Sedgewick (1983)] can be used, respectively. The following result was proved in [Arai (2007a)]:

Proposition 6.11. *For any* $a \in A$,
$$\begin{cases} \operatorname{Inv}(f_a, K) \subset |\operatorname{Inv} \mathcal{G}(F, K, A)| \\ \operatorname{Inv}(Df_a, L) \subset |\operatorname{Inv} \mathcal{G}(DF, L, A)|. \end{cases} \quad (6.105)$$

Furthermore, if $\mathcal{R}(f_a) \subset Int(K)$ holds for all $a \in A$, then we have
$$\mathcal{R}(f_a) \subset |\operatorname{Scc} \mathcal{G}(F, K, A)| \tag{6.106}$$
for all $a \in A$.

The algorithm used to prove the quasi-hyperbolicity in [Arai (2007a)] is based on the subdivision algorithm given in [Dellnitz & Junge (2002)], i.e., if the proof of quasi-hyperbolicity fails, then the subdivision of all the cubes in K and L is needed to give a better approximation of the invariant set, and repeat the step until we succeed with the proof. Indeed, let A be a fixed set of parameters. The role of the Arai algorithm is to check if $\mathcal{R}(f_a)$ is quasi-hyperbolic for all $a \in A$. The principle of this algorithm is as follows: If the set A contains a non-quasi-hyperbolic parameter value, then the Arai algorithm never stops.

Algorithm A (for proving quasi-hyperbolicity for all $a \in A$)

(1) Find K such that $\mathcal{R}(f_a) \subset Int(K)$ holds for all $a \in A$, and let $L := K \times [-1, 1]^n$.
(2) Compute $\operatorname{Scc} \mathcal{G}(F, K, A)$, and replace K with $|\operatorname{Scc} \mathcal{G}(F, K, A)|$.
(3) Replace L with $L \cap (K \times [-1, 1]^n)$.
(4) Compute $\operatorname{Inv} \mathcal{G}(DF, L, A)$.
(5) If $|\operatorname{Inv} \mathcal{G}(DF, K, A)| \subset K \times \int [-1, 1]^n$, then stop.
(6) Otherwise, replace L with $|\operatorname{Inv} \mathcal{G}(DF, L, A)|$, and refine the decomposition of K and L by bisecting all cubes in them. Then go to step 2.

The following result was proved in [Arai (2007a)]:

Theorem 6.17. *If Algorithm A stops, then f_a is quasi-hyperbolic on $\mathcal{R}(f_a)$ for every $a \in A$.*

Proof. Assume that Algorithm A stops, and choose $a \in A$. Let
$$N_a = L \cap (\mathcal{R}(f_a) \times [-1, 1]^n). \tag{6.107}$$
Thus the set N_a contains the zero-section of $T\mathcal{R}(f_a)$. Thus it suffices to show that N_a is an isolating neighborhood with respect to $Df_a : T\mathcal{R}(f_a) \to T\mathcal{R}(f_a)$. Because the algorithm stops, one has
$$Inv(Df_a, N_a) \subset |Inv\mathcal{G}(DF, L)| \subset K \times Int([-1, 1]^n). \tag{6.108}$$
Thus the inclusion $Inv(Df_a, N_a) \subset N_a \subset \mathcal{R}(f_a) \times [-1, 1]^n$ implies that
$$Inv(Df_a, N_a) \subset \mathcal{R}(f_a) \times Int([-1, 1]^n). \tag{6.109}$$
Finally, $Inv(Df_a, N_a) \subset Int(N_a)$ since $\mathcal{R}(f_a) \times Int([-1, 1]^n)$ is the interior of N_a with respect to $T\mathcal{R}(f_a)$. $\square$

To apply Algorithm A for a large family of diffeomorphism, the algorithm should involve an automatic selection of parameter values in the set A using the same subdivision algorithm to realize such a procedure, *i.e.*, the decomposition of A into a finite union of elements (denoted by $\mathcal{A}$) of $\mathbb{IF}^r$, and remove cubes in which the hyperbolicity is proved. In this case, various *selection rules* can be applied, and the effectiveness of a rule depends on the dynamical system under consideration. An example of such a rule is the selection of the set A_i such that the sets N_i and K_i consist of a smaller number of cubes. Using this rule, the computations are concentrated on parameter values on which the computation is relatively fast. Generally, this rule works sufficiently well and reasonably fast. To avoid the problem in which the computations using this rule are focused only on parameters with a small invariant set, one can use the number of cubes multiplied by the subdivision depth of A_i instead of the number of cubes itself, or one can distribute computational effort equally across the whole parameter space by simply selecting all cubes in $\mathcal{A}$ sequentially.

Algorithm B (adaptive selection of quasi-hyperbolic parameters)

(1) Find K such that $\mathcal{R}(f_a) \subset K$ holds for all $a \in A$.
(2) Let $\mathcal{A} = \{A_0\}$ where $A_0 = A$ and $K_0 = K$, $L_0 = K_0 \times [-1,1]^n$.
(3) Choose a cube $A_i \in \mathcal{A}$ according to the "selection rule".
(4) Apply step 2, 3, and 4 of Algorithm A with $A = A_i$, $K = K_i$, and $L = L_i$.
(5) If $|\operatorname{Inv} \mathcal{G}(DF, L_i)| \subset Int(L_i)$, then remove A_i from $\mathcal{A}$, and go to step 2.
(6) Otherwise, bisect A_i into two cubes A_j and A_k. Remove A_i from $\mathcal{A}$, and add A_j and A_k to $\mathcal{A}$. Put $K_j = K_k = K_i$ and $L_j = L_k = \operatorname{Inv} \mathcal{G}(DF, L_i)$, and then go to step 2.

The effectiveness of Algorithm B is that it does not stop if there is a non-quasi-hyperbolic parameter in the set A, and Theorem 6.10 implies that if the cube A_i is removed, then A_i consists of quasi-hyperbolic parameter values. An example of the application of this algorithm to the Hénon map is given in Sec. 10.2.2.

6.11 Exercises

Exercise 118. (a) Show that the stable and unstable submanifolds E^s, E^u of a map f are Df-invariant.

(b) Show that the stable and unstable subspaces E^s and E^u of a map f depend continuously on the point x and they are invariant and interchanged when one passes from a map to its inverse.

Exercise 119. (a) Show that there must be at least three holes for a compact surface hyperbolic attractor and that it looks locally like a Cantor set × an interval.

(b) Show that hyperbolic attractors have dense periodic points and a point with a dense orbit.

Exercise 120. Show that the Smale horseshoe and the Plykin attractors are locally maximal attractors.

Exercise 121. (a) Show that if a set Λ is a hyperbolic set with nonempty interior for a map f, then f is Anosov if it is transitive, locally maximal, and Ω is a surface.

(b) Show that locally maximal transitive hyperbolic sets have the following properties: Shadowing property introduced in Sec. 5.3, structural stability introduced in Sec. 4.1, Markov partitions, and the SRB measures (for attractors) introduced in Sec. 3.1.2.

Exercise 122. Show that the homoclinic tangencies are absent for Anosov diffeomorphisms.

Exercise 123. Show that not every manifold admits an Anosov diffeomorphism or flow.

Exercise 124. Show that hyperbolicity implies coexistence of highly complicated long-term behavior, sensitive dependence on initial conditions, and overall stability of the orbit structure.

Exercise 125. Show that all Anosov systems are expansive.

Exercise 126. Give an example of a Kupka-Smale diffeomorphism.

Exercise 127. Show that for any $r \geq 1$, the set of Kupka-Smale G_{KS} diffeomorphisms is a dense subset of $Diff^r(\Omega)$ equipped by the C^r-topology.

Exercise 128. Show that the Smale–Williams solenoid introduced in Sec. 6.4.3 is Lyapunov-stable.

Exercise 129. Show that in two dimensions, only the torus supports Anosov diffeomorphisms and all are topologically conjugate to hyperbolic toral automorphisms.

Exercise 130. Construct a hyperbolic flow using the special flows method introduced in Sec. 6.2.7 for the Arnold cat map (6.18).

Exercise 131. (a) Show that maps induced on $\mathbb{T}^n$ by a hyperbolic linear map L of $\mathbb{R}^n$ with integer entries and determinant ± 1 are Anosov diffeomorphisms.

(b) Show that there is an Euclidean norm in $\mathbb{R}^n$ that makes L contracting in $E^s(L)$ and expanding in $E^u(L)$, thus there exists an invariant splitting into subspaces parallel to $E^s(L)$ and $E^u(L)$.

Exercise 132. Show that any transitive Anosov diffeomorphism has the following properties:
 (a) The periodic points are dense.
 (b) There exists a point whose orbit is dense.
 (c) There are many ergodic invariant probability measures with full support.

Exercise 133. (a) Show that he map defined by the matrix $\begin{pmatrix} 1 & 1 \\ 1 & 0 \end{pmatrix}$ has an entropy given by $\log \frac{3+\sqrt{5}}{2}$.

(b) Show that a toral automorphism map defined by the matrix $\begin{pmatrix} 1 & 1 \\ 1 & 0 \end{pmatrix}$ has the following properties: The periodic points are dense. There exists a point whose orbit is dense. There are many ergodic invariant probability measures with full support.

Exercise 134. Show that the map (6.6) is hyperbolic under some suitable conditions on the matrix A.

Exercise 135. (a) Show that for the Anosov automorphism $f : \mathbb{T}^n \to \mathbb{T}^n$, one has $\Omega(f) = \mathbb{T}^n$.

(b) Show that Anosov diffeomorphisms on $\mathbb{T}^n$ with $n \geq 2$ are structurally stable diffeomorphisms on a compact manifold whose nonwandering set contains infinitely many points. See [Mather (1967)].

Exercise 136. Show that the solenoid attractor defined by equation (6.24) has the following properties:
 (a) The forward orbit under f of every point in J accumulates on Λ.
 (b) The restriction of f to the attractor Λ is transitive.
 (c) The set of periodic points of f is dense in Λ.
 (d) The set Λ has a dense subset of periodic orbits and also a dense orbit. In this case, every point in a neighborhood of Λ converges to Λ.

Exercise 137. Depict the so-called *Witorius–Van Danzig solenoid* introduced in Sec. 6.2.3. See [Smale (1967), Arnold & Avez (1968), Arnold (1983), Shilnikov (2002)].

Exercise 138. Prove Proposition 6.2. See [Smale (1967), Shilnikov (2002)].

Exercise 139. Show that the DA (*derived from Anosov*)-diffeomorphism (6.22) is hyperbolic.

Exercise 140. Show that the map (6.5) is expanding if the spectrum of the integer matrix A lies strictly outside the unit circle.

Exercise 141. (a) Show the behavior of the manifolds of a saddle for the cat map (6.18) and for the modified cat map (6.19).
  (b) Illustrate the robust hyperbolic attractor for $\delta < \frac{\pi}{2}$.
  (c) Calculate $P(\phi)$ (see Sec. 6.9.1) versus $0 \leq \phi \leq 90$, $\phi_{\min}$ versus $0 \leq \delta \leq 0.2$, and λ_1 versus the parameter $0 \leq \delta \leq 0.2$.

Exercise 142. Show that for the Blaschke product map (6.11), one of the following three mutually exclusive cases holds:
  (a) B has all its fixed points on $\mathbb{T}$. There is exactly one of them z_0 that is a sink, and the other n are expanding.
  (b) B has $n-1$ fixed points on $\mathbb{T}$, all expanding. It has one fixed point inside the disk which is a sink, and one outside. These two fixed points are related by the formula $z_0 \to \frac{1}{z_0}$ (hence they lie on the same ray passing through the origin).
  (c) B has all its fixed points on $\mathbb{T}$. There is one that is an indifferent saddle-node fixed point, $B(z_0) = z_0$ and $B'(z_0) = 1$.
  (d) Show that in all three cases, there is an open set of points in the disk which tends to z_0 under iteration of B.
  (e) Show that the sequence $B^{(n)}(z)$ is uniformly bounded in the unit disk, *i.e.*, it is a normal family.

Exercise 143. Show that the map (6.13) is hyperbolic.

Exercise 144. Prove Proposition 6.4.

Exercise 145. Show that the basis sets of Smale systems may be of the following three types: attractors, repellers, and saddles.

Exercise 146. Show that saddle basis sets are one-dimensional in the case of flows and null-dimensional in the case of cascades. See [Bowen (1970)].

Exercise 147. Prove Theorem 6.5.

Exercise 148. Show that the (6.8) displays expanding solenoids or a hyperbolic attractor. See [Williams (1967-1974)].

Exercise 149. Show that the map (6.9) is hyperbolic, where A is a matrix with integer entries other than 1, $\det |A| = 1$, the eigenvalues of A do not lie on the unit circle, and $f(\theta)$ is a periodic function of period-1 with

$$\theta' = A\theta + f(\theta) \,(\mathrm{mod}\, 1) \qquad (6.110)$$

Exercise 150. (a) Show that the Bernoulli map (6.13) implies sensitivity to initial conditions. See [Sinai (1979), Devaney (1989), Ott (1993), Katok & Hasselblatt (1999)].

$$\phi_{n+1} = 2\phi_n (\mathrm{mod}\, 2\pi) \qquad (6.111)$$

(b) Show that the Bernoulli map (6.13) is expanding and it transforms the angle variable ϕ is a nonuniform way, but in any case it must be monotonic and possess the characteristic topological property, in particular, the map (6.13) displays *homogeneous chaotic dynamics*.

(c) Show that the Bernoulli map (6.13) is an exactly solvable model of deterministic chaos.

(d) Show that the *Frobenius–Perron operator* for the Bernoulli map (6.13) is solvable. See [Driebe (1999)].

(e) Show that the eigenvalues of the Bernoulli map (6.13) are multiples of $\frac{1}{2}$ and the eigenfunctions are the *Bernoulli polynomials*. See [Gaspard (1992)].

Exercise 151. Show that the Arnold cat map (6.18) has the following properties:
(a) It is the Anosov torus $\mathbb{T}^2$.
(b) It is conservative.
(c) It has a hyperbolic chaotic attractor.
(d) It has two Lyapunov exponents expressed by eigenvalues of the matrix associated with it, *i.e.*, $\Lambda_1 = \frac{\ln(3+\sqrt{5})}{2} = 0.9624$ and $\Lambda_2 = -\frac{\ln(3+\sqrt{5})}{2} = -0.9624$, and their sum vanishes.
(e) The second iteration of the Fibonacci map yields the Arnold cat map (6.18).

Exercise 152. Make a study of the common properties of the three types of chaotic attractors of dynamical systems.

Exercise 153. Show that hyperbolic attractors satisfy the following three conditions:

(a) A hyperbolic attractor consists of a continuum of *unstable leaves*, or curves, which are dense in the attractor and along which close trajectories exponentially diverge.

(b) A hyperbolic attractor (in the neighborhood of each point) has the same geometry defined as a product of the Cantor set and an interval.

(c) A hyperbolic attractor has a neighborhood foliated into *stable leaves* along which the close trajectories converge to the attractor. See [Afraimovich (1989-1990)].

(d) Show that properties (a) and (b) hold under perturbations.

(e) The structure of a hyperbolic attractor is homogeneous.

(f) Dissipative hyperbolic systems exhibit strange chaotic attractors with strong chaotic properties.

(g) Hyperbolic chaotic attractors are structurally stable (robust) under small changes in their governing equations. See [Katok & Hasselblatt (1995), Afraimovich & Hsu (2003), Kuznetsov (1998)].

(h) The absence of homoclinic tangencies implies robust hyperbolicity.

Exercise 154. Show that the map φ defined by (6.20) has the following characteristics:

(a) The diffeomorphism $\varphi|_S$ is structurally stable because φ is hyperbolic.

(b) The attractor $\mathbb{S}$ defined by (6.21) is chaotic and has sensitive dependence on initial conditions in the $\mathbb{S}^1$ direction under repeated application of the map φ.

(c) The attractor $\mathbb{S}$ is not a manifold.

(d) The map φ can be rewritten in the form f given by (6.24).

(e) The forward orbit under f of every point in J accumulates on $\Lambda = \cap_{n \geq 0} f^n(J)$.

(f) The restriction of f to the attractor Λ is transitive.

(g) The set of periodic points of f is dense in Λ.

(h) The set Λ has a dense subset of periodic orbits and also a dense orbit. In this case, every point in a neighborhood of Λ converges to Λ.

Exercise 155. Explain why the Smale–Williams attractor cannot be produced by a continuous flow in three-dimensional phase space.

Exercise 156. Show that the Plykin attractor is a diffeomorphism of a two-dimensional torus projected onto a two-dimensional sphere having four repelling fixed points in its simplest case.

Exercise 157. (a) Show that the interval $[0,1]$ is invariant under the logistic map (6.25).

(b) Show that if $\mu > 4$, there is no invariant interval.

(c) Show that the set Ω defined by (6.27) is a closed invariant set and its dynamics are topologically conjugate to a one-sided shift on two symbols and it is hyperbolic for $4 < \mu \leq 2 + \sqrt{5}$. See [Guckenheimer (1979), van Strien (1981), Misiurewicz (1981), de Melo & van Strien (1993), Robinson (1995), Katok & Hasselblatt (1997)].

(d) Show Lemma 6.1.

(e) Show that the logistic map (6.25) is topologically conjugate to the quadratic map family given by (6.29) according to (6.30).

(f) Show that the map f_r has the following properties: The most interesting dynamics of (f_r) lie in an interval $I_r = [-a, a]$ where $x = -a$ is the fixed point of f_r given by (6.31). If $r < 2$, then one has $I_r \subset [-1, 1]$. The conjugating function, p, satisfying $F_\mu = p^{-1} \circ f_r \circ p$ is affine with $p(x) = Ax + B$. The conjugating function p maps the interval $[0, 1]$ onto I_r, and Ω and Ω_n map onto the corresponding sets for f_r denoted Λ and Λ_n.

(g) Show that if $r = 2$, then the change of variable h defined by (6.34) conjugates the quadratic map f_2 to the tent map g_2 defined by (6.35).

(h) Show relation (6.36) for $\theta \in h^{-1}(\Lambda)$.

(i) Show relation (6.37) for the map g_r.

Exercise 158. Let $f : U \setminus \Gamma \to f(U \setminus \Gamma) \subset U$ be the C^2-diffeomorphism defined in Sec. 6.6.

(a) Show that the Lozi and Belykh attractors satisfy the assumption about the differentiability up to the singularity curves for both f and f^{-1}.

(b) Show that the D set defined in (6.51) is invariant under both f and f^{-1}. See [Sataev (1992)].

(c) Show that some (but not all) Lorenz, Lozi, and Belykh attractors satisfy Condition A1, but in some cases, a weaker assumption than Condition A1 is sufficient. See [Pesin (1992), Afraimovich et al. (1995)].

(d) Show that Condition A1 implies Condition A2 with $K_0 = 2^{\frac{1}{\tau}}$ and that Condition A2 holds for a single, sufficiently large value of $m \geq 1$. See [Sataev (1992)].

(e) Show that Condition A2 implies Conditions A3 and A4. See [Afraimovich et al. (1995)].

(f) Show that the subsets defined in (6.58) satisfy the following properties: The sets $\hat{D}^+_{\varepsilon,l}, D^{\pm}_{\varepsilon,l}$, and D^0_ε are closed. $D^0_\varepsilon = D^+_\varepsilon \cap D^-_\varepsilon$. The set D^+_ε

is f-invariant, D_ε^- is f^{-1}-invariant, and D_ε^0 is both f and f^{-1} invariant. $D_\varepsilon^0 \subset D$ for any $\varepsilon > 0$.

(g) Show that limit (6.59) exists, is positive, continuous on $D_{s,l}^-$, and uniformly bounded away from zero and infinity on D_ε^- because f is smooth up to S^+.

Exercise 159. (a) Show that the map (6.74) is a strong contraction along x and it is close to the degenerate map (6.75).

(b) Show that if $m = 1$, then the closure $cl\left(W_{L_0}^u\right)$ is a two-dimensional torus, and it is smooth if (6.74) is a diffeomorphism. If $m = -1$, then $cl\left(W_{L_0}^u\right)$ is a Klein bottle, and also smooth if (6.74) is a diffeomorphism. In Theorem 6.15, the set $cl\left(W_{L_0}^u\right)$ is not a manifold. In the case of $\mathbb{R}^n$ ($n \geq 4$), the constant m may be any integer.

(c) Prove Theorem 6.16.

(d) Show that the Poincaré map (6.79) is close to the shortened map (6.80), and deduce that Poincaré map (6.79) is an Anosov map.

Exercise 160. Explain the notion of density of hyperbolicity and homoclinic bifurcations in dynamical systems.

Exercise 161. Apply the numerical procedure introduced in Sec. 6.9.1 for the Lozi map (8.88).

Exercise 162. Find the Poincaré map for the forced damped pendulum given by (6.94). See [Grebogi et al. (1987c)].

Exercise 163. Compare Algorithm A and Algorithm B defined in Sec. 6.10.

Exercise 164. (Open problem) Determine whether the universal cover of a manifold that admits an Anosov diffeomorphism must be $\mathbb{R}^n$ for some n.

Exercise 165. (Open problem) Prove Conjecture 6.1.

Exercise 166. (Open problem)

(a) Are one-dimensional dynamics generally hyperbolic? See [Smale (1998)].

(b) Can a complex polynomial T be approximated by one of the same degree with the property that every critical point tends to a periodic sink under iteration? Equivalently, can a polynomial map $T : \mathbb{C} \to \mathbb{C}$ be approximated by one that is hyperbolic? See [Smale (1998)].

(c) Can a smooth map $T : [0; 1] \to [0; 1]$ be C^r approximated by one that is hyperbolic for all $r > 1$? See [Smale (1998)].

Exercise 167. (Open problem) [Gorodnik (2007)]

(a) Is every Anosov diffeomorphism topologically conjugate to a hyperbolic automorphism on an infranilmanifold. See [Franks (1968), Newhouse (1970), Manning (1974), Verjovsky (1974), Farrell & Jones (1978), Franks & Williams (1979), Ghys (1986), Flaminio & Katok (1991), Benoist & Labourie (1993), Ghys (1995), Kalinin & Sadovskaya (2003)].

(b) Show that expanding maps are structurally stable. See [Shub (1978)].

Chapter 7

Robust Chaos in Hyperbolic Systems

In this chapter, we give several examples of realistic models describing hyperbolic (robust) chaos. Indeed, in Sec. 7.1.1, we present a method for the construction of the Smale–Williams attractor introduced in Sec. 6.4.3 for a three-dimensional map along with a theoretical and numerical verification of its hyperbolicity. In Sec. 7.1.4, an example of a flow system is presented with an attractor concentrated mostly on the surface of a two-dimensional torus, the dynamics of which is governed by the Arnold cat map introduced in Sec. 6.2.7. In Sec. 7.1.5, a model for the Bernoulli map (6.13) is constructed using the Poincaré map method discussed in Sec. 1.1 with a hyperchaotic attractor (one with two positive Lyapunov exponents) in a perturbed heteroclinic cycle. The Poincaré map defined by a three-dimensional flow constructed as a bursting neuron model exhibiting a Plykin-like attractor (a strange hyperbolic attractor) is given in Sec. 7.1.6. At the end of the chapter, a set of exercises and open problems are listed with some references.

7.1 Modeling hyperbolic attractors

Recall that in strange attractors of the hyperbolic type, all orbits in phase space are of the saddle type, and the invariant sets of trajectories approach the original one in forward or backward time, *i.e.*, the stable and unstable manifolds intersect transversally. Generally, most known physical systems do not belong to the class of systems with hyperbolic attractors[1] [Anishchenko & Strelkova (1997)]. The type of chaos in them is

---

[1] Although the hyperbolic theory of dynamical systems is widely used for characterizing chaotic behavior of realistic nonlinear systems, it has never been applied to any physical object.

characterized by chaotic trajectories and a set of stable orbits of large periods, not observable in computations because of their extremely narrow domains of attraction. These attractors are called *quasi-attractors*, introduced in Chapter 10. Hyperbolic strange attractors are robust (structurally stable) [Mira (1997)]. Thus, both from the point of view of fundamental studies and of applications, it would be interesting to find physical examples of hyperbolic chaos, *i.e.*, noise generators and transmitters in chaos-based communications. One of the methods used to prove the robustness of a chaotic attractor is the construction of a system with robustness of the corresponding Poincaré mapping, *i.e.*, if a system is known that has robust chaos, then it is possible to construct another model that also has robust chaos. This method was applied to a hyperbolic-type system [Kuznetsov & Seleznev (2006)]. Another example of this method was used to conclude that there is no robust chaos in the quasi-attractor-type systems [Mira (1997)] introduced in Chapter 10. The method most used for such a construction is based on the use of coupled self-sustained oscillators with alternating excitation with a numerical study[2] to visualize diagrams illustrating the phase transfer [Hunt (2000), Belykh *et al.* (2005), Kuznetsov (2005), Kuznetsov & Seleznev (2006), Isaeva *et al.* (2006), Kuznetsov & Pikovsky (2007), Kuznetsov (2008)] where additional coupling permits one to transfer the phases simultaneously from one partner to another to obtain a desired chaotic map on a circle or on a torus (robust hyperbolic). One of the constructed systems (**model A** in [Kuznetsov & Pikovsky (2007)]) is an example of a system of minimal possible dimension (four) possessing the attractor of the Smale–Williams type in its 3-D Poincaré map. **Models A,B,C,D** in [Kuznetsov & Pikovsky (2007)] consist of two or three oscillators excited alternately due to the appropriate mutual amplitude nonlinearity. The basic tools used here are the Poincaré map technique introduced in Sec. 1.1, iteration diagrams illustrating the phase transformation in the course of the excitation transfer, calculation of the Lyapunov exponents introduced in Sec. 3.1.5, and characterization of the resulting processes with correlation functions introduced in Sec. 3.3.

7.1.1 *Modeling the Smale–Williams attractor*

In this section, the Smale–Williams attractor was constructed for a three-dimensional map, namely the composed equations [Kuznetsov & Seleznev

---

[2]This includes the Lyapunov spectrum, where the positive exponent is compared to that of the underlying chaotic map.

(2006)] given by

$$\begin{cases} x' = -2\pi u + (h_1 + A_1 \cos 2\pi\tau/N) x - \frac{1}{3}x^3 \\ u' = 2\pi (x + \varepsilon_2 y \cos 2\pi\tau) \\ y' = -4\pi v + (h_2 - A_2 \cos 2\pi\tau/N) y - \frac{1}{3}y^3 \\ v' = 4\pi (y + \varepsilon_1 x^2), \end{cases} \qquad (7.1)$$

which was introduced for the first time in [Kuznetsov (2005)]. The system (7.1) is a nonautonomous nonlinear system consisting of two coupled van der Pol oscillators whose frequencies are ω_0 and $2\omega_0$. The system (7.1) exhibits a Smale–Williams type strange attractor when it is represented by a 4-D stroboscopic Poincaré map. In this case, hyperbolicity was verified numerically by analyzing the distribution of the angle φ (see Sec. 6.9.1) between the stable and unstable subspaces of the manifolds of the resulting chaotic invariant set.

The system (7.1) has been constructed as a laboratory device [Kuznetsov & Seleznev (2006)], and experimental and numerical solutions were found. This example of a physical system with a hyperbolic chaotic attractor is of considerable significance since it opens the possibility for real applications of the hyperbolic theory of dynamical systems.

Equations (7.1) are obtained by applying the so-called *equations of Kirchhoff* [Archibald (1988)], where the variables x and u are normalized voltages and currents in the corresponding LC circuit shown in Fig. 2 in [Kuznetsov & Seleznev (2006)] for the first self-oscillator (U_1 and I_1, respectively), and y and v are normalized voltages and currents in the second oscillator (U_2 and I_2). Time is normalized to the period of oscillations of the first LC oscillator, and the parameters A_1 and A_2 determine the amplitude of the slow modulation of the parameter $h_{1,2} \pm A_{1,2} \cos \frac{2\pi\tau}{N}$ responsible for the Andronov-Hopf bifurcation in both self-oscillators. The parameters h_1 and h_2 determine a map of the mean value of this parameter from the bifurcation threshold, and ε_1 and ε_2 are coupling parameters. From system (7.1), it is clear that each subsystem is described with time-varying coefficients. The normalization (7.4) below implies that the circular frequencies are $2\pi^3$ and 4π.

The Kirchhoff's law and the equation relating the current through the coil of inductance $L_{1,2}$ to the voltage across it for each subsystem gives the

---

[3] *i.e.*, It corresponds to the period $\Delta\tau = 1$.

following system:

$$\begin{cases} C_1 \frac{dU_1}{dt} + I_1 - \frac{U_1}{R_1} + f(U_1) + U_1\left(g_1 - k_1 a \cos \frac{\omega_0 t}{N}\right) = 0, \\ L_1 \frac{dI_1}{dt} = U_1 + k_2 U_2 \cos \omega_0 t \\ C_2 \frac{dU_2}{dt} + I_2 - \frac{U_2}{R_2} + f(U_2) + U_2\left(g_2 + k_2 a \cos \frac{\omega_0 t}{N}\right) = 0, \\ L_2 \frac{dI_2}{dt} = U_2 + k_1 U_1^2 \end{cases} \quad (7.2)$$

where the current through the nonlinear circuit component consisting of semiconductor diodes as a function of the voltage across it is given by

$$f(U) \approx \alpha U + \beta U^3. \quad (7.3)$$

The coefficients $k_{1,2}$ characterize the coupling between the subsystems (U_1, I_1) and (U_2, I_2), and the factor $g \pm ka \cos\left(\frac{\omega_0 t}{N}\right)$ represents the conductance of the field-effect transistor controlled by the ac gate voltage $\pm a \cos\left(\frac{\omega_0 t}{N}\right)$.

Note that system (7.1) was obtained from (7.2) using the dimensionless variables

$$\begin{cases} \tau = \frac{\omega_0 t}{2\pi} \\ x = U_1 \sqrt{\frac{6\pi\beta}{\omega_0 C_1}}, u = I_1 \sqrt{\frac{6\pi\beta}{\omega_0^3 C_1^3}}, \\ y = U_2 \sqrt{\frac{6\pi\beta}{\omega_0 C_2}}, v = I_2 \sqrt{\frac{3\pi\beta}{2\omega_0^3 C_2^3}} \end{cases} \quad (7.4)$$

and parameters

$$\begin{cases} A_1 = \frac{2\pi k_1 a}{\omega_0 C_1}, A_2 = \frac{2\pi k_2 a}{\omega_0 C_2} \\ h_1 = \frac{2\pi}{\omega_0 C_1}\left(\frac{1}{R_1} - \alpha - g_1\right) \\ h_2 = \frac{2\pi}{\omega_0 C_2}\left(\frac{1}{R_2} - \alpha - g_2\right) \\ \varepsilon_1 = k_1 \sqrt{\frac{\omega_0 C_1^2}{6\pi\beta C_2}}, \varepsilon_2 = k_2 \sqrt{\frac{C_2}{C_1}}. \end{cases} \quad (7.5)$$

Note that the circuit diagram representing the electronic device corresponding to system (7.1) was realized with the following assumptions:

(1) Both oscillators 1 and 2 contain a coil of inductance $L_{1,2}$ and a capacitor of capacitance $C_{1,2}$ making up an oscillating circuit, hence $\omega_0 = \frac{1}{\sqrt{L_1 C_1}}, 2\omega_0 = \frac{1}{\sqrt{L_2 C_2}}$.
(2) Negative-resistance components with resistances $-R_{1,2}$ based on operational amplifiers were introduced and can be treated as constant parameters in the entire operating voltage ranges of the corresponding oscillators 1 and 2.
(3) To ensure an increase in energy loss with oscillation amplitude, a nonlinear conductance was implemented in a circuit component consisting of two oppositely poled parallel arrays of series-connected semiconductor diodes.

(4) A field-effect transistor is used as an almost linearly conducting component, with drain current controlled by the gate voltage slowly varying as a periodic function of time with period $T = \frac{2\pi N}{\omega_0}$, where N is an integer.

In this case, the excitation is alternately transferred between oscillators 1 and 2 because of the following:

(a) One oscillator is active during successive half-periods of its variation, while the other is idle, and *vice versa*.
(b) A second-harmonic signal, which triggers the active oscillator 2 into oscillation in a frequency range around $2\omega_0$ was generated by the voltage squarer A_1 in oscillator 1.
(c) Detector A_2 heterodynes the output of oscillator 2 with an auxiliary signal of frequency ω_0 to produce a difference-frequency signal resonant with the frequency of oscillator 1, which triggers oscillator 1 when it becomes active.

Now, if one assume that the signal generated by the active oscillator 1 has a phase

$$\varphi : U_1 \propto \cos(\omega_0 t + \varphi), \qquad (7.6)$$

then one has the following:

(1) The output of voltage squarer A_1 contains the second harmonic $\cos(2\omega_0 t + 2\varphi)$ with phase 2φ.
(2) The generated signal U_2 has the phase 2φ, when oscillator 2 becomes active.
(3) After the signal U_2 is heterodyned with the auxiliary signal by means of detector A_2, the resulting single-frequency signal has the same phase 2φ.
(4) The signal generated by oscillator 1 during the next half-period of its operation has the phase 2φ.

Hence the phases φ_n of the signals generated by oscillator 1 during subsequent half-periods can be represented approximately by the sawtooth map (6.13) which exhibits chaotic dynamics. The state of system (7.1) can be described at any instant by the voltages and currents in both oscillating circuits, *i.e.*, $V = \{U_1, I_1, U_2, I_2\}$, and if the vector $V = V_n$ is given at $t = nT$, then the state variables at the next point of stroboscopic section $V_{n+1} = F(V_n)$ is determined uniquely by the dynamics. Thus the dynamics

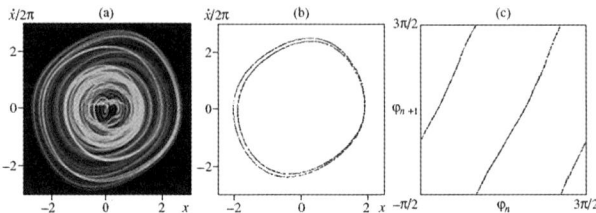

Fig. 7.1 Chaotic attractor obtained from system (7.1) for $N = 8, A_1 = 1.5, A_2 = 6, \varepsilon_{1,2} = 0.1$, and $h_{1,2} = 0$. (a) Portrait projected onto the x, x'-plane of oscillator 1; (b) stroboscopic section at $\tau_n = nN$, (c) first-return map for the phase of oscillator 1 [Kuznetsov and Seleznev, 2006].

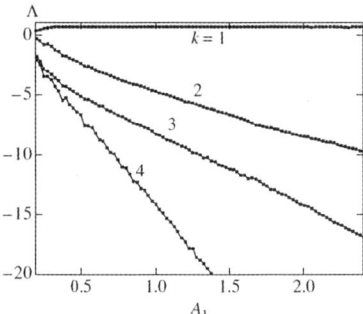

Fig. 7.2 Lyapunov spectrum of the stroboscopic map versus A_1 for $N = 8, A_2 = 4A_1, \varepsilon_{1,2} = 0.1$, and $h_{1,2} = 0$. The largest exponent is consistent with the estimated $\Lambda_1 \approx \ln 2$ [Kuznetsov and Seleznev, 2006].

of system (7.1) can be studied by analyzing the sequence of instantaneous states V_n generated by F at the period T. Geometrically, consider a 4-D solid toroid (direct product of a 1-D circle with a 3-D ball). Hence at each step of the stroboscopic map $V_{n+1} = F(V_n)$, the solid toroid is stretched to twice its original length, contracted in the transverse direction, folded in half, and squeezed into its original volume, because in the 4-D state space of system (7.1), the eigendirection associated with the phase φ is expanding and the remaining three are contracting. The above process behaves exactly as the Smale–Williams solenoid. Hence system (7.1) has a hyperbolic chaotic attractor, i.e., it is robust. For $N = 8, A_1 = 1.5, A_2 = 6, \varepsilon_1 = \varepsilon_2 = 0.1$, and $h_1 = h_2 = 0$. Figure 7.1(a) shows the projection onto the x, x'-plane of the phase portrait of the chaotic attractor. Figure 7.1(b) represents the stroboscopic section of the attractor at successive instants $\tau_n = nN$. Figure 7.1(c) shows the first-return map, i.e., phases of the signal generated by oscillator 1 at τ_n and τ_{n+1}, where their abscissa and ordinate are calculated by the following relation:

$$\varphi = \begin{cases} \arctan\left(-\frac{x'}{2\pi x}\right), x > 0 \\ \pi + \arctan\left(-\frac{x'}{2\pi x}\right), x < 0. \end{cases} \quad (7.7)$$

The first-return map appears to be topologically equivalent with minor discrepancy[4] to the sawtooth map (6.13). To confirm the chaoticity of system (7.1), the Lyapunov spectrum shown in Fig. 7.2 was obtained using the Bennetin's algorithm [Benettin et al. (1980), Christiansen & Rugh (1997)], i.e., using the following algorithm:

(1) Compute system (7.1) simultaneously with the corresponding linearized equations for perturbations,

$$\begin{cases} \tilde{x}' = -2\pi\tilde{u} - x^2\tilde{x} + \left(h_1 + A_1\cos\frac{2\pi\tau}{N}\right)\tilde{x}, \\ \tilde{u}' = 2\pi\left(\tilde{x} + \varepsilon_2\tilde{y}\cos 2\pi\tau\right), \\ \tilde{y}' = -4\pi\tilde{v} - y^2\tilde{y} + \left(h_2 - A_2\cos 2\frac{2\pi\tau}{N}\right)\tilde{y}, \\ \tilde{v}' = 4\pi\left(\tilde{y} + 2\varepsilon_1 x\tilde{x}\right). \end{cases} \quad (7.8)$$

(2) Perform a Gram–Schmidt orthonormalization at each integration step.
(3) Orthogonalize and normalize the perturbation vectors.
(4) Average the growth rates of the sum of logarithms of norms.

For $N = 8, A_1 = 1.5, A_2 = 6, \varepsilon_1 = \varepsilon_2 = 0.1$, and $h_1 = h_2 = 0$, one has $\Lambda_1 \approx 0.69 \approx \ln 2$, $\Lambda_2 \approx -6.64$, $\Lambda_3 \approx -11.12$, and $\Lambda_4 \approx -22.24$. This implies that the stroboscopic map F is chaotic[5]. The Kaplan–Yorke dimension (3.28) of the stroboscopic section of the attractor is $D = 1.29$ which agrees with the Grassberger–Procaccia correlation dimension $D \approx 1.4$. Note that the respective Lyapunov exponents λ_k for the system (7.1) and the Lyapunov exponents Λ_k of the map F satisfy the relation

$$\lambda_k = N^{-1}\Lambda_k. \quad (7.9)$$

The hyperbolicity of system (7.1) can be deduced from the fact that the largest Lyapunov exponent Λ_1 shown in Fig. 7.2 has an approximately constant value of about $\ln 2$ over the interval of A_1, and the others components

---

[4]This effect arises from the definition of the phase, and the inaccurate qualitative derivation of Eq. (7.1), but if the period ratio N is large, then better agreement can be achieved.
[5]The absence of the zero Lyapunov exponent is natural for maps and nonautonomous continuous-time flows.

are monotonically varying functions. The noticeable decrease of Λ_1 toward the left endpoint of the interval signifies a deviation from hyperbolicity. Analogous dynamics are obtained for other integer values of N, for example $N = 4$. Figure 6(a) in [Kuznetsov & Seleznev (2006)] shows the fractal fine structure between the attractor and the Smale–Williams solenoid depicted in Fig. 7.1(a). This implies that the resulting attractor is structurally stable under changes in system (7.1).

To test the hyperbolicity conjecture of the resulting attractor from system (7.1), Fig. 7 in [Kuznetsov & Seleznev (2006)] shows the spectra of chaotic signals generated by both oscillators 1 and 2. The spectra were calculated using a fast Fourier transform algorithm. These spectra are continuous and localized around $\omega_0 = 2\pi$ and $2\omega_0 = 4\pi$, respectively. It is easy to see that they do not exhibit pronounced multiple peaks characteristic of the spectra of nonhyperbolic attractors. The observed narrow peaks in the low-frequency tail of the spectrum of oscillator 2 are attributed to the heterodyning effect of the voltage squarer. To verify the hyperbolicity of the attractor, the numerical procedure given in Sec. 6.9.1 [Lai et al. (1993), Anishchenko et al. (2000)] was applied. Because the only unstable manifold of system (7.1) is one-dimensional and the stable one is three-dimensional, the following modified procedure was used:

(1) Generate a representative orbit $\{x(\tau), u(\tau), y(\tau), v(\tau)\}$ over a sufficiently long time interval on the attractor by computing system (7.7).
(2) Compute system (7.8) for perturbations of the orbit forward in time.
(3) Normalize the vector $a(\tau) = \{\tilde{x}(\tau), \tilde{u}(\tau), \tilde{y}(\tau), \tilde{v}(\tau)\}$ at each step to preclude divergence.
(4) Compute three replicas of system (7.8) backwards in time to find three vectors $b(\tau), c(\tau)$, and $d(\tau)$.
(5) Perform Gram Schmidt orthonormalization of the vectors at each integration step to avoid divergence and predominance of one of the vectors.
(6) Calculate at each point $\tau_n = nN$ of the stroboscopic section the vector $a_n = a(\tau_n)$ and the span of $\{b_n, c_n, d_n\} = \{b(\tau n), c(\tau n), d(\tau n)\}$ corresponding to the unstable direction and the 3-D stable manifold, respectively.
(7) Evaluate the angle α between the unstable and stable manifolds as follows: (a) Determine a vector v_n transversal to the 3-D stable manifold by solving the linear system of equations $v_n b_n = 0$, $v_n c_n = 0$, and $v_n d_n = 0$. (b) Calculate the angle $\beta_n \in \left[0, \frac{\pi}{2}\right]$ between v_n and a_n from $\cos\beta = \frac{|v_n a_n|}{|v_n||a_n|}$. (c) Set $\alpha_n = \frac{\pi}{2} - \beta_n$.

The result of the above procedure is shown in Fig. 8 in [Kuznetsov & Seleznev (2006)], where the histograms of the computed α_n are shown. The histograms of angle α between stable and unstable subspaces for system (7.1) for hyperbolic attractors corresponding to $A_1 = 1.5, A_2 = 6, \varepsilon_1 = \varepsilon_2 = 0.1$, and $h_1 = h_2 = 0$, and (a) $N = 8$, (b) $N = 4$ are shown in Fig. 8(a) and 8(b) in [Kuznetsov & Seleznev (2006)]. The histograms for a nonhyperbolic attractor corresponding to $N = 8, A_1 = 0.2, A_2 = 0.8, \varepsilon_{1,2} = 0.1$, and $h_{1,2} = 0$ is shown in Fig. 8(c) in [Kuznetsov & Seleznev (2006)].

Figure 8 in [Kuznetsov & Seleznev (2006)] shows clearly that both distributions are well separated from zero. Hence the hyperbolicity of the attractor was guaranteed from the test while the same test shown in Fig. 8(c) in [Kuznetsov & Seleznev (2006)] indicates nonhyperbolic dynamics for the set of parameters $N = 8, A_1 = 0.2, A_2 = 0.8, \varepsilon_{1,2} = 0.1$, and $h_{1,2} = 0$.

For the experimental results, the circuit design shown in Fig. 2 in [Kuznetsov & Seleznev (2006)] was implemented in a laboratory device with the values[6] $C_1 = 20$ nF, $C_2 = 5$ nf, $L_{1,2} = 1$ H, $f_1 = \frac{\omega_0}{2\pi} = 1090$ Hz, $f_2 = 2f_1 = 2180$ Hz and the following considerations:

(a) The negative-resistance amplifier was implemented using a 140UD26 operational amplifier.
(b) The nonlinear conductance was implemented using KD102 diodes.
(c) The time-varying conductance was introduced using KP303G field-effect transistors.
(d) The nonlinear components responsible for the coupling between the subsystems were based on 525PS2 analog frequency multipliers.
(e) The output voltages U_1 and U_2 were fed into a measuring device[7] or into a computer by an ADM12-3 analog-to-digital (A/D) converter with 12-bit resolution and a maximum sampling frequency of 3 MHz.
(f) The functions U'_1 and U'_2 were generated using a standard analog differential amplifier consisting of a 500 pF capacitor and a 62 kΩ resistor combined with a 140UD26 operational amplifier.

In this case, the experimental system exhibits chaotic oscillations shown in Fig. 10(a) in [Kuznetsov & Seleznev (2006)] for $N = 4$ photographed with an exposure time of a few seconds to capture a sufficiently large number of recurrent loops of a trajectory on the attractor. This attractor agrees with the above one shown in Fig. 6 in [Kuznetsov & Seleznev (2006)], and it

---
[6]Here, the ferrite-core coils having equal inductances L_1 and L_2, and $f_{1,2}$ are the free-running frequencies of the oscillators.
[7]Oscilloscope or spectrum analyzer.

is similar to the analogous portrait of the Smale–Williams solenoid. Here, the Grassberger–Procaccia correlation dimension calculated by processing a time series generated at a sampling rate of 200 kHz is $d \approx 2.3$, which agrees reasonably with numerical results. Furthermore, the largest Lyapunov exponent of the stroboscopic map evaluated by processing a time series sampled with the period $T = \frac{2\pi N}{\omega_0}$ is $\Lambda \approx 0.73$, which agrees reasonably with the estimated value of $\Lambda \approx \ln 2$.

Due to the topological equivalence of the first-return maps to the sawtooth map (6.13), these maps has hyperbolic attractors. Hence their hyperbolicity can be extracted from experimental data. Indeed, the first-return maps calculated by substituting into (7.7) the near-maximum values of $U_1'(t)$ sampled with a period $T = \frac{2\pi N}{\omega_0}$ and the derivatives of $U_1'(t)$ at the same instants produced by the differential amplifier are shown in Figs. 9(b) and 9(d) in [Kuznetsov & Seleznev (2006)].

As a concluding remark, the numerical and experimental results obtained above strongly suggest the similarity of system (7.1) to Smale–Williams type dynamics. However, the determination of the Lyapunov spectrum or the verification of the hyperbolicity of such an attractor cannot be verified by experiment.

7.1.2 Testing hyperbolicity of system (7.1)

The amplitude equations for the system (7.1) were introduced in [Kuptsov et al. (2008)] to prove that the resulting attractor of the system of amplitudes is also uniformly hyperbolic and to discuss some attributes of its hyperbolic dynamics. The universality of the amplitude equations implies that system (7.1) model the dynamics of various physical systems [Kuznetsov & Seleznev (2006)].

To derive a system of amplitude equations corresponding to system (7.1), the standard technique was used. Assume the solution of system (7.1) has oscillations with frequencies ω_0 and $2\omega_0$, respectively, and with slow varying complex amplitudes

$$\begin{cases} x(\tau) = a(\tau) e^{i\omega_0 \tau} + a^*(\tau) e^{-i\omega_0 \tau} \\ y(\tau) = b(\tau) e^{2i\omega_0 \tau} + b^*(\tau) e^{-2i\omega_0 \tau} \end{cases} \quad (7.10)$$

where the upper index "*" means complex conjugation and τ indicates time. The derivatives of the complex amplitudes a and b should satisfy additional conditions

$$\begin{cases} ae^{i\omega_0 \tau} + a^* e^{-i\omega_0 \tau} = 0 \\ be^{2i\omega_0 \tau} + b^* e^{-2i\omega_0 \tau} = 0. \end{cases} \quad (7.11)$$

Because a and b are supposed to be slow, the resulting equations using (7.11) and substituting (7.10) into (7.1) and averaging over the period $\frac{2\pi}{\omega_0}$ have the following form:

$$\begin{cases} a' = (\frac{A}{2})a\cos\left(\frac{2\pi\tau}{P}\right) - |a|^2\frac{a}{2} - \frac{i\varepsilon b}{4\omega_0} \\ b' = -(\frac{A}{2})a\cos\left(\frac{2\pi\tau}{P}\right) - |b|^2\frac{b}{2} - \frac{i\varepsilon a^2}{4\omega_0}. \end{cases} \quad (7.12)$$

Equations (7.12) allows the following rescaling of the time variable and parameters:

$$t = \frac{\tau}{2}, T = \frac{P}{2}, \epsilon = \frac{\varepsilon}{2\omega_0} \quad (7.13)$$

Hence the resulting system of amplitude equations is given by

$$\begin{cases} a' = Aa\cos\left(\frac{2\pi t}{T}\right) - |a|^2 a - i\epsilon b \\ b' = -Ab\cos\left(\frac{2\pi t}{T}\right) - |b|^2 b - i\epsilon a^2. \end{cases} \quad (7.14)$$

Define phases within the interval $[0, 2\pi)$ as follows: $\phi = \arg a$ and $\psi = \arg b$. If one supposes that the first oscillator is excited and its amplitude $|a|$ is high, then the second one is suppressed and its amplitude $|b|$ is small. In this case, the phases can be changed only as a result of interaction between subsystems because the coefficients in Eqs. (7.14) are real except for the coupling. Hence the phase of a remains constant during the excitation stage because if a is excited, then $|b|$ is small and its action on a is negligible. After a half period $\frac{T}{2}$, the oscillator b inherits a doubled phase of a at the threshold of its own excitation because the opposite influence from a to b is high and the coupling term is proportional to a^2. Hence the phase also gets a shift $-\frac{\pi}{2}$ due to the imaginary unit in the coupling term. This means that the roles of the subsystems a and b are exchanged, i.e., the phase of b remains constant and the first oscillator a doubles its phase during the period T. When $A = 3, T = 5$, and $\epsilon = 0.05$, the alternate excitation of the subsystems a and b demonstrate the chaotic nature of the dynamics. Also the time dependence of the variable a shows the existence of a series of spikes corresponding to the stages of the excitation of a. Thus it is possible to derive a map for a series of phases $\phi_n = \arg a(nT)$ that are measured over the time step T

$$\phi_{n+1} = 2\phi_n - \pi \,(\text{mod}\, 2\pi) \quad (7.15)$$

up to an additive constant. Map (7.15) is a Bernoulli map (6.13) that displays homogeneous chaotic dynamics, i.e., the rate of exponential divergence of two close trajectories is identical at each point of the phase space and equal to ln 2. The diagram of ϕ_{n+1} versus ϕ_n shows such a correspondence with the Bernoulli map (6.13). The most important point is the topological equivalence between the empirical map (7.15) and the Bernoulli map (6.13). One full circle that passes a preimage implies two passes of a circle for an image. The preimages imply a phase change of 2π and 4π for images. Thus the presented attractor from system (7.1) is robust and hyperbolic because for $A = 3, T = 5$, and $\epsilon = 0.05$, the Lyapunov exponents of the corresponding Poincaré map are $\Lambda_1 \approx 0.691 \simeq \ln 2, \Lambda_2 \approx -4.06, \Lambda_3 \approx -6.48$, and $\Lambda_4 \approx -9.06$. The first component Λ_1 is with a good agreement with the one for the Bernoulli map (6.13). This component is almost independent on A, and the others vary rather smoothly, without peaks and valleys.

At the instant $t_n = nT$, the state of system (7.14) is given by a vector

$$x_n = \{\operatorname{Re} a(t_n), \operatorname{Im} a(t_n), \operatorname{Re} b(t_n), \operatorname{Im} b(t_n)\}. \tag{7.16}$$

If x_n is an initial state, then the integration of (7.14) over the period T gives a new vector

$$x_{n+1} = \{\operatorname{Re} a(t_{n+1}), \operatorname{Im} a(t_{n+1}), \operatorname{Re} b(t_{n+1}), \operatorname{Im} b(t_{n+1})\} \tag{7.17}$$

which is determined by the vector x_n. Hence the Poincaré map of (7.14) that operates in $\mathbb{R}^4$ is given by:

$$x_{n+1} = T(x_n). \tag{7.18}$$

Geometrically, each vector x_n belongs to a 4-D hyperplane $t = t_n = nT$, which is a section of a flow of trajectories in the 5-D extended phase space $\{\operatorname{Re} a, \operatorname{Im} a, \operatorname{Re} b, \operatorname{Im} b, t\}$. Periodicity of the phase in t implies that these hyperplanes can be identified, and map (7.18) maps the 4-D hyperplane $\{\operatorname{Re} a, \operatorname{Im} a, \operatorname{Re} b, \operatorname{Im} b\}$ onto itself. According to (7.15), the map T expands a volume element in a direction associated with the phase ϕ, while three others directions are contracting. Due to the periodicity of the phase, the 4-D toroid denoted by U can be considered as a direct product of a 1-D circle and a 3-D ball. Hence one iteration of T corresponds to a longitudinal stretch and a transverse contraction of the 4-D toroid followed by its kinking and insertion into the initial area as a double loop. This exactly corresponds to the Smale–Williams procedure embedded into a 4-D rather then a 3-D state space as is usual. In this case, the toroid U with its interior $IntU$ is an absorbing domain for the map T, i.e., $T(U) \subset IntU$ and any attractor A of the map T can be expressed as

$$A = \cap_{n=1}^{\infty} T^n(U). \tag{7.19}$$

Let us define a torus in space $\{\operatorname{Re} a, \operatorname{Im} a, \operatorname{Re} b, \operatorname{Im} b\}$ as
$$(|a| - r)^2 + |b|^2 = (\chi d)^2 \tag{7.20}$$
where r is the radius of the 1-D circle, and χd is the radius of the 3-D ball that appears in a radial section of the torus in a hyperplane perpendicular to the circle. The values r and d should be found for the attractor to fit inside the torus U. For $A = 3, T = 5$, and $\epsilon = 0.05$, a large number of points belonging to the attractor A of the map (7.18) were computed. Their projections onto the plane $\{\operatorname{Re} a, \operatorname{Im} a\}$ fall somewhere in the vicinity of a circle whose radius $r = r(|a|) = 0.631166$. The role of χ is to scan points both on the surface ($\chi = 1$) and inside the absorbing domain U, $0 \leq \chi \leq 1$. However, the torus (7.20) can also be defined parametrically as
$$\begin{cases} \operatorname{Re} a = (\chi d \cos\theta + r)\cos\phi \\ \operatorname{Re} b = \chi d \sin\theta \cos\psi \\ \operatorname{Im} a = (\chi d \cos\theta + r)\sin\phi \\ \operatorname{Im} b = \chi d \sin\theta \sin\psi, \end{cases} \tag{7.21}$$
where ϕ and ψ are the phases of a and b, respectively. The computation of d, requires the following steps:

(1) For $\chi = 1$ consider a series of tori with identical radii r parameterized by a section radius dx that grows from 0 in small steps.
(2) For each dx, cover the surface of a corresponding torus by a mesh using parametric Eqs. (7.21) with a step $\frac{2\pi}{50}$ for the angle variables ϕ, ψ, and θ.
(3) Iterate the map (7.18) once from each node of the mesh. Thus the torus dx generates a set of *image* tori because an image point falls on the surface ($\chi = 1$) of a new torus whose section radius is equal to $(|a| - r)^2 + |b|^2$, where $|a|$ and $|b|$ are here related to the image point.
(4) Take as an absorbing domain the worst case where the image torus has the largest section radius dy and obtain the function $dy = F(dx)$. In this case, F grows monotonically and intersects the line $dy = dx$ at $dx \approx 0.0918$. Now, take a torus whose section radius d is about 1% above the intersection point[8].

Finally, the set of parameters of the system (7.1) and corresponding parameters of the absorbing domain U are given by
$$A = 3, T = 5, \epsilon = 0.05, r = 0.631166, d = 0.0927. \tag{7.22}$$

---

[8]In this case, all tori to the right from the point $dx \approx 0.0918$ can be taken as an absorbing domain U for the map T because their surfaces and interiors are mapped into their interiors.

Verification of hyperbolicity of the attractor of the system (7.14) was done using the method developed in [Kuznetsov & Sataev (2007)]. This method is based on the theorem for expanding and contracting cones [Schuster (1984), Katok & Hasselblatt (1995), Hunt (2000)] and some numerical calculations.[9] Theorem 7.1 below deals with the Poincaré map diffeomorphism $T(x)$ of class C^∞ in the Euclidian space $\mathbb{R}^4$. Let $\mathbf{DT}_x = \{\frac{\partial x'_i}{\partial x_j}\}, i,j = 0,1,2,3$ be the Jacobian matrix of the Poincaré map $x' = T(x)$ at x, and DT_x^{-1} is the derivative matrix for the inverse map $T^{-1}(x)$. If δx (vectors δx form a tangent space V_x associated with x) is the vector of a small perturbation to x, then in a linear approximation the evolution of a perturbed state $x + \delta x$ corresponds to a transformation of the vector δx according to a linear mapping $\delta x' = \mathbf{DT}_x \delta x$. Hence the following result was proved in [Schuster (1984), Katok & Hasselblatt (1995), Hunt (2000)]:

Theorem 7.1. *Suppose that a diffeomorphism T of class C^∞ maps a bounded domain $U \subset \mathbb{R}^4$ into itself: $T(U) \subset IntU$, and $A \subset IntU$ is an invariant set for the diffeomorphism. The set A will be uniformly hyperbolic if there exists a constant $\gamma > 1$ and the following conditions hold: (1) The expanding and contracting cones S_x^γ and C_x^γ may be defined in the tangent space V_x at each $x \in A$, such that $\|\mathbf{DT}_x u\| \geq \gamma \|u\|$ for all $u \in S_x^\gamma$, and $\|\mathbf{DT}_x^{-1} v\| \geq \gamma \|v\|$ for all $v \in C_x^\gamma$; moreover, for all $x \in A$, they satisfy $S_x^\gamma \cap C_x^\gamma = \emptyset$ and $S_x^\gamma + C_x^\gamma = V_x$. (2) The cones S_x^γ are invariant with respect to action of DT, and C_x^γ are invariant with respect to action of $\mathbf{DT}^{-1}$, i.e., for all $x \in A$, $\mathbf{DT}_x(S_x^\gamma) \subset S_{T(x)}^\gamma$ and $\mathbf{DT}_x^{-1}(C_x^\gamma) \subset C_{T^{-1}(x)}^\gamma$.*

Note that the conditions of Theorem 7.1 are valid for all points on the attractor $A \subset U$ if they are valid for the absorbing domain U. The following procedure was performed for a verification of the conditions of the Theorem 7.1:

(1) Choose the vector $x = \{\text{Re}\,a, \text{Im}\,a, \text{Re}\,b, \text{Im}\,b\} \in U$, and solve Eqs. (7.14) numerically on the interval $t \in [0, T]$ to get the image $x' = T(x)$.

(2) Initialize the four sets of equations for small perturbations given by

$$\begin{cases} \delta x'_0 = A\cos(\frac{2\pi t}{T})\delta x_0 - (3x_0^2 + x_1^2)\delta x_0 - 2x_0 x_1 \delta x_1 + \epsilon \delta x_3 \\ \delta x'_1 = A\cos(\frac{2\pi t}{T})\delta x_1 - (x_0^2 + 3x_1^2)\delta x_1 - 2x_0 x_1 \delta x_0 - \epsilon \delta x_2 \\ \delta x'_2 = -A\cos(\frac{2\pi t}{T})\delta x_2 - (3x_2^2 + x_3^2)\delta x_2 + w_1 \\ \delta x'_3 = -A\cos(\frac{2\pi t}{T})\delta x_3 - (x_2^2 + 3x_3^2)\delta x_3 + w_2 \\ w_1 = -2x_3 \delta x_2 x_3 + 2\epsilon(x_0 \delta x_1 + x_1 \delta x_0) \\ w_2 = -2x_2 \delta x_3 x_2 - 2\epsilon(x_0 \delta x_0 - x_1 \delta x_1) \end{cases} \quad (7.23)$$

---

[9]This is the suggestion of Sinai [Sinai (1979)] for the verification of hyperbolicity for the attractor of a Poincaré map.

with unit vectors $(1,0,0,0)$, $(0,1,0,0)$, $(0,0,1,0)$, and $(0,0,0,1)$, respectively, and solve these equations simultaneously with the original system (7.1). From the resulting four vector-columns, compose the matrix $\mathbf{U} = \mathbf{DT}_x$. Here, $x_0 = \operatorname{Re} a$, $x_1 = \operatorname{Im} a$, $x_2 = \operatorname{Re} b$, $x_3 = \operatorname{Im} b$, and $\delta x_0, \delta x_1, \delta x_2, \delta x_3$ are small perturbations to these values.

(3) Iterate the Poincaré map T one time from x so that any perturbation vector u transforms to $u' = \mathbf{U}u$, and one has $\|u'\|^2 = u^T \mathbf{U}^T \mathbf{U} u$, $u = \mathbf{U}^{-1} u'$ and $\|u\|^2 = u'^T \mathbf{U}^{-1,T} \mathbf{U}^{-1} u'$ (T is the transposition). Hence the preimage of u' is related to the expanding cone S_x^γ if $\|u'\| \geq \gamma \|u\|$, or

$$\begin{cases} u'^T \mathbf{H}_\gamma u' \leq 0 \\ \mathbf{H}_\gamma = \mathbf{U}^{-1,T} \mathbf{U}^{-1} - \gamma^{-2}. \end{cases} \quad (7.24)$$

(4) Start from $x' = T(x)$, and obtain a vector u' that transforms to $u'' = \mathbf{U}' u'$, and $\|u''\|^2 = u'^T \mathbf{U}'^T \mathbf{U}' u'$. The expanding cone $S_{T(x)}^\gamma$ at $x' = T(x)$ is determined by an inequality $\|u''\| \geq \gamma \|u'\|$, or

$$\begin{cases} u'^T \mathbf{H}'_\gamma u' \geq 0 \\ \mathbf{H}'_\gamma = \mathbf{U}'^T \mathbf{U}' - \gamma^{-2}. \end{cases} \quad (7.25)$$

In this case, the condition $\mathbf{DT}_x(S_x^\gamma) \subset S_{T(x)}^\gamma$ of Theorem 7.1 is formulated in terms of two quadratic forms: If (7.24) holds, then (7.25) holds also. To simplify the analysis with the quadratic form $u'^T \mathbf{H}'_\gamma u'$, it is necessary to reduce it to its canonical form using a coordinate change. Indeed, the matrix $\mathbf{U}'^T \mathbf{U}'$ is real and symmetric, and an orthonormal basis of eigenvectors d_0, d_1, d_2, d_3 is chosen with the matrix $D = \{d_0, d_1, d_2, d_3\}$ as a diagonalizer, i.e.,

$$\mathbf{D}^T \mathbf{U}'^T \mathbf{U}' \mathbf{D} = \{\Lambda_i \delta_{ij}\}, i, j = 0, 1, 2, 3. \quad (7.26)$$

Assume that $\Lambda_0 \leq \Lambda_1 \leq \ldots \leq \Lambda_4$. In this case, there is one stretching and three contracting directions because $\Lambda_0^2 > 1$ and $\Lambda_{1,2,3}^2 < 1$. Let γ be a parameter selected such that $\Lambda_0^2 > \gamma^2$, $\Lambda_{1,2,3}^2 < \gamma^2$. The transformation $\mathbf{D}$ transforms the matrix $\mathbf{H}'_\gamma$ into a diagonal one, i.e.,

$$\mathbf{D}^T \mathbf{H}'_\gamma \mathbf{D} = \mathbf{D}^T \left(\mathbf{U}'^T \mathbf{U}' - \gamma^2 \right) \mathbf{D} = \left\{ \left(\Lambda_i - \gamma^2 \right) \delta_{ij} \right\}. \quad (7.27)$$

By additional dilatation (compression)

$$S = \{s_i^{-1} \delta_{ij}\}, s_0 = \sqrt{\Lambda_0 - \gamma^2}, s_{1,2,3} = \sqrt{\gamma^2 - \Lambda_{1,2,3}^2} \quad (7.28)$$

one has

$$\mathbf{H}'_\gamma = \mathbf{S}^T \mathbf{D}^T \left(\mathbf{U}'^T \mathbf{U}' - \gamma^2 \right) \mathbf{D} \mathbf{S} = \sigma_i \delta_{ij}, \sigma_0 = 1, \sigma_{1,2,3} = -1. \quad (7.29)$$

The same transformation $\mathbf{D}$ applied to the matrix $\mathbf{H}_\gamma = \mathbf{U}^{-1,T}\mathbf{U}^{-1} - \gamma^{-2}$ gives

$$\tilde{H}'_\gamma = \mathbf{S}^T\mathbf{D}^T(\mathbf{U}^{-1,T}\mathbf{U}^{-1} - \gamma^{-2})\mathbf{D}\mathbf{S} = \{h_{ij}\}, \text{ where } h_{ij} = h_{ji}. \quad (7.30)$$

The condition (7.25) for which a vector $c = (1, c_1, c_2, c_3)$ is in the expanding cone $S^\gamma_{T(x)}$ is given by

$$c^T \tilde{H}'_\gamma c \geq 0, \text{ or } c_1 + c_2 + c_3 \leq 1, \quad (7.31)$$

which corresponds to the interior of the unit ball in the 3-D space $\{c_1, c_2, c_3\}$. The condition (7.24) for which the preimage of the vector $c = (1, c_1, c_2, c_3)$ belongs to the expanding cone S^γ_x is

$$c^T \tilde{H}'_\gamma c \leq 0 \text{ or } h_{00} + \sum_{\alpha=1}^{3}(h_{0\alpha}c_\alpha + h_{\alpha 0}c_\alpha) + \sum_{\alpha,\beta=1}^{3} h_{\alpha\beta}c_\alpha c_\beta \leq 0, \quad (7.32)$$

which determines the interior of a certain ellipsoid in the space $\{c_1, c_2, c_3\}$. In this case, the inclusion $\mathbf{DT}_x(S^\gamma_x) \subset S^\gamma_{T(x)}$ holds if the ellipsoid is placed inside the unit ball. The coordinates for the center of the ellipsoid can be obtained from the equations

$$\sum_{\beta=1}^{3} h_{\alpha\beta}\bar{c}_\beta = -h_{\alpha 0}, \alpha = 1, 2, 3, \quad (7.33)$$

and the distance of this point from the center of the unit ball is given by

$$\rho = \sqrt{\bar{c}_1^2 + \bar{c}_2^2 + \bar{c}_3^2}. \quad (7.34)$$

If the origin was transformed to the point (c_1, c_2, c_3), then the equation for the surface of the ellipsoid becomes

$$\begin{cases} \sum_{\alpha,\beta=1}^{3} h_{\alpha\beta}\tilde{c}_\alpha\tilde{c}_\beta = R^2 \\ \tilde{c}_\alpha = c_\alpha - \bar{c}_\alpha, \\ R^2 = -h_{00} - \sum_{\alpha=1}^{3}(h_{0\alpha}\bar{c}_\alpha + h_{\alpha 0}\bar{c}_\alpha) - \sum_{\alpha,\beta=1}^{3} h_{\alpha\beta}\bar{c}_\alpha\bar{c}_\beta. \end{cases} \quad (7.35)$$

Now consider a symmetric 3×3 matrix $\mathbf{h} = (h_{\alpha\beta})$. In this case, the equation for the ellipsoid surface takes the form

$$l_1\xi_1 + l_2\xi_2 + l_3\xi_3 = R^2, \quad (7.36)$$

where l_1, l_2, l_3 are eigenvalues of the matrix $\mathbf{h}$, and the largest principal semi-axis of the ellipsoid is expressed by the smallest eigenvalue

$$r_{\max} = \frac{R}{\sqrt{l_{\min}}}. \quad (7.37)$$

This equation was obtained in the diagonal representation obtained with an orthogonal coordinate transformation $(\tilde{c}_1, \tilde{c}_2, \tilde{c}_3) \to (\xi_1, \xi_2, \xi_3)$. Hence the ellipsoid is positioned inside the unit ball when

$$f = r_{\max} + \rho < 1. \tag{7.38}$$

Inequality (7.38) completes[10] the procedure for verification of the expanding cones inclusion for the point x taking into account the fact that the cones S^γ and $C^{1/\gamma}$ are complimentary sets[11]: $S^\gamma \cup C^{1/\gamma} = \mathbb{V}$, the above procedure, applied to the points of the absorbing domain U with $\gamma < 1$ is equivalent to verification of the condition for the contracting cones in the domain $x \in T^2(U)$ with the parameter $\gamma' = \frac{1}{\gamma} > 1$, $\mathbf{DT}_x^{-1}\left(S_x^{\frac{1}{\gamma}}\right) \subset S_{T^{-1}(x)}^{\frac{1}{\gamma}}$.

7.1.3 Numerical verification of the hyperbolicity of system (7.1)

Numerical tests for the above method for the Poincaré map of the system (7.14) confirms that Theorem 7.1 is fulfilled and the chaotic attractor is indeed hyperbolic. To achieve this numerical test, the following tasks must be taken into account:

(1) The algorithms for computation of the eigenvalues of a real symmetric matrix, for matrix inversion, and for the solution of a set of linear algebraic equations can be found in [Press *et al.* (1992)].
(2) At each point, the matrix $\mathbf{U}_x$ is computed to find its eigenvalues Λ_i.
(3) The expanding and contracting cones are defined if the eigenvalues of the matrix $\mathbf{U}_x^T \mathbf{U}_x$ satisfy $\Lambda_0^2 > 1$ and $\Lambda_1^2 < 1$, respectively, and the computations show that both inequalities are valid in the entire absorbing domain U for the parameters set (7.23).
(4) In Eqs. (7.21), the parameters χ and ψ are constant while ϕ and θ are varied from 0 to 2π with a step $\frac{2\pi}{50}$. For the set (7.23) at $\chi = 1$ (the surface of the absorbing domain U) and at $\psi = 0$, one has $\Lambda_0^2 > 1$, while $\Lambda_1^2 < 1$ and $\Lambda_{2,3}^2 < 1$. For $\chi = 1, \psi = 0, \gamma = 1.1$, and the set (7.23),

---

[10]*i.e.*, checking condition (7.38) inside U for two parameters γ and $1/\gamma$ implies that both conditions for expanding and for contracting cones are valid in the domain $A \subset T^2(U)$.
[11]Here S^γ corresponds to the cone of vectors that either expand or contract but no stronger than by factor γ. Furthermore, the cones S^γ and C^γ have a common border only at $\gamma = 1$, while for $\gamma > 1$ they do not intersect as required by the condition of Theorem 7.1: $S_x^\gamma \cap C_x^\gamma = \emptyset$.

the function f is smooth and varies sufficiently slowly[12] and is always less then 0.5 so that there are no peaks which can be candidates for violation of the inequality (7.38) between the checked values of ϕ and θ. Thus inequality (7.38) is valid not only on the surface of the absorbing domain U, but also for $0 \leq \chi \leq 1$ and far beyond. Furthermore, the inequality $f_{\max} < 1$ holds for a wide range of γ both below and above the point $\gamma = 1$, and this confirms the location of expanding and contracting cones in the absorbing domain U.

As a final remark, the reader can well understand the above procedure when solving the exercises at the end of this chapter. For further information, see [Kuznetsov & Sataev (2006)]. Due to some complications of high-order dynamical systems as seen for system (7.1), four examples of uniformly hyperbolic, smooth, low-order, autonomous dynamical systems are presented in [Kuznetsov & Pikovsky (2007)]. The approach used is similar to the one given in [Kuznetsov (2005)]. These models are termed by the authors as **A, B, C,** and **D**. **Model A** is the flow model of minimal dimension (four) that capable of modeling the Smale–Williams type attractor introduced in Sec. 6.4.3. This model consists of two asymmetrically coupled oscillators with amplitudes similar to those in the predator-prey model. These amplitudes become large or small alternately. In this case, the excitation transfer is accompanied with a doubling of the phase variable, namely, it corresponds to the Bernoulli map (6.13) or expanding circle map (6.8) [Gaspard (2005)]. **Model A** has a Smale–Williams type attractor in its three-dimensional associated Poincaré map. The **model A** is given by

$$\begin{cases} x_1' = \omega_0 y_1 + \left(1 - a_2 + \tfrac{1}{2}a_1 - \tfrac{1}{50}a_1^2\right) x_1 + \varepsilon x_2 y_2 \\ y_1' = -\omega_0 x_1 + \left(1 - a_2 + \tfrac{1}{2}a_1 - \tfrac{1}{50}a_1^2\right) y_1 \\ x_2' = \omega_0 y_2 + (a_1 - 1) x_2 + \varepsilon x_1 \\ y_2' = -\omega_0 x_2 + (a_1 - 1) y_2 \end{cases} \quad (7.39)$$

where ε is the coupling constant and $a_1 = x_1^2 + y_1^2, a_2 = x_2^2 + y_2^2$. For system (7.39), several different initial conditions lead to the same attractor when $\omega_0 = 2\pi$, and $\varepsilon = 0.3$, i.e., the observed chaotic attractors appear structurally stable when $\varepsilon \in [0.0004, 0.3]$ and other parameters (about $\pm 10\%$ of the presented value) because nothing changes regarding its nature, its shape, and the approximate value of the largest Lyapunov exponent for the associated Poincaré map. In the course of transition of the second oscillator

---

[12]It is shown numerically that f depends essentially on ϕ only, while the dependence on θ and ψ is very weak.

to the active stage, it is forced by the term x_1 and adopts the phase of the first oscillator. In turn, the first oscillator at the transition to the active stage is forced by the term $x_2 y_2$. Hence, in accordance with this mechanism, the phase doubling will occur during transfer of the excitation. To demonstrate that the phase doubling holds for system (7.39), the Poincaré map method was used in the four-dimensional phase space $\{x_1, y_1, x_2, y_2\}$ as follows:

(1) Define the surface
$$S = x_1^2 + y_1^2 - x_2^2 - y_2^2 = 0 \tag{7.40}$$
that corresponds to the case where the instantaneous amplitudes for the two oscillators are equal.
(2) Consider the intersections of the surface S with trajectories of system (7.39) in the direction of the increasing value of S.
(3) Choose a 3-D initial point v_n on the surface S.
(4) The next intersection of the surface S in the same direction gives the image vector v_{n+1} and defines the three-dimensional Poincaré map
$$v_{n+1} = T(v_n). \tag{7.41}$$
(5) Define the phases of the second oscillator as
$$\varphi = \arg(x_2 + iy_2) \tag{7.42}$$
at successive intersections of the surface $S = 0$.
(6) Plot the phases of the second oscillator $y_{1,2}$ on the diagram φ_{n+1} versus φ_n.

For $\omega_0 = 2\pi$ and $\varepsilon = 0.3$, the average time interval between the intersections of the surface $S = 0$ is approximately $T_{av} = 7.245$. This is a characteristic time interval between the bursts for a typical time dependence for the variables $x_{1,2}$, $y_{1,2}$ obtained by computer solution of Eq. (7.39). Furthermore, the diagram φ_{n+1} versus φ_n looks topologically similar to the expanding circle map (6.8). Indeed, both portraits possess a Cantor-like transversal structure. Lyapunov exponents are computed using the Benettin method [Benettin et al. (1992)]. The results are $\lambda_1 = 0.0918 \pm 0.0002 > 0$, $\lambda_2 = -0.000004 \pm 0.00003$, $\lambda_3 = -4.074 \pm 0.002$, and $\lambda_4 = -4.3936 \pm 0.0005$. In this case, the positive exponent λ_1 indicates chaos, and the near zero exponent λ_2 indicates a perturbation associated with a shift along the orbit. The exponents $\lambda_{3,4} < 0$, and they are the cause for the approach of trajectories of system (7.39) to their attractor. For the

Poincaré map T, the nonzero Lyapunov exponents are obtained using the relation $\Lambda = \lambda T_{av}$. Here, $\rho_1 = \exp[\lambda_1 T_{av}] = 1.945 \simeq 2$ is close to the value expected from the approximation based on the expanding circle map (6.8). The Kaplan–Yorke dimension for system (7.39) is $D_L \simeq 2.0225$, and for the attractor of the Poincaré map T one has $d = D_L - 1 = 1.0225$.

The same approach and results can be obtained from **model B** which displays in the 5-D Poincaré map a Smale–Williams attractor in a perturbed heteroclinic cycle

$$\begin{cases} x_1' = \omega_0 y_1 + \left(1 - a_1 - \tfrac{1}{2}a_2 - 2a_3\right) x_1 + \varepsilon x_2 y_2 \\ y_1' = -\omega_0 x_1 + \left(1 - a_1 + \tfrac{1}{2}a_2 - 2a_3\right) y_1 \\ x_2' = \omega_0 y_2 + \left(1 - a_2 + \tfrac{1}{2}a_3 - 2a_1\right) x_2 + \varepsilon x_3 y_3 \\ y_2' = -\omega_0 x_2 + \left(1 - a_2 + \tfrac{1}{2}a_3 - 2a_1\right) y_2 \\ x_3' = \omega_0 y_3 + \left(1 - a_3 + \tfrac{1}{2}a_1 - 2a_2\right) x_3 + \varepsilon x_1 y_1 \\ y_3' = -\omega_0 x_3 + \left(1 - a_3 + \tfrac{1}{2}a_1 - 2a_2\right) y_3, \end{cases} \quad (7.43)$$

where $a_i = x_i^2 + y_i^2$, $i = 1, 2, 3$, and ε is the coupling constant. Major properties of system (7.43) with $\omega_0 = 1$ and $\varepsilon = 0.03$ are listed below:

(1) Only one attractor was obtained when testing different initial conditions as in the case of **model A**.
(2) The doubling phase mechanism, was demonstrated using the Poincaré map, defined in the six-dimensional phase space $\{x_1, y_1, x_2, y_2, x_3, y_3\}$ due to the symmetry of the system (7.43)[13] with some additional modification and three surfaces given by

$$\begin{cases} S_1 = x_1^2 + y_1^2 - x_2^2 - y_2^2 = 0 \\ S_2 = x_2^2 + y_2^2 - x_3^2 - y_3^2 = 0 \\ S_1 = x_3^2 + y_3^2 - x_1^2 - y_1^2 = 0, \end{cases} \quad (7.44)$$

and successive intersections of these surfaces in the direction of growth of S_j were examined. The intersection of $S_j = 0$ given in the order $S_1 \to S_3 \to S_2 \to \ldots$ with a trajectory of system (7.43) indicates the condition that an amplitude of the rising oscillator (x_j, y_j) overtakes that of the previously excited partner. If $v_n \in \mathbb{R}^5$ is a point on a surface $S_j = 0, j = 1, 2, 3$, then the next intersection of a surface S_j by the trajectory of system (7.43) started at v_n defines the Poincaré map (7.41). Hence the Poincaré map corresponds to a threefold iteration of the map T.

---

[13] In this case, it is possible to identify the spaces of images and pre-images under T because of the cyclic symmetry of system (7.43).

(3) The average time interval between the intersection of the surfaces $S_j = 0$, is $T_{av} = 10.96$.
(4) The Lyapunov exponents for the system (7.43) are $\lambda_1 = 0.06305 \pm 0.00004$, $\lambda_2 = -0.000001 \pm 0.00001$, $\lambda_3 = -0.2101 \pm 0.0002$, $\lambda_4 = -0.2259 \pm 0.0004$, $\lambda_5 = -0.3395 \pm 0.0006$, and $\lambda_6 = -1.9999349 \pm 0.0000005$. The Kaplan–Yorke dimension is $D = 2.30$ and $d = D - 1 = 1.30$.
(5) The autocorrelation function for the observable $X = x_1 + x_2 + x_3$ shows that the correlations decay rather fast[14], and there exists a slowly decaying nearly periodic tail resulting from the fact that phases of the oscillators are strongly chaotic.

Another example of maps that can display a Smale–Williams type strange attractor can be found in [Kuznetsov & Ponomarenko (2008)] where a radiotechnical circuit generating chaotic signals has been proposed and implemented based on the nonautonomous *delay-feedback van der Pol oscillator* (dFV) given by

$$x'' - \left(A\cos\Omega t - x^2\right)x' + \omega_0^2 x = \varepsilon x\left(t-\tau\right).x'\left(t-\tau\right)\cos\omega_0 t \quad (7.45)$$

where x is the dynamic variable of (7.45), A is the amplitude, ω_0 (it is possible to assume that ω_0 is an integer multiple of Ω, so that the external action is on the whole periodic) is the working frequency of the vdP oscillator (7.45) with $\Omega << \omega_0$, and the factor ε determines the small amplitude of the delay feedback. The signal of this circuit alternatively exhibits states of excitation and decay. This is a result of the periodic variation of a parameter responsible for the bifurcation creating a limit cycle. In this case, the phase variable exhibits doubling at every excitation stage, which implies that the phase obeys the circle-stretching mapping, *i.e.*, the Bernoulli map (7.15) with chaotic dynamics.

7.1.4 *Modeling the Arnold cat map*

In [Isaeva et al. (2006)], an example of a flow system is presented with an attractor concentrated mostly at a surface of a two-dimensional torus, the dynamics on which is governed by the Arnold cat map [Arnold (1988), Anishchenko et al. (2002)],

$$\begin{cases} p_{n+1} = p_n + q_n \pmod{1} \\ q_{n+1} = p_n + 2q_n \pmod{1}, \end{cases} \quad (7.46)$$

---

[14]Generally, correlations in continuous-time systems with hyperbolic attractors must not demonstrate a perfect decay.

or its modification with added nonlinear terms on the right-hand side of the equations. This system illustrates a realistic example of dynamics approximately described by a hyperbolic map on a torus, *i.e.*, an attractor in the eight-dimensional phase space of a chosen Poincaré map of a 9-D flow. The 8-D map can be reduced approximately to a two-dimensional map for the phases of one pair of the oscillators due to the strong phase space volume compression.

To a good approximation, a set of two phase variables φ_x and φ_y observed stroboscopically follows the Arnold cat map (6.18). These two pairs of oscillators become active in turn as they pass the excitation from one to the other due to the slow modulation of the parameters responsible for the self excitation and because of the appropriately introduced coupling.

The main idea is adopted from a recent work of [Kuznetsov (2005)] introduced in Sec. 7.1.1.

Lemma 7.1. *The Arnold cat map (6.18) can be represented as a twofold composition of a simpler map given by*

$$\begin{cases} p_{n+1} = q_n \pmod{1} \\ q_{n+1} = p_n + q_n \pmod{1}. \end{cases} \qquad (7.47)$$

Proof. Use the relation $\begin{pmatrix} 1 & 1 \\ 1 & 2 \end{pmatrix} = \begin{pmatrix} 0 & 1 \\ 1 & 1 \end{pmatrix} \begin{pmatrix} 0 & 1 \\ 1 & 1 \end{pmatrix}$. □

To implement the corresponding dynamics physically, the following system composed of a set of four coupled nonautonomous van der Pol oscillators was proposed in [Isaeva *et al.* (2006)]:

$$\begin{cases} x'' - \left[A \cos \frac{2\pi t}{T} - x^2\right] x' + \omega_0^2 x = \varepsilon z \cos \omega_0 t \\ y'' - \left[A \cos \frac{2\pi t}{T} - y^2\right] y' + \omega_0^2 y = \varepsilon w \\ z'' - \left[-A \cos \frac{2\pi t}{T} - z^2\right] z' + 4\omega_0^2 z = \varepsilon xy \\ w'' - \left[-A \cos \frac{2\pi t}{T} - w^2\right] w' + \omega_0^2 w = \varepsilon x \end{cases} \qquad (7.48)$$

with the following assumptions:

(1) The variables $x, y,$ and w have equal basic frequencies ω_0, and z has a frequency $2\omega_0$.
(2) The control parameter responsible for the Andronov-Hopf bifurcation in autonomous partial systems is forced to oscillate slowly with some period $T \gg \frac{2\pi}{\omega_0}$.

(3) For one half period, the first pair of oscillators (x, y) is active, while the second pair (z, w) is passive[15]. For the other half period, the situation is reversed.
(4) The product xy, contains a component close to the doubled basic frequency, and the oscillators x and y affect the oscillator z by xy. This term serves as a priming for the oscillator z as it becomes active due to the parameter variation. The oscillator w accepts excitation from the oscillator x at the same time. Hence the excitation returns to the first pair when the active stage for the second pair of oscillators ends.
(5) The oscillator z affects the oscillator x by the coupling term $z \cos \omega_0 t$, and the oscillator y accepts excitation from the oscillator w.
(6) To make the external driving periodic, assume that the period of the slow parameter modulation contains an integer number of periods of the auxiliary signal $N_0 = \frac{2\pi}{\omega_0}$.

Using a rough approximation for the phases of the oscillators, it was shown that the dynamics of (7.48) follow the one of (7.1). Indeed, if the first and the second oscillators on an active stage of the process have some initial phases φ_x and φ_y, then assume

$$\begin{cases} x \simeq \cos(\omega_0 t + \varphi_x) \\ y \simeq \cos(\omega_0 t + \varphi_y). \end{cases} \quad (7.49)$$

For the active stage, the third oscillator z inherits the phase

$$\varphi_z \simeq \varphi_x + \varphi_y + const, \quad (7.50)$$

and the fourth oscillator accepts the phase of the first one,

$$\varphi_w \simeq \varphi_x + const, \quad (7.51)$$

because the last component in the expression is of a frequency close to that of the oscillator z and effectively excites it, and at the beginning of the next stage of activity for the oscillators x and y, the coupling term in the first equation $z \cos \omega_0 t$ excites the oscillator x, and it acquires the phase

$$\varphi'_x \simeq \varphi_z + const \simeq \varphi_x + \varphi_y + const. \quad (7.52)$$

Finally, the oscillator y inherits the phase from

$$w : \varphi'_y \simeq \varphi_w + const \simeq \varphi_x + const. \quad (7.53)$$

---

[15]The word 'active' here means that is above the self-oscillation threshold, and the word 'passive' means that is below the self-oscillation threshold.

The formulas for xy and $z\cos\omega_0 t$ are given by

$$\begin{cases} xy \simeq \frac{1}{2}\left[\cos\left(\varphi_x - \varphi_y\right) + \cos\left(2\omega_0 t + \varphi_x + \varphi_y\right)\right] \\ z\cos\omega_0 t \simeq \frac{1}{2}\left[\cos\left(3\omega_0 t + \varphi_x + \varphi_y\right) + \cos\left(\omega_0 t + \varphi_x + \varphi_y\right)\right]. \end{cases} \quad (7.54)$$

Therefore, an approximate mapping for the phases confirming that the stroboscopic dynamics observed with the basic time interval $2T$ for the phases of the oscillators (x, y) follow at least in the considered approximation the Arnold cat map $(6.18)^{16}$ which displays a strange chaotic attractor,

$$\begin{cases} \varphi'_x = \varphi_x + \varphi_y + const \\ \varphi'_y = \varphi_x + const. \end{cases} \quad (7.55)$$

This result was confirmed by plotting the variables x, y, z, and w versus time in the system (7.48) at parameter values

$$\omega_0 = 2\pi, N_0 = T = 20, A = 2, \varepsilon = 0.4. \quad (7.56)$$

To demonstrate the correspondence of the dynamics to the Arnold cat map (6.18), the following method was applied:

(1) Find the phases for the first and the second oscillators x, y in the middle of an excitation stage using numerical integration of system (7.48) and the relations[17]

$$\begin{cases} \varphi_x = \arg\left(x - \frac{ix'}{\omega_0}\right) \\ \varphi_y = \arg\left(y - \frac{iy'}{\omega_0}\right). \end{cases} \quad (7.57)$$

(2) If the point $(q, p) = \left(\frac{\varphi_x}{2\pi}, \frac{\varphi_y}{2\pi}\right)$ falls in the area of the cat face drawn in the unit square, then mark a dot on the diagram and the dots for the images after the time intervals $2T$ and $4T$ on two subsequent plots, respectively.

(3) Repeat step 2. In this case, a great many dots accumulate, and one can observe the cat face on the plot.

For a better description of the dynamics of system (7.48), consider the Poincaré map defined by

(1) Consider the vector $X_n = \left\{x, \frac{x'}{\omega_0}, y, \frac{y'}{\omega_0}, z, \frac{z'}{2\omega_0}, w, \frac{w'}{\omega_0}\right\}$ as a certain state of the system (7.48) at $t_n = 2nT$.

(2) Integrate system (7.48) with the initial state X_n.

---

[16]Up to some additive constants.
[17]Note that when the first pair of oscillators (x, y) is passive, the respective amplitudes are small, and the phases φ_x and φ_y are not well defined on the whole time interval T.

(3) After time interval $2T$, a new vector X_{n+1} is determined uniquely by the vector X_n.

(4) Define the function $T:[X_n]^{18} \to [X_n]$ by the expression

$$X_{n+1} = T(X_n). \tag{7.58}$$

Thus the map T defined by (7.58) is a diffeomorphism in $\mathbb{R}^8$, a one-to-one differentiable map of class C^∞ because the right-hand side of (7.48) is smooth and bounded. The Lyapunov exponents for the system (7.48) with the parameters (7.56) are given by:

$$\begin{cases} \Lambda_1 = 0.962, \Lambda_2 = -0.970 \\ \Lambda_3 = -15.525, \Lambda_4 = -19.074 \\ \Lambda_5 = -20.053, \Lambda_6 = -21.315 \\ \Lambda_7 = -32.444, \Lambda_8 = -32.898. \end{cases} \tag{7.59}$$

For $N_0 = 6$ and $\omega_0 = 2\pi$, $T = 6$, $A = 6.5$, and $\varepsilon = 0.4$, the system (7.48) has a chaotic attractor, and the Lyapunov exponents are

$$\begin{cases} \Lambda_1 = 0.961, \Lambda_2 = -1.021 \\ \Lambda_3 = -11.256, \Lambda_4 = -12.908 \\ \Lambda_5 = -13.794, \Lambda_6 = -15.886 \\ \Lambda_7 = -23.663, \Lambda_8 = -24.464. \end{cases} \tag{7.60}$$

Hence the linear toral map is not a good approximation of the dynamics of (7.48). Thus it is more appropriate to use the following Anosov map with added nonlinear terms:

$$\begin{cases} p_{n+1} = p_n + q_n + f_1(p_n, q_n) \pmod{1} \\ q_{n+1} = p_n + 2q_n + f_2(p_n, q_n) \pmod{1} \end{cases} \tag{7.61}$$

where $f_1(p_n, q_n)$ and $f_2(p_n, q_n)$ are some smooth functions of period-1 with respect to both arguments (p_n, q_n). To find the best approximation to $f_{1,2}$ the following method was used:

(1) Represent the functions $f_{1,2}$ as Fourier expansions over two arguments,

$$\begin{cases} f_{1,2}(p_n, q_n) = \sum_{n=0}^{M} \sum_{m=-M}^{M} [a_{n,m} + b_{n,m}] \\ a_{n,m} = A_{n,m}^{1,2} \cos(2\pi(np + mq)) \\ b_{n,m} = B_{n,m}^{1,2} \sin(2\pi(np + mq)). \end{cases} \tag{7.62}$$

---

[18]The set $[X_n]$ is an eight–dimensional space. Geometrically, in the 9-D extended phase space $\{X_n, t\}$ of system (7.58) one has a cross section of the flow by the 8-D hyperplanes $t = t_n = 2nT$, that can be identified due to the periodicity of the phase space in t.

(2) Calculate the stroboscopic sequences p_n, q_n at $t_n = 2nT$.
(3) Accumulate sufficiently long series $f_{1,n}$ and $f_{2,n}$ given by

$$\begin{cases} f_{1,n} = p_{n+1} - p_n - q_n \,(\text{mod}\, 1) \\ f_{2,n} = q_{n+1} - p_n - 2q_n \,(\text{mod}\, 1) \,. \end{cases} \qquad (7.63)$$

(4) Estimate using a least squares method, the coefficients $A_{n,m}^{1,2}$ and $B_{n,m}^{1,2}$ to obtain the best approximation for $f_{1,n}$ and $f_{2,n}$ by their Fourier expansions (7.62) at some finite M.

Finally, the fractal invariant measure on the attractor of system (7.48) has the following characteristics:

(1) The correlation dimensions is $D_2 = 1.93$ for the set of parameters (7.56), and it is distinct from the value 2, which implies a uniform distribution of the invariant density over the surface of the 2-D torus. This dimension was calculated using the Grassberger-Procaccia algorithm [Grassberger & Procaccia (1983)] by computing two-component time series composed of data for phases (7.57) from 10^5 iterations of the Poincaré map (7.58) obtained from a numerical solution of Eqs. (7.48).
(2) The Kaplan–Yorke formula is given by $D_L = 1 + \frac{\Lambda_1}{|\Lambda_2|} \simeq 1.94$ using the Lyapunov exponents (7.60). This value of D_L is in good agreement with the Grassberger-Procaccia result. For the parameter set (7.56), one has $D_2 = 1.98$ and $D_L = 1.99$, much closer to 2 for the attractor used to illustrate the cat map dynamics.
(3) The uniform hyperbolicity for the attractor in the Poincaré map (7.58) is not easily checked because of the high dimension of its phase-space. A new method checking this property can be found in [Kuznetsov (2007)]. See Sec. 7.1.2.
(4) A circuitry realization on a base of coupled nonautonomous self-oscillators is possible for the system (7.48) using the electronic device given for a system modeling the Smale–Williams solenoid [Kuznetsov (2005)], and it is represented in Fig. 2 in [Kuznetsov & Seleznev (2006)].

A second example that permits modeling the Arnold cat map (6.18) is **model C** [Kuznetsov & Pikovsky (2007)] in a perturbed heteroclinic cycle given by

$$\begin{cases} x_1' = \omega_0 y_1 + \left(1 - a_1 - \frac{1}{2}a_2 - 2a_3\right)x_1 + \varepsilon\left(x_2 y_3 + x_3 y_2\right) \\ y_1' = -\omega_0 x_1 + \left(1 - a_1 - \frac{1}{2}a_2 - 2a_3\right)y_1 \\ x_2' = \omega_0 y_2 + \left(1 - a_2 - \frac{1}{2}a_3 - 2a_1\right)x_2 + \varepsilon\left(x_1 y_3 + x_3 y_1\right) \\ y_2' = -\omega_0 x_2 + \left(1 - a_2 - \frac{1}{2}a_3 - 2a_1\right)y_2 \\ x_3' = \omega_0 y_3 + \left(1 - a_3 - \frac{1}{2}a_1 - 2a_2\right)x_3 + \varepsilon\left(x_2 y_1 + x_1 y_2\right) \\ y_3' = -\omega_0 x_3 + \left(1 - a_3 - \frac{1}{2}a_1 - 2a_2\right)y_3. \end{cases} \quad (7.64)$$

For system (7.64), three oscillators become excited in the nonperiodic order $1 \to 3 \to 2 \to 1 \to ...$ because the symmetrized observables $X = x_1 + x_2 + x_3$ and $Y = y_1 + y_2 + y_3$ display oscillations with irregular phase and amplitude modulation. This can also be seen from the diagram of the phase portrait on the plane (X, Y).

For system (7.64), the Poincaré map is defined in the same way in the six-dimensional phase space $\{x_1, y_1, x_2, y_2, x_3, y_3\}$ using the three surfaces determined by Eq. (7.44). Successive intersections of these surfaces by a trajectory were examined for the vector $v_n \in \mathbb{R}^5$ associated with a point on the surface $S_j = 0$, and a mapping (7.41) was constructed. The image under T is a vector v_{n+1} on the next intersected surface $S_k = 0$, which follows in accordance with the order $S_1 \to S_3 \to S_2 \to S_1$. In this case, the transversal structure of the attractor is distinguishable. The phases of the oscillations $\varphi_n = \arg(x_j + iy_j)$ were defined for a sequence of successive intersections of the surfaces $S_j = 0$, and they obey (with good accuracy) the Fibonacci map

$$\varphi_{n+1} = \varphi_n + \varphi_{n-1} + const. \quad (7.65)$$

Since the second iteration of the Fibonacci map (7.65) yields the Arnold cat map (6.18), the action of the map T displays a traditional picture of the cat face, *i.e.*, the attractor in the Poincaré map is concentrated close to a two-dimensional torus which agrees with the fact that the sum of the largest two nonzero Lyapunov exponents is very small, $\lambda_1 + \lambda_3 \simeq 0.00054$. The Lyapunov exponents for **model C** (7.64) are $\lambda_1 = 0.04786 \pm 0.00009$, $\lambda_2 = -0.000001 \pm 0.00001$, $\lambda_3 = -0.04840 \pm 0.00005$, $\lambda_4 = -0.3184 \pm 0.0004$, $\lambda_5 = -0.3673 \pm 0.0004$, and $\lambda_6 = -2.0000207 \pm 0.0000002$, and the average time between the intersection of the surface $S_j = 0$ at the given parameters is approximately $T_{av} = 9.943$. Hence the Lyapunov exponents for the map T are $\rho_1 = \exp[\lambda_1 T_{av}] \simeq 1.6094$ and $\rho_3 \simeq 0.618$. The squared numbers $\rho_1^2 \simeq 2.59$ and $\rho_3^2 \simeq 0.382$ are in a good agreement with the eigenvalues for the Arnold cat map (6.18) of 2.618 and 0.382. The Lyapunov dimension of the attractor is nearly $D \simeq 3$, and the dimension of the attractor in the Poincaré map is $d \simeq 2$. In fact, the correlations decay rapidly without any visible tail for the observable $X = x_1 + x_2 + x_3$.

7.1.5 Modeling the Bernoulli map

In this section, **model D** [Kuznetsov & Pikovsky (2007)] was constructed using the same method as discussed above, but its associated Poincaré map has a hyperchaotic (with two positive Lyapunov exponents) attractor in a perturbed heteroclinic cycle given by

$$\begin{cases} x_1' = 3\omega_0 y_1 + \left(1 - a_1 - \tfrac{1}{2}a_2 - 2a_3\right) x_1 + \varepsilon_1 x_3^2 y_3 \\ y_1' = -3\omega_0 x_1 + \left(1 - a_1 - \tfrac{1}{2}a_2 - 2a_3\right) y_1 \\ x_2' = \omega_0 y_2 + \left(1 - a_2 - \tfrac{1}{2}a_3 - 2a_1\right) x_2 + \varepsilon_2 \left(x_1 y_3 + x_3 y_1\right) \\ y_2' = -\omega_0 x_2 + \left(1 - a_2 - \tfrac{1}{2}a_3 - 2a_1\right) y_2 \\ x_3' = \omega_0 y_3 + \left(1 - a_3 - \tfrac{1}{2}a_1 - 2a_2\right) x_3 + \varepsilon_3 x_2 \\ y_3' = -\omega_0 x_3 + \left(1 - a_3 - \tfrac{1}{2}a_1 - 2a_2\right) y_3. \end{cases} \quad (7.66)$$

System (7.66) is not symmetric as are **models A, B, C** given above. Here $\varepsilon_{1,2,3}$ are the coupling constants. For $\omega_0 = 1$, $\varepsilon_1 = \varepsilon_3 = 0.004$, and $\varepsilon_2 = 0.1$ and various initial conditions, the system (7.66) displays always only one attractor (it is structurally stable). For system (7.66), the partial oscillators become excited in the order $1 \to 3 \to 2 \to 1....$ at the moments in time when the conditions (7.44) are fulfilled just as for **models B, C**. The phases φ_j of the rising oscillators are defined by $\varphi_j = \arg\left(x_j + iy_j\right)$.

The transformation $\bar{\varphi}_1 \to \bar{\varphi}_3 \to \bar{\varphi}_2 \to \bar{\varphi}_1 \to \bar{\varphi}_3 \to \bar{\varphi}_2 \to ...$ adopts the phase

$$\bar{\varphi}_1 = 3\varphi_3 + const \quad (7.67)$$

because oscillator 1 is driven by a force (that contains the third harmonic) proportional to $x_3^2 y_3$. The role of the third harmonic is that it is in resonance with the oscillator 1 and stimulates its excitation at the beginning of the epoch of activity. Oscillator 3 is driven in resonance by a force proportional to x_2 in the course of the excitation, and hence it adopts the phase from the oscillator 2,

$$\bar{\varphi}_3 = \varphi_2 + const. \quad (7.68)$$

Finally, oscillator 2 is driven by a signal proportional to $x_1 y_3 + x_3 y_1$, and it is excited in a nonresonant manner at the beginning of the active epoch[19]. Oscillator 2 adopts the phase

$$\bar{\varphi}_2 = \varphi_1 + \varphi_3 + const. \quad (7.69)$$

---
[19]For that, a relatively large value of ε_2 was assigned.

Observe that these relations are equivalent to (6.65) up to a constant term. The computed Lyapunov exponents for the model (7.66) are $\lambda_1 = 0.01878 \pm 0.00006$, $\lambda_2 = 0.00621 \pm 0.00002$, $\lambda_3 = 0.00000 \pm 0.000003$, $\lambda_4 = -0.36803 \pm 0.0003$, $\lambda_5 = -0.43133 \pm 0.0003$, and $\lambda_6 = -2.000026 \pm 0.000002$, and the Lyapunov dimension of the attractor estimated from the Kaplan–Yorke formula is 3.068. The presence of two positive Lyapunov exponents confirms the *hyperchaotic* nature of the dynamics of model (7.66) if one considers the overall transformation[20] with the surface of section S_1 as the Poincaré map. Then the average time between successive crossings is $T_{av} \simeq 41.475 = 3(13.825)$, and the Lyapunov exponents for the Poincaré map are $\rho_1 \simeq 2.179 \simeq \frac{\sqrt{13}+1}{2} \simeq 2.303$ and $\rho_2 \simeq 1.294 \simeq \frac{\sqrt{13}-1}{2} \simeq -1.303$, which are in good agreement with the eigenvalues of the hyperchaotic map (6.65). The necessary diagrams demonstrate the chaotic properties of system (7.66).

Another example for modeling the Bernoulli doubling map (6.13) is given in [Kuznetsov & Pikovsky (2008)] where they propose a device with possible experimental implementations based on a Q-switched self-sustained oscillator with two nonlinear delayed feedback loops. In this case, the phase[21] difference between the two neighboring pulses obeys the Bernoulli doubling map (6.13) which displays a hyperbolic (robust and structurally stable) chaotic attractor. This is a result of the presence of the delay that makes the phase space of the system infinite-dimensional from the mathematical point of view, but the delays can be arranged rather easily in a physical implementation that is mostly appropriate for a realization with lasers [Siegman (1986)]. The so-called *active Q-switch* is shown in Fig. 7.3, *i.e.*, the block diagram of the system, where the modulation is typically done by applying an external signal by an acousto-optic or electro-optic modulator. In this case, it is possible to control explicitly the periodicity of the pulse train contrary to the case of a passive Q-switch where one can confirm the appearance of pulsation because of the instability of the continuous generation (saturable absorber). The block diagram of the system contains the following elements: (a) The main active element. (b) A self-sustained oscillator with an operating frequency ω. In this case, this system works as follows: The oscillator becomes periodically active with a period $\tau = \frac{2\pi}{\Omega}$, and the stages of activity are interrupted by intervals of silence as shown in Fig. 7.4(a). Let the indices $..., n-1, n, n+1, ...$ denote successive pulses of oscillation created by a slow external modulation of the

---

[20] Because all consecutive transformations $\varphi_1 \to \varphi_3 \to \varphi_2 \to \varphi_1$ are different.
[21] This phase is determined by the phases from two previous activity stages through the chaotic Bernoulli doubling map.

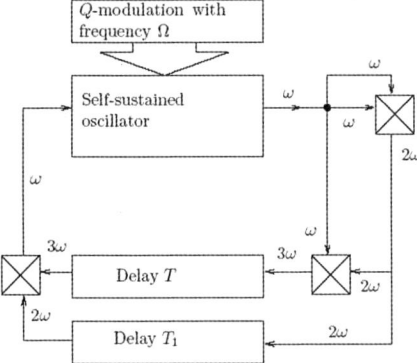

Fig. 7.3 Block diagram of the Q-switched oscillator, where crossed boxes indicate nonlinear elements-quadratic multipliers, and separate blocks indicate a modulator with frequency Ω. The feedback loop includes two delay units with times T and T_1. Reused with permission from [S. P. Kuznetsov and A. Pikovsky, Hyperbolic chaos in the phase dynamics of a Q-switched oscillator with delayed nonlinear feedbacks, EPL 84 (2008) 10013].

activity/losses with frequency $\Omega \ll \omega$. In this case, the phase of each new pulse is arbitrary if there are no additional feedback loops. In the actual case, the sequence of pulses with undetermined phases is modified by the delayed feedback loops, taking into account the following tasks about the qualitative explanation of the phase transformation:

(1) The oscillatory process was represented as
$$x(t) = \sum_n f_n(t) \cos(\omega t + \varphi_n), \qquad (7.70)$$
where f_n is the amplitude and φ_n is the phase of the nth pulse.

(2) The time constants of the delay loops, T (less than τ) and T_1 (less than $T + \tau$) are chosen such that the $(n-1)^{th}$ pulse φ_{n-1} delayed by time T_1 and the nth pulse φ_n delayed by time T overlap during the time interval when the intensity of the $(n+1)^{th}$ pulse φ_{n+1} starts to grow.

(3) The second harmonic of the process $x(t)$ was sent using two multipliers with quadratic nonlinearity in the input of the delay line T_1, and one has a process proportional to
$$f_{n-1}^2 \cos(2\omega t + 2\varphi_{n-1} + const). \qquad (7.71)$$

(4) The third harmonic of the process $x(t)$ was sent using two multipliers with quadratic nonlinearity in the input of the delay line T, and one has a process proportional to
$$f_n^3 \cos(3\omega t + 3\varphi_n + const). \qquad (7.72)$$

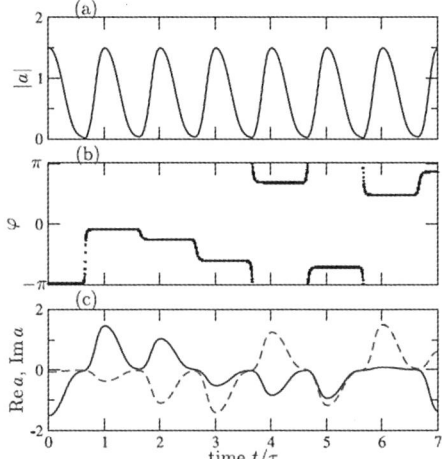

Fig. 7.4 Modulus of the amplitude (a), and the phase $\varphi = \arg(a)$ of the oscillator (b). The real and imaginary parts of the complex amplitude (c) versus time, obtained for $\Omega = 1, \gamma_0 = 0.2, \gamma_1 = 2, \epsilon = 0.1, T = 5$, and $T_1 = 11$. Reused with permission from [Kuznetsov, S. P. and Pikovsky, A., Hyperbolic chaos in the phase dynamics of a Q-switched oscillator with delayed nonlinear feedbacks, EPL 84 (2008) 10013].

(5) All other harmonics possible at the outputs of the delay lines are supposed to be filtered out.
(6) The two processes (7.71) and (7.72) are mixed in the quadratic multiplier, and the resulting component at frequency ω is proportional to the process

$$f_n^3 f_{n-1}^2 \cos(\omega t + 3\varphi_n - 2\varphi_{n-1} + const) \qquad (7.73)$$

which forces the self-sustained oscillator under consideration at the beginning of the activity stage $(n + 1)$ and serves as a *germ* for the excitation of the oscillator with the relation

$$\varphi_{n+1} = 3\varphi_n - 2\varphi_{n-1} + const, \qquad (7.74)$$

which can interpreted in terms of the one-dimensional dynamics using the phase difference between the successive pulses given by

$$\vartheta_n = \varphi_n - \varphi_{n-1}. \qquad (7.75)$$

Hence the new variable ϑ_n behaves in accordance with the expanding map of a circle, the Bernoulli doubling map (6.13). A practical algorithm for the calculation of the phase differences ϑ_n is to calculate the phases φ_n taken stroboscopically at times $\tau, 2\tau, 3\tau...$, and from them the new variable ϑ_n can be calculated from (7.75).

Thus the delay differential equation corresponding to the block diagram of Fig. 7.3 can be represented by means of the slowly varying complex amplitude $x = \text{Re}\left[a(t)e^{i\omega t}\right]$ as follows:

$$\frac{da}{dt} = (\gamma_0 + \gamma_1 \cos \Omega t - |a|^2)a + \epsilon[a(t-T)]^3[a^*(t-T_1)]^2 \qquad (7.76)$$

where a^* is the complex conjugate and ϵ is the feedback strength with the following assumptions about the retarded terms in the right-hand part of (7.76) (selected in accordance with the signal transformations in the feedback loops):

(a) The oscillator is assumed to be of the Stuart–Landau type, and the modulation of the parameter responsible for its excitation follows the expression

$$\Gamma(t) = \gamma_0 + \gamma_1 \cos \Omega t. \qquad (7.77)$$

In this case, the stages of activity correspond to $\Gamma(t) > 0$, and those of silence to $\Gamma(t) < 0$.

(b) The resonant term[22] at frequency ω was chosen in the set of all possible terms appearing from multiplication of the signal of the third harmonic with delay time T and the signal of the second harmonic with delay time T_1. This resonant term is proportional to $[a(t-T)]^3[a^*(t-T_1)]^2$.

Numerical simulation of the system dynamics can be summarized as follows: For $\Omega = 1, \gamma_0 = 0.2, \gamma_1 = 2, \epsilon = 0.1, T = 5$, and $T_1 = 11$, Fig. 7.4(a) shows that the amplitudes of the pulses are nearly equal, while the phases are different from pulse to pulse as shown Fig. 7.4(b). Fig. 7.5 shows that a stroboscopic mapping of the phase difference ϑ_n is in a good correspondence with the theoretical prediction (6.13). Fig. 7.6 illustrates similarity of the attractor in (7.76) to the Smale–Williams solenoid. Fig. 7.7 shows the seven largest Lyapunov exponents in dependence on the feedback strength ϵ for $\gamma_0 = 0.2$ and $\gamma_1 = 2$, where it is clear that there is no chaos for $\epsilon < 0.05$, and there is hyperbolic chaos for $\epsilon > 0.065$ with one positive Lyapunov exponent in a good agreement with what one expects from (6.13). A transitional region occurs for $0.05 < \epsilon < 0.065$, where the largest Lyapunov exponent is positive but less than the value following from (6.13). Furthermore, there is a zero Lyapunov exponent corresponding to the invariance of (7.76) in respect to the phase shift, and all other exponents are negative. The Lyapunov dimension of the attractor $\epsilon \in [0.065, 0.018]$ is between 1.5 and 2.5. Figure 7.8 shows qualitatively similar behaviors to that

---
[22]Practically nonresonant terms are neglected because they do not affect the oscillator.

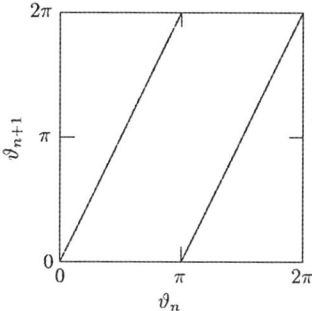

Fig. 7.5 Iteration diagrams for the phase differences ϑ_n obtained from numerical simulation (7.76) for $\Omega = 1$, $\gamma_0 = 0.2$, $\gamma_1 = 2, \epsilon = 0.1, T = 5$, and $T_1 = 11$. Reused with permission from [Kuznetsov, S. P. and Pikovsky, A., Hyperbolic chaos in the phase dynamics of a Q-switched oscillator with delayed nonlinear feedbacks, EPL 84 (2008) 10013].

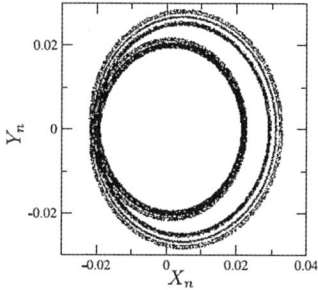

Fig. 7.6 Similarity of the attractor in (7.76) to the Smale–Williams solenoid introduced in Sec. 6.4.3 drawn in projection on the plane (X_n, Y_n), where $X_n + iY_n = a(n\tau + \tau_0)a^*((n-1)\tau + \tau_0), \tau = \frac{2\pi}{\Omega}, \tau_0 = \frac{8\tau}{11}$ for $\Omega = 1$, $\gamma_0 = 0.2$, $\gamma_1 = 2, \epsilon = 0.1, T = 5$, and $T_1 = 11$. Reused with permission from [Kuznetsov, S. P. and Pikovsky, A., Hyperbolic chaos in the phase dynamics of a Q-switched oscillator with delayed nonlinear feedbacks, EPL 84 (2008) 10013].

of Fig. 7.7 for the dependence on the modulation level γ_1 for fixed values of $\epsilon = 0.1$ and $\gamma_0 = 0.2$. In particular, there is a large range $1.8 < \gamma_1 < 5$ in which hyperbolic chaos appears. At all values of the parameters presented in Figs. 7.7 and 7.8, only a single attractor was observed with no multistability because the phase space is largely determined by the nondelayed dynamics[23] and the delayed feedback is relatively weak. Thus multistability can appear only for rather large values of the feedback parameter ϵ.

---

[23]Generally a multistability may appear in systems with delay.

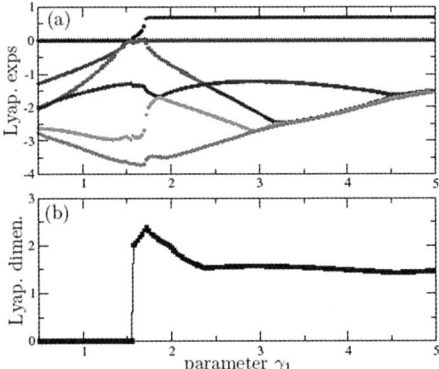

Fig. 7.7 (a) The seven largest Lyapunov exponents normalized by period τ in (7.76) as functions of the real feedback strength $\epsilon \in [0, 0.18]$ for $\Omega = 1$, $\gamma_0 = 0.2$, $\gamma_1 = 2$, $T = 5$, and $T_1 = 11$. (b) The Lyapunov dimension calculated without the zero Lyapunov exponent corresponding to the phase invariance. Reused with permission from [Kuznetsov, S. P. and Pikovsky, A., Hyperbolic chaos in the phase dynamics of a Q-switched oscillator with delayed nonlinear feedbacks, EPL 84 (2008) 10013].

From an experimental point of view, the above Q-switched self-sustained oscillator was realized using laser experiments, and a proper observable $\rho(t)$ was obtained by looking at the interference between two successive pulses. In this case, the sum of the pulse fields given by

$$\rho(t) = a(t) + a(t - \tau) \qquad (7.78)$$

illustrates a strong irregular variability of intensity as shown in Fig. 7.8. If the amplitude A of the field $a(t)$ is periodic and only the phase changes, then one can write

$$\begin{cases} \rho(t) = A(t)(e^{i\varphi(t)} + e^{i\varphi(t-\tau)}) \\ |\rho(t)| = |A|.|1 + e^{i(\varphi(t) - \varphi(t-\tau))}| = |A|.rn \\ rn = \sqrt{1 + \cos(\varphi(t) - \varphi(t-\tau))} \end{cases} \qquad (7.79)$$

where the second formula of (7.79) has a strong dependence on the phase difference. Since

$$\begin{cases} |\rho_n|^2 = 2|A_n|^2(1 + \cos \vartheta_n) \\ A_n \approx const, \end{cases} \qquad (7.80)$$

then the dynamics of the observable ρ can also be characterized by the stroboscopic nonlinearly transformed Bernoulli doubling map (7.15). If the constant in (7.15) vanishes, then this stroboscopic map is the logistic map (6.25) in the regime of full chaos as shown in Fig. 7.9. To conclude this section, we give some comparisons between several proposed schemes for

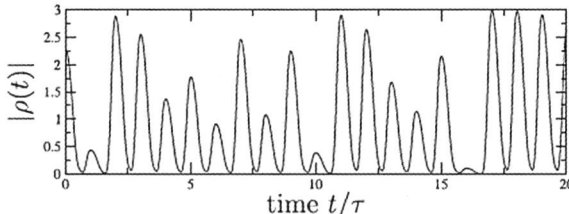

Fig. 7.8 Chaotically varying pulse amplitudes shown by the observable $|\rho(t)| = |a(t) + a(t - \tau)|$ versus time for $\Omega = 1$, $\gamma_0 = 0.2$, $\epsilon = 0.1, T = 5$, and $T_1 = 11$. Reused with permission from [Kuznetsov, S. P. and Pikovsky, A., Hyperbolic chaos in the phase dynamics of a Q-switched oscillator with delayed nonlinear feedbacks, EPL 84 (2008) 10013].

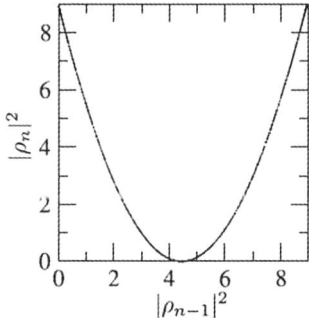

Fig. 7.9 The stroboscopic map of the variable $|\rho_n|^2$ for $\Omega = 1$, $\gamma_0 = 0.2$, $\gamma_1 = 2, \epsilon = 0.1, T = 5$, and $T_1 = 11$. Reused with permission from [Kuznetsov, S. P. and Pikovsky, A., Hyperbolic chaos in the phase dynamics of a Q-switched oscillator with delayed nonlinear feedbacks, EPL 84 (2008) 10013].

realizing hyperbolic chaos in practice: The scheme proposed in Sec. 7.1.5 for the Bernoulli map (6.13) has several advantages from the point of view of a possible experimental realization compared with those given in the other sections of this chapter. Indeed, this scheme has the following advantages:

(1) It was realized by the use of a single oscillatory element. In this case, the device is simple in comparison to those with two or more active elements.
(2) It exploits a resonant transmission of the excitation from the previous activity stages to the next one.

The last item implies that this system can be relevant for a wide range of oscillators, *i.e.*, from very fast systems like lasers to extremely slow ones like

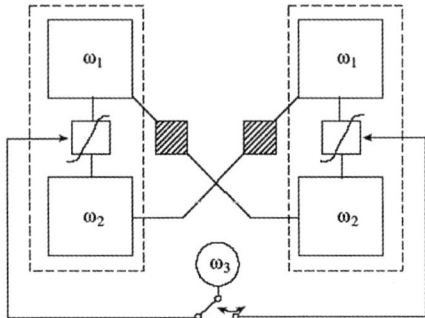

Fig. 7.10 Parametric chaos generator block diagram. Blocks labeled ω_1 and ω_2 represent the respective oscillators with these free-running frequencies, those labeled with a waveform are reactive coupling elements characterized by a parameter oscillating at the driving carrier frequency ω_3, and batched squares are signal sources. [Kuznetsov, 2008].

chemical reactions. In conclusion, we note that in [Kuznetsov (2008)], the feasibility of a parametric generator of hyperbolic chaos was discussed. In particular, this generator was used for systems consisting of two subsystems, and each of them is a pair of parametrically coupled oscillators whose free-running frequencies ω_1 and ω_2 differ by factor of two as in the previous cases discussed in this chapter. The block diagram is shown in Fig. 7.10.

7.1.6 Modeling Plykin's attractor

In this section, we discuss one example of the problem of modeling Plykin attractor introduced in Sec. 6.4.4 using a dynamical system. Indeed, in [Hunt (2000)], an artificial 3-D flow was constructed that has Plykin's attractor as represented by the corresponding Poincaré map. This example is very difficult to be implemented in a physical system. In [Belykh et al. (2005)], this problem was solved by showing that the Poincaré map defined by a 3-D flow constructed as a bursting neuron model exhibits a Plykin-like attractor (a strange hyperbolic attractor). In fact, this work is the first such example of a realistic system. The method of analysis is based on the use of a geometrical approach and the derivation of a 2-D flow-defined Poincaré map giving an accurate account of the system's dynamics. The analysis of bifurcation phenomena of the neuron system under consideration shows that in some parameter ranges this system undergoes bifurcations causing transitions between tonic spiking and bursting. In this case, the 2-D Poincaré map becomes a map of a disk with several (at least three) periodic holes, and this map can display the Plykin example of a planar hyperbolic attractor. Hence the corresponding attractor of the 3-D neuron model appears to be hyperbolic due to the existence of a two-loop (secondary) homoclinic bifurcation of a saddle.

The dynamical system that can model a Plykin attractor is a models of bursting neurons given in [Hindmarsh & Rose (1984), Rinzel (1987)] and represented by the singular perturbed system

$$\begin{cases} x' = X(x,z) \\ z' = \mu(Z(x) - z - \delta), \end{cases} \quad (7.81)$$

where $x = (x_1, x_2)$ is two fast variables x_1 and x_2, with the first component x_1 representing the membrane potential and the second component x_2 representing a fast current. The scalar variable z is associated with a slow current. The functions $Z : \mathbb{R}^2 \to \mathbb{R}^1$ with $Z(0) = 0$ and $X : \mathbb{R}^2 \to \mathbb{R}^2$ are assumed to be smooth. The parameter μ is small, and δ is a bifurcation parameter. For $\mu = 0$, the fast reduced subsystem

$$\begin{cases} x' = X(x,z) \\ z = const \end{cases} \quad (7.82)$$

defines a z-parameter family of two-dimensional phase portraits. Hence system (7.82) represents the usual adiabatic approach in which the reduced system ($\mu = 0$) accounts for the fast dynamics, and variations of z describe the slow dynamics for $\mu > 0$. The functions $X(x,z)$ and $Z(x)$ satisfy a large number of typical assumptions about equilibrium points, stable limit cycles, homoclinic orbits, and the nature of the stable and unstable separatrices. The dynamics of the full system (7.81) essentially depends on the position of the separating surface $\kappa = \{z = Z(x) - \delta\}$ and the small parameter μ. The study of the dynamics of the full system (7.81) is done by constructing a global[24] cross-section D_0 of the fast system by connecting the left and right branches of a set of equilibria where the unstable and stable manifolds transversally intersect the global cross-section D_0. Thus the Poincaré map f was constructed on the basis of the typical assumptions for the system (7.81). Hence f is a global Poincaré return map and represents the complete set of attractors and bifurcations of system (7.81). Finally, it was shown in [Belykh et al. (2005)] that the Poincaré return map f matches the Plykin example of a planar hyperbolic attractor, which implies that system (7.81) displays robust chaos.

---

[24]$i.e.$, the vector field must be transversal to D, and any trajectory starting from D must return to this cross-section.

7.2 Exercises

Exercise 168. (a) Show that in the 4-D state space of system (7.1), the eigendirection associated with the phase φ is expanding and the remaining three are contracting.

(b) Show that the respective Lyapunov exponents λ_k for the system (7.1) and the Lyapunov exponents Λ_k of the stroboscopic map F satisfy the relation $\lambda_k = N^{-1}\Lambda_k$.

(c) Show that for $N = 4, A_1 = 1.5, A_2 = 6, \varepsilon_1 = \varepsilon_2 = 0.1$, and $h_1 = h_2 = 0$, the Lyapunov exponents of the stroboscopic map associated with system (7.1) are given by $\Lambda_1 \approx 0.69, \Lambda_2 \approx -2.40, \Lambda_3 \approx -4.24$, and $\Lambda_4 \approx -6.85$.

(d) Using the plot of the Lyapunov exponent, show that the system (7.1) has a nonhyperbolic attractor for $N = 8, A_1 = 0.2, A_2 = 0.8, \varepsilon_{1,2} = 0.1$, and $h_{1,2} = 0$.

Exercise 169. Prove Lemma 7.1 using the algebraic relation $\begin{pmatrix} 1 & 1 \\ 1 & 2 \end{pmatrix} = \begin{pmatrix} 0 & 1 \\ 1 & 1 \end{pmatrix} \begin{pmatrix} 0 & 1 \\ 1 & 1 \end{pmatrix}$.

Exercise 170. Transform the system (7.48) to an 8-D system.

Exercise 171. Prove the relations (7.54).

Exercise 172. With the normalization $q = \frac{\varphi_x}{2\pi}$ and $p = \frac{\varphi_y}{2\pi}$, show that the map (7.55) is exactly the map (7.47) up to some additive constants.

Exercise 173. Show and compare the transformation of the area in a plane of phase variables in the system (7.47) and the Arnold cat map (7.18) after time intervals $2T$ and $4T$ for $\omega_0 = 2\pi, N_0 = T = 20, A = 2$, and $\varepsilon = 0.4$.

Exercise 174. Using theorems of existence, uniqueness, continuity, and differentiability of solutions of differential equations, show that the Poincaré map T is a diffeomorphism in $\mathbb{R}^8$, a one-to-one differentiable map of class C^∞.

Exercise 175. (a) Calculate all the Lyapunov exponents given by (7.59) using an appropriately adapted algorithm of Benettin [Benettin et al. (1980)] based on simultaneous solution of (7.48) and its perturbed 8-D linearized

equations given by
$$\begin{cases} \tilde{x}'' + 2xx'\tilde{x} - \left[A\cos\frac{2\pi t}{T} - x^2\right]\tilde{x}' + \omega_0^2\tilde{x} = \varepsilon\tilde{z}\cos\omega_0 t \\ \tilde{y}'' + 2yy'\tilde{y} - \left[A\cos\frac{2\pi t}{T} - y^2\right]\tilde{y}' + \omega_0^2\tilde{y} = \varepsilon\tilde{w} \\ \tilde{z}'' + 2zz'\tilde{z} - \left[-A\cos\frac{2\pi t}{T} - z^2\right]\tilde{z}' + 4\omega_0^2\tilde{z} = \varepsilon\left(\tilde{x}y + x\tilde{y}\right) \\ \tilde{w}'' + 2ww'\tilde{w} - \left[-A\cos\frac{2\pi t}{T} - w^2\right]\tilde{w}' + \omega_0^2\tilde{w} = \varepsilon\tilde{x} \end{cases} \quad (7.83)$$

using $2T$ as the unit time in the definition of the Lyapunov exponents, and at the time instants $t = 2nT$ ($n = 1, 2, ...$) perform the Gram–Schmidt orthogonalization and normalization for eight vectors $\left\{\tilde{x}, \frac{\tilde{x}'}{\omega_0}, \tilde{y}, \frac{\tilde{y}'}{\omega_0}, \tilde{z}, \frac{\tilde{z}'}{2\omega_0}, \tilde{w}, \frac{\tilde{w}'}{\omega_0}\right\}$ and their mean rates of growth.

(b) Plot all eight Lyapunov exponents of the system (7.48) versus the amplitude A of the slow modulation for $\omega_0 = 2\pi$, $N_0 = T = 20$, $\varepsilon = 0.4$, and show that the largest two exponents are almost constant over a wide interval of the parameter variation and are close to the values characteristic of the Arnold cat map (6.18). The other exponents are large negative, and they correspond to strong compression of the phase volume along six of the eight dimensions of the phase space of the Poincaré map (6.18).

(c) Show that qualitatively similar dynamical behavior is observed for $N_0 = 6$ and $\omega_0 = 2\pi$, $T = 6$, $A = 6.5$, and $\varepsilon = 0.4$.

Exercise 176. Plot a 3-D view of the functions expressed by Eq. (7.62) at $M = 12$, i.e., in the spaces $(p, q, f_{1,n})$ and $(p, q, f_{2,n})$.

Exercise 177. Find possible bifurcation scenarios for the onset of hyperbolic strange attractors in systems **A**, **B**, **C**, and **D** given in this chapter.

Exercise 178. To demonstrate that the component a in system (7.12) has almost constant phase when being excited, but the phase is doubled at every new stage of the excitation, it is necessary to perform the following tasks:

(a) Plot the alternate excitation and decay of the components a and b of equations (7.12).

(b) Plot the time dependence of the real and imaginary parts of a.

(c) Plot the time dependence of the complex variable a for $A = 3, T = 5$ and $\epsilon = 0.05$ and show the existence of a series of spikes corresponding to the stages of the excitation of a.

(d) Confirm the alternate excitation of the subsystems a and b by showing numerically that the projections of spikes on the plane $(\operatorname{Re} a, \operatorname{Im} a)$ are almost straight lines.

(e) Plot a numerical solution of Eqs. (7.14) using a fifth-order Runge–Kutta method at $A = 3, T = 5$, and $\epsilon = 0.05$, and show that there is an alternate excitation of the subsystems a and b, *i.e.*, at each time the amplitudes $|a|$ and $|b|$ attains their maximum, the height of the maximum is a bit different from the previous maximum.

(f) Compute the map for the phase $\phi_n = \arg a(nT)$ measured over the time step T at $A = 3, T = 5$, and $\epsilon = 0.05$, and show the correspondence with the Bernoulli map (7.15), *i.e.*, in the plane $\frac{\phi_n}{2\pi} - \frac{\phi_{n+1}}{2\pi}$.

(g) Plot the Lyapunov exponents λ_k, Λ_k versus parameter A for the system (7.14) and system (7.1) at $T = 5$ and $\epsilon = 0.05$ and at $P = 10, \omega_0 = 2\pi$, and $\epsilon = 0.2$, respectively.

(h) Draw a 3-D view of the absorbing domain U of the map (7.18) with an enclosed attractor A projected from 4-D space $\{\operatorname{Re} a, \operatorname{Im} a, \operatorname{Re} b, \operatorname{Im} b\}$ along axis $\operatorname{Im} b$ onto the plane section $\psi = 0$.

(i) Plot the first and second eigenvalues of the matrix $\mathbf{U}_x^T \mathbf{U}_x$ versus the two angle variables $0 \leq \phi, \theta \leq 2\pi$ at the boundary of the absorbing area and at $\psi = 0$.

(j) To show that $f < 1$ in the inequality (7.38) on the surface of the absorbing domain U at $\psi = 0$, $\gamma = 1.1$ and at the parameter values $A = 3, T = 5, \epsilon = 0.05, r = 0.631166$, and $d = 0.0927$. Plot the function f versus the angles $0 \leq \phi, \theta \leq 2\pi$.

(k) Plot the maximum of $f(\phi, \psi, \theta)$ versus χ at $\gamma = 1.1$ and $A = 3, T = 5, \epsilon = 0.05, r = 0.631166$, and $d = 0.0927$.

(l) Plot the maximum of $f(\phi, \psi, \theta, \chi)$ versus γ and $A = 3, T = 5, \epsilon = 0.05, r = 0.631166$, and $d = 0.0927$.

Exercise 179. (a) Show typical time dependencies for the variables $x_{1,2}, y_{1,2}$ obtained from a computer solution of Eq. (7.39) for $\omega_0 = 2\pi$ and $\varepsilon = 0.3$.

(b) Using the Poincaré map defined for the **model A**, show that the diagram φ_{n+1} versus φ_n looks topologically similar to the expanding circle map (6.8).

(c) Show that both portraits possess a Cantor-like transversal structure.

(d) Show that the attractor of the map T is just a hyperbolic chaotic attractor of the Smale–Williams type.

Exercise 180. (a) Show typical time dependencies for x_1, x_2, and x_3, obtained from a computer solution of Eq. (7.43) when $\omega_0 = 1$ and $\varepsilon = 0.03$.

(b) Plot the autocorrelation function of system (7.43) for the observable $X = x_1 + x_2 + x_3$.

Exercise 181. (a) Show that for $\omega_0 = 1$ and $\varepsilon = 0.03$, different initial conditions of **model C** (7.64) display almost the same attractor.

(b) Plot the evolution of the variables x_1, x_2, and x_3 as functions of time for system (7.64) after the decay of transients.

(c) Draw the diagram of the phase portrait on the plane $(X = x_1 + x_2 + x_3, Y = y_1 + y_2 + y_3)$.

(d) Draw the projections of the attractor in the Poincaré map onto three planes of the variables (x_j, y_j) corresponding to three partial oscillators.

(e) Show that the phases of the oscillations $\varphi_n = \arg(x_j + iy_j)$ defined for a sequence of successive intersections of the surfaces $S_j = 0$ obey (with a good accuracy) the Fibonacci map (7.65).

(f) Plot of the autocorrelation function for the observable $X = x_1 + x_2 + x_3$, and show that the correlations decay rapidly without any visible tail.

Exercise 182. (a) Illustrate the time evolution of the oscillators (7.66).

(b) Illustrate the phase portrait for the symmetrized observables $X = x_1 + x_2 + x_3$ and $Y = y_1 + y_2 + y_3$.

(c) Plot the autocorrelation function for the observable $X = x_1+x_2+x_3$.

Exercise 183. (Open problems)

(a) Numerical computations show that the hyperbolic strange attractors A,B,C,D introduced in this chapter does not exists in some parameters ranges. In these ranges a complex objects was observed and it is still cannot classify and characterize. See [Kuznetsov & Sataev (2007)].

(b) Rigorous mathematical proof using computer verification of conditions of hyperbolicity in terms of cones or in an absorbing domain containing the attractor of **models A,B,C,D**. See [Kuznetsov & Sataev (2007)].

(c) Generalization of the heteroclinic cycle construction for systems A,B,C,D of more than three oscillators. See [Kuznetsov & Sataev (2007)].

(d) The problem of the transition to hyperbolic chaos under parameter variations is not yet resolved neither in mathematical nor in physics. See [Kuznetsov (2008)].

Exercise 184. (Open problems)

(a) Show that the corresponding attractor of the 3-D neuron model (7.81) is hyperbolic when its 2-D Poincaré map displays a planar hyperbolic Plykin attractor. See [Belykh et al. (2005)].

(b) Examine further properties of the new bifurcation phenomenon, *i.e.*, the existence of a two-loop (secondary) homoclinic bifurcation of a saddle in the study of the 2-D Poincaré map associated with the 3-D neuron model (7.81). See [Belykh et al. (2005)].

Chapter 8

Lorenz-Type Systems

In this chapter, we present some recent results about Lorenz-type systems. In Sec. 8.1, an overview outlines the major properties of such systems along with one definition resulting from several observations of the dynamics of the standard Lorenz system which is presented on Sec. 8.2 from the standpoint of its existence and structure. In Sec. 8.3, we present another example of Lorenz-type systems, *i.e.*, expanding and contracting Lorenz attractors. Section 8.4 deals with pseudo-hyperbolic theory. In particular, we discuss the essence of *wild strange attractors* along with their dynamical properties. In Sec. 8.5, one example (the Lozi map) of Lorenz-type attractors realized in two-dimensional maps is presented with some properties and notes. Some exercises and open problems concerning Lorenz-type systems are given in Sec. 8.6.

8.1 Lorenz-type attractors

Recall that the Lorenz-type attractors are almost not structurally stable (see Definition 4.3), although their homoclinic and heteroclinic orbits are structurally stable (hyperbolic), and no stable periodic orbits appear under small parameter variations, as for example in the Lorenz system [Lorenz (1963)]. These attractors are considered as examples of *truly* strange attractors [Shilnikov (1980), Williams (1977), Cook & Roberts (1970)]. Furthermore, there are finitely many dynamical systems that are Lorenz-type systems and are closest in their structure and properties to robust hyperbolic attractors (see Chapter 6). For example, in the Lorenz attractor, all trajectories are saddles and a variation of parameters does not create stable points or cycles [Bykov & Shilnikov (1989), Afraimovich (1984-1989-1990)]. Furthermore, the Lozi and Belykh attractors [Lozi (1978), Belykh

(1982-1995)] are examples of such Lorenz-type attractors realized in two-dimensional (2-D) maps. For these attractors, at least one of the three conditions for hyperbolicity given in Definition 6.1(d) is violated. For the Lorenz attractor, the second condition is not valid. In fact, it was proved that the chaotic attractor in the Lorenz system corresponding to a certain set of parameter values exhibits the key properties of hyperbolic attractors, which confirms that the system is not completely hyperbolic in the sense of Definition 6.1(d). For this reason, the Lorenz system is sometimes called *quasi-hyperbolic* [Afraimovich *et al.* (1977), Mischaikow & Mrozek (1995-1998)]. Their difference from the robust hyperbolic attractors introduced in Chapter 6 is a local violation of homogeneity due to singular phase trajectories belonging to another saddle type such as saddle equilibrium states, which have a different dimension of manifolds, or separatrix circuits like the Lorenz attractor, which includes a denumerable set of separatrix loops of the saddle equilibrium state as shown in [Williams (1977), Afraimovich *et al.* (1977), Shilnikov (1980)]. The homogeneity of the attractor can also be violated as a result of the birth of nonrobust homoclinic trajectories. In fact, an attractor is a Lorenz-type if it has nondangerous trajectories with a zero measure on the attractor, *i.e.*, their appearance and disappearance should not lead to the birth of stable trajectories and affect the structure of the chaotic hyperbolic set. In fact, Lorenz-like attractors can appear from singular cycles, and they are characterized in a robust way by the presence of infinitely many periodic orbits in any neighborhood of a singularity. In this situation, an adequate notion of hyperbolicity with singularities called *singular hyperbolicity* was introduced by Morales, Pacífico, and Pujals using the concept of *dominated splitting* over an invariant set which contains a singularity, *i.e.*, a splitting on the tangent bundle.

From the above notes, the Lorenz-like attractors can be defined as follows:

Definition 8.1. A Lorenz-like attractor is an attractor with the following characteristics:

(1) It is a robust, transitive attractor which is not hyperbolic.

(2) The origin $(0, 0, 0)$ accumulates hyperbolic periodic orbits.

(3) The attractor has sensitive dependence on initial conditions (or chaotic).

For recent results about the topic of Lorenz-type attractors, see [Komuro (1984), Morales *et al.* (1998), Anishchenko *et al.* (2002), Klinshpont *et al.* (2005), Bautista & Morales (2006), Klinshpont (2006), Alves *et al.* (2007),

Araujo & Pacifico (2007), Arroyo & Pujals (2007), Araujo et al. (2007), Araújo & Pacifico (2008)].

Let Λ be an attractor of a Lorenz-like system in a compact, boundaryless three-manifold M. Let $C^1(M)$ denote the set of C^1 vector fields on M endowed with the C^1 topology. In [Araújo & Pacifico (2008)], an overview of some recent results on the dynamics of Lorenz-like attractors is given where the following details are in Chapters 8 and 9:

(1) There is an invariant foliation (recall Definition 5.2) whose leaves are forward contracted by the flow.
(2) There is a positive Lyapunov exponent (recall Definition 3.11) at every orbit.
(3) They are expansive and thus sensitive to initial data.
(4) They have zero volume if the flow is C^2.
(5) There is a unique physical measure (recall Definition 3.2) whose support is the whole attractor and which is an equilibrium state with respect to the center-unstable Jacobian.

8.2 The Lorenz system

The best known example of a Lorenz-type system is the original Lorenz system [Lorenz (1963)] given by

$$\begin{cases} x' = \sigma(y-x) \\ y' = rx - y - xz \\ z' = -bz + xy, \end{cases} \quad (8.1)$$

which is the first known case of a fully persistent[1] and transitive attractor that is not hyperbolic [Lorenz (1963)] for all sufficiently small variations of the parameters $\sigma = 10, b = \frac{8}{3}$ and $r = 28$ where system (8.1) has a chaotic attractor with periodic and homoclinic orbits that are everywhere dense. The attractor is structurally unstable because a saddle equilibrium state is embedded with a one-dimensional unstable manifold on the attractor. This implies that under small smooth perturbations, stable periodic orbits do not arise, and the strange attractors are obtained through a finite number of bifurcations, namely, a route consists of three steps only because the Lorenz system is symmetric under the transformation $(x, y, z) \to (-x, -y, z)$. Generally, for $\sigma = 10, b = \frac{8}{3}$, the Lorenz system displays the following

---

[1] It exists for all sufficiently small variations of the parameters $\sigma = 10, b = \frac{8}{3}$, and $r = 28$.

attractors: For $r \in [-\infty, 1]$, the point $(0,0,0)$ is an attracting equilibrium. For $r \in [1, 13.93]$ the two other symmetric equilibria are attracting, and the origin is unstable. For $r \in [13.93, 24.06]$, there is a transient to chaos, *i.e.*, there are chaotic orbits but no chaotic attractors. For $r \in [24.06, 24.74]$, a chaotic attractor coexists with attracting two symmetric equilibria. For $r \in [24.74, 28.0]$ a chaotic attractor exists, but the two symmetric equilibria are no longer attracting.

Equations (8.1) have proved to be very resistant to rigorous analysis and also present obstacles to numerical study. A very successful approach was taken in [Guckenheimer & Williams (1979), Afraimovich *et al.* (1982)] where they constructed so-called *geometric models* introduced in Sec. 8.2.2 below (these models are flows in three dimensions) for the behavior observed by Lorenz for which one can rigorously prove the existence of a robust attractor. Another approach through *rigorous numerics* [Hasting & Troy (1992), Hassard, *et al.* (1994), Mischaikow & Mrozek (1995-1998)] showed that Eq. (8.1) exhibits a suspended Smale horseshoe as shown in Sec. 8.2.1. In particular, they have infinitely many closed solutions. A computer-assisted proof of chaos for the Lorenz equations is given in [Franceschini *et al.* (1993), Mischaikow & Mrozek (1995), Galias & Zgliczynski (1998), Tucker (1999), Stewart (2000), Sparrow (1982)]. For the proof of Mischaikow and Mrozek, see on outline in Sec. 8.2.1, and for the proof of Tucker see Sec. 9.2.

8.2.1 *Existence of Lorenz-type attractors*

In this section, we present two methods for proving rigorously the existence of the Lorenz attractor shown in Fig. 8.1. The first is due to Shilnikov and investigates the cases where a dynamical system can display a Lorenz-type attractor. The second method is due to Mischaikow and Mrozek and deals with the rigorous proof of chaos in the Lorenz system (8.1) with $(\sigma, r, b) = (45, 54, 10)$ using essentially the method of fixed point index introduced in Sec. 1.4.1.

For clarity of the presentation, we introduce the following concepts about the so-called *strong stable manifold* and *strong unstable manifold*. Indeed, let M be a closed three-manifold, and f denotes a C^r flow in M, $r \geq 1$. If $d(.,.)$ is the metric in M, and $\delta > 0$, we denote

$$B_\delta(A) = \{x \in M : d(x, A) < \delta\}. \tag{8.2}$$

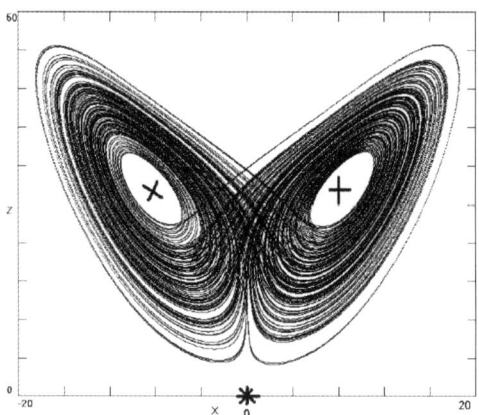

Fig. 8.1 The Lorenz chaotic attractor obtained from (8.1) with $\sigma = 10, r = 28$, and $b = \frac{8}{3}$. The origin $P_0 = (0,0,0)^T$ is shown using the symbol $*$, and the other equilibria are shown using symbols $\times$ and $+$.

Given $p \in M$ and $\epsilon > 0$, we define

$$\begin{cases} W_f^{ss}(p) = \{x : d(f_t(x), f_t(p)) \to 0, t \to \infty\} \\ W_f^{uu}(p) = \{x : d(f_t(x), f_t(p)) \to 0, t \to -\infty\} \\ W_f^{ss}(p, \epsilon) = \{x : d(f_t(x), f_t(p)) \leq \epsilon, \forall t \geq 0\} \\ W_f^{uu}(p, \epsilon) = \{x : d(f_t(x), f_t(p)) \leq \epsilon, \forall t \leq 0\}. \end{cases} \quad (8.3)$$

Then one has the following definition:

Definition 8.2. (a) The set $W_f^{ss}(p)$ is called the stable set of p.
(b) The set $W_f^{uu}(p)$ is called the unstable set of p.
(c) The set $W_f^{ss}(p, \epsilon)$ is called the local stable set of p.
(d) The set $W_f^{uu}(p, \epsilon)$ is called the local unstable set of p.

We also define

$$\begin{cases} W_f^s(p) = \bigcup_{t \in \mathbb{R}} W_f^{ss}(f_t(p)) \\ W_f^u(p) = \bigcup_{t \in \mathbb{R}} W_f^{uu}(f_t(p)). \end{cases} \quad (8.4)$$

In this case, using the stable manifold theorem given in [Hirsch et al. (1977)], it follows that if H is a hyperbolic set (recall Definition 6.1(d)) of f and $p \in H$, then the sets $W_f^{ss}(p), W_f^s(p), W_f^{uu}(p), W_f^u(p), W_f^{ss}(p, \epsilon)$, and $W_f^{uu}(p, \epsilon)$ are C^r-submanifolds of M, and $\dim W_f^{ss}(p) = \dim(E^s)$ and $\dim W_f^{uu}(p) = \dim(E^u)$. In this case, we have the following definition:

Definition 8.3. The sets $W^{ss}_f(p)$ and $W^{uu}_f(p)$ are called the *strong* stable and the strong unstable manifolds of p, respectively.

Now consider a finite-number parameter family of vector fields defined by the system of differential equations

$$x' = f(x, \mu) \tag{8.5}$$

where $(x, \mu) \in (\mathbb{R}^{n+1}, \mathbb{R}^m)$ and $f(x, \mu)$ is a C^r-smooth function of x and μ. Then one has the following result proved in [Shilnikov (1981)]:

Theorem 8.1. *Assume that the following two conditions hold: (a) System (8.5) has an equilibrium state $O(0,0)$ of the saddle type. The eigenvalues of the Jacobian at $O(0,0)$ satisfy $\operatorname{Re}\lambda_n < ... < \operatorname{Re}\lambda_2 < \lambda_1 < 0 < \lambda_0$. (b) The separatrices Γ_1 and Γ_2 of the saddle $O(0,0)$ returns to the origin as $t \to +\infty$. Then for $\mu > 0$, there exists an open set V whose boundary contains the origin such that in V system (8.5) possesses the Lorenz attractor in the following three cases:* **Case 1.** *(a) Γ_1 and Γ return to the origin tangentially to each other along the dominant direction corresponding to the eigenvalue λ_1 (b) if*

$$\begin{cases} \frac{1}{2} < \gamma < 1 \\ \nu_i = -\frac{\operatorname{Re}\lambda_i}{\lambda_0} > 1 \\ \gamma = -\frac{\lambda_1}{\lambda_0}. \end{cases} \tag{8.6}$$

Case 2. *(a) Γ_1 and Γ belong to the nonleading manifold $W^{ss}_f(0,0) \in W^s_f(0,0)$ and enter the saddle along the eigendirection corresponding to the real eigenvector λ_2. (b) Condition (8.6)* **Case 3.** *(a) $\Gamma_{1,2} \in W^{ss}_f(0,0)$, (b) $\gamma = 1$, (c) $\lambda_{1,2} \neq 0$, and $|\lambda_{1,2}| < 2$. In this case $m = 3$ and $\mu_3 = \gamma - 1$.*

For the **Case 3**, Shilnikov in [Shilnikov (1986-1993)] shows that both subclasses (a) and (c) are realized in the Shimizu–Morioka model in which the appearance of the Lorenz attractor and its disappearance through bifurcations of lacunae are explained. Some systems of type (a) were studied in [Rychlik (1990)], and those of type (c) were studied in [Robinson (1989)]. Furthermore, a proposed scheme for the classification of Lorenz-type strange attractors is given and proved in [Shilnikov (1981)] as follows:

Proposition 8.1. *The Lorenz-type strange attractors have a complete topological invariant. Geometrically, two Lorenz-like attractors are topologically equivalent if the unstable manifolds of both saddles behave similarly*[2].

---

[2]The similarity was given in terms of kneading invariants [Milnor and Thurston (1977)] when studying continuous (discontinuous), monotonic mappings on an interval.

The method of analysis is based on the fact that the corresponding Poincaré map for system (8.1) is reduced to the form

$$\begin{cases} \bar{x} = F(x,y) \\ \bar{y} = G(y) \end{cases} \quad (8.7)$$

with a continuous right-hand side, apart from the discontinuity line $y = 0$, and G is piecewise monotonic due to the existence of a foliation. Using the technique of taking the inverse spectrum, the map (8.7) can be reduced to a one-dimensional map $\bar{y} = G(y)$, and it is shown in [Guckenheimer & Williams (1979)] that a pair of the kneading invariants is a complete topological invariant for the associated two-dimensional maps if

$$\inf |G'| > 1. \quad (8.8)$$

More details can be found in [Shilnikov (1981-1986-1993a), Shilnikov et al. (1993)]. The proof of Mischaikow and Mrozek given in [Mischaikow & Mrozek (1995-1998)] uses the technique given in [Hastings & Troy (1992)] for obtaining rigorous results concerning the global dynamics of nonlinear systems, especially the fact that the Lorenz Eqs. (8.1) exhibit chaotic dynamics, combined with abstract existence results based on the Conley index theory introduced in Sec. 1.4.1 with finite computer-assisted computations necessary to verify the assumptions of the theorems cited below. The advantages of this technique can be summarized as follows:

(1) This technique is applicable to concrete differential equations such as the Lorenz system (8.1).
(2) In terms of semi-conjugacies (see Sec. 1.4.1), this technique provides a relatively strong description of global dynamics of differential equations.
(3) The necessary computer-assisted computations of this technique are small enough to be performed on any computer.

Let $f : \mathbb{R}^n \to \mathbb{R}^n$ be a homeomorphism. For a good description of this method, we need the following definitions (compare with Definition 1.14):

Definition 8.4. (a) For $N \subset \mathbb{R}^n$, the maximal invariant set of N is defined by

$$Inv(N, f) = \{x \in N : f^n(x) \in N, \forall n \in \mathbb{Z}\}. \quad (8.9)$$

(b) A compact set N is called an isolating neighborhood if $Inv(N; f) \subset intN$. (c) An invariant set S of f is said to be isolated if there is an isolating neighborhood N such that $Inv(N, f) = S$.

To determine the Conley index (recall Definition 1.13) of an isolated invariant set for a map, one begins with an index pair (N, L), where N is an isolating neighborhood, L is its exit set, and the map f induces a homomorphism called the *index map* on the cohomology of the index pair, i.e., $I_f^* : H^*(N, L) \to H^*(N, L)$.

Definition 8.5. The cohomological Conley index of an isolating neighborhood under f is given by

$$Con^*(Inv(N; f)) = (CH^*(Inv(N; f)), X^*(Inv(N; f))) \quad (8.10)$$

where $CH^*(Inv(N, f))$ is the graded module obtained by quotienting $H^*(N, L)$ by the generalized kernel of I_f^* and $X^*(Inv(N; f))$ is the induced graded module automorphism on $CH^*(Inv(N; f))$.

If $P(\mathbb{R}^n)$ is the power set of $\mathbb{R}^n$, then some definitions on real functions are also needed:

Definition 8.6. (a) A multivalued map from $\mathbb{R}^n$ to itself is a function $F: \mathbb{R}^n \to P(\mathbb{R}^n) - \{\emptyset\}$.

(b) A continuous function $f : \mathbb{R}^n \to \mathbb{R}^n$ is a *selector* for F if $f(x) \in F(x)$ for all $x \in \mathbb{R}^n$.

(c) A multivalued function F is an extension of a continuous function $f : \mathbb{R}^n \to \mathbb{R}^n$ if $f(x) \in F(x)$ for all $x \in \mathbb{R}^n$.

(d) Let F be a multivalued function on $\mathbb{R}^n$. For $A \subset \mathbb{R}^n$, let $F(A) = \cup_{x \in A} F(x)$ and define $F^{n+1}(A) = F(F^n(A))$. In this way F recursively defines a multivalued discrete semi-dynamical system on $\mathbb{R}^n$.

To define the Conley index theory for such systems, consider a set $B \subset \mathbb{R}^n$, its *inverse image* given by $F^{-1}(B) = \{x \in \mathbb{R}^n : F(x) \subset B\}$, and its *weak inverse image* given by $F^{*-1}(B) = \{x \in \mathbb{R}^n : F(x) \cap B \neq \emptyset\}$. Given $N \subset \mathbb{R}^n$, then one has the following definitions:

Definition 8.7. (a) The invariant set of N is given by

$$Inv(N, F) = \{\exists \gamma_x : \mathbb{Z} \to N : \gamma_x(0) = 0, \gamma_x(n+1) \in F(\gamma(n))\} \quad (8.11)$$

for $x \in N$.

(b) The *diameter* of F over N is the number[3]

$$diam_N F = \sup_{x \in N} \{\|z - y\| : z, y \in F(x)\}. \quad (8.12)$$

(c) N is called *an isolating neighborhood under F* if

$$B(Inv(N, F), diam_N F) \subset int(N). \quad (8.13)$$

---
[3] The diameter of F over its domain is also called the size of F.

Note that some conditions (admissibility) must be met to define the Conley index for F in N, and in this case it suffices to see that convex-valued maps with continuous selectors are admissible.

Due to technical computations, it is convenient to use the following special form of an isolating neighborhood:

Definition 8.8. The set N is an isolating block for F if

$$B\left(F^{*-1}(N) \cap N \cap F(N), diam_N F\right) \subset int(N). \tag{8.14}$$

In this case, the notion of an isolating block uses only a finite number of iterates of F (one forward and one backward), contrary to the notion of an isolating neighborhood.

The main theorem that proves that Lorenz system (8.1) contains chaotic dynamics is given by [Mischaikow & Mrozek (1992)]:

Theorem 8.2. *Let*

$$P = \{f(x,y,z) : z = 53\}. \tag{8.15}$$

For all parameter values in a sufficiently small neighborhood of $(\sigma, r, b) = (45, 54, 10)$, *there exists a Poincaré section* $N \subset P$ *such that the Poincaré map g induced by (8.1) is Lipschitz and well defined. Furthermore, there exists a* $d \in \mathbb{N}$ *and a continuous surjection* $\rho : Inv(N, g) \to \Sigma_2$ *such that*

$$\rho \circ g^d = \sigma \circ \rho \tag{8.16}$$

where $\sigma : \Sigma_2 \to \Sigma_2$ *is the full shift dynamics on two symbols.*

The following important remarks can be made for this case:

(1) Theorem 8.2 provides a description of the dynamics in a neighborhood of an explicitly presented parameter value, i.e., $(\sigma, r, b) = (45, 54, 10)$, and this choice of parameters was made to minimize the necessary computations because analogous computations for the classical choice of parameters $(\sigma, r, b) = \left(10, 28, \frac{8}{3}\right)$ are more complex.
(2) Theorem 8.2 shows only the existence of an unstable invariant set which maps onto a horseshoe, and there is no sufficiently strong abstract result that proves that the whole attractor of Lorenz system (8.1) is chaotic because the unstable invariant set may lie within a strange attractor.

The proof of Theorem 8.2 has the following five distinct components:

The first step is concerned with the use of the Conley index theory to find algebraic invariants which guarantee the structure of the global dynamics of the Lorenz system (8.1), in this case the semi-conjugacy to the full two shift, *i.e.*, isolating neighborhoods and chaos. Indeed, let $N = N_0 \cup N_1$ be an isolating neighborhood under f where N_0 and N_1 are disjoint compact sets. For $k; l = 0; 1$, let

$$\begin{cases} N_{kl} = N_k \cap f(N_l), S_k = Inv((N_{kk}, f)) \\ S_{lk} = Inv(N_{kk} \cup N_{kl} \cup N_{ll}, f). \end{cases} \quad (8.17)$$

Hence one has the following result proved in [Mischaikow & Mrozek (1995-1998)]:

Theorem 8.3. *Assume that*

$$Con^n(S_k) = \begin{cases} (Q, id) & \text{if } n = 1 \\ 0 & \text{otherwise} \end{cases} \quad (8.18)$$

and that $X^*(S_{lk})$ is not conjugate to $X^*(S_k) \oplus X^*(S_l)$. Then there exists a $d \in N$ and a continuous surjection $\rho : Inv(N, f) \to \Sigma_2$ such that

$$\rho \circ f^d = \sigma \circ \rho \quad (8.19)$$

where $\sigma : \Sigma_2 \to \Sigma_2$ is the full shift dynamics on two symbols.

The second step of the proof is concerned with the extension of the algebraic invariants to multivalued maps providing the following index theory for multivalued maps:

Theorem 8.4. *Let F be an admissible multivalued map and f a selector of F. If N is an isolating neighborhood for F, then it is an isolating neighborhood for f. Furthermore,*

$$Con^*(Inv(N, f)) \approx Con^*(Inv(N, F)). \quad (8.20)$$

Theorem 8.4 implies that if the Poincaré map g given by the Lorenz Eqs. (8.1) is replaced by an a multivalued map extension G such that $g(x) \in G(x)$ and N is an isolating neighborhood for G, then any index information obtained for G is valid for g. The next typical convergence theorem in this step states that any index information of g can be determined by a sufficiently small multivalued map extension G.

Theorem 8.5. *Let N be an isolating neighborhood for $f : \mathbb{R}^n \to \mathbb{R}^n$, a Lipschitz continuous function. Let $\{F_n\}$ be a family of extensions of f such that $F_n \to f$. Then for n sufficiently large, N is an isolating neighborhood for F.*

The third step of the proof needs a combination of the above-mentioned algebraic invariants and a theory of finite representable multivalued maps, which bridge the gap between the continuous dynamics of the Lorenz Eqs. (8.1) and the finite dynamics of the computer. Hence they are more readily attainable than simplicial approximations. To obtain cohomological information generated by the continuous map g, it is necessary to use a computer[4], and this is guaranteed due to the simplicial approximation theorem since g is Lipschitz, but the computations must be performed in which simplicial approximation apply to all nearby maps, including the original map g.

In the proof of Theorem 8.2, the following technical aspects were chosen to facilitate and reduce the computations:

(1) The set of representable real numbers is given by the standard floating point representable coordinates.
(2) The representable sets in $\mathbb{R}^3$ were chosen by selecting a compact set $M \subset \mathbb{R}^3$ within which the dynamics of the Lorenz system (8.1) occur, where a set D of representable vectors and a representable number were used to cover the set. M is the collection of cubic balls $B(d, \eta)$, $d \in D^5$, and in this case a representable set is any union of a subcollection of these cubes.
(3) Double precision arithmetic and the standard fourth-order Runge–Kutta method with a step size $\frac{100}{2^{20}}$ were used to integrate the Lorenz system (8.1). This resulted in the growth factor[6] for Eq. (8.1) and in the function G described in (8.1) being approximately 10^6. The number of initial grid points needed to fulfill condition (8.14) by the associated multivalued map G is proportional to the square of the growth factor. To determine a significantly smaller growth factor, 23 intermediate cross-sections, labeled Ξ_k, $k = -1, ...22$, were introduced, with the original Poincaré section appearing as $P = \Xi_{-1} = \Xi_{22}$. Using the technique described above, 21 finite representable multivalued maps $G_i : \Xi_{i-1} \to \Xi_i$ were obtained. Each multivalued map is an extension of the flow defined map $g_i : \Xi_{i-1} \to \Xi_i$ determined by the Lorenz Eqs. (8.1). In this case, $G = G_{21} \circ G_{20} \circ ... \circ G_0$. The calculation was performed beginning with approximately 700 000 cubes covering N_1[7].

---

[4]Which works only with a finite set of objects.
[5]In this case the sup norm was chosen to increase the efficiency of the algorithm.
[6]Ratio of the size of a value to the size of the grid.
[7]Due to the symmetry of the Lorenz equations (8.1), it was not necessary to compute G on the set N_0.

The growth factor in this case was approximately 30, which resulted in the diameter of $G(q)$ being less than 0.044 for all $q \in Q$. The condition

$$B\left(G^{*-1}(N) \cap N \cap G(N), 0.044\right) \subset int(N) \quad (8.21)$$

is fulfilled, then N is an isolating block under G, and N is an isolating neighborhood of G, and hence by Theorem 8.5, N is an isolating neighborhood of g. To prove the existence of a sequence F_n of finite representable extensions of f we introduce the following definition:

Definition 8.9. Let $M_0 \subset M$. A representable multivalued map on M_0 is a multivalued map $F : M_0 \to P(M)$ such that the set $\{F(x) : x \in M_0\}$ is a finite collection of representable sets.

Let $f : M_0 \to M$ be a Lipschitz continuous function with Lipschitz constant L. Assume that for every representable vector d one can compute a representable vector $f_0(d)$ such that $\|f(d) - f_0(d)\| < \delta$, where δ is a given representable number. Defining $F(x)$ as the smallest convex representable set which contains $B(f_0(x), \delta + L\eta)$, one obtains a multivalued map $F : M_0 \to M$ such that $f(x) \in F(d)$ for all $x \in B(d, \eta)$. Then

$$F^u(x) = \cup_{\|d-x\| \leq \eta} F(d), \text{ and } F^l(x) = \cap_{\|d-x\|} F(d). \quad (8.22)$$

F^l is convex valued because it is easy to see that $F^u(x)$ and $F^l(x)$ are finite, representable, and respectively upper and lower semi-continuous extensions of f. Letting $\delta \to 0$ results in the following proposition proved in [Mischaikow & Mrozek (1995-1998)]:

Proposition 8.2. *Let $M_0 \subset \mathbb{R}^n$ be a compact set, and let $f : M_0 \to \mathbb{R}^n$ be a Lipschitz continuous function. Then there exists a sequence F_n of finite representable extensions of f such that $F_n \to f$ as $n \to +\infty$.*

The fourth step of the proof consists of numerical computations of the finite multivalued map of interest. A return map which strongly suggests the existence of horseshoe dynamics in the Lorenz system (8.1) was obtained for $(\sigma, r, b) = (45, 54, 10)$. The existence of an invariant set which is conjugate to the full shift dynamics on two symbols is guaranteed because the two rectangles R_0 and R_1 are in the plane P, which appear to cross themselves and each other transversally. The proof of semi-conjugacy described in Theorem 8.2 was obtained by computing a representable multivalued admissible extension G of the Poincaré map g such that $N = N_0 \cup N_1$[8] is an isolating neighborhood under G.

---
[8] The rectangles N_0 and N_1 were taken to be within R_0 and R_1 to simplify computation.

Ideally, the multivalued extension G must calculated from the center of each cube in the grid of representable sets using a numerical integration of the Lorenz Eqs. (8.1) and incorporating all errors into the size of the assigned value. In this case, technical difficulties arise due to the exponential growth of these errors caused from the following four sources where in every case rigorous error bounds can be obtained:

(i) The approximate arithmetic of the machine.

(ii) The numerical procedure used to integrate the equations.

(iii) Extending the value from the center of a cube to the whole cube by means of Lipschitz constant estimates. In this case the error estimates were based on the Gronwall inequality, the local Lipschitz constants, and logarithmic norms.

(iv) Estimating the point of intersection of the trajectory with the cross-section from two consecutive steps of the numerical method. In this case, the errors were estimated using the second-order Taylor expansion of the solution.

The fifth step of the proof is concerned with the calculation of the combinatorial computations of the Conley index for the multivalued map G. For this purpose, one first needs to find an index pair (N, L) that can be determined by inspection because the only cohomology group of interest is $H^1(N, L)$, and then to determine the index map defined in Definition 8.5

$$I_G^* : H^*(N, L) \to H^*(N, L). \tag{8.23}$$

For more general problems, the map G is a multivalued simplicial map because D can be viewed as the set of vertices of a simplicial decomposition of P. Hence finite computation permits one to find I_G. From I_G one computes the Conley index $Con^*(Inv(N, G))$ and hence by Theorem 8.5 the index of interest $Con^*(Inv(N, g))$ that leads to the algebraic hypotheses of Theorem 8.4, and hence Theorem 8.2 follows. For the more detail about the above five steps of the proof, see [Mischaikow & Mrozek (1998)].

8.2.2 Geometric models of the Lorenz equation

The example of a singular axiom A vector field without cycles and with a singular basic set equivalent to the Lorenz attractor introduced in [Guckenheimer (1976)] belongs to the interior $C^r(M)$ of the set of C^r vector fields in M whose critical elements are hyperbolic (recall Definition 6.1(d)). This example consists of a vector field on a compact manifold which does not lie in the closure of Ω-stable (see the end of Sec. 4.1.2) or axiom A vector fields (recall Definition 6.3). The importance of this example is that it has additional instability properties, not topologically Ω-stable, and the violation of axiom A occurs differently than in the previous examples which have non-isolated singularities in nonattracting parts of the nonwandering set (recall Definition 6.1(a)) [Abraham & Marsden (1967), Abraham (1972)]. Furthermore, it confirms that it is not true that the singularities of a vector field are isolated in its nonwandering set. This example is based on numerical studies of the Lorenz system (8.1) with the classical bifurcation values, *i.e.*, $\sigma = 10, r = 28$, and $b = \frac{8}{3}$. The construction of such an example is given by the following steps:

(1) Define a C^∞ vector field f using the coordinates (x, y, z) in $\mathbb{R}^3$ in a bounded region. Inside this region there is a compact invariant 2-D set A which is an attractor[9].
(2) Assume that vector field f has the following properties:
   (a) The vector field f has three singular points,
   $$p = (0,0,0), q_\pm = (\pm 1, \pm\frac{1}{2}, 1), \quad (8.24)$$
   where p is a saddle with a two-dimensional stable manifold $W^s(p)$, and $q_\pm$ are saddle points with one-dimensional stable manifolds $W^s(q_\pm)$. The negative eigenvalues of f at $q_\pm$ have large absolute values, and the remaining ones are complex with eigenspaces spanned by $\frac{\partial}{\partial y}$ and $\frac{\partial}{\partial z}$. The real parts of these eigenvalues are small.
   (b) The rectangle
   $$\{(x, y, z) : x = 0, -1 \le y \le 1, 0 \le z \le 1\} \quad (8.25)$$
   is contained in $W^s(p)$.
   (c) The stable eigenvectors of f at p are $\frac{\partial}{\partial y}$ with an eigenvalue of large absolute value and $\frac{\partial}{\partial z}$ with an eigenvalue of small absolute value.
   (d) The unstable manifold $W^u(p)$ contains the segment from $(-1, 0, 0)$ to $(1, 0, 0)$ and has an eigenvalue of intermediate absolute value. The segments from $(\pm 1, -1, 1)$ to $(\pm 1, 1, 1)$ are contained in $W^s(q_\pm)$.
(3) Consider the square
$$R = \{(x, y, z) : -1 \le x \le 1, -1 \le y \le 1, z = 1\} \quad (8.26)$$
and its Poincaré return map θ. In this case, the map θ is not defined when f is ± 1 or 0 since these points lie in the stable manifold of one of the singular points, and the orbits in R for $f = \pm 1$ never leave R while those for $f = 0$ never return. At all other points of R, θ is defined.

---
[9]*i.e.*, In the sense that A has a fundamental system of neighborhoods, each of which is forward invariant under the flow of f.

(4) Let

$$\begin{cases} R_+ = R \cap \{(x,y,z) : 0 < x < 1\} \\ R_- = R \cap \{(x,y,z) : -1 < x < 0\}. \end{cases} \quad (8.27)$$

Define $\theta_\pm$ to be the restriction of θ to $R_\pm$. Assume the following:
(a) There are functions[10] $f_\pm, g_\pm$ and a number $\alpha > 1$ with the properties that

$$\begin{cases} \theta_\pm(x,y) = (f_\pm(x), g_\pm(x,y)) \\ 0 < \frac{\partial g_\pm}{\partial y} < \frac{1}{2} \\ \frac{\partial f_\pm}{\partial x} > \alpha. \end{cases} \quad (8.28)$$

(b) The numbers $\rho_\pm = \lim_{x \to 0} f_\pm(x)$ have the properties

$$\begin{cases} \rho_+ < 0 \\ \rho_- > 0 \\ \theta_-(\rho_+) < 0 \\ \theta_+(\rho_-) > 0. \end{cases} \quad (8.29)$$

(c) The first intersections of $W^u(p)$ with R occur at the points with $x = \rho_\pm$.
(d) The images of $g_\pm$ are contained in the intervals $\left[\frac{\pm 1}{4}, \frac{\pm 3}{4}\right]$.

By solving a linear system of differential equations near the saddle point p, it easy to show that at the point p, one has $\lim_{x \to 0} \frac{\partial g_\pm(x,y)}{\partial y} = 0$ and $\lim_{x \to 0} \frac{\partial f_\pm(x,y)}{\partial x} = \infty$. Since the trajectories of $R_\pm$ come arbitrarily close to p, then the return maps $\theta_\pm$ acquire singularities like a power of f. If one assumes that the vector field f is extended to a vector field on a compact three-manifold M, then the following results were proved in [Guckenheimer (1976)]:

Theorem 8.6. *(a) There is a neighborhood $\mathbf{U}$ of f in the space of C^r vector fields on M $(r \geq 1)$ and a set $\mathbf{V}$ of second category in $\mathbf{U}$ such that if $g \in \mathbf{V}$, then g has a singular point which is not isolated in its nonwandering set.*
(b) The vector field f has a neighborhood $\mathbf{U}$ in the space of C^r vector fields on M $(r > 1)$ with the property that if $\mathbf{V} \subset \mathbf{U}$ is an open set in the space of C^r vector fields, then there are vector fields in $\mathbf{V}$ whose nonwandering sets are not homeomorphic to each other.

---
[10] Note that this assumption does not remain after perturbation of f to a compact three-manifold M. These functions are introduced to simplify the discussion and are not essential properties of f.

Note that Theorem 8.6(b) states that f is not in the closure of the set of topologically Ω-stable vector fields. The proof of Theorem 8.6 is based on the description (in terms of *symbolic dynamics* in [Smale (1965)]) of the nonwandering set $\Omega(f)$ of the vector field f. In fact, the geometric Lorenz attractor described above resembles the horseshoe (see Sec. 1.3) in some aspects; each is the closure of its periodic orbits, transitive, and has sensitive dependence on initial conditions.

8.2.2.1 *Chaoticity of the geometric Lorenz attractor*

The chaoticity of the geometric Lorenz attractor described above can be seen from the calculation of its Hausdorff dimension (see Sec. 3.1.3) given in [Lizana & Mora (2008)] where the lower bound for this dimension was estimated for the geometric Lorenz attractor Λ (with flow L) in terms of the eigenvalues of the singularity and the symbolic dynamic associated with the geometrical distribution with the attractor in the homoclinic case, *i.e.*, when both branches of the unstable manifold of the unique singularity O meet its stable manifold. Indeed, let $F(x,y) = (f(x), g(x,y))$ be the Poincaré return map F induced by the flow of L for some cross-section surface, and let Λ_F be the hyperbolic attractor of F. The following result was proved in [Lizana & Mora (2008)]:

Theorem 8.7. *For the homoclinic case, there exists a $0 < \gamma < 1$ such that*

$$\dim_H(\Lambda_F) \geq 1 + \frac{\ln \rho(A)}{\ln\left(\frac{1}{\gamma}\right)} > 1, \qquad (8.30)$$

where $\rho(A)$ is the spectral radius of the matrix A, a $(0,1)$-matrix which describes the geometric distribution of F in the y-direction.

The proof of Theorem 8.7 is based on relations between a constructed geometric Lorenz attractor and shift map of finite type.

The following corollary of Theorem 8.7 was given also in [Lizana & Mora (2008)]:

Corollary 8.1. *For the homoclinic case, the Hausdorff dimension of the geometric Lorenz attractor Λ satisfies the following bound:*

$$\dim_H(\Lambda) \geq 2 + \frac{\ln \rho(A)}{\ln\left(\frac{1}{\gamma}\right)} > 2. \qquad (8.31)$$

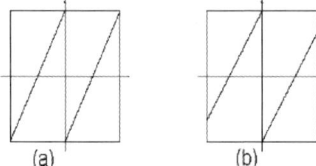

Fig. 8.2 (a) The baker's map (8.32). (b) A tired baker's map (8.33).

8.2.2.2 *Transitivity of the Lorenz attractor*

It was proved in [Tucker (1996b)] that some of the one-dimensional Poincaré maps of the geometric Lorenz model are not good models of the real flow. The method of analysis is the construction of a family of tired baker's maps and the proof that uniform expansion is not a sufficient condition for topological transitivity. Indeed, the baker's map $B : [-1,1] \to [-1,1]$ is given by

$$B(x) = sgn(x)(2|x| - 1) \tag{8.32}$$

in which the interval $I = [-1,1]$ should be thought of as a piece of dough the baker cuts I in half and then stretches both halves by a factor and finally puts one of the resulting pieces on top of the other. The map (8.32) defines a completely mixed dough, *i.e.*, the map B is topologically transitive on I [Glendinning (1994)].

Definition 8.10. Consider a map $f : I \to I$ where I is any compact interval. f is topologically transitive on I if for any two open sets $U, V \subset I$ there exists an $n \geq 0$ such that $f^n(U) \cap V \neq \emptyset$.

The *tired baker's maps* (TBMs) $(B_s)_{1<s<2}$ defined in [Tucker (1996b)] were obtained by assuming that the baker gets tired stretching the two halves of the dough by a factor of two all day long, and instead stretches them by a factor $s \in (1,2]$. Thus the tired baker's maps (TBMs) are defined by the following definition (see Fig. 8.2):

Definition 8.11. We have

$$B_s(x) = sgn(x)(s|x| - 1). \tag{8.33}$$

In this case, the map B_s is topologically transitive on I for some values of s, and it is uniformly expanding on I if $s > 1$. Using Definition 8.11 and [Parry (1979)], one can obtain the following result proved in [Tucker (1996b)]:

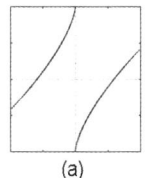 

(a)   (b)

Fig. 8.3 (a) Conditions (1)–(4) satisfied. (b) Conditions (1)–(3), but not (4), satisfied.

Theorem 8.8. *B_s is topologically transitive on I if and only if $s \in \left[-\sqrt{2}, \sqrt{2}\right]$.*

As invariance and expansion are trivially fulfilled, one has the following corollary:

Corollary 8.2. *B_s is chaotic on I if and only if $s \in \left[-\sqrt{2}, \sqrt{2}\right]$.*

Many simplifying assumptions have been made about one-dimensional Poincaré maps of the geometric Lorenz model described in the beginning of this section. In [Tucker (1996b)], a way of overcoming these simplifications is shown for more truthful conditions on the Poincaré maps. Indeed, the geometric Lorenz model requires roughly the following conditions of the class of maps $f_a : [-a, a] \to [-a, a]$ with

$$f_a(x) = sgn\,(x)\,(f\,(|x|) - a)\,. \tag{8.34}$$

When $a > c$ and $(\sigma, b, r) = \left(10, \tfrac{8}{3}, 28\right)$, with all $x > 0$, f satisfies the following:

(1) $f(x) = kx^v + O(x^v)$, where $k > 0$ and $v \in (0, 1)$,
(2) $f''(x) < 0$,
(3) $f(c) = 2c$ for some $c > 0$,
(4) $f'(x) > \sqrt{2}$.

In this case and for the usual parameter values, condition (4) does not hold. A weaker version of (4) is given by

(4̃) $\left(f^2\right)'(x) > \sqrt{2}, s < \sqrt{2}$,

and in this case, one has Poincaré maps C^0-close to the TBMs defined by (8.33). These Poincaré maps are very poor models of the Lorenz flow. Figure 8.3 shows the *limit image* (which is the attractor lifted onto the two branches of the map) of two similar maps illustrating the problem. Figure 8.3(b) shows the gaps created by the mechanism resulting from the

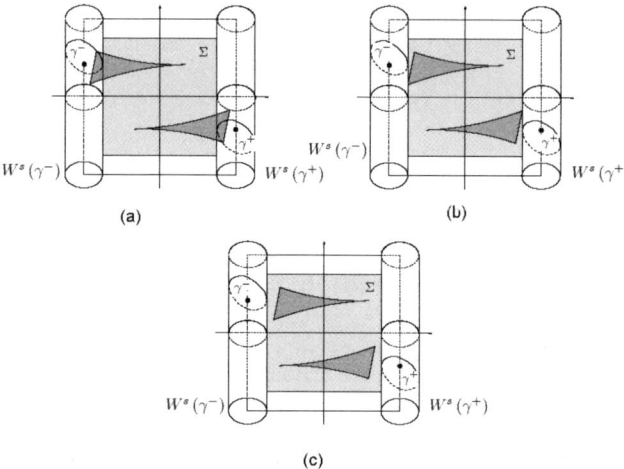

Fig. 8.4 At $(\sigma, b, r) \approx (10, \frac{8}{3}, 24.06)$, the cusps leave the stable cylinders, and the invariant set becomes attracting for the Lorenz system (8.1).

use of a slightly less restrictive version of condition (4) which gives false results. Furthermore, the use of milder conditions guaranteeing topological transitivity are presented in [Robinson (1984), Tucker (1996a)] giving a worse description of the real dynamics of the Lorenz Eq. (8.1) if they are not uniformly expanding.

In [Tucker (1996b)], a completely new way of solving this problem was provided by considering a different parameter range $(\sigma, b, r) \approx (10, \frac{8}{3}, 24.06)$ than the classical one where the invariant set of the geometric Lorenz system goes from being an unstable horseshoe to being a stable attractor as shown in Fig. 8.4. Here the top bottom cusp is the first return of the left right half of the rectangle Σ which is transversal to the flow, and the invariant set of this return map becomes attracting just as the cusps leave the two cylinders representing the stable manifolds of the two unstable closed orbits γ^+ and γ^-. Note that for a certain small neighborhood of $(\sigma, b, r) \approx (10, \frac{8}{3}, 24.06)$, conditions (1) and (4) can be dropped almost entirely without any loss of topological transitivity as shown in [Tucker (1996a)]. Thus the above problem was solved by considering the following conditions on the map (8.33):

($\tilde{1}$) $f(0) = 0$, and $f'(0^+) > 0$,
($\tilde{2}$) $f''(x) < 0$, for all $x > 0$,

($\tilde{3}$) $f(c) = 2c$ for some $c > 0$,
($\tilde{4}$) $f'(c) \geq 1$.

Thus for a larger than c, the one-dimensional model with conditions ($\tilde{1}$)–($\tilde{4}$) exhibits a horseshoe, and this confirms the results of Mischaikow and Mrozek described in Sec. 8.2.1. Furthermore, it was shown in [Tucker (1996a)] that by a perturbation argument, the map (8.33) can be turned into a two-dimensional map, and if $R_{a,b}$ corresponds to a Poincaré map of a three-dimensional flow, then the 2-D map induces an attracting set $\mathcal{A}$ in an open neighborhood of the parameters (a,b) in which $R_{a,b}$ is topologically transitive on $\mathcal{A}$ and has positive Lyapunov exponents for all $x \in \mathcal{A}$.

8.2.2.3 Singular horseshoe

Some examples of a structural stable vector field on the unit disk $\mathbb{D}^3 \subset \mathbb{R}^3$ are tangent to the boundary of $\mathbb{D}^3$ and have a nonhyperbolic nonwandering set[11]. In [Labarca & Pacifico (1986)], an example of structural stability on manifolds with a boundary was introduced and called a *singular horseshoe*. Indeed, let M be a compact manifold with a boundary $\partial M \neq \emptyset$. Let $C^1(M, \partial M)$ be the space of C^1 vector fields on M that are tangent to the boundary ∂M. Then the following result was proved in [Labarca & Pacifico (1986)]:

Theorem 8.9. *Let $\mathbb{D}^3$ be the unit disk in $\mathbb{R}^3$. There exists a structurally stable vector field $f \in C^1(M, \partial M)$ whose nonwandering set $\Omega(f)$ is non-hyperbolic.*

The basic idea of such an example is the construction of a vector field $f \in C^1(M, \partial M)$ with a saddle connection (p, σ) along the boundary of $\mathbb{D}^3$, where $p \in \partial \mathbb{D}^3$ is a singularity with a one-dimensional unstable manifold contained in $\partial \mathbb{D}^3$ and σ is a saddle type closed orbit contained in $\partial \mathbb{D}^3$ such that $W^u(p) \subset W^s(\sigma) \subset \partial \mathbb{D}^3$. Hence this produces a persistent cycle evolving p and σ, because there are two orbits of transversal intersection between $W^u(p)$ and $W^s(\sigma)$. Note that the singular horseshoe is a first return map F defined on a cross section at $q \in \sigma$ associated with this cycle, and it is a modification of Smale's horseshoe introduced in Sec. 1.3.

---

[11]As a criterion, the existence of such a singularity implies the nonhyperbolicity of the nonwandering set, and maybe the structural stability without singularities implies the hyperbolicity of the nonwandering set.

8.2.2.4 Lorenz-like families with criticalities

In [Luzzatto & Viana (2000)], a study of the interaction between *singular behavior* (corresponding to trajectories near equilibria) and *critical behavior* (near folding regions) was given using a one-dimensional map called *Lorenz-like families with criticalities* as models of rich, nonsmooth dynamics in one-dimension.

Note that the Lorenz-like families with criticalities φ_a defined below in (8.35) is inspired by the properties of the original Lorenz flow (8.1), in particular, the study of the bifurcation which occurs as the parameter crosses the value $a = c$, and the persistence (see Sec. 2.3) of the Lorenz attractor for all $a < c$. Indeed, let $\{\varphi_a\}$ be one-parameter families of real maps of the form

$$\varphi_a(x) = \begin{cases} \varphi(x) - a & \text{if } x > 0 \\ -\varphi(-x) + a & \text{if } x < 0 \end{cases} \quad (8.35)$$

where $\varphi : \mathbb{R}^+ \to \mathbb{R}^+$ is smooth and satisfies the following properties that makes it uniformly expanding for all parameters up to c using the so-called *measure theoretic persistence* of positive Lyapunov exponents proved by Jakobson in [Jakobson (1981)] for maps in the quadratic family $f_a(x) = 1 - ax^2$ close to parameter values $\bar{a}$ satisfying the conditions of [Misiurewicz (1981)]:

L1 $\varphi(x) = \psi(x^\lambda)$ for all $x > 0$, where $0 < \lambda < \frac{1}{2}$ and ψ is a smooth map defined on $\mathbb{R}$ with $\psi(0) = 0$ and $\psi'(0) \neq 0$;

L2 There exists some $c > 0$ such that $\varphi'(c) = 0$;

L3 $\varphi''(x) < 0$ for all $x > 0$.

L4 Let $x_{\sqrt{2}}$ denote the unique point in $(0, c)$ such that $\varphi'(x_{\sqrt{2}}) = \sqrt{2}$, then we suppose

$$0 < \varphi_a(x_{\sqrt{2}}) < \varphi_a(a) < x_{\sqrt{2}} \quad (8.36)$$

for all $a \in (a_2, c]$.

L5 $\left|(\varphi_c^2)'(x)\right| > 2$ for all $x \in [-c, c] \setminus \{0\}$ such that $|\varphi_c(x)| \in [x_{\sqrt{2}}, c]$.

Hence the following results were proved in [Luzzatto & Viana (2000)]:

Proposition 8.3. *Given any $a \in [a_1, c]$, one has*
1. *The interval $[-a, a]$ is forward invariant and $\varphi|_{[-a,a]}$ is transitive.*
2. *We have*

$$\left|(\varphi_a^n)'(x)\right| \geq \min\left\{\sqrt{2}, |\varphi_a'(x)|\right\} \left(\sqrt{2}\right)^{n-1} \quad (8.37)$$

for all $x \in [-a, a]$ such that $\varphi_a^j(x) \neq 0$ for every $j = 0, 1, ..., n-1$.

Theorem 8.10. *Let $\{\varphi_a\}$ be a Lorenz-like family satisfying conditions L1–L5. Then there are $\sigma > 0$ and $\mathcal{A}^+ \subset \mathbb{R}$ such that*

$$\left|\left(\varphi_a^j\right)'(c_1(a))\right| \geq e^{\sigma j} \tag{8.38}$$

for all $a \in \mathcal{A}^+$ and $j \geq 1$ and

$$\lim_{\varepsilon \to 0} \frac{m\left(\mathcal{A}^+ \cap [e, e+\varepsilon]\right)}{\varepsilon} = 1. \tag{8.39}$$

(m=Lebesgue measure on $\mathbb{R}$).

Moreover, there is a $\sigma_1 > 0$ such that if $a \in \mathcal{A}^+$ then for m-almost all $x \in [-a, a]$ we have

$$\limsup_{n \to \infty} \frac{1}{n} \left|\log\left(\varphi_a^n\right)'(x)\right| \geq \sigma_1. \tag{8.40}$$

8.2.2.5 Topological classification of geometrical Lorenz attractors

The problem of topological classification of geometrical Lorenz attractors was considered in [Klinshpont et al. (2005), Klinshpont (2006)] as follows: Consider the model of the Lorenz attractor suggested in [Williams (1979)], and let L_1 be a branched manifold and a semi-flow φ_t, $t > 0$ on L_1 resulting from the factorization of a neighborhood of the Lorenz attractor by the one-dimensional invariant stable foliation transversal to the trajectories of the Lorenz flow. Let $\tilde{L} = \lim_{\leftarrow}(L_1, \varphi_t, t > 0)$ be the inverse limit. In this case, the branches of the negative semi-trajectories of the semi-flow φ_t are the elements $\hat{z} \in \hat{L}$, and a flow $\hat{\varphi}_t$ is naturally defined on the space $\hat{L}$. Thus the pair $\left(\hat{L}, \hat{\varphi}_t\right)$ is called the *geometrical Lorenz attractor*. Let $\hat{O}$ be the fixed point of the flow $\hat{\varphi}_t$, $\hat{W}^u(\hat{O})$ be the unstable manifold of the point $\hat{O}$, and $A = Cl(\hat{W}^u(\hat{O})) \setminus \hat{W}^u(\hat{O})$. The concepts of symbolic prehistory and their equivalence are given by the following definition:

Definition 8.12. (a) A symbolic prehistory $\alpha(\hat{z})$ is a symbolic sequence $\{\alpha_i\}$ characterizing the behavior of the negative semi-trajectory $\tilde{z}$.

(b) We call two prehistories $\alpha(\hat{z}) = \{\alpha_i\}_{i>1}$ and $\alpha(\hat{w}) = \{\gamma_i\}_{i>1}$ equivalent if their *tails* coincide, that is there is a k, l such that for any $m \geq 0$: $\alpha_{k+m} = \gamma_{l+m}$.

Let Ω be the set of symbolic prehistories of points of A, and let $\hat{\Omega}$ be the set of equivalence classes. For each class $\beta \in \hat{\Omega}$, consider the set $C(\beta) = \{\hat{z} : \alpha(\hat{z}) \in \beta\}$. Thus we have the following definition:

Definition 8.13. We call the set of cardinals $X = \{card(C(\beta) : \beta \in \hat{\Omega}\}$ the Lorenz manuscript.

The work of Klinshpont is based on the concept of *Lorenz manuscript* or topological invariant given by Definition 8.13. This notion is the principal cause of the existence of an uncountable set of nonhomeomorphic geometrical Lorenz attractors. Hence the following result was proved in [Klinshpont et al. (2005), Klinshpont (2006)]:

Theorem 8.11. *1. X is a topological invariant, that is if $\hat{L}_1$ is homeomorphic to $\hat{L}_2$, then $X_1 = X_2$;*

2. for any sequence $\{n_i\}$ of natural numbers there exists a pair $\left(\hat{L}, \hat{\varphi}_t\right)$ such that $X = \{n_1, n_2, ..., N_0, c\}$.

Using Theorem 8.11, several new examples of attractors, which are suspensions of an expanding map of an interval with n points of discontinuity, were constructed. In fact, all topological types of these attractors can be obtained by small C^1 perturbation of map of an interval.

Finally, we give the main references that deal with the study of the dynamics of Lorenz maps defined that show the properties of Lorenz system (8.1). For this purpose, the reader can be see [Guckenheimer (1976), Guckenheimer & Williams (1979), Gambaudo & Tresser (1985), Hubbard & Sparrow (1990), Brucks et al. (1991), Glendinning & Sparrow (1993), Campbell et al. (1996), Galeeva et al. (1997), Labarca & Moreira (2001-2003-2006)] and references therein.

8.2.3 Structure of the Lorenz attractor

The description of the structure of the Lorenz attractor (8.1) was done in [Afraimovich et al. (1977-1983)] as follows: Let B denote the Banach space of C^r-smooth dynamical systems X ($r \geq 1$) with the C^r-topology on a smooth three-dimensional manifold M. Let $W^s = W^s(X)$ denote the stable two-dimensional manifold of the saddle. Let $W^u = W^u(X)$ denote the unstable manifold of O and two trajectories $\Gamma_{1,2} = \Gamma_{1,2}(X)$ originating from it. Assume that in the domain $U \subset B$, each system X has an equilibrium state O of the saddle type, i.e., $\lambda_1 < \lambda_2 < 0 < \lambda_3$ holds for the roots $\lambda_i = \lambda_i(X)$, $i = 1, 2, 3$ of the characteristic equation at O, and the saddle value $\sigma(X) = \lambda_2 + \lambda_3 > 0$. Here it is assumed that for a certain local map $V = \{(x_1, x_2, x_3)\}$, containing O, X can be written in the form

$$x_i = \lambda_i x_i + P_i(x_1, x_2, x_3), i = 1, 2, 3 \qquad (8.41)$$

because in this case it is known that both W^s and W^u depend smoothly on X on each compact subset. Suppose that the system $X_0 \subset U$ satisfies

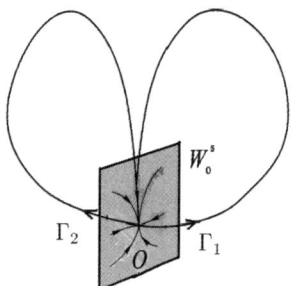

Fig. 8.5   Illustration of a homoclinic butterfly.**

the following conditions (see Fig. 8.7):

(1) $\Gamma_i(X_0) \subset W^s(X_0), i = 1,2$, i.e., $\Gamma_i(X_0)$ is doubly asymptotic to O.
(2) $\Gamma_1(X_0)$ and $\Gamma_2(X_0)$ approach O tangentially to one another.

Hence the following lemma was obtained:

Lemma 8.1. *(a) If $\lambda_1 < \lambda_2$, then the nonleading manifold W_O^{ss} of W_O^s, consisting of O and the two trajectories tangential to the axis x_1 at the point O, divides W_O^s into two open domains, W_+^s and W_-^s.*

(b) If $\Gamma_i(X_0) \subset W_+^s(X_0)$, then Γ_i is tangent to the positive semi-axis x_2.

(c) If v_1 and v_2 are sufficiently small neighborhoods of the separatrix butterfly $\bar{\Gamma} = \overline{\Gamma_1 \cup O \cup \Gamma_2}$ and $M_>$ is the connection component of the intersection of $W_+^s(X_0)$ with v_i which contains $\Gamma_i(X_0)$, then M_i is a two-dimensional C^0-smooth manifold homeomorphic either to a cylinder or to a Möbius band.

Assume that the *separatrix values* $A_1(X_0)$ and $A_2(X_0)$ are not equal to zero. Then the above assumptions lead to the following results:

Lemma 8.2. *The system X_0 belongs to the bifurcation set B_1^2 of codimension-two, and B_1^2 is the intersection of two bifurcation surfaces B_1^1 and B_2^1 each of codimension-one, where B_i^1 corresponds to the separatrix loop $\bar{\Gamma}_i = \overline{O \cup \Gamma_i}$.*

Consider a two-parameter family of dynamical systems $X(\mu), \mu = (\mu_1, \mu_2), |\mu| < \mu_0, X(0) = X_0$, such that $X(\mu)$ intersects with B_1^2 only along X_0 and only for $\mu = 0$. Assume also that the family $X(\mu)$ is transverse to B_1^2, i.e., if we recall Definition 5.1, then transversality here means that for the system $X(\mu)$, the loop $\Gamma_1(X(\mu))$ deviates from $W_+^s(X(\mu))$ by a value on the order of μ_1, and the loop $\Gamma_2(X(\mu))$ deviates from $W_+^s(X(\mu))$ by a value of the order of μ_2. In [Shilnikov (1970)], the above assumptions imply that in the transition to a system close to X_0 the separatrix loop can generate only one periodic orbit, which is of the saddle type. Furthermore, assume that the loop $\Gamma_1(X_0) \cup O$ generates a periodic orbit L_1 for $\mu_1 > 0$ and $\Gamma_2(X_0) \cup O$ generates the periodic orbit L_2 for $\mu_2 > 0$. Let U_0 be the corresponding domain in U, which is the intersection of the stability regions for L_1 and L_2, i.e., the domain in which the periodic orbits L_1 and L_2 are structurally stable (recall Definition 4.3). Let W_i^s be a stable manifold of L_i for the system $X \subset U_0$ and the unstable one be W_i^u. Then one has the following lemma:

Lemma 8.3. *(a) If $A_i(X_0) > 0$, then W_i^u is a cylinder. (b) If $A_i(X_0) < 0$, then W_i^u is a Möbius band. (c) If M is an orientable manifold, then W_i^s is a cylinder if $A_i(X_0) > 0$. Otherwise W_i^s is a Möbius band.*

The study of the signs of the separatrix values gives the following three main cases: **Case A** (orientable) $A_1(X_0) > 0$, $A_2(X_0) > 0$, **Case B** (semi-orientable) $A_1(X_0) > 0$, $A_2(X_0) < 0$, **Case C** (nonorientable) $A_1(X_0) < 0$, $A_2(X_0) < 0$. The domain U_0 contains two bifurcation surfaces B_3^1 and B_4^1 in each of the above three cases:

Lemma 8.4. *(a) In Case A, B_3^1 corresponds to the inclusion $\Gamma_1 \subset W_2^s$, and B_4^1 corresponds to the inclusion $\Gamma_2 \subset W_1^s$, (b) In Case B, B_3^1 corresponds to the inclusion $\Gamma_1 \subset W_1^s$, and B_4^1 corresponds to the inclusion $\Gamma_2 \subset W_1^s$; (c) In Case C, (c-1) The generated orbits L_1 and L_2 arise. (c-2) There also arises a saddle periodic orbit L_3 which makes one revolution along $\Gamma_1(X_0)$ and $\Gamma_2(X_0)$. (c-3) If both W_i^u are Möbius bands, $i = 1, 2$, then the unstable manifold W_3^u of the periodic orbit L_3 is a cylinder, and in this case the inclusions $\Gamma_1 \subset W_2^s$ and $\Gamma_2 \subset W_3^s$ correspond to the surfaces B_3^1 and B_4^1, respectively.*

Now, suppose that B_3^1 and B_4^1 intersect transversely over the bifurcation set B_2^2, i.e., the curves B_3^1 and B_4^1 intersect at some point $\mu_1 = (\mu_{11}, \mu_{12})$. Let U_1 be the domain lying between B_3^1 and B_4^1. Assume that for each $X \in U$ there exists a transversal D with the following properties:

(1) The Euclidean coordinates (x, y) can be introduced on D such that

$$D = \{(x, y) : |x| \leq 1, |y| < 2\}. \tag{8.42}$$

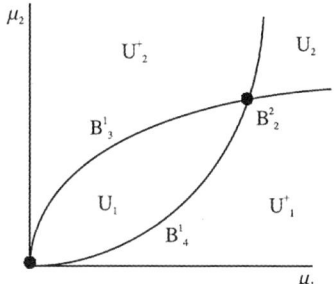

Fig. 8.6 Schematic bifurcation diagram for the Lorenz attractor (8.1) in the plane (μ_1, μ_2).**

(2) The equation $y = 0$ describes a connection component S of the intersection $W_O^s \cap D$ such that no ω-semi-trajectory that begins on S possesses any point of intersection with D for $t > 0$.

(3) The mapping $T_1(X) : D_1 \to D$ and $T_2(X) : D_2 \to D$ are defined along the trajectories of the system X, where

$$\begin{cases} D_1 = \{(x,y) : |x| \leq 1, 0 < y \leq 1\} \\ D_2 = \{(x,y) : |x| \leq 1, -1 \geq y < 1\}, \end{cases} \quad (8.43)$$

and $T_i(X)$ is written in the form,

$$\begin{cases} \bar{x} = f_i(x, y) \\ \bar{y} = g_i(x, y), \end{cases} \quad (8.44)$$

where $f_i, g_i \in C^r, i = 1, 2$.

(4) f_i and g_i admit continuous extensions on S, and

$$\begin{cases} \lim_{y \to 0} f_i(x, y) = x_i^{**} \\ \lim_{y \to 0} g_i(x, y) = y_i^{**} \end{cases}, i = 1, 2. \quad (8.45)$$

(5)

$$\begin{cases} T_1 D_1 \in P_{i_1} = \{(x,y) : \frac{1}{2} \leq x \leq 1, |y| < 2\} \\ T_2 D_2 \in P_{i_2} = \{(x,y) : -1 \leq x \leq -\frac{1}{2}, |y| < 2\}. \end{cases} \quad (8.46)$$

(6) If $\|.\| = \sup_{(x,y) \in D \setminus S} |.|$, then let $T(X) \equiv T_i(X)|_{D_i}$, $(f, g) \equiv (f_i, g_i)$ on D_i, $i = 1, 2$. Let us impose the following restrictions on $T(X)$:

$$\begin{cases} \|f_x\| < 1 \\ 1 - \|(g_y)^{-1}\| \cdot \|f_x\| > 2\sqrt{\|(g_y)^{-1}\| \|g_x\| \|(g_y)^{-1} \cdot f_y\|} \\ \|(g_y)^{-1}\| < 1 \\ \|(g_y)^{-1} \cdot f_y\| \cdot \|g_x\| < (1 - \|f_x\|)(1 - \|(g_y)^{-1}\|). \end{cases} \quad (8.47)$$

Hence one has the following result:

Lemma 8.5. *In a small neighborhood of S, the following representation of the behavior of trajectories near W_O^s is valid:*

$$\begin{cases} f_1 = x_1^{**} + \varphi_1(x,y)y^\alpha \\ g_1 = y_1^{**} + \psi_1(x,y)y^\alpha \\ f_2 = x_2^{**} + \varphi_2(x,y)(-y)^\alpha \\ g_2 = y_2^{**} + \psi_2(x,y)(-y)^\alpha, \end{cases} \quad (8.48)$$

where $\varphi_1, ..., \psi_2$ are smooth with respect to x, y for $y \neq 0$, and $T_i(x)$ satisfies (8.47) for sufficiently small y.

Let $A_1(X)$ be the limit of φ_1, and $A_2(X)$ be one of ψ_2. By analogy with $A_1(X_0)$ and $A_2(X_0)$ introduced above, one can obtain the following definition:

Definition 8.14. *The functionals $A_1(X)$ and $A_2(X)$ are called the separatrix values[12].*

Thus the following result was obtained:

Lemma 8.6. *(a) For a system lying in a small neighborhood of the system X, all the conditions 1–6 are satisfied near S, and the concept of orientable, semi-orientable, and nonorientable cases can be extended to any system $X \in U$. (b) The point P_i with the coordinates (x_i^{**}, y_i^{**}) is the first point of intersection of $\Gamma_i(X)$ with D.*

Consider the constant q defined by

$$\begin{cases} q = \dfrac{1 + \|f_x\| \|(g_y)^{-1}\| + S_{x,y}}{2\|(g_y)^{-1}\|} \\ S_{x,y} = \sqrt{1 - \|(g_y)^{-1}\|^2 \|f_x\| - 4\|g_x\| \|(g_y)^{-1}\| \|(g_y)^{-1} \cdot f_y\|} \end{cases} \quad (8.49)$$

Hence one has the following result:

Lemma 8.7. *If (8.49) holds, then $q > 1$, and all the periodic points are of the saddle type.*

If Σ denotes the closure of the set of points of all the trajectories of the mapping $T(X)$, which are contained entirely in D, the set Σ is described most simply in the domain U_1. Here the following theorems hold [Afraimovich *et al.* (1977)]:

---
[12] For simplicity, assume that $A_{1,2}(X)$ do not vanish.

Theorem 8.12. *(a) If $X \in U_1$, then $T(X)|_\Sigma$ is topologically conjugated with the Bernoulli scheme (σ, Ω_2) with two symbols. (b) The system $X \in U_2$ has a two-dimensional limiting set Ω, which satisfies the following conditions: 1. The set Ω is structurally unstable. 2. $[\Gamma_1 \cup \Gamma_2 \cap O] \subset \Omega$. 3. Structurally stable periodic orbits are everywhere dense in Ω. 4. Under perturbations of X, periodic orbits in Ω disappear as a result of matching to the saddle separatrix loops Γ_1 and Γ_2.*

To reformulate the properties of Ω in terms of mappings, let a domain $\tilde{D} \subset D$ defined by the following: For the **Case A**,

$$\tilde{D} = \{(x,y) \in D_1 \cup D_2 : y_2(x) < y < y_1(x)\}; \tag{8.50}$$

and for the **Case B**,

$$\tilde{D} = \{(x,y) \in D_1 \cup D_2 : y_{12}(x) < y < y_1(x)\}, \tag{8.51}$$

where $y = y_{12}(x), |x| \leq 1$ denotes a curve in D whose image lies on the curve $y = y_1(x)$, and finally in **Case C**, let $\tilde{\Sigma}$ be the closure of points of all the trajectories of the mapping $T(X)$, which are entirely contained in D. Thus one has the following [Afraimovich et al. (1977)]:

Theorem 8.13. *Let $X \in U_2$. Then (a) $\tilde{\Sigma}$ is compact, one-dimensional, and consists of two connection components in **Cases A** and **C**, and of a finite number of connection components in **Case B**. (b) $\tilde{D}$ is foliated by a continuous stable foliation H^+ into leaves satisfying the Lipschitz conditions, along which a point is attracted to $\tilde{\Sigma}$; inverse images of the discontinuity line $S : y = 0$ (with respect to the mapping $T^k, k = 1, 2, ...$) are everywhere dense in $\tilde{D}$. (c) There exits a sequence of $T(X)$-invariant null-dimensional sets Δ_k, $k \in \mathbb{Z}_+$ such that $T(X)|_{\Delta_k}$ is topologically conjugated with a finite topological Markov chain with a nonzero entropy, the condition $\Delta_k \in \Delta_{k+1}$ being satisfied, and $\Delta_k \to \tilde{\Sigma}$ as $k \to \infty$. (d) The nonwandering set $\Sigma_1 \in \tilde{\Sigma}$ is a closure of saddle periodic points of $T(X)$, and either $\Sigma_1 = \tilde{\Sigma}$ or $\Sigma_1 = \Sigma^+ \cup \Sigma^-$, where: (d-1) The set Σ^- is null-dimensional and is an image of the space Ω^- of a certain TMC (G^-, Ω^-, σ) under the homeomorphism $\beta : \Sigma^- \to \Sigma^-$ which conjugates $\sigma|_{\Omega^-}$ and $T(X)|_{\Sigma^-}$; $\Sigma^- = \cup_{m=1}^{l(X)} \Sigma_m^-, l(X) < \infty$, where $T(X)(\Sigma_m^-) = \Sigma_m^-$, $\Sigma_{m_1}^- \cap \Sigma_{m_2}^- = \emptyset$ for $m_1 = m_2$ and $T(X)|_{\Sigma_m^-}$ is transitive; (d-2) The set Σ^+ is compact, one-dimensional and (d-3) If $\Sigma^+ \cap \Sigma^- = \emptyset$, then Σ^+ is an attracting set in a certain neighborhood; (d-4) If $\Sigma^+ \cap \Sigma^- \neq \emptyset$, then $\Sigma^+ \cap \Sigma^- = \Sigma_m^+ \cap \Sigma_m^-$ for a certain m, and this intersection consists of periodic points of no more than two periodic orbits, and (d-4-1) If Σ_m^- is* finite, Σ^+ is ω-limiting for all the trajectories in a certain neighborhood; (d-4-2) If Σ_m^- is infinite, Σ^+ is not locally maximal but is ω-limiting for all the trajectories in $\tilde{D}$ excluding those asymptotic to $\Sigma^-\backslash\Sigma^+$.

Finally, we discuss a recent result about the existence of periodic orbits in the Lorenz system (8.1).

8.2.3.1 Existence of short periodic orbits for the Lorenz system

Applying the method described in Sec. 1.2.1, Galias and Tucker [Galias & Tucker (2008)] proved rigorously the existence of short periodic orbits for the Lorenz system (8.1) with $(\sigma, r, b) = \left(10, 28, \frac{8}{3}\right)$ under the diagonal form of the system given by

$$\begin{cases} x_1' = \lambda_1 x_1 - k_1(x_1 + x_2)x_3 \\ x_2' = \lambda_2 x_2 + k_1(x_1 + x_2)x_3 \\ x_3' = \lambda_3 x_3 + (x_1 + x_2)(k_2 x_1 + k_3 x_2). \end{cases} \quad (8.52)$$

Eq. (8.52) can be obtained using a linear change of variables of system (8.1). The advantage of this form is that the invariant manifolds of the origin $P_0 = (0,0,0)$ are tangent to the coordinate axes x_1', x_2', x_3'. The constants in system (8.52) are given by

$$\begin{cases} u = \sqrt{(s+1)^2 + 4s(r-1)} \\ k_1 = \frac{s}{u} \approx 0.2886 \\ k_2 = \frac{(s-1+u)}{(2s)} \approx 2.1828 \\ k_3 = \frac{(s-1-u)}{(2s)} \approx -1.2828 \\ \lambda_1 = \frac{(-s-1+u)}{2} \approx 11.8277 \\ \lambda_2 = \frac{(-s-1-u)}{2}/ \approx -22.8277 \\ \lambda_3 = -q \approx -2.6667. \end{cases} \quad (8.53)$$

A trajectory of the Lorenz system is shown in Fig. 8.1. For $(\sigma, r, b) = \left(10, 28, \frac{8}{3}\right)$, the Lorenz system has three equilibria. The origin $P_0 = (0,0,0)^T$ shown in Fig. 8.1 using the symbol $*$ has one positive eigenvalue λ_1 and two negative eigenvalues λ_2 and λ_3 as defined in (8.53). The other equilibria are shown in Fig. 8.1 using symbols $\times$ and $+$ are

$$P_{\pm} = \left(\mp\frac{\lambda_2\sqrt{q(r-1)}}{u}, \pm\frac{\lambda_1\sqrt{q(r-1)}}{u}, r-1\right). \quad (8.54)$$

Numerically, the equilibria are $P_{\pm} \approx (\pm 5.589, \pm 2.896, 27)$ and have a pair of complex eigenvalues with positive real parts $\mu_{1,2} \approx 0.094 \pm 10.19j$ and one real negative eigenvalue $\mu_3 \approx -13.854$. To find a trapping region and average return time for the system (8.52), the Poincaré map P was defined by the section

$$\Sigma = \{x = (x_1, x_2, x_3) : x_3 = 27, x_3' < 0\}. \tag{8.55}$$

The analysis is based on the following tasks:

(1) The fact that there are trajectories in the Lorenz attractor which pass arbitrarily close to the origin and for which the return time is arbitrarily long, implies that it is not possible to find the trapping region by a direct integration of the differential equation (8.52).
(2) The trapping region was found using a modified Euler method with rigorous error bounds.
(3) A partitioning technique (when a box has expanded enough, it is partitioned, and the subboxes are then treated separately) was used to reduce expansion along the trajectories and reduce the wrapping effect.
(4) For trajectories passing close to the origin, a cube was defined around the origin, and computations were interrupted if the trajectory hits the cube. Hence the normal coordinates was changed, and the exit of the trajectory was explicitly computed.
(5) A rectangle can pass through the cube in two different ways. If the box intersects the stable manifold of the origin, it is split along the line of intersection and exits the cube in two pieces. Otherwise, the box flows out in one piece.
(6) The original coordinates were used in computations again after leaving the cube.

The details of the above tasks can be found in [Tucker (1999)]. Practically, the trapping region is composed of 14 518 boxes of size $\frac{1}{2^7} \times \frac{1}{2^7}$. There are 514 126 connections in the corresponding graph. Using the fact that the average return time τ_{aver} between two crossings is larger than $\frac{t_n}{n}$ for each n, for $n = 10\,000$, one obtains $\tau_{aver} > 0.6397$. Then the length of any periodic orbit corresponding to a period-n cycle of P is larger than $0.6397n$. In this case, there are no period-1 cycles in the graph, hence there are no period-1 orbits. A set composed of 86 boxes was used to study the existence of period-2 orbits. Finding all period-2 orbits is unsuccessful for this set due to long return times and too strong stretching. Hence more sections for the Poincaré map P were introduced to make the return time shorter. Let us now consider the Poincaré map P_1 corresponding to $\Sigma = \Sigma_1 \cup \Sigma_2$, where

$$\begin{cases} \Sigma_1 = \{x : x_3 = 27\} \\ \Sigma_2 = \{x : x_3 = 14\}. \end{cases} \quad (8.56)$$

Clearly, period-2 orbits of the original map P correspond to period-8 orbits of the map P_1. A graph has 182 boxes and 638 connections was used to generate the graph for P_1 (with higher accuracy) and one finds 26 086 period-8 cycles in the graph. The next step is to make a partition for each box into 16×16 smaller boxes, and generate the graph for this division, which gives a covering composed of 216 boxes of size $\frac{1}{2^{11}} \times \frac{1}{2^{11}}$ with 724 connections. Coming back to the original box size gives 16 boxes and 24 connections. In this case, there are 36 self-symmetric cycles of length 8. The Krawczyk operator is evaluated for each cycle, and it is verified that only one of them corresponds to a periodic orbit. For period-3 and period-4 orbits, similar computations have been performed using a set containing all period-3 cycles composed of 254 boxes. The sections

$$\begin{cases} \Sigma_1 = \{x : x_3 = 27\} \\ \Sigma_2 = \{x : x_3 = 18\} \end{cases} \quad (8.57)$$

were introduced, and a graph composed of 636 cycles and 2748 connections was obtained (boxes of size $\frac{1}{2^{13}} \times \frac{1}{2^{13}}$ were considered to reduce the number of cycles). The resulting graph has 460 boxes and 1500 connections, and by increasing the box size, a set of 38 boxes of size $\frac{1}{2^9} \times \frac{1}{2^9}$ with 54 connections was obtained. There are 72 period-12 cycles in the graph, and two of them (symmetric to each other) correspond to periodic orbits. For period-4 orbits, there are 654 boxes covering period-4 orbits of P. Using sections

$$\begin{cases} \Sigma_1 = \{x : x_3 = 27\} \\ \Sigma_2 = \{x : x_3 = 19\}, \end{cases} \quad (8.58)$$

one obtain a graph composed of 2220 boxes with 15 516 connections. Using the box size $\frac{1}{2^{17}} \times \frac{1}{2^{17}}$, one obtains 3166 boxes with 21 450 connections. By increasing the box size, one obtains a set of 72 boxes of size $\frac{1}{2^{10}} \times \frac{1}{2^{10}}$ with 90 connections. In this case, three (a pair of orbits symmetric to each other and a self-symmetric orbit) of 188 period-16 cycles correspond to periodic orbits. Hence the length of each orbit found is shorter than 3.1, and the length of any other periodic orbit is larger than $5 \times 0.6397 = 3.1985$. Thus the periodic orbits found are the shortest due to relation (8.55).

8.3 Expanding and contracting Lorenz attractors

In this section, we discuss expanding and contracting Lorenz attractor systems with their method of construction. These attractors are examples of Lorenz-type attractors realized in 3-D continuous-time systems. Indeed, in [Morales et al. (2006)], the so-called *contracting Lorenz attractor* in the unfolding of certain resonant double homoclinic loops in dimension-three was obtained from the geometric Lorenz attractor by replacing the usual expanding condition $\lambda_2 < \lambda_3 < 0 < \lambda_1$ at the origin by a contracting condition $\lambda_3 + \lambda_1 < 0$ as given in [Rovella (1993)]. This work answers a question posed in [Robinson (2000)] about contracting attractors in a positive Lebesgue measure set of parameters. In [Morales et al. (2005)], a generalization of the work of Robinson given in [Robinson (2000)] in which the existence of Lorenz attractors in the unfolding of resonant double homoclinic loops in dimension-three was studied. The analysis is based on two methods, the first of which is the search for *attractors* instead of the *weak attractors* used in [Robinson (2000)], and the second way is to enlarge considerably the region in the parameter space corresponding to flows allowing expanding Lorenz attractors.

Let f_η be a family of $C^r, r \geq 1$ vector fields on $\mathbb{R}^3$, and let Q_η be a hyperbolic singularity. $W^u(Q_\eta, \eta)$ is a one-dimensional unstable manifold tangent to the eigenvector v^u, $W^s(Q_\eta, \eta)$ is a two-dimensional stable manifold tangent to the eigenvector v^s, $W^{ss}(Q_\eta, \eta)$ is a one-dimensional strong stable manifold tangent to the eigenvector v^{ss}, $W^{cu}(Q_\eta, \eta)$ is a two-dimensional extended stable manifold tangent to the eigenvectors v^s and v^u, and $\Gamma^\pm$ are two branches of

$$\Gamma = W^u\left(Q_{\eta_0}, \eta_0\right) \subset \left\{Q_{\eta_0}\right\} \cup \Gamma^+ \cup \Gamma^-. \tag{8.59}$$

Let $q^\pm(t)$ be a parametrization of the solution along $\Gamma^\pm$ and $div_2(q^\pm(t))$ be the Jacobian of f_{η_0} at t restricted to $T_{\Gamma^\pm} W^{cu}$. Define the change in area within the plane

$$P(q) = T_q W^{cu}\left(Q_{\eta_0}, \eta_0\right) \text{ for } q \in \Gamma \tag{8.60}$$

along the whole length of $\Gamma^\pm$ as the quantity $C_{\eta_0}^\pm$ given by

$$C_{\eta_0}^\pm = \exp\left(\int_{-\infty}^{\infty} div_2\left(q^\pm(t)\right) dt\right). \tag{8.61}$$

Assume that the family f_η satisfies the following conditions:

(A1) For every η, f_η has a hyperbolic singularity Q_η such that the eigenvalues of $Df_\eta(Q_\eta)$ are real with
$$\lambda_{ss}(\eta) < \lambda_s(\eta) < 0 < \lambda_u(\eta), \qquad (8.62)$$
and with eigenvalues of v^{ss}, v^s, and v^u, respectively.

(A2) For the bifurcation value η_0, there is a double homoclinic connection with the unstable manifold of Q_{η_0} contained in the stable manifold, but outside the strong stable manifold, Γ given by (8.59).

(A3) For η_0, the central manifold $W^{cu}\left(Q_{\eta_0}, \eta_0\right)$ is transverse to the stable manifold $W^s\left(Q_{\eta_0}, \eta_0\right)$ along Γ.

(A4) We assume that
$$\begin{cases} \lambda_{ss}\left(\eta_0\right) - \lambda_s\left(\eta_0\right) + \lambda_u\left(\eta_0\right) < 0 \\ \lambda_{ss}\left(\eta_0\right) < 2\lambda_s\left(\eta_0\right). \end{cases} \qquad (8.63)$$

(A5) Assume that
$$B = \frac{C_{\eta_0}^+ + C_{\eta_0}^-}{C_{\eta_0}^+ C_{\eta_0}^-} > 1. \qquad (8.64)$$

(A6) There is a one-to-one resonance between the unstable and weakly stable eigenvalue for η_0 given by
$$\lambda_u\left(\eta_0\right) + \lambda_s\left(\eta_0\right) = 0. \qquad (8.65)$$

(A7) Let $N \subset C^1\left(\mathbb{R}^3\right)$ be the three-submanifold defined by conditions (A1)–(A6). We assume that the family $\{f_\eta\}$ is transverse to N at η_0.

Then the following result was proved in [Morales et al. (2005)]:

Theorem 8.14. *Let $\{f_\eta\}$ be a C^k-parameterized family of C^r-vector fields $(r, k \geq 3)$ satisfying (A1)–(A7). Then there is an open set $\mathcal{O}$ in the parameter space with $\eta_0 \in Cl(\mathcal{O})$ such that f_η has an expanding Lorenz attractor for all $\eta \in \mathcal{O}$.*

The proof of Theorem 8.14 is done using the so-called *rescaling techniques*[13] introduced in [Palis & Takens (1993)] to obtain convergence to non-continuous maps after the reduction of the dynamics to a one-dimensional Poincaré map and the ideas given in [Morales & Pujals (1997)] for the study of the certain heteroclinic connections involving saddle-node periodic orbits. In conclusion, we note the expanding Lorenz attractor is the prototype example of a singular basic set which is not hyperbolic.

---
[13]Note that rescaling techniques in the unfolding of homoclinic loops were used in [Naudot (1996)] to obtain convergence to the Hénon map (6.88).

8.4 Wild strange attractors and pseudo-hyperbolicity

In this section, we present a class of dynamical systems with strange attractors presented in [Turaev & Shilnikov (1996)]. These attractors are almost stable, chain-transitive closed sets that contain wild hyperbolic sets. Because the main interest of this section is the investigation of the essence of the so-called *wild strange attractor,* we need the following definition [Newhouse (1972)]:

Definition 8.15. A hyperbolic set is called *wild* if the system possesses an absorbing area embracing the hyperbolic basis set in which the stable and unstable subsets may touch each other.

In other words, a transitive hyperbolic set is said to be wild if its stable and unstable manifolds have a tangency that cannot be removed by a small perturbation. In particular, if a Poincaré mapping corresponding to a continuous system admits an absolutely continuous contractive invariant foliation and the corresponding quotient mapping expands in volume and is injective, then the chain-transitive attractor A is unique and coincides with the set of points accessible from a set L. In this case, the attractor A is wild if the system belongs to a Newhouse domain, *i.e.*, a C^r-open set of systems for which L is contained in a wild hyperbolic set [Palis & Viana (1994)].

In this case, the following results hold:

(1) Either the system itself or a close one has a saddle periodic orbit with a nontransversal homoclinic trajectory along which the stable and unstable manifolds of the cycle have in general a quadratic tangency.
(2) The saddle value $|\lambda\gamma|$ must be less then 1, where λ and γ are the multipliers of the saddle periodic orbit. This condition is always true when the divergence of the vector field is negative in the absorbing area. Hence near the given system, there will exist so-called *Newhouse regions*[14] [Newhouse (1979)] in the space of the dynamical system, *i.e.*, regions of dense structural instability.
(3) Since systems with a countable set of periodic orbits of arbitrarily high degrees of degeneracies are dense in the Newhouse regions [Gonchenko

---

[14] A system in the Newhouse region has countably many stable periodic orbits which cannot principally be separated from the hyperbolic subset. If this hyperbolic set contains a saddle periodic orbit with the saddle value exceeding 1, then there is a countable set of repelling periodic orbits next to it, and whose closure is not separable from the hyperbolic set.

et al. (1993a-b)], then if the divergence of the vector field is sign-alternating in the absorbing area, the dynamics become exotic and require infinitely many continuous topological invariant-moduli needed for the proper description of the system in the Newhouse regions.

Now, let X be a smooth $(C^r, r \geq 4)$ flow in $\mathbb{R}^n (n \geq 4)$ having an equilibrium state O of a saddle-focus type with characteristic exponents $\gamma, -\lambda \pm i\omega, -\alpha_1, \dots, -\alpha_{n-3}$, where $\gamma > 0, 0 < \lambda < \operatorname{Re} \alpha_j, \omega = 0$.
Suppose that [Ovsyannikov & Shilnikov (1986-1991)]

$$\gamma > 2\lambda. \tag{8.66}$$

Then one has the following result:

Theorem 8.15. *If X has a homoclinic loop, i.e., one of the separatrices of O returns to O as $t \to +\infty$, then no stable periodic orbit could appear.*

Consider the coordinates $(x, y, z) \in \mathbb{R}^1 \times \mathbb{R}^2 \times \mathbb{R}^{n-3}$, and assume the following:

(1) The equilibrium state is at the origin.
(2) The one-dimensional unstable manifold of O is tangent to the x-axis and the $(n-1)$-dimensional stable manifold is tangent to $\{x = 0\}$.
(3) The coordinates $y_{1,2}$ correspond to the leading exponents $\lambda \pm i\omega$, and the coordinate z corresponds to the nonleading exponent α.
(4) The flow X possesses a cross-section of the form

$$\Pi : \{\|y\| = 1, \|z\| \leq 1, |x| \leq 1\}. \tag{8.67}$$

(5) The stable manifold W^s is tangent to $\{x = 0\}$ at O. Thus it is given locally by an equation of the form $x = h^s(y, z)$ where h^s is a smooth function $h^s(0,0) = 0$, $(h^s)'(0,0) = 0$ at least when $(\|y\| \leq 1, \|z\| \leq 1)$ and $|h^s| < 1$.

Then one has the following:

Lemma 8.8. *The surface Π is a cross-section for W^s_{loc}, and the intersection of W^s_{loc} with Π has the form $\Pi_0 : x = h_0(\varphi, z)$ where φ is the angular coordinate $y_1 = \|y\| \cos\varphi, y_2 = \|y\| \sin\varphi$, and h_0 is a smooth function $-1 < h_0 < 1$. One can make $h_0 \equiv 0$ by a coordinate transformation, and we assume that is done.*

If we suppose that all the orbits starting on $\Pi \backslash \Pi_0$ return to Π, then the Poincaré map is defined by $T_+ : \Pi^+ \to \Pi, T_- : \Pi^- \to \Pi$, where

$$\begin{cases} \Pi_+ = \Pi \cap \{x > 0\} \\ \Pi_- = \Pi \cap \{x < 0\}. \end{cases} \quad (8.68)$$

In this case the following is evident:

Lemma 8.9. *If P is a point on Π with coordinates (x, φ, z), then $\lim_{x \to -0} T_-(P) = P_-^1$ and $\lim_{x \to -0} T_+(P) = P_+^1$ where P_-^1 and P_+^1 are the first intersection points of the one-dimensional separatrices of O with Π.*

Therefore, the maps T_+ and T_- are defined as

$$T_-(\Pi_0) = P_-^1, T_+(\Pi_0) = P_+^1. \quad (8.69)$$

Evidently, from Fig. 8.7 one has the following result:

Lemma 8.10. *The cylinder region*

$$D = \{\|y\| \leq 1, \|z\| \leq 1, |x| \leq 1\} \quad (8.70)$$

with two glued handles surrounding the separatrices filled by the orbits starting on Π (plus the point O and its separatrices) is an absorbing domain for the system X, i.e., the orbits starting in ∂D enter D and stay there for all positive values of time t.

It is necessary to assume that the semi-flow is pseudo-hyperbolic in D. Here we give the definitions due to Shilnikov [Shilnikov (2002)], and it is stronger than is usually done in [Hirsch *et al.* (1977)] because it prevents the appearance of stable periodic orbits.

Definition 8.16. A semi-flow is called pseudo-hyperbolic if the following two conditions hold:

A. At each point of the phase space, the tangent space is uniquely decomposed (and this decomposition is invariant with respect to the linearized semi-flow) into a direct sum of two subspaces N_1 and N_2 (continuously depending on the point) such that the maximal Lyapunov exponent in N_1 is strictly less than the minimal Lyapunov exponent in N_2 at each point M for any vectors $u \in N_1(M)$ and $v \in N_2(M)$

$$\limsup_{t \to +\infty} \frac{1}{t} \ln \frac{\|u_t\|}{\|u\|} < \liminf_{t \to +\infty} \frac{1}{t} \ln \frac{\|v_t\|}{\|v\|}, \quad (8.71)$$

where u_t and v_t denote the shift of the vectors u and v by the semi-flow linearized along the orbit of the point M.

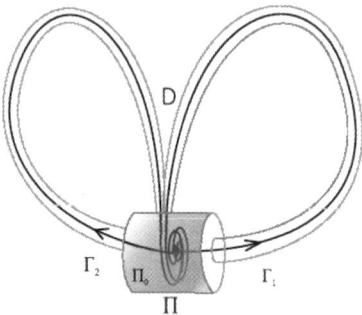

Fig. 8.7 Schematic illustration of the construction of the wild strange attractor.**

B. The linearized flow restricted to N_2 is volume expanding, $V_t \geq const \cdot e^{\sigma t} V_0$ with some $\sigma > 0$. Here, V_0 is the volume of any region in N_2, and V_t is the volume of the shift of this region by the linearized semi-flow.

This property is stable with respect to small smooth perturbations of the system [Hirsch et al. (1977)], i.e., the invariant decomposition of the tangent space is not destroyed by small perturbations, and the spaces N_1 and N_2 depend continuously on the system. The property of volume expansion in N_2 is also stable with respect to small perturbations. The additional condition **B** is essential because wild strange attractors need the fact that the linearized semi-flow is exponentially contracting in N_1. Furthermore, Definition 8.16 is quite broad because it is possible to assume for hyperbolic flows that $(N_1, N_2) = (N^s, N^u \oplus N_0)$ or $(N_1, N_2) = (N^s \oplus N_0, N^u)$ where N^s, N^u and N_0 are the stable, unstable invariant subspaces, and a one-dimensional invariant subspace spanned by the phase velocity vector, respectively.

An example of such a situation is the geometrical Lorenz model presented in Sec. 8.2.2 and studied in [Afraimovich et al. (1977-1983)], or [Guckenheimer & Williams (1979)], where N_1 is tangent to the contracting invariant foliation of codimension-two and the expansion of areas in a two-dimensional subspace N_2 is provided by the property that the Poincaré map is expanding in a direction transverse to the contracting foliation. Furthermore, Lorenz attractors, wild spiral attractors, and their periodic perturbations are examples of pseudo-hyperbolic attractors.

Because condition **A** of Definition 8.16 implies that for vectors of N_2 that contract, that contraction must be weaker than those on N_1. Then one has the following definition:

Definition 8.17. The set $N_1 = N^{ss}$ is called the *strong stable subspace*, and $N_2 = N^c$ the *center subspace*.

Lemma 8.11. *If (a) If N_1 has codimension-three, then $\dim N_1 = n-3$ and $\dim N_2 = 3$, and the linearized flow (at $t \geq 0$) is exponentially contracting on N_1. (b) The coordinates (x, y, z) in $\mathbb{R}^n$ are such that at each point of D, the space N^{ss} has a nonzero projection onto the coordinate space z, and N^c has a nonzero projection onto the coordinate space (x, y). Then the pseudo-hyperbolicity conditions **A** and **B** are satisfied at the point O.*

Proof. We remark that the space N^{ss} coincides with the coordinate space z, and N^c coincides with the space (x, y), and we use condition (8.66) which guarantees the expansion of volumes in the invariant subspace (x, y). In a small neighborhood of O, the pseudo-hyperbolicity of the linearized flow is inherited by the orbits. In this case, it is necessary that this property would extend into a non-small neighborhood D of O. For that, recall that the exponential contraction in N^{ss} implies the existence of an invariant contracting foliation N^{ss} with C^r-smooth leaves which are tangent to N^{ss} [Hirsch et al. (1977)], and the foliation is absolutely continuous [Afraimovich et al. (1983)]. After a factorization along the leaves, the region D becomes a branched manifold [Williams (1979)][15]. □

The property of pseudo-hyperbolicity is inherited by the Poincaré map $T \equiv (T_+, T_-)$ on the cross-section Π. Here, we have the following:

Lemma 8.12. *(a) There exists a foliation with smooth leaves of the form $(x, \varphi) = h(z)|_{-1 \leq z \leq 1}$, where the derivative $h'(z)$ is uniformly bounded, which possesses the following properties: the foliation is invariant in the sense that if l is a leaf, then $T_+^{-1}(l \cap T_+(\Pi_+ \cup \Pi_0))$ and $T_-^{-1}(l \cap T_-(\Pi_- \cup \Pi_0))$ are also leaves of the foliation (if they are not empty sets), the foliation is absolutely continuous in the sense that the projection along the leaves from one two-dimensional transversal to another increases or decreases the area in a finite number of times and the coefficients of expansion or contraction of areas are bounded away from zero and infinity, the foliation is contracting in the sense that if two points belong to one leaf, then the distance between the iterations of the points with the map T tends to zero exponentially;*

(b) The quotient maps $\tilde{T}_+$ and $\tilde{T}_-$ are area-expanding.

---

[15] Because the set D is bounded, and the quotient-semi-flow expands in volumes.

Assume the following conditions:
Condition: Let us write the map T as

$$(\bar{x}, \bar{\varphi}) = g(x, \varphi, z), \bar{z} = f(x, \varphi, z), \tag{8.72}$$

where f and g are functions smooth at $x \neq 0$ and discontinuous at $x = 0$,

$$\begin{cases} \lim_{x \to -0}(g, f) = (x_-, \varphi_-, z_-) \equiv P_-^1 \\ \lim_{x \to +0}(g, f) = (x_+, \varphi_+, z_+) \equiv P_+^1. \end{cases} \tag{8.73}$$

Let

$$\det \frac{\partial g}{\partial(x, \varphi)} \neq 0. \tag{8.74}$$

Denote

$$\begin{cases} A = \frac{\partial f}{\partial z} - \frac{\partial f}{\partial(x,\varphi)} \left(\frac{\partial g}{\partial(x,\varphi)}\right)^{-1} \frac{\partial g}{\partial z} \\ B = \frac{\partial f}{\partial(x,\varphi)} \left(\frac{\partial g}{\partial(x,\varphi)}\right)^{-1} \\ C = \left(\frac{\partial g}{\partial(x,\varphi)}\right)^{-1} \frac{\partial g}{\partial z}, D = \left(\frac{\partial g}{\partial(x,\varphi)}\right)^{-1}. \end{cases} \tag{8.75}$$

If

$$\lim_{x \to 0} C = 0, \lim_{x \to 0} \|A\| \|D\| = 0 \tag{8.76}$$

$$\sup_{p \in \Pi \setminus \Pi_0} \sqrt{\|A\| \|D\|} + \sqrt{\sup_{p \in \Pi \setminus \Pi_0} \|B\| \sup_{p \in \Pi - \Pi_0} \|C\|} < 1, \tag{8.77}$$

then the map has a continuous invariant foliation with smooth leaves of the form $(x, \varphi) = h(z)|_{-1 \leq z \leq 1}$ where the derivative $h'(z)$ is uniformly bounded. If, additionally,

$$\sup_{P \in \Pi \setminus \Pi_0} \|A\| + \sqrt{\sup_{P \in \Pi \setminus \Pi_0} \|B\| \sup_{P \in \Pi - \Pi_0} \|C\|} < 1, \tag{8.78}$$

then the foliation is contracting and if, moreover, for some $\beta > 0$ the functions $A|x|^{-\beta}, D|x|^{\beta}, B, C$ are uniformly bounded and Hölder continuous, and $\frac{\partial \ln \det D}{\partial z}$ and $\frac{\partial \ln \det D}{\partial(x,\varphi)} D|x|^{\beta}$ are uniformly bounded, then the foliation is absolutely continuous. The additional condition

$$\frac{1 - \sqrt{\sup_{P \in \Pi \setminus \Pi_0} \|B\| \sup_{P \in \Pi \setminus \Pi_0} \|C\|}}{\sup_{P \in \Pi \setminus \Pi_0} \sqrt{\det D}} = q > 1 \tag{8.79}$$

guarantees that the quotient map $\tilde{T}$ expands areas.

It follows from [Ovsyannikov & Shilnikov (1986-1991)] that:

Theorem 8.16. *If the equilibrium state is a saddle-focus, then the Poincaré map near $\Pi_0 = \Pi \cap W^s$ is written in the following form under some appropriate choice of the coordinates:*

$$(\bar{x}, \bar{\varphi}) = Q_\pm (Y, Z), z = R_\pm (Y, Z). \tag{8.80}$$

Here

$$\begin{cases} Y = |x|^\rho \begin{pmatrix} \cos(\Omega \ln |x| + \varphi) & \sin(\Omega \ln |x| + \varphi) \\ -\sin(\Omega \ln |x| + \varphi) & \cos(\Omega \ln |x| + \varphi) \end{pmatrix} + \Psi_1(x, \varphi, z) \\ Z = \Psi_2(x, \varphi, z), \end{cases} \tag{8.81}$$

where $\rho = \frac{\lambda}{\gamma} < \frac{1}{2}$, $\Omega = \frac{\omega}{\gamma}$, and, for some $\eta > \rho$,

$$\begin{cases} \left\| \frac{\partial p^+|q|\Psi_i}{\partial x^p \partial (\varphi, z)^q} \right\| = O(|x|\eta - p) \\ 0 \leq p + |q| \leq r - 2. \end{cases} \tag{8.82}$$

The functions $Q_\pm R_\pm$ in (8.80) ("+" corresponds to $x > 0_-$ in the map T_+, and "−" corresponds to $x < 0_-$ in the map T_-) are smooth functions in a neighborhood of $(Y, Z) = 0$ for which the Taylor expansion can be written as

$$\begin{cases} Q_\pm = (x_\pm \varphi_\pm) + a_\pm Y + b_\pm Z + ... \\ R_\pm = z_\pm + c_\pm Y + d_\pm Z + ... \end{cases} \tag{8.83}$$

Lemma 8.13. *(a) If the point O is a saddle-focus satisfying (8.66), then if $a_+ \neq 0$ and $a_- \neq 0$, the map T satisfies conditions (8.76) with $\beta \in (\rho, \eta)$.*

(b) The map (8.80),(8.81),(8.83) is continued onto the whole cross-section Π.

(c) There are no stable periodic orbits in D, and any orbit in D has a positive maximal Lyapunov exponent, i.e., there is a strange attractor.

Proof. (a) Use conditions (8.80)–(8.83).

(b) Analogues of conditions (8.74),(8.77),(8.78), and (8.79) are fulfilled where the supremum should be taken not over $|x| \leq 1$ but over a small x.

(c) The expansion of volumes by the quotient-semi-flow restricts the possible types of limit behavior of orbits. □

An example is given by the following map:

$$\begin{cases} \bar{x} = 0.9|x|^\rho \cos(\ln |x| + \varphi) \\ \bar{\varphi} = 3|x|^\rho \sin(\ln |x| + \varphi) \\ \bar{z} = (0.5 + 0.1z|x|^\eta) \operatorname{sgn} x \end{cases} \tag{8.84}$$

where $0.4 = \rho < \eta$.

To prove the existence of a wild set, recall some definitions and simple facts from topological dynamics. Let $X_t P$ be the time-t shift of a point P by the flow X.

Definition 8.18. (a) For a given $\epsilon > 0$ and $\tau > 0$, an (ϵ, τ)-orbit is a sequence of points $P_1, P_2, ..., P_k$ such that P_{i+1} is at a distance less than ϵ from $X_t P_i$ for some $t > \tau$.

(b) A point Q is called (ϵ, τ)-attainable from P if there exists an (ϵ, τ)-orbit connecting P and Q, and it called attainable from P if, for some $\tau > 0$, it is (ϵ, τ)-attainable from P for any ϵ (this definition, obviously, does not depend on the choice of $\tau > 0$).

(c) A set C is attainable from P if it contains a point attainable from P.

(d) A point P is called chain-recurrent if it is attainable from $X_t P$ for any t.

(e) A compact invariant set C is called chain-transitive if for any points $P \in C$ and $Q \in \bar{C}$ (the compliment of C) and for any $\epsilon > 0$ and $\tau > 0$, the set C contains an (ϵ, τ)-orbit connecting P and Q.

(f) A compact invariant set C is called orbitally stable if for any its neighborhood U there is a neighborhood $V(C) \subset U$ such that the orbits starting in V stay in U for all $t \geq 0$.

(g) An orbitally stable set is called completely stable if for any of its neighborhood $U(C)$ there exist an $\epsilon_0 > 0$, $\tau > 0$ and a neighborhood $V(C) \subset U$ such that the (ϵ_0, τ)-orbits starting in V never leave U.

(h) We call the set A of the points attainable from the equilibrium state O, the attractor of the system X.

(l) A set A is called epsilon-stable if, for any ε, every two points of A can be joined by an ε-trajectory contained entirely in A, and for any neighborhood U of A, there exists an ε such that the ε-trajectories starting from A never leave U (the epsilon stability property can also be formulated as follows: A is an intersection of embedded absorbing domains).

Thus one can obtain the following lemma:

Lemma 8.14. *(a) All points of a chain-transitive set are chain-recurrent.*

(b) The set C is orbitally stable if and only if $C = \cap_{j=1}^{\infty} U_j$, where $(U_j)_j$ is a system of embedded open invariant (with respect to the forward flow) sets.

(c) The set C is completely stable if the sets U_j are not just invariant but are absorbing domains, i.e., the orbits starting on ∂U_j enter inside U_j for a time interval not greater than some τ_j.[16]

---

[16] In this situation, (ϵ, τ)-orbits starting on ∂U_j lie always inside U_j if ϵ is sufficiently small and $\tau \geq \tau_j$.

Definition 8.18(h) is justified by the following theorem:

Theorem 8.17. *The set A is chain-transitive, completely stable, and attainable from any point of the absorbing domain D.*

Now consider a one-parameter family X_μ of such systems. Then one has the following result:

Theorem 8.18. *Assume that X_μ has a transversal homoclinic loop of the saddle-focus O at $\mu = 0$, i.e., one of the separatrices (say, Γ_+) returns to O as $t \to +\infty$. Then there exists a sequence of intervals Δ_i (accumulated at $\mu = 0$) such that when $\mu \in \Delta_i$, the attractor A_μ contains a wild set (nontrivial transitive closed hyperbolic invariant set whose unstable manifold has points of tangency with its stable manifold). Furthermore, for any $\mu^* \in \Delta_i$, for any system C^r-close to a system X_μ, its attractor A also contains the wild set.*

From [Gonchenko et al. (1992-1993)], it is possible to state the following result:

Theorem 8.19. *The systems whose attractors contain structurally nontransversal homoclinic trajectories as well as structurally unstable periodic orbits of higher orders of degeneracies are dense in the given regions in the space of dynamical systems.*

In [Turaev & Shilnikov (2008)], more discussions of the pseudo-hyperbolic attractor theory introduced in [Turaev & Shilnikov (1996)] are given by means of the following assumptions:

Property 1: All orbits in $\mathcal{D}$ are unstable; each of them has a positive maximal Lyapunov exponent

$$\lambda_{\max}(x) = \limsup_{t \to \infty} \frac{1}{t} \ln \|DX_t(x)\| > 0. \tag{8.85}$$

Property 2: The pseudo-hyperbolicity conditions are not violated under small smooth perturbations of the system. Moreover, the subspaces N_1 and N_2 change continuously under such perturbations.

Property 3: There exists a unique invariant, absolutely continuous, contractive foliation with C^r-smooth leaves tangent to N_1 on $\mathcal{D}$.

Note that Property 1 comes from Definition 8.18, and Properties 2 and 3 are proved in a standard way [Anosov (1967)]. Furthermore, pseudo-hyperbolic attractors include hyperbolic and Lorenz-type attractors, and new examples were constructed by applying the following two useful theorems, which are implied by Property 2:

Theorem 8.20. *(a) If a system of the form $y' = Y(y)$ is pseudo-hyperbolic on an absorbing domain $\mathcal{D}$ and a function $p(y,\theta)$ is periodic in θ and small together with its derivatives, then the system $y' = Y(y) + p(y,\theta), \theta' = 1$ is pseudo-hyperbolic on $\mathcal{D} \times \mathbb{S}^1$.*

(b) If a system of the form $y' = Y(y)$ is pseudo-hyperbolic on an absorbing domain $\mathcal{D}$ and functions $p(y,z)$ and $q(y,z)$ are small together with their derivatives, then the system $y' = Y(y) + p(y,z), z' = Y(z) + q(y,z)$ is pseudo-hyperbolic on $\mathcal{D} \times \mathcal{D}$.

Using the fact that not all systems have transitive asymptotically stable attractors, for example the Lorenz geometric model [Afraimovich *et al.* (1982)], Turaev and Shilnikov in [Turaev & Shilnikov (2008)] proved the following result by considering that an attractor is a chain-transitive and epsilon-stable set:

Theorem 8.21. *Suppose that $\mathcal{D}$ contains a finite set of points $Q_1, Q_2, \ldots, Q_m$ such that, for any point $P \in D$, some of the points Q_j are accessible from P (i.e., for any $\varepsilon > 0$, there exists an ε-trajectory joining P to Q_j), but Q_j is not accessible from Q_i for any $i \neq j$. Then it contains precisely m attractors $A_1, A_2, \ldots, A_m$, where each A_j is precisely the set of points accessible from Q_j.*

Theorem 8.21 can be considered as a fairly constructive criterion for the finiteness of the number of attractors in the case of pseudo-hyperbolic flows.

Now, if a flow in $\mathcal{D}$ has a secant, that is, a finite number of smooth $(n-1)$-surfaces $\Pi_1, \Pi_2, \ldots, \Pi_l$ transversal to the flow and such that the positive semi-trajectory of any point from $\mathcal{D}$ attains one of the surfaces $\Pi_1, \Pi_2, \ldots, \Pi_l$ at some moment, this implies that on $\Pi_1 \cup \Pi_2 \cup \ldots \cup \Pi_l$, a Poincaré mapping is defined and the corresponding quotient mapping expands the $(k-1)$-volume due to Property 3 and the nature of the flow on $N_{1,2}$. Suppose that $\mathcal{D}$ contains finitely many equilibrium states $O_1, O_2, \ldots, O_m$ and/or periodic orbits $L_1, L_2, \ldots, L_p$ in which the secant minus finitely many pieces of the stable manifolds of the orbits $O_1, O_2, \ldots, O_m$ and $L_1, L_2, \ldots, L_p$ decomposes into connected components, on each of which the quotient mapping is injective. Then some iteration of any open domain under the action the Poincaré mapping must intersect the stable manifold of one of the orbits $O_1, O_2, \ldots, O_m$ and $L_1, L_2, \ldots, L_p$, and at least one of these orbits is accessible from any point of the domain $\mathcal{D}$.

Property 4: The union of stable manifolds of the equilibrium states $O_1, O_2, \ldots, O_m$ and the periodic orbits $L_1, L_2, \ldots, L_p$ from Theorem 8.21 implies the following assertion:

Theorem 8.22. *The domain $\mathcal{D}$ contains finitely many attractors, and each of them is the set of points attainable from one of the points O_j or from one of the periodic orbits L_j.*

Assume that $m = 1$, i.e., the class of pseudo-hyperbolic systems with one saddle O with one-dimensional unstable manifold that contain the Lorenz geometric model [Afraimovich et al. (1982)] and the model has a wild spiral attractor [Turaev & Shilnikov (1998)].

Theorem 8.23. *If the above condition holds, then the map has a continuous invariant foliation with smooth leaves of the form $(x, \varphi) - h(z)\big|_{\|z\| \leq 1}$ where the derivative $h'(z)$ is uniformly bounded. If (8.78) holds, then the foliation is contracting and if, moreover, for some $\beta > 0$ the functions $A|x|^{-\beta}, D|x|^{\beta}, B, C$ are uniformly bounded and Hölder continuous, and $\frac{\partial \ln \det D}{\partial z}$ and $\frac{\partial \ln \det D}{\partial(x,\varphi)} D|x|^{\beta}$ are uniformly bounded, then the foliation is absolutely continuous. If, in addition, (8.79) holds, then the quotient map $\tilde{T}$ expands in m-volume.*

Note that Theorem 8.23 generalizes Theorem 8.16. The case of $m = 1$ corresponds to the Lorenz geometric model given in Sec. 8.2.2. In this case, if $g > \frac{1+\sqrt{5}}{2}$ or if $q > \sqrt{2}$ such that the mapping is symmetric with respect to the change $x \to -x$, then A is also a maximal attractor, and the intersection $A \cap \Pi$ coincides with the nonwandering set Ω of the Poincaré mapping, it is one-dimensional and transitive, and the saddle periodic points and structurally stable homoclinic points are dense in it.

For an arbitrary q, the set Ω contains a nonempty zero-dimensional invariant hyperbolic set Σ_0 on which the mapping T is conjugate to a TMC with finitely many symbols in which the set $\Sigma_1 = cl(\Omega \backslash \Sigma_0)$ is one-dimensional and transitive, and the saddle periodic and structurally stable homoclinic points are dense in it. If $\Sigma_1 \cap \Sigma_0 = \emptyset$, then $A \cap \Pi = \Sigma_1$, and A is an asymptotically stable transitive attractor. If $\Sigma_1 \cap \Sigma_0 \neq \emptyset$ [Afraimovich et al. (1982)], then the chain-transitive attractor A is nontransitive because

Σ_1 is unstable, and the system has no stable transitive sets at all, and $A \cap \Pi$ contains the set of points accessible from Σ_1.

Now, let r be the degree of smoothness of the system, and let $r_0 \geq 1$ be the maximum integer strictly less than $\frac{\gamma}{|\lambda_1+\lambda_2+...+\lambda_m|}$.

Theorem 8.24. *If $s = r$ for $m = 1$ and $s = min(r, r_0)$ for $m \geq 2$, then both separatrices of the point O can be closed by an arbitrarily small (in the C^s-norm) perturbation of the vector field.*

For the Lorenz geometric model, the C^1-version of this theorem was proved in [Afraimovich et al. (1982)]. For $m \geq 2$, and O a saddle-focus, it was shown in [Turaev & Shilnikov et al. (1998)] that, if a C^r-smooth ($r \geq 2$) pseudo-hyperbolic system of the form described above has a homoclinic loop, then any neighborhood of this system in the C^r-topology contains C^r-open domains such that the attractor A is wild, i.e., it contains a wild hyperbolic set together with its unstable manifold for systems from these domains.

A new class of wild attractor arising from a periodic perturbation of pseudo-hyperbolic systems with an equilibrium state was presented in [Turaev & Shilnikov (2008)] as follows: Let $u = (x, y, z)$, and assume the system

$$u' = U(u) \tag{8.86}$$

satisfies the above conditions in the absorbing domain $\mathcal{D}$. Let $p(u, \theta, \mu)$ be a periodic function in θ and bounded together with its derivatives. For small μ in the periodically perturbed system

$$\begin{cases} u' = U(u) + \mu p(u, \theta, \mu) \\ \theta' = 1, \end{cases} \tag{8.87}$$

the equilibrium state O system (8.86) corresponds to a saddle periodic orbit L_μ with an unstable two-manifold and stable n-manifolds. Thus the following result was proved in [Turaev & Shilnikov (2008)] to solve a problem about periodic perturbations of Lorenz-type attractors[17]:

Theorem 8.25. *For all small μ, system (8.87) has a unique chain-transitive attractor A in the absorbing domain $\mathcal{D} \times \mathbb{S}^1$, which coincides with the set of points attainable from L_μ (the attractor A contains, in particular, the unstable two-manifold $W^u(L_\mu)$ and its closure). If, in the unperturbed system (8.86), the point O has a homoclinic loop, then the addition p can*

---
[17]This problem arises in studying local bifurcations of periodic orbits [Shilnikov et al. (1993), Gonchenko et al. (2005a)].

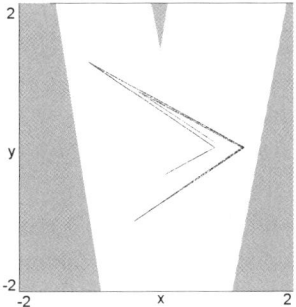

Fig. 8.8 The Lozi attractor (8.88) with its basin of attraction (in gray) obtained for $a = 1.7$ and $b = 0.3$.

be chosen so that the one-parameter family (8.87) contains a countable set of intervals of values of μ accumulating to $\mu = 0$ at which L_μ is contained in a wild hyperbolic set, and the attractor A is wild.

A generalized version of Theorem 8.25 can be proved if the system has a saddle periodic orbit L in the absorbing domain.

8.5 Lorenz-type attractors realized in two-dimensional maps

In this section, we present an example of a Lorenz-type attractor realized in two-dimensional maps, namely, the Lozi map attractor. From Sec. 12.1.4, it is known that *dangerous* tangencies are an essential characterization of the dynamics of the Hénon map (6.88). However, this phenomena can be avoided using piecewise-smooth functions [Banerjee et al. (1998)]. According to this principle, the Lozi map [Lozi (1978)] shown in Fig. 8.8 and given by

$$f_{a,b}(x, y) = \begin{pmatrix} 1 - a|x| + y \\ bx \end{pmatrix} \qquad (8.88)$$

avoids this type of tangency, where a and b are real parameters. The map (8.88) is one of the central subjects of study in dynamical system theory [Lozi (1978), Misiurewicz (1979), Palis & de Melo (1982), Rychlik (1983), Collet & Levy (1984), Young (1985), Brucks et al. (1991), Pesin (1992), Sataev (1992), Brucks & Misiurewicz (1996), Ishii (1997(a-b)-1998), Bruin (1998), Cao & Liu (1998), Young (1998b), Brucks & Buczolich (2000),

Brucks & Bruin (2004), Kiriki & Soma (2007), Zeraoulia (2011)] and references therein. Misiurewicz in [Misiurewicz (1979)] showed that the map $f_{a,b}$ admits a unique strange attractor $\Lambda_{a,b}$ if (a,b) belongs to the open set $\mathcal{M}$ defined by the inequalities

$$\begin{cases} 0 < b < 1, a > b+1, 2a+b < 4 \\ a\sqrt{2} > b+2, b < \frac{a^2-1}{2a+1}. \end{cases} \quad (8.89)$$

Furthermore, it was shown in [Kiriki (2004)] that there exists an open set $\mathcal{O}$ in the parameter space such that, for almost every parameter in $\mathcal{O}$, the forward limit set of a point on the y-axis, which is a singularity in a trapping region, coincides with the strange attractor. This fact generalizes the corresponding result about *turning orbits* in the dynamical core of tent maps on $\mathbb{R}$ by Brucks and Misiurewicz. Indeed, for any $(a,b) \in \mathcal{M}$, the map $f_{a,b}$ has a saddle fixed point contained in the first quadrant given by

$$P_{a,b} = \left(\frac{1}{1+a-b}, \frac{b}{1+a-b}\right). \quad (8.90)$$

The unstable set $W^u(P_{a,b})$ of $P_{a,b}$ contains the line segment that connects $P_{a,b}$ and the point

$$z_{a,b} = \left(\frac{2+a+\sqrt{a^2+4b}}{2+2a-2b}, 0\right) \quad (8.91)$$

on the x-axis. Thus *a trapping region* of (8.88) is the triangle $T_{a,b}$ with vertices $z_{a,b}$, $f_{a,b}(z_{a,b})$ and $f^2_{a,b}(z_{a,b})$ such that $f_{a,b}(T_{a,b}) \subset T_{a,b}$. Hence it was shown in [Misiurewicz (1979)] that the Lozi attractor is given by

$$\Lambda_{a,b} = \bigcap_{i \geq 0} f^i_{a,b}(T_{a,b}) = cl(W^u(P_{a,b})). \quad (8.92)$$

If $\mathcal{Y}_{a,b}$ denotes the segment of the y-axis in $T_{a,b}$, the following result was proved in [Kiriki (2004)]:

Theorem 8.26. *There exists an open set $\mathcal{O} \subset \mathcal{M}$ whose closure contains $(2,0)$ such that, for Lebesgue almost every*

$$((a,b),z) \in \{\ ((a,b),z)(a,b) : (a,b) \in \mathcal{O}, z \in \mathcal{Y}_{a,b}\ \}, \quad (8.93)$$

the ω-limit set $\omega(z, f_{a,b})$ coincides with the Lozi attractor $\Lambda_{a,b}$.

In conclusion, we note that the Lozi strange attractor has almost hyperbolic structure (recall Definition 6.1(d)), off the y-axis where the Lozi maps (8.88) are not differentiable. However, by the *influence of the singularities* on the y-axis, the dynamics of the Lozi maps (8.88) are quite delicate. The Lozi map (8.88) has the attractor shown in Fig. 8.8, which has properties close to the Lorenz attractor given by (8.1) (see Sec. 9.2). Furthermore, the Lozi family is a two-parameter family of piecewise affine, uniformly hyperbolic maps (recall Definition 6.3) on $\mathbb{R}^2$ with strange attractors.

8.6 Exercises

Exercise 185. (a) Show that the Belykh and Lozi attractors are examples of Lorenz-type systems realized in two-dimensional maps. See [Lozi (1978), Belykh (1982-1995)].

(b) Determine which condition of Definition 6.1(d) is violated for the Belykh and Lozi attractors.

(c) Determine a set of parameter values exhibiting the key properties of hyperbolic attractors of Definition 6.1(d) for the Belykh, Lozi, and Lorenz attractors. See [Afraimovich *et al.* (1977), Mischaikow & Mrozek (1995-1998)].

(d) Give two examples of Lorenz-like attractors rather than Belykh, Lozi, and Lorenz attractors.

Exercise 186. Explain the major differences between Lorenz-type systems and the robust hyperbolic attractors introduced in Chapter 6.

Exercise 187. (a) Show that the Lorenz attractor obtained from (8.1) for $\sigma = 10, b = \frac{8}{3}$, and $r = 28$ is a fully persistent and transitive attractor which is not hyperbolic.

(b) In this case, show that periodic and homoclinic orbits are everywhere dense.

(c) Show that the Lorenz attractor is structurally unstable and that it is obtained through a finite number of bifurcations.

(d) Show that for $\sigma = 10$ and $b = \frac{8}{3}$, the Lorenz system displays the following attractors: For $r \in [-\infty, 1]$, the point $(0, 0, 0)$ is an attracting equilibrium. For $r \in [1, 13.93]$ the two other symmetric equilibria are attracting, and the origin is unstable. For $r \in [13.93, 24.06]$, there is a transient to chaos. For $r \in [24.06, 24.74]$, a chaotic attractor coexists with two symmetric attracting equilibria. For $r \in [24.74, 28.0]$, a chaotic attractor exists, but the two symmetric equilibria are no longer attracting.

(e) Explain the fact that equation (8.1) is very resistant to rigorous analysis and also presents obstacles to numerical study.

(f) Show that the Lorenz system (8.1) satisfies the conditions of Theorem 8.1.

Exercise 188. Show that if H is a hyperbolic set of a vector field f and $p \in H$, then the sets $W_f^{ss}(p)$, $W_f^{s}(p), W_f^{uu}(p)$, $W_f^{u}(p)$, $W_f^{ss}(p, \epsilon)$ and $W_f^{uu}(p, \epsilon)$ are C^r-submanifolds of M, and $\dim(W_f^{ss}(p)) = \dim(E^s)$ and $\dim(W_f^{uu}(p)) = \dim(E^u)$. See [Hirsch *et al.* (1977)].

Exercise 189. Show that for the **Case 3** of Theorem 8.1, both subclasses (a) and (c) are realized in the Shimizu–Morioka model in which the appearance of the Lorenz attractor and its disappearance through bifurcations of lacunae are explained. See [Shilnikov (1986-1993), Robinson (1989), Rychlik (1990)].

Exercise 190. Show that Theorem 8.6(b) states that f is not in the closure of the set of topologically Ω-stable vector fields.

Exercise 191. (a) Show that the map (8.32) defines a completely mixed dough, *i.e.*, the map B is topologically transitive on I. See [Glendinning (1994)].

(b) Show that the map B_s defined by (8.33) is topologically transitive on I for some values of $s > \sqrt{2}$ and that it is uniformly expanding on I if $s > 1$.

(c) Show that B_s is topologically transitive on I if and only if $s \in \left[-\sqrt{2}, \sqrt{2}\right]$.

(d) Show that B_s is chaotic on I if and only if $s \in \left[-\sqrt{2}, \sqrt{2}\right]$.

Exercise 192. (a) Show that almost all 1-D Poincaré maps defined in the literature give less description of the real dynamics of the Lorenz system (8.1).

(b) Explain the role of the TBMs defined by (8.33).

Exercise 193. (a) Explain the arguments used in the definition of the Lorenz-like families with criticalities φ_a defined below in (8.35). See [Luzzatto & Viana (2000)].

(b) Show that inequality (8.36) implies that given any y with $|y| \in \left(x_{\sqrt{2}}, a\right]$, there exists a unique $x \in [-a, a]$ such that $\varphi_a(x) = y$, and x and y have apposite signs, and hence $|x| < x_{\sqrt{2}}$.

(c) Show that the properties L1–L5 are satisfied by a nonempty open set of one-parameter families, where openness is meant with respect to the C^2 topology in the space of real maps ψ.

(d) Show that for any $a \in [a_1, c]$, one has that $[-a, a]$ is forward invariant and $\varphi|_{[-a,a]}$ is transitive and (8.37) holds for all $x \in [-a, a]$ such that $\varphi_a^j(x) \neq 0$ for every $j = 0, 1, ..., n-1$. See [Luzzatto & Viana (2000)].

Exercise 194. (a) Find a linear change of variables that transforms the Lorenz system (8.1) to the diagonal form given by equation (8.52).

(b) Find the trapping region for the Poincaré map $P : x_3 = 27$ defined for the Lorenz diagonal form given by equation (8.52).

(c) Find the lower bound t_n of return times for P^n for $n = 1, 2, 5, 10, 100$, and 10000.

(d) Find regions containing all periodic orbits with period $p = 2, 3, 4$, $x_3 = 27$, and draw short periodic orbits for each case.

Exercise 195. Give two examples of expanding and contracting Lorenz attractor systems, and analyze their properties.

Exercise 196. Give two examples of wild strange attractors and analyze their properties.

Exercise 197. Show that Definition 8.16 is stronger than the usual case given in [Hirsch et al. (1977)].

Exercise 198. Show that the geometrical Lorenz model presented in Sec. 8.2.2, Lorenz attractors, wild spiral attractors, and their periodic perturbations satisfy the conditions of Definition 8.16.

Exercise 199. Show that Property 1 comes from Definition 8.16.

Exercise 200. (Open problem) Is the dynamics of the ordinary differential equations of Lorenz given by (8.1) that of the geometric Lorenz attractor of Williams, Guckenheimer and Yorke discussed in Sec. 8.2.2? See [Smale (1998)].

Chapter 9

Robust Chaos in the Lorenz-Type Systems

In this chapter, we discuss robust chaos in the Lorenz-type system in the sense of Definition 2.5. In Sec. 9.1, we give an overview of the classical results concerning robust Lorenz-type attractors. In Sec. 9.2, the well-known results on structural stability and robustness of the standard Lorenz attractor governed by Eq. (8.1) are listed to make clear the very close properties to the hyperbolic systems introduced in Chapter 6. Section 9.3 deals with robust chaos in the 2-D Lozi discrete mapping (8.88) as an example of Lorenz-type attractors in this case. Note that the method of analysis is based on the contents of Chapters 3, 6, and 8. In Sec. 9.4, a set of exercises is given to fix ideas about the topic of robust chaos in the Lorenz-type systems.

9.1 Robust chaos in the Lorenz-type attractors

Smale's conjecture known for flows on disks [Andronov & Pontryagin (1937)] and orientable surfaces, especially for diffeomorphisms of the circle [Peixoto (1962)], suggests that every system can be approximated by a hyperbolic one. This conjecture is not true in general due to counter examples given in [Smale (1967), Abraham & Smale (1968), Simon (1972)] and references therein. A special case is the Lorenz attractor (8.1), which is not hyperbolic and still robust, *i.e.*, totally persistent under small perturbations of the initial flow. A number of specific Lorenz-like attractors (recall Definition 8.1) are found, and it was proved mathematically that they are robustly transitive and not hyperbolic and have sensitive dependence on initial conditions [Afraimovich *et al.* (1977), Guckenheimer & Williams (1979)]. These attractors can also appear in three-dimensional families of cubic differential equations — not quadratic like the Lorenz Eq. (8.1) [Robinson (1989), Rychlik (1990)].

A new kind of attractor in three dimensions, the contracting Lorenz attractor or Lorenz–Rovella attractor, which is probability persistent but not robust, was obtained by Rovella in [Rovella (1993)] based upon a previous work by Arneodo, Coullet and Tresser [Arnéodo et al. (1981)]. It contains a hyperbolic singularity with real eigenvalues (but now the sum of any two eigenvalues is negative), and it is probability persistent in terms of Lebesgue probability or just probability persistent, but not robust.

As a general result, it was proved in [Araujo, et al. (2005)] that the so-called *singular-hyperbolic attractor* (or Lorenz-like attractor) of a three-dimensional flow is chaotic in two different strong senses. Firstly, the flow is expansive; if two points remain close for all times, possibly with time reparametrization, then their orbits coincide. Secondly, there exists a physical (or Sinai–Ruelle–Bowen) measure (recall Definition 3.2) supported on the attractor whose ergodic basin covers a full Lebesgue (volume) measure subset of the topological basin of attraction. In particular, these results show that both the flow defined by the Lorenz Eqs. (8.1) and the geometric Lorenz flows introduced in Sec. 8.2.2 are expansive. Furthermore, it was proved in [Tucker (2002a)] that this flow exhibits a singular-hyperbolic attractor.

Another proof of the robustness of the Lorenz attractor is given in [Franceschini et al. (1993)] where the chaotic attractors of the Lorenz system associated with $r = 28$ and $r = 60$ were characterized in terms of their unstable periodic orbits and eigenvalues. While the Hausdorff dimension (see Sec. 3.1.3) is approximated with very good accuracy in both cases, the topological entropy (recall Definition 3.5) was computed exactly only for $r = 28$. A general method for proving the robustness of chaos in a set of systems called C^1-*robust transitive sets with singularities* for flows on closed three-manifolds is given in [Morales, et al. (2004)]. The elements of the set C^1 are partially hyperbolic with a volume-expanding central direction and are either attractors or repellers. In particular, any C^1-robust attractor with singularities for flows on closed three-manifolds always have an invariant foliation (recall Definition 5.2(a)) whose leaves are forward contracted by the flow and has a positive Lyapunov exponent at every orbit, showing that any C^1-robust attractor resembles the geometric Lorenz attractor introduced in Sec. 8.2.2. A new topological invariant (see Sec. 8.2.2) (Lorenz manuscript) leading to the existence of an uncountable set of topologically various attractors is proposed in [Klinshpont, et al.

(2005)] where a new definition of the hyperbolic properties of the Lorenz system (8.1) close to singular hyperbolicity is introduced, as well as a proof that small nonautonomous perturbations do not lead to the appearance of stable solutions.

Other than the Lorenz attractor, there are some works that focus on the proof of chaos and its robustness property in 3-D continuous systems, for example the C^1 set introduced in [Morales, et al. (2004)], and a characterization of maximal transitive sets with singularities for generic C^1-vector fields on closed three-manifolds in terms of homoclinic classes associated with a unique singularity is given and applied to some special cases.

9.2 Robust chaos in Lorenz system

It was shown in [Guckenheimer & Williams (1979)] that the Lorenz system (8.1) is robust in the sense that there is a structurally stable (recall Definition 4.3) two-parameter family of flows containing the geometric Lorenz flow, *i.e.*, the geometric Lorenz flow is structurally stable of codimension-two. In this case, any perturbations of the geometric Lorenz flow is topologically conjugate on a neighborhood of the attractor to a nearby member of this family. The fact that the Lorenz system (8.1) is invariant under rotation of $\mathbb{R}^3$ around the z-axis by π implies that the geometric Lorenz flow (see Sec. 8.2.2) is structurally stable and of codimension-one rather than two.

Thus the following result was proved in [Guckenheimer & Williams (1979)]:

Theorem 9.1. *There is an open set $\mathfrak{U}$ in the space of all vector fields in $\mathbb{R}^3$ and a continuous mapping h of $\mathfrak{U}$ into a two-dimensional disk, such that:*

(A) Each $f \in \mathfrak{U}$ has a 2-dimensional Lorenz attractor.

(B) f and $g \in \mathfrak{U}$ are topologically conjugate by a homeomorphism close to the identity if and only if they have the same image under h.

The proof of Theorem 9.1 is based on the following two things:

(1) The construction of a suspension [Smale (1967)] and inverse limits [Williams (1975)] in dynamical systems.
(2) Results on the bifurcations of maps of the unit interval given in [Guckenheimer (1977)].

In [Tucker (1999)], a rigorous proof was provided that the geometric model does indeed (see Sec. 8.2.2) give an accurate description of the

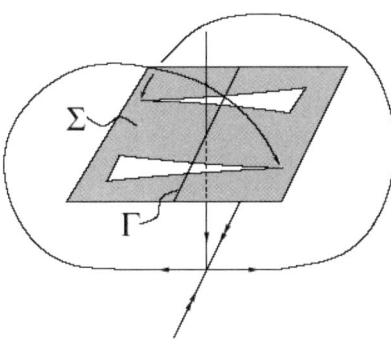

Fig. 9.1 The return map acting on Σ.

dynamics of the Lorenz Eqs. (8.1), *i.e.*, it supports a strange attractor as conjectured by Lorenz in 1963. This conjecture was listed by Steven Smale as one of several challenging mathematical problems for the 21st century [Smale (1998-2000)]. See Exercise 200. Also a proof that the attractor is robust, *i.e.*, it persists under small perturbations of the coefficients in the underlying differential equations was given. This proof is based on a combination of *normal form theory* and rigorous numerical computations. The robust chaotic Lorenz attractor is shown in Fig. 8.1.

The novelty of this method of proof lies on the construction of an algorithm in a C-program which, if successfully executed, proves the existence of the strange attractor. The source codes and the list of initial data that are used in the proof are available in the literature.

For $\sigma = 10, b = \frac{8}{3}$, and $r = 28$ the system (8.1) displays a strange attractor denoted by A. The geometric model described in Sec. 8.2.2 does indeed give an accurate description of the dynamics of (8.1). Indeed, using a Poincaré section, the flow of (8.1) can be reduced to a *first return map* R acting on the section

$$\Sigma \subset \{z = r - 1\} \qquad (9.1)$$

as schematically illustrated in Fig. 9.1. The map R is not defined on the line $\Gamma = \Sigma \cap W^s(0)$ because for these points, R tends to the origin and never returns to the section Σ. In this case, the return times are not bounded due to the fixed points at the origin, and this is overcome by introducing a local change of coordinates (8.52) as done in [Tucker (1999)] as follows: There exists a compact set $N \subset \Sigma$ such that $N \setminus \Gamma$ is *forward invariant* under R, *i.e.*, $R(N \setminus \Gamma) \subset int(N)$. This implies that the flow has an attracting set A with a large basin of attraction. The new cross-section of the attracting set of the new system is given by

$$A \cap \Sigma = \cap_{n=0}^{\infty} R^n(N) = \Lambda \qquad (9.2)$$

in which on the set N, there exists a cone field $\mathfrak{C}$ which is mapped strictly into itself by DR, i.e., for all

$$x \in N, DR(x).\mathfrak{C}(x) \subset \mathfrak{C}(R(x)), \qquad (9.3)$$

the cones of $\mathfrak{C}$ are centered along two curves which approximate Λ, and each cone has an opening of 10^0. Furthermore, the tangent vectors in $\mathfrak{C}$ are eventually strongly expanded under the action of DR; there exists a $C > 0$ and $\lambda > 1$ such that for all $v \in \mathfrak{C}(x), x \in N$, one has

$$|DR^n(x)v| \geq C\lambda^n |v|, n \geq 0. \qquad (9.4)$$

Hence the map R is topologically transitive on Λ. Thus the following result was proved in [Tucker (1999)]:

Theorem 9.2. *For the classical parameter values, the Lorenz Eqs. (8.1) support a robust strange attractor A. Furthermore, the flow admits a unique SRB measure (recall Definition 3.2) μ_X with $\mathrm{supp}(\mu_X) = A$.*

The proof of Theorem 9.2 is based on following result: Let M be a manifold and $f : M \to M$ be a C^1 diffeomorphism. Let N be a trapping region[1] for f. Let $T_N M = F^s \oplus F^u$ be a continuous splitting approximating $E^s \oplus E^u$. Given $\alpha \geq 0$, the stable and unstable cones are defined by

$$\begin{cases} \mathfrak{C}_x^s = \{v_1 + v_2 \in F_x^s \oplus F_x^u : |v_2| \leq \alpha |v_1|\} \\ \mathfrak{C}_x^u = \{v_1 + v_2 \in F_x^s \oplus F_x^u : |v_2| \geq \alpha |v_1|\}. \end{cases} \qquad (9.5)$$

Thus the following theorem provides a practical way of proving that a set is hyperbolic[2]:

Theorem 9.3. *Let N be a trapping region for a C^1 diffeomorphism f. Suppose that there exists a continuous splitting $T_N M = F^s \oplus F^u$, and that there are constants $\alpha \geq 0$ and $C > 0$, and $\sigma > 1$ so that*

$$\begin{cases} Df_x^{-1}.\mathfrak{C}_x^s(\alpha) \subset \mathfrak{C}_{f^{-1}(x)}^s(\alpha) \\ Df_x.\mathfrak{C}_x^u(\alpha) \subset \mathfrak{C}_{f(x)}^u(\alpha) \\ \|Df_x^{-n}|\mathfrak{C}_x^s(\alpha)|\| \geq C\sigma^n \\ \|Df_x^{n}|\mathfrak{C}_x^u(\alpha)|\| \geq C\sigma^n \end{cases} \qquad (9.6)$$

for every $x \in N$. Then $\Lambda = \bigcap_{i=0}^{\infty} f^i(N)$ is hyperbolic for f.

---
[1] The set N is a trapping region for f if $f(N) \subset \mathrm{int}(N)$.
[2] Because in practice it is impossible to find explicitly the invariant set N and the splitting $T_N M = E^s \oplus E^u$ except in the most trivial cases.

Fig. 9.2 Projection of the Lorenz attractor (8.1) on the plane (x, z) (black color) and the basin of its attraction (gray color) in the Poincaré section $y = 0$ for $\sigma = 10, b = \frac{8}{3}$, and $r = 28$. Reused with permission from Anishchenko, V. and Strelkova, G., Discrete Dynamics in Nature and Society (1998). Copyright 1998, Hindawi Publishing Corporation.

It is clear that the hypotheses of Theorem 9.3 are open in the C^1 topology which proves that hyperbolicity is a robust property. Especially, if g is C^1 close to f, then $\Lambda = \bigcap_{i=0}^{\infty} f^i(N)$ is a hyperbolic set for g.

To show that the Lorenz attractor (8.1) with $\sigma = 10, b = \frac{8}{3}$, and $r = 28$ has the typical properties of hyperbolic systems introduced in Chapter 6, we present in what follows some results and statistical computations leading to the fact that the Lorenz attractor is robust in the sense of Definition 2.5. Indeed, the Lorenz attractor (8.1) with $\sigma = 10, b = \frac{8}{3}$, and $r = 28$ is the attracting set consisting of the phase trajectories characterized by the individual exponential instability that do not depend on initial conditions and do not change under small variations of parameters. Figure 9.2 shows a projection of the Lorenz attractor on the plane (x, z) and the basin of its attraction. For the Lorenz attractor (8.1) with the classical values of parameters and $r = 26$, the Lyapunov exponents (LCE) are $\lambda_1 = 0.9, \lambda_2 = 0$, and $\lambda_3 = -14.57$, and the Lyapunov's dimension is $D = 2.06$ in which the fractional part is close to 0 as a result of the strong contraction of the flow in the dissipative system (8.1) for which $divF = -(\sigma + b + r)$, i.e., the Poincaré map of the Lorenz attractor is very close to a one-dimensional map. Furthermore, the set of nonsingular orbits of the system includes only points, cycles, and the Lorenz attractor because $divF$ does not depend on phase variables and the birth of the regime of two-frequency quasi-periodic oscillations is impossible. The bifurcation diagram of system (8.1) is shown

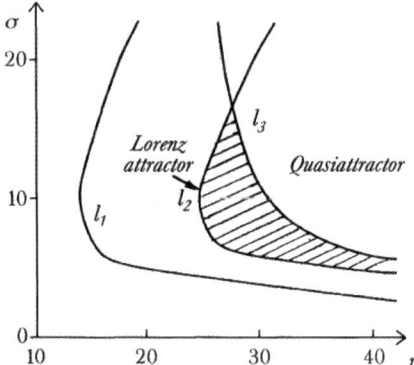

Fig. 9.3 Bifurcation diagram of the Lorenz system (8.1) in the plane of parameters r, σ for $b = 8/3$, l_1 is the line of the existence of a symmetrical separatrix loop for the zero equilibrium state; l_2 is the line at the birth of the Lorenz attractor; l_3 is the line of the bifurcation transition to quasi-attractors. Reused with permission from Anishchenko, V. and Strelkova, G., Discrete Dynamics in Nature and Society (1998). Copyright 1998, Hindawi Publishing Corporation.

in Fig. 9.3 [Bykov & Shilnikov (1989)], where the shaded region corresponds to the existence of the Lorenz attractor, while outside of this region the properties of the chaotic attractor are different, and system (8.1) displays a quasi-attractor (see Chapter 10) after the line $l_1 l_3$ in Fig. 9.3 which demonstrates the *transition Lorenz attractor-quasi-attractor*. In fact, as the parameters are varied, the Lorenz system (8.1) exhibits a bifurcational transition to a nonhyperbolic attractor [Anishchenko et al. (1994)]. The fact that the Lorenz attractor is robust (no bifurcations of the attractor occur) can be explained by the LCE spectrum. Indeed, this spectrum does not change under variation of initial conditions, which means that this attractor is the only one and the basin of its attraction is the entire phase space as shown in Figs. 9.3 and 9.4. Furthermore, this spectrum does not significantly change with variation of the parameters in the region of Lorenz attractor existence (see Fig. 9.4). On the other hand, the autocorrelation function (see Sec. 3.3.1) and power spectrum of the Lorenz attractor (8.1) presented in Fig. 9.6 are typical for intermixing systems. Indeed, the autocorrelation function is decreases almost exponentially with increasing time as shown in Fig. 9.5, and the power spectrum is a continuously decreasing function of frequency and does not contain pronounced peaks at any characteristic frequencies as shown in Fig. 9.6. These properties do not significantly change in the presence of additive (or multiplicative) noise with

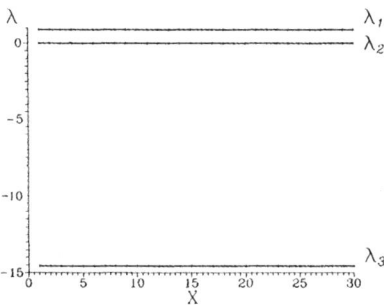

Fig. 9.4 Dependence of the LCE spectrum of the Lorenz attractor on the values of the x coordinate for $r = 28$, $\sigma = 10$, and $b = 8/3$. Reused with permission from Anishchenko, V. and Strelkova, G., Discrete Dynamics in Nature and Society (1998). Copyright 1998, Hindawi Publishing Corporation.

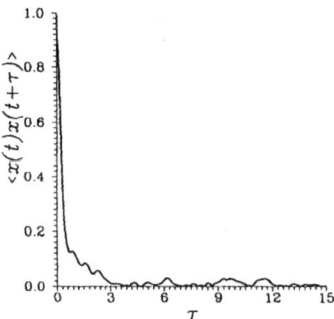

Fig. 9.5 Autocorrelation function for the Lorenz attractor (8.1) for $r = 28$, $\sigma = 10$, and $b = 8/3$. Reused with permission from Anishchenko, V. and Strelkova, G., Discrete Dynamics in Nature and Society (1998). Copyright 1998, Hindawi Publishing Corporation.

small intensity as shown in Fig. 9.7 [Anishchenko (1995)]. Figure 9.8(a) shows the probability distribution (see Sec. 6.9.1) of angles $p(\phi)$ for the Lorenz attractor (8.1) [Muller et al. (1997)]. It is clear that the probability of homoclinic tangency is exactly zero $[p(\phi) = 0]$. From Fig. 9.8(b), it is also clear that the effect of homoclinic tangency emerges where the Lorenz attractor exists. The above analysis shows strongly that the Lorenz attractor (8.1) demonstrates practically all properties and qualities of robust hyperbolic attractors (see Chapter 6) in a finite range of values of its control parameters [Anishchenko & Strelkova (1998)]. Essentially, the stable and unstable manifolds of the attractor trajectories intersect transversally in some ranges of parameters as shown in [Anishchenko et al. (1995)].

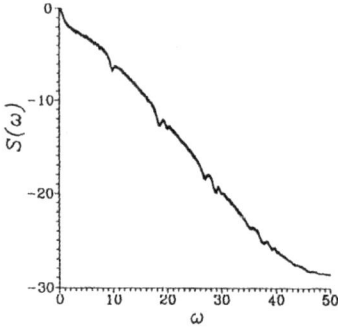

Fig. 9.6 Power spectrum for the Lorenz attractor (8.1) for $r = 28$, $\sigma = 10$, and $b = 8/3$. Reused with permission from Anishchenko, V. and Strelkova, G., Discrete Dynamics in Nature and Society (1998). Copyright 1998, Hindawi Publishing Corporation.

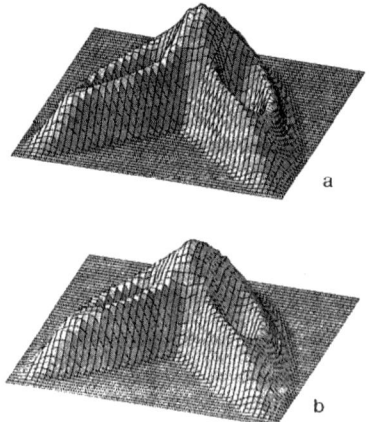

Fig. 9.7 Stationary two-dimensional probability density $\rho(x, z)$ of the Lorenz attractor (8.1) for $r = 28$, $\sigma = 10$, and $b = 8/3$. (a) In the absence of noise, and (b) in the presence of additive white noise with intensity $d = 0.8$. Reused with permission from Anishchenko, V. and Strelkova, G., Discrete Dynamics in Nature and Society (1998). Copyright 1998, Hindawi Publishing Corporation.

(1) The Lorenz system (8.1) has no stable regular attractors when any perturbations are added even in the presence of a denumerable set of separatrix loops of equilibrium states.
(2) The Lorenz system (8.1) has bifurcations, and no other stable attracting subsets appear for any perturbations of the system parameters in a finite range.

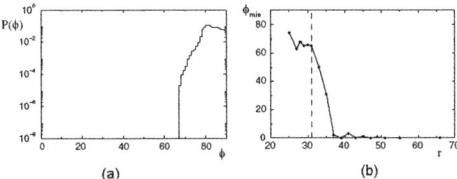

Fig. 9.8 Calculation results for Lorenz system (8.1) at $\sigma = 10$ and $b = 8/3$. (a) Probability distribution of angles for the Lorenz attractor at $r = 27$, and (b) the minimum angle $\phi_{\min}$ as a function of the parameter r. The vertical line marks the theoretically determined onset of the transition from the Lorenz attractor to nonhyperbolic attractors.

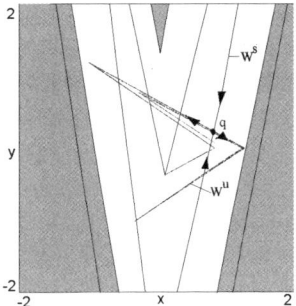

Fig. 9.9 Stable and unstable manifolds of saddle points in the Lozi map (8.88) for $a = 1.7$ and $b = 0.3$.

(3) The Lorenz system (8.1) has small changes in the attractor's structure for any slight perturbation or for external noise of small intensity.
(4) The Lorenz system (8.1) admits the construction of a smooth function of distribution density[3] $\rho(x, y, z)$, in which small perturbations cause small changes in the probability measure. See [Kifer (1974)].

9.3 Robust chaos in 2-D Lorenz-type attractors

An example of a 2-D Lorenz-type attractor realized in a discrete mapping is the Lozi map (8.88). The manifolds of chaotic orbits of this map are transversal, i.e., they have the same qualitative behaviors of manifolds as the saddle cycles shown in Fig. 9.9. Using for the Lozi map (8.88) the same algorithm as in the case of the Hénon map (6.88), one obtains Fig. 9.10(a) that shows the distribution $P(\phi)$ of the angle ϕ between the manifolds of

---
[3] The probability measure of the attractor.

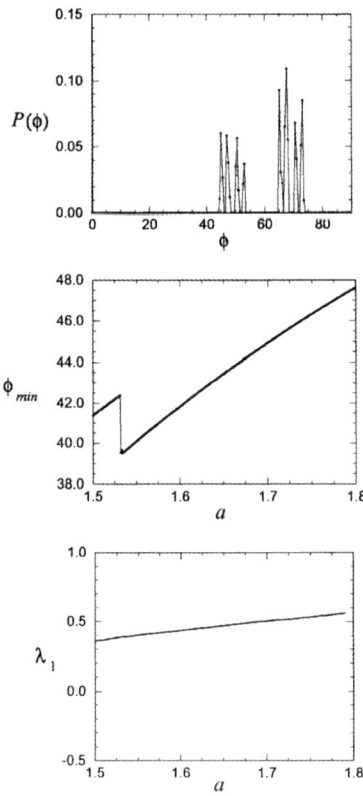

Fig. 9.10 Calculation results of the characteristics for the Lozi attractor (8.88). (a) A distribution of the probabilities of the angle ϕ between stable and unstable manifolds for $a = 1.7$ and $b = 0.3$, (b) a plot of the minimal angle $\phi_{\min}$ versus parameter a for $b = 0.3$, and (c) the dependence of the largest Lyapunov exponent on a for $b = 0.3$. Reused with permission from Anishchenko, V. S., Vadivasova, T. E., Strelkova, G. I., and Kopeikin, A. S., Discrete Dynamics in Nature and Society (1998). Copyright 1998, Hindawi Publishing Corporation.

a chaotic orbit of the Lozi map (8.88). Hence there is some minimal and bounded values $\phi_{\min}$ of the angle ϕ obtained for a with $b = 0.3$ as shown in Fig. 9.10(b). Indeed, for $1.5 \leq a \leq 1.8$, one has $\phi_{\min} > 40$ degrees and is never equal to 0. Generally, in 2-D invertible dissipative maps, the following properties are characteristic of Lorenz-type attractors [Anishchenko et al. (1998)]:

(1) Transversality is a necessary condition for the attractor to be a Lorenz-type attractor.

(2) The probability density of the angle between separatrixes of a chaotic orbit on the attractor is strictly equal to zero in the neighborhood of a zero value of the angle, i.e., $p(\phi) = 0$ for $\phi \simeq 0$, and $p(\phi) > 0$ for $0 < \phi < \phi_0$ (see Fig. 9.10(a)).

(3) The dependence of the largest Lyapunov exponent λ_1 is a smooth positive definite function of the parameter in the region where the attractor exists, i.e., $\lambda_1(a) > 0$, $a_1 < a < a_2$. (see Fig. 9.10).

(4) The power spectrum of Lorenz-type attractors depends smoothly on the frequency and does not contain any pronounced peaks.

9.4 Exercises

Exercise 201. Explain a method for approximating a system by a hyperbolic one. See [Smale (1967), Abraham & Smale (1968), Simon (1972)].

Exercise 202. Show that the contracting Lorenz attractor or Lorenz–Rovella attractor are probability persistent in terms of Lebesgue probability but are not robust. See [Rovella (1993)].

Exercise 203. Prove Theorem 9.1. See [Guckenheimer & Williams (1979)].

Exercise 204. (a) By using an appropriate program, show that the Lorenz system (8.1) is robust for $\sigma = 10, b = \frac{8}{3}$, and $r = 28$.
(b) Use the same program for other values suggesting robust chaos.

Exercise 205. (a) Show that the *first return map* R defined in Sec. 9.1 is not defined on the line $\Gamma = \Sigma \cap W^s(0)$.
(b) Show that the return times are not bounded.
(c) Explain a method for overcoming this difficulty.

Exercise 206. Show that hyperbolicity is a robust property.

Exercise 207. (a) Show that the Lorenz attractor (8.1) with $\sigma = 10, b = \frac{8}{3}$, and $r = 28$ is characterized by individual exponential instability that does not depend on initial conditions and does not change under small variation of parameters.
(b) Calculate the Lyapunov exponents (LCEs) for the Lorenz attractor (8.1) with $\sigma = 10, b = \frac{8}{3}$, and $r = 28$.
(c) Calculate the Lyapunov dimension.
(d) Show that the Poincaré map of the Lorenz attractor is very close to a one-dimensional map.

(e) Show that the set of nonsingular orbits of the system includes only points, cycles, and the Lorenz attractor.

(f) Show that the birth of the regime of two-frequency quasi-periodic oscillations is impossible.

(g) Find explicitly the expressions for the lines l_1, l_2, and l_3 in Fig. 9.3. See [Bykov & Shilnikov (1989)].

(h) Illustrate the transition Lorenz attractor to quasi-attractor.

(i) Find regions of robust Lorenz attractors other than those shown in Fig. 9.4.

(j) Show that the probability of homoclinic tangency is exactly zero $[p(\phi) = 0]$ in Fig. 9.8 (a) where the Lorenz attractor exists.

(k) Show by numerical calculation that the Lorenz system (8.1) has not stable regular attractors when any perturbations are added, even in the presence of a denumerable set of separatrix loops of equilibrium states.

(l) Show by numerical calculation that the Lorenz system (8.1) has bifurcations and that no other stable attracting subsets appear for any perturbations of the system parameters in a finite range.

(m) Show by numerical calculation that the Lorenz system (8.1) has small changes in the attractor's structure for any slight perturbation or with external noise of small intensity.

(n) Show by numerical calculation that the Lorenz system (8.1) admits the construction of a smooth function of distribution density $\rho(x, y, z)$.

Exercise 208. (a) Show that for the Lozi map (8.88), manifolds of chaotic orbits of this map are transversal.

(b) Find other parameter values for the Lozi map (8.88) in which the property in (a) holds.

(c) Calculate the distribution $P(\phi)$ of the angle ϕ between the manifolds of a chaotic orbit of the Lozi map (8.88) for the set of parameters values obtained in (b).

Exercise 209. Show by examples that the following properties are characteristic for the Lorenz-type attractors in a 2-D invertible dissipative map:

(a) Transversality is a necessary condition for the attractor to be a Lorenz-type attractor.

(b) The probability density of the angle between separatrices of a chaotic orbit on the attractor is strictly equal to zero in the neighborhood of a zero value of the angle.

(c) The dependence of the largest Lyapunov exponent λ_1 is a smooth positive definite function of the parameter in the region where the attractor exists.

(d) The power spectrum of Lorenz-type attractors depends smoothly on the frequency and does not contain any pronounced peaks. See [Anishchenko et al. (1998)].

Chapter 10

No Robust Chaos in Quasi-Attractors

In this chapter, we discuss some relevant results and properties of quasi-attractors as the third type in the classification of strange attractors of dynamical systems introduced in Sec. 6.3. Indeed, Sec. 10.1 gives concepts and properties of these attractors. Section 10.2 gives some well-known results for the Hénon map such as the existence of transversal homoclinic points that prove that this map is a quasi-attractor. The uniform hyperbolicity is discussed for both the Hénon map and other Hénon-like maps in Sections 10.2.1 and 10.2.2. Using the procedure described in Sec. 6.9.1, we give a proof that the Hénon attractor is a quasi-attractor in Sec. 10.2.3. As another example of systems with quasi-attractors, the so-called *Strelkova–Anishchenko map* and the *Anishchenko–Astakhov oscillator* are described in Sections 10.3 and 10.4, respectively. In Sec. 10.5, we describe in some detail Chua's circuit that is the famous example of a system displaying a quasi-attractor, this including the existence of homoclinic and heteroclinic orbits and the dynamic of this circuit near them, along with a short description of the subfamilies of the double scroll family. In Sec. 10.5.2, the so-called *geometric model* is also introduced with some details. The particularity of this model is that it is different from the Lorenz-type (see Chapter 8) or quasi-attractors discussed in this chapter. This new type contains unstable points, in addition to the Cantor set structure of hyperbolic points. In particular, the corresponding two-dimensional Poincaré map has a strange attractor with no stable orbits. In Sec. 10.6 are some exercises and open problems about quasi-attractors, in particular for the Hénon and Hénon-like attractors.

10.1 Quasi-attractors, concepts, and properties

Recall that the *quasi-attractors* are the limit sets enclosing periodic orbits of different topological types (for example, stable and saddle periodic orbits) and structurally unstable orbits. It is important to note that most known chaotic attractors are quasi-attractors [Afraimovich & Shilnikov (1983), Lichtenberg & Lieberman (1983), Rabinovich & Trubetskov (1984), Schuster (1984), Neimark & Landa (1989), Anishchenko (1990-1995)], *i.e.*, they are neither robust hyperbolic nor almost hyperbolic attractors. Some applications of quasi-attractors in neurodynamics can be found in [Fujii *et al.* (2007a-b)]. These attractors have the following important properties:

(1) They separate loops of saddle-focuses or homoclinic orbits of saddle cycles at the point of tangency of their stable and unstable manifolds because they enclose nonrobust singular trajectories that are *dangerous*.
(2) A map of *Smale's horseshoe-type* (see Sec. 1.3) appears in the neighborhood of their trajectories. This map contains both a nontrivial hyperbolic subset of trajectories and a denumerable subset of stable periodic orbits (see Sec. 5.3.2 describing Shilnikov's theorems) [Shilnikov (1963), Gavrilov & Shilnikov (1972-1973)] and Newhouse's theorem [Newhouse (1980)].
(3) The quasi-attractor is the unified limit set of the whole attracting set of trajectories including a subset of both chaotic and stable periodic trajectories which have long periods and a weak and narrow basin of attraction and stability region. This is a result of the effect that this attractor is *holed* by a set of basins of attraction (see Sec. 2.1.3) of different periodic orbits[1].
(4) The basins of attraction of stable cycles are very narrow[2].
(5) Some orbits do not ordinarily reveal themselves in numerical simulations except for some quite large stability windows where they are clearly visible.

The above properties imply that quasi-attractors have a very complex structure of embedded basins of attraction in terms of initial conditions and a set of bifurcation parameters of nonzero measure. The homoclinic tangency

---

[1]More generally, this implies the co-existence of a denumerable set of different chaotic and regular attractors in a bounded element of the system phase space volume when the bifurcation parameters are fixed.
[2]This implies that these basins are mixed in the phase space and that it is not always possible to detect them in numerical experiments.

(recall Definition 1.9) of stable and unstable manifolds of saddle points in the Poincaré section (see Sec. 1.1) is the principal cause of this complexity as shown in [Gavrilov & Shilnikov (1972-1973), Afraimovich (1984-1989-1990)]. On the other hand, it is well known that basin boundaries (see Sec. 2.1.3) arise in dissipative dynamical systems when two or more attractors are present. In such situations, each attractor has a basin of initial conditions that lead asymptotically to that attractor. The sets that separate different basins are called the *basin boundaries*. In some cases, the basin boundaries can have very complicated fractal structure and hence pose an additional impediment to predicting long-term behavior. Hence for the quasi-attractors, the basins of attraction of co-existing limit sets can have fractal boundaries and occupy very narrow regions in phase space as mentioned above. However, a rigorous mathematical description of quasi-attractors is still an open problem because almost all nonhyperbolic attractors are obscured by noise.

Some new results concerning quasi-attractors can be found in [Luchinski & Khovanov (1999)] where the authors investigate numerically the noise-induced escape from the basin of attraction of a strange attractor in a periodically excited nonlinear oscillator. The result is that escape occurs in two steps. The first is transfer of the system from the strange attractor to a close-lying saddle cycle along several optimal trajectories, and the second is a subsequent fluctuation-induced transfer from the basin of attraction of the strange attractor along a single optimal trajectory. In [Ding (2005)], it was shown that a quasi-attractor is the limit point of attractors with respect to the Hausdorff metric (see Sec. 3.1.3), and if a component of an attractor is not an attractor, then it must be a real quasi-attractor. The properties of borderlines in discontinuous conservative systems were studied in [Wang & Fang (2006)]. In particular, it was shown that a chaotic quasi-attractor is confined by the preceding lower order images of the borderline, *i.e.*, there exists a forbidden zone (which is the sub-phase space of one side of the first image of the borderline) in which any orbit cannot visit. In this case, each order of the images of the forbidden zone can be qualitatively divided into two sub-phase regions, one of which is the escaping region and the other is the so-called *dissipative region* where the phase space contraction occurs. The necessary and sufficient condition for the existence of a quasi-attractor was given in [Zuo & Wang (2007)] by generalizing some results of Conley. The method of analysis is based on the connection among the attractor, the attractor neighborhood, and the domain of its influence.

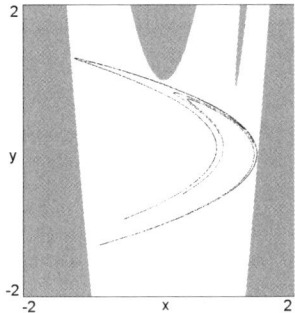

Fig. 10.1 The Hénon attractor with its basin of attraction (in gray) obtained for $a = 1.4$ and $b = 0.3$.

In the remainder of this chapter, we present some examples of quasi-attractors, including the Hénon map, the Anishchenko–Astakhov oscillator, the Strelkova–Anishchenko map, and Chua's circuit. For each example, we state the most important results concerning their nature, especially the different methods used for proving that these systems are quasi-attractors, and hence that there is no robust chaos in them.

10.2 The Hénon map

In this section, we discuss some relevant results about the homoclinic tangency in the Hénon map [Hénon (1976)] given by

$$h(x,y) = \begin{pmatrix} 1 - ax^2 + y \\ bx \end{pmatrix} \qquad (10.1)$$

as the principal cause of its complexity and nonrobustness, *i.e.*, the Hénon map (10.1) is a quasi-attractor. The map (10.1) which is the subject of many works that deal with its important and unusual properties as a simplified model of the Poincaré map (see Sec. 1.1) for the Lorenz model (8.1). In this section, several forms of the Hénon map are used to state the corresponding results, and the reader can deduce these results for the map (10.1) using an affine or linear transformation. The Hénon attractor with its basin of attraction (in gray) obtained for $a = 1.4$ and $b = 0.3$ is shown in Fig. 10.1. First, note that the Hénon map (10.1) has the following properties:

(1) It has two fixed points, it is invertible, it is conjugate to its inverse, and its inverse is also a quadratic map. The parameter b is a measure of the rate of area contraction (dissipation).

(2) It is the most general 2-D quadratic map with the property that the contraction is independent of the variables x and y.
(3) It reduces to the quadratic map when $b=0$, which is conjugate to the logistic map (6.25).
(4) It has bounded solutions over a range of a and b values, and a portion of this range (about 6%) yields chaotic solutions.
(5) For $a = 1.4$ and $b = 0.3$, the Hénon map (10.1) has the chaotic attractor as shown in Fig. 10.1 with a correlation dimension of 1.25 ± 0.02 [Grassberger & Procaccia (1983)] and a capacity dimension of 1.261 ± 0.003 [Russell, et al (1980)]. A numerical algorithm for estimating generalized dimensions for large negative q for the Hénon map (10.1) was given in [Pastor-Satorras & Riedi (1996)] because the standard fixed-size box-counting algorithms are inefficient for computing generalized fractal dimensions in the range of $q < 0$.

Most results on the dynamical behavior for the Hénon map (10.1) are done numerically. See [Zeraoulia & Sprott (2010)] for more detail.

10.2.1 Uniform hyperbolicity of the Hénon map

In this section, we discuss the concept of uniform hyperbolicity of the Hénon map. Indeed, there are many known results that give sufficient conditions for hyperbolicity, for example see [Hirsch & Pugh (1970), Newhouse & Palis (1971), Moser (1973), Palis & Takens (1993), Robinson (1999)]. In [Newhouse (2004)], some conditions for dominated and hyperbolic splittings on compact invariant sets for a diffeomorphism in terms of its induced action on a cone field and its complement were given. These results were applied to prove the well-known theorems for hyperbolicity of the set of bounded orbits in real and complex Hénon mappings using complex methods [Bedford & Smillie (2006c)]

First, we need the following definitions and concepts: Consider a diffeomorphism f on the manifold M and a compact invariant set Λ which admits a dominated splitting in terms of E_{1x} and E_{2x}.

Definition 10.1. (a) Let $E \subset \mathbb{R}^n$ be a proper subspace, i.e., $0 < \dim E < n$. Let F be a complementary subspace, i.e., $\mathbb{R}^n = E \oplus F$. Then the *standard unit cone* determined by the subspaces E and F is the set

$$K_1(E, F) = \{(v_1, v_2) : v_1 \in E, v_2 \in F, \text{ and } |v_2| \leq |v_1|\}. \qquad (10.2)$$

(b) A *cone* in $\mathbb{R}^n$ with core E, denoted $C(E)$, is the image $T(K_1(E,F))$ where $T: \mathbb{R}^n \to \mathbb{R}^n$ is a linear automorphism such that $T(E) = E$. In particular, in $\mathbb{R}^n$ a cone C is a set $C(E)$ for some proper subspace E of $\mathbb{R}^n$. (c) A *cone field* $C = \{C_x\}$ on Λ is a collection of cones $C_x \subset T_x M$ for $x \in M$. (d) The cone field C_x has a constant orbit core dimension on Λ if $\dim E_x = \dim E_{f(x)}$ for all $x \in \Lambda$ where $E_x, E_{f(x)}$ are the cores of $C_x, C_{f(x)}$, respectively.

Now, given such a cone field $C = \{C_x\}_{x \in M}$, and let
$$\begin{cases} m_{C,x} = m_{C,x}(f) = \inf_{v \in C_x \setminus \{0\}} \frac{|Df_x(v)|}{|v|} \\ m'_{C,x} = m'_{C,x}(f) = \inf_{v \notin C_{f(x)}} \frac{|Df^{-1}_{f(x)}(v)|}{|v|}. \end{cases} \quad (10.3)$$
Then one has the following definition:

Definition 10.2. (a) The real quantity $m_{C,x}$ is called the *minimal expansion* of f on C_x or of Df on C_x.

(b) The real $m'_{C,x}$ is called the *minimal co-expansion* of f on C_x or of Df on C_x.

(c) The *domination coefficient* of f on C is given by
$$m_d(C) = m_d(C, f) = \inf_{x \in \Lambda} m_{C,x} \cdot m'_{C,x}. \quad (10.4)$$

(d) The map f is *dominating* on C over Λ (or C is *a dominating cone field* on Λ) if C has constant orbit core dimension and
$$m_d(C) > 1. \quad (10.5)$$

(e) The map f is *strongly dominating* on C over Λ if C has constant orbit core dimension and
$$\left(\inf_{x \in \Lambda} m_{C,x}\right) \cdot \left(\inf_{x \in \Lambda} m'_{C,x}\right) > 1. \quad (10.6)$$

Thus the following results were proved in [Newhouse (2004)]:

Theorem 10.1. *(a) Suppose that f is dominating on C over Λ. Then there is a unique Df-invariant splitting $T_\Lambda M = E_1 \oplus E_2$ such that for all $x \in \Lambda$, we have $E_{1x} \subset Cx$ and $E_{2x} \subset T_x M \setminus C_x$. Further, if f is strongly dominating on C over Λ, then the functions $x \to E_{1x}, x \to E_{2x}$ are continuous in x.*

(b) A necessary and sufficient condition for Λ to be a uniformly hyperbolic set for f is that there is an integer $N > 0$ and a cone field C with constant orbit core dimension over Λ such that f^N is both expanding and co-expanding on C.

The proof of Theorem 10.1 is based on the following results also proved in [Newhouse (2004)]:

Proposition 10.1. *(a) Suppose that f is dominating on C over Λ. Then C is an f-invariant cone field. That is, for $x \in \Lambda$, we have*

$$Df_x(C_x) \subset C_{f(x)}. \tag{10.7}$$

(b) A sufficient condition for f to have a dominated splitting over Λ is that there is an integer $n_0 > 0$ such that f^{n_0} has a strongly dominated cone field C over Λ.

The preceding results were applied to both the real and complex Hénon map (10.1), *i.e.*, with respect to the localization of the parameters a and b either in $\mathbb{R}$ or in $\mathbb{C}$. This gives another simple proof of hyperbolicity on the set of bounded orbits in the Hénon family (10.1) for the real case in Sec. 6.9.2 for $|a|$ sufficiently large depending on $|b|$. The most important result here is that the same estimate holds in the complex case, and the arguments used for proof do not depend on the real geometry as in the case of the proof of Devaney and Nitecki given in Sec. 10.2.2. Better estimates can be obtained using complex methods, as was done in Proposition 7.4.6 in [Morosawa *et al.* (2000)] where it was proved that the bounded orbits form a topological horseshoe for

$$a > 2(1 + |b|)^2. \tag{10.8}$$

Thus the following result was proved also in [Newhouse (2004)]:

Theorem 10.2. *Consider the real or complex Hénon family $H_{a,b} = \begin{pmatrix} a + by - x^2 \\ x \end{pmatrix}$ with $0 < |b| \leq 1$. Let $\Lambda_{a,b}$ denote the set of points with bounded orbits. Assume that*

$$\begin{cases} |a| > \frac{(5+2\sqrt{2})(1+|b|^2)}{4} \\ 0 < |b| \leq 1. \end{cases} \tag{10.9}$$

In the complex case or the real case with $a > 0$, we have that $\Lambda_{a,b}$ is a nonempty compact invariant uniformly hyperbolic set. In the real case with $a < 0$, the set $\Lambda_{a,b}$ is empty.

Another proof of the uniform hyperbolicity, *i.e.*, the existence of many regions of hyperbolic parameters in the parameter plane, is done for the Hénon map $H_{a,b}(x,y)$ in [Arai (2007)(a)] using the rigorous computational method introduced in Sec. 6.10 for testing uniform hyperbolicity. In this section, this method is applied to the chain recurrent set

$\mathcal{R}(H_{a,b}) = \frac{1+|b|+\sqrt{(1+|b|^2)+4a}}{2}$ of the Hénon family, where an *a priori* knowledge of the size of $\mathcal{R}(H_{a,b})$ is needed. The method of analysis is based on the application of Theorem 6.1. Indeed, the dynamics restricted to $\mathcal{R}(H_{a,b})$ is chain recurrent, as proved in [Arai (2007(a)].

Lemma 10.1. *The chain recurrent set $\mathcal{R}(H_{a,b})$ is contained in $S(a,b) = \{(x,y) \in \mathbb{R}^2, |x| \leq R(a,b), |x| \leq R(a,b)\}$, and $H_{a,b}$ restricted to $\mathcal{R}(H_{a,b})$ is chain recurrent.*

Proof. If $x \notin S(a,b)$, then for small $\varepsilon_0 > 0$ such that if $\varepsilon < \varepsilon_0$, then all ε-chains through x must diverge to infinity, and hence x cannot be chain recurrent as shown in Corollary 2.7 of [Bedford & Smillie (1991)]. Using the same method in [Robinson (1999)] for the compact case, it follows that the set $\mathcal{R}(H_{a,b})$ is chain recurrent because it is still possible to choose a compact neighborhood S' of $S(a,b)$ and $\varepsilon_0 > 0$ such that if $\varepsilon < \varepsilon_0$ then all ε-chains from $x \in \mathcal{R}$ to x must be contained in S'. □

For an arbitrary polynomial map of $\mathbb{R}^n$, in particular for the Hénon map $H_{a,b}(x,y)$, the condition given in Sec. 6.10 can be verified using rigorous interval arithmetic on a CPU that satisfies the IEEE754 standard for binary floating-point arithmetic. In the case of the Hénon map (10.1), it suffices to consider the case $b \in [-1,1]$ because its inverse is conjugate to the Hénon map $H_{\frac{a}{b^2},\frac{1}{b}}$, whose Jacobian is $\frac{1}{b}$, and the hyperbolicity of a diffeomorphism is equivalent to that of the inverse map. Using the results of [Devaney & Nitecki (1979)] described in Sec. 10.2.2, it is sufficient to choose the case $(a,b) \in [-1,12] \times [-1,1]$ because otherwise the set $\mathcal{R}(H_{a,b})$ is hyperbolic or empty. Let $A = [-1,12] \times [-1,1]$, $K = [-8,8] \times [-8,8]$, and $L = K \times [-1,1]^2$. Then Lemma 10.1 implies that $\mathcal{R}(H_{a,b}) \subset \int K$ holds for all $(a,b) \in A$. With this initial data, Theorem 10.2 is proven by applying **Algorithm B** described in Sec. 6.10. Theorem 10.3 below was obtained by fixing $b = -1$ and starting the computation with $A = [4,12]$ with the same sets K and L in the case of Theorem 10.3 below. The method requires 1000 hours of computation using a 2 GHz PowerPC 970 CPU to achieve Theorem 10.3 below and 260 hours for Theorem 10.4 below. Then the results of [Arai (2007a)] are given by the following:

Theorem 10.3. *There exists a set $P \subset \mathbb{R}^2$, which is the union of 8943 closed rectangles, such that if $(a,b) \in P$, then $\mathcal{R}(H_{a,b})$ is uniformly hyperbolic. The set P is illustrated in Fig. 10.2 (shaded regions).*

The set P is called a *"plateau"* because the hyperbolicity of the chain recurrent set $\mathcal{R}(H_{a,b})$ implies its Ω-stability (see Sec. 4.1.2), and in this case, on each connected component of P, no bifurcation occurs in $\mathcal{R}(H_{a,b})$, and hence numerical invariants such as the topological entropy, the number of periodic points, *etc.*, are constant on it. On the other hand, Theorem 10.3 does not determine regions for nonhyperbolic parameters, and more computations for the one-parameter area-preserving Hénon family $H_{a,-1}$ gives a set P' of uniformly hyperbolic parameters such that $P \subset P'$. Namely, the following result was proved in [Arai (2007a)]:

Theorem 10.4. *If a is in one of the following closed intervals,*

[4.5383300781250, 4.5385742187500],	[4.5388183593750, 4.5429687500000],
[4.5623779296875, 4.5931396484375],	[4.6188964843750, 4.6457519531250],
[4.6694335937500, 4.6881103515625],	[4.7681884765625, 4.7993164062500],
[4.8530273437500, 4.8603515625000],	[4.9665527343750, 4.9692382812500],
[5.1469726562500, 5.1496582031250],	[5.1904296875000, 5.5366210937500],
[5.5659179687500, 5.6077880859375],	[5.6342773437500, 5.6768798828125],
[5.6821289062500, 5.6857910156250],	[5.6859130859375, 5.6860351562500],
[5.6916503906250, 5.6951904296875],	[5.6999511718750, ∞),

then $\mathcal{R}(H_{a,-1})$ is uniformly hyperbolic.

Some important remarks about Theorem 10.4 are the following:

(a) Theorem 10.4 justifies the conjecture of Davis *et al.* [Davis *et al.* (1991)] due to the presence of the three intervals where the Hénon map (10.1) is uniformly hyperbolic.

(b) Comparison of Fig. 10.2 with the bifurcation diagrams of the Hénon map (10.1) obtained numerically in [El Hamouly & Mira (1981), Sannami (1989-1994)] show a very close similarity between the boundary of P and these bifurcation curves.

In conclusion, we note that it was shown in [Cao *et al.* (2008)] that the Hénon map is nonhyperbolic (it has a tangency) if the parameter is on the boundary of the full horseshoe plateau [Bedford & Smillie (2005-2006)]. From the above results in this section and Theorem 10.4, one has that $H_{a,-1}$ has a tangency when a is close to 5.699 951 171 875. Hence the following theorem was proved in [Arai (2007a)] using the rigorous computational method developed in [Arai & Mischaikow (2006)]:

Proposition 10.2. *There exists an $a \in [5.6993102, 5.6993113]$ such that $H_{a,-1}$ has a homoclinic tangency with respect to the saddle fixed point on the third quadrant.*

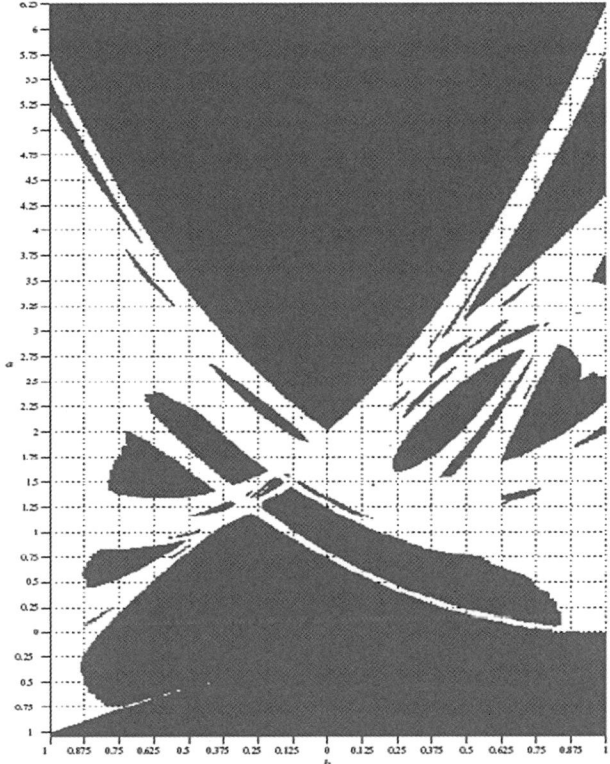

Fig. 10.2 Uniformly hyperbolic plateaus for the Hénon map $H_{a,b}$. This figure is provided by Zin Arai.

Consequently, Theorem 10.4 and Proposition 10.2 yield the following result proved also in [Arai (2007a)]:

Corollary 10.1. *When we decrease $a \in \mathbb{R}$ of the area-preserving Hénon family $H_{a,-1}$, the first tangency occurs in $[5.6993102, 5.6999)$.*

10.2.2 Hyperbolicity of Hénon-like maps

Hoensch in [Hoensch (2008)] establishes conditions under which certain C^2 diffeomorphisms exhibit hyperbolic invariant sets. This result can be apply for an abstract class of maps that contains the Hénon family

$$H(x,y) = (ax(1-x) - by, x) \qquad (10.10)$$

and small C^2 perturbations of them called *Hénon-like diffeomorphisms* within a specified parameter ranges in the ab-plane.

The result of Hoensch that works for an abstract class of diffeomorphisms, for which Hénon-like maps provide an example class, differs from the works given in [Bedford & Smillie (2005)] and [Cao et al. (2008)] in the following respects:

(1) The methods used in [Bedford & Smillie (2005)] does not apply by its nature to perturbations of the Hénon map.
(2) The method in [Cao, et al. (2008)] does not *a priori* work with perturbations of the Hénon map.

Hence the main results proved in [Hoensch (2008)] for Hénon-like maps is given by the following:

Theorem 10.5. *There exists a $b_0 > 0$ such that for each $0 < b \leq b_0$ there is a unique $r = r(b)$ with the property that,*
(a) if $r > r(b)$, the nonwandering set $\Omega(H)$ is uniformly hyperbolic, and
(b) if $r = r(b)$, there is a quadratic homoclinic tangency between the stable and the unstable manifolds of the fixed point $(0,0)$. This homoclinic tangency causes a loss of hyperbolicity of the nonwandering set $\Omega(H)$. However, the set $\Omega(H) \backslash \mathcal{O}(q)$ is uniformly hyperbolic in the sense that a first-return map on $\Lambda \backslash \mathcal{O}(q)$ is uniformly hyperbolic. ($\mathcal{O}(q)$ is the orbit of the homoclinic tangency.).

These statements are also true for C^2-perturbations of the Hénon map because, for these perturbations, $r > r(b)$ corresponds to the situation of the map being "before the first tangency," and $r = r(b)$ corresponds to the situation of the map being "at the first tangency." These two situations are illustrated in Fig. 10.3.

The ideas of the proof of Theorem 10.5 are based on the geometry exhibited by the invariant manifolds of the two fixed points of the Hénon map and Hénon-like maps with small positive Jacobian determinant and the proposed abstract model, which permits one to give sufficient conditions for the existence of a quadratic tangency, where the sets $l_{1,1}^s, l_{1,2}^s, l_1^u$, and l_2^u in Fig. 10.3 are some special elementary curves defined for Hénon-like maps.

10.2.3 Hénon attractor is a quasi-attractor

The above analysis in Sections 10.2.1 and 10.2.2 and the theoretical proof given in [Newhouse (1980)] confirm that the Hénon attractor is an example

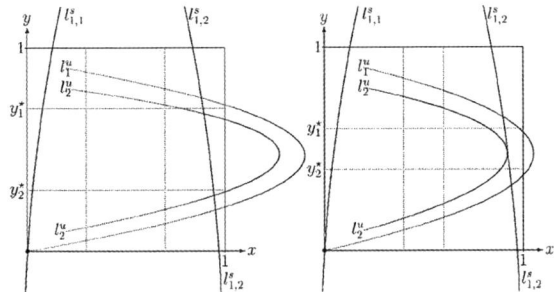

Fig. 10.3 Parts of the invariant manifolds before the first tangency when $r > r(b)$) (left) and parts of the invariant manifolds at the first tangency when $r = r(b)$ (right). Reused with permission from U.A. Hoensch, Nonlinearity (2008). Copyright 2008, IOP Publishing Ltd.

of a quasi-attractor, *i.e.*, the homoclinic tangencies are everywhere dense in the parameter space, and the quasi-attractor is a typical limit set for the Hénon map (10.1). This result is the basis of another numerical proof that the Hénon map given by

$$\begin{cases} x_{n+1} = a - x_n^2 + y_n \\ y_{n+1} = bx_n \end{cases} \quad (10.11)$$

is a quasi-attractor [Anishchenko *et al.* (1998)]. Indeed, Fig. 10.4 shows the behavior of the stable and the unstable manifolds of the period-1 cycle obtained from the Hénon map (10.11) for $a = 1.3$ and $b = 0.3$. This feature is qualitatively the same for manifolds of saddle cycles of other periods. For the case of 2-D maps, manifolds of saddles are 1-D curves, and chaotic attractors are located along the unstable manifolds of saddle cycles repeating their form. For the case of a 2-D dissipative map, the unstable manifolds of saddles and the chaotic attractor must be *packed* in some bounded region of the phase plane. Thus it follows that for smooth 2-D maps, the unstable manifolds inevitably undergo a bending in the form of *horseshoe* which leads to *dangerous* tangencies between the stable and unstable manifolds and to the quasi-attractor, respectively. To determine whether the attractor governed by the Hénon map (10.11) is a quasi-attractor or a hyperbolic attractor, one can use the algorithm described in Sec. 3.3.1. The calculation of the following quantities is sufficient for this purpose:

(1) Calculate the angles ϕ between manifolds W^s and W^u of a chaotic orbit for different points on the attractor, and the analyze their statistics.

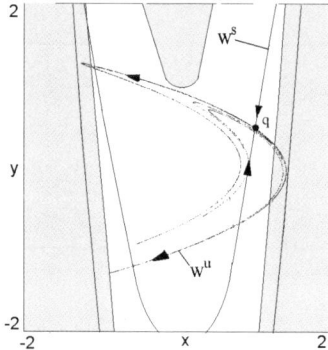

Fig. 10.4 Stable and unstable manifolds of saddle points in the Hénon map (10.11) for $a = 1.3$ and $b = 0.3$.

(2) Calculate the distribution of the probabilities of the angle between the manifolds of a chaotic trajectory $P(\phi)$ on the quasi-attractor shown in Fig. 10.5(a). Recall that the definition of $P(\phi)$ is given in Sec. 3.3.2.
(3) Calculate the probability $P^{\delta\varphi}$ that the angle φ falls within a small neighborhood of zero ($\delta\varphi = 1$ degree) as a function of the control parameter a. This indicates that the hyperbolic attractors are not characteristic of the map (10.11). The principle of such an algorithm is that if at a certain value of the parameter a the trajectory does not have the unstable manifold, then $P^{\delta\varphi} = 0$.
(4) Calculate the dependencies on control parameters of the Lyapunov exponents for chaotic orbits.

Hence the results are as follows:

(1) From Fig. 10.5(a), it is clear that in the neighborhood of zero, the probability $P(\phi)$ of the angle ϕ is finite. This implies the presence of tangency points of manifolds, and hence a nonrobust homoclinic curve of saddle cycles exists along which manifolds of the cycles approach each other tangentially.
(2) Figure 10.5(b) shows the dependence of the probability $P^{\delta\varphi}$ as a function of a. Hence, except of a set of a values corresponding to $P^{\delta\varphi}(a) \neq 0$ falling within the neighborhood of zero, there is a denumerable set of a values for which $P^{\delta\varphi}(a) = 0$ that corresponds to windows of stability for periodic orbits.

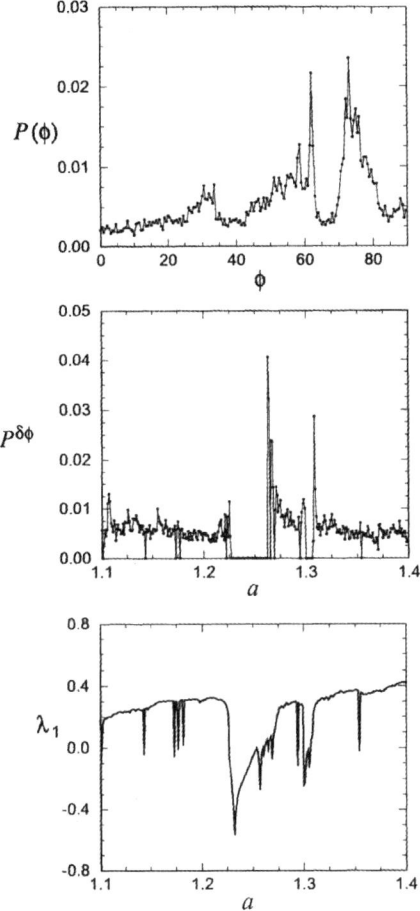

Fig. 10.5 Calculated results for the characteristics for the Hénon attractor given by (10.11). (a) A distribution of the probabilities of the angle ϕ between stable and unstable manifolds for $a = 1.179$ and $b = 0.3$, (b) the probability that the angle ϕ falls within the interval $0 < \phi < 1$ ($\delta\varphi = 1$ degree) versus the parameter a for $b = 0.3$, and (c) dependence of the largest Lyapunov exponent on the parameter a for $b = 0.3$. Reused with permission from Anishchenko, V. and Strelkova, G., Discrete Dynamics in Nature and Society (1998). Copyright 1998, Hindawi Publishing Corporation.

(3) Figure 10.5(c) displays the dependence of the largest Lyapunov exponent of system (10.11) on the variable a and shows a set of jumps to regions of negative value, which correspond to periodic windows. This property is typical of quasi-attractors.

10.3 The Strelkova–Anishchenko map

The so-called *Strelkova–Anishchenko map* [Strelkova & Anishchenko (1997)] also displays a quasi-attractor, and it is a discrete dynamical system in the form of two coupled logistic maps given by

$$\begin{cases} x_{n+1} = 1 - \alpha x_n^2 + \gamma(y_n - x_n) \\ y_{n+1} = 1 - \alpha y_n^2 + \gamma(x_n - y_n). \end{cases} \quad (10.12)$$

The map (10.12) has the following properties:

(1) For $\alpha = 0.9$ and $\gamma = 0.285$, the map (10.12) displays a hyperchaotic regime whose basin of attraction is a bounded rhombus on the parameter plane (x, y) as shown in Fig. 10.6(a).
(2) When one changes a parameter, the number of co-existing attractors increases, and the structure of their basins becomes more complicated. The results are shown in Fig. 10.6(b).
(3) The map (10.12) has the following *reproducibility from initial conditions property*: If one chooses a small box denoted by 2 in Fig. 10.6(b) as a region of uncertainty in initial conditions from the basin of attraction of the quasi-attractor and examines its evolution in time, then a combination of all three attractors of the system (10.12) are obtained and one of the three coexisting regimes will dominate randomly.

10.4 The Anishchenko–Astakhov oscillator

The so-called *Anishchenko–Astakhov oscillator* [Anishchenko (1990-1995)][3] is a three-dimensional differential equation system with two bifurcation parameters. This system is a typical system whose chaotic dynamics fully illustrate Shilnikov's theorem about the properties of dynamical systems with a saddle-focus separatrix loop of the equilibrium state,

$$\begin{cases} x' = mx + y - xz \\ y' = -x \\ z' = -gz + gl(x)x^2, \end{cases} \quad (10.13)$$

where

$$l(x) = \begin{cases} 1, x > 0 \\ 0, x \leq 0. \end{cases} \quad (10.14)$$

---

[3]Or a *modified oscillator with inertial nonlinearity*.

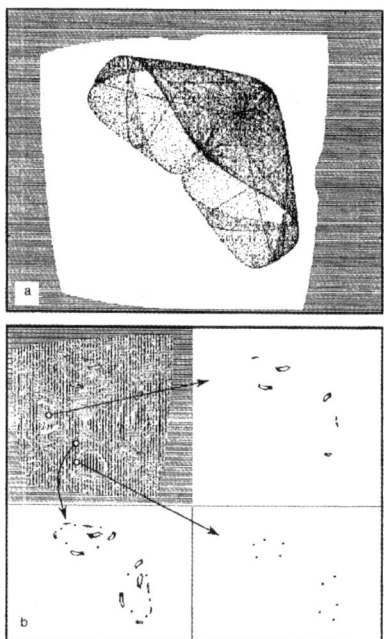

Fig. 10.6 Hyperchaos in the Strelkova–Anishchenko map (10.12) for parameters $\alpha = 0.9$ and $\gamma = 0.285$. (a) Co-existing attractors and their basins of attraction in system (10.12) for $\alpha = 0.78$ and $\gamma = 0.2876$. (b) The regions (in arrows) of uncertainty $1, 2, 3$ in initial conditions leading to the corresponding limit sets. Reused with permission from Anishchenko, V. and Strelkova, G., Discrete Dynamics in Nature and Society (1998). Copyright 1998, Hindawi Publishing Corporation.

For $m = 1.42$ and $g = 0.097$, the LCE spectrum as a function of initial conditions is shown in Figs. 10.7, 10.8, and 10.9, where for $-2.0 < x < 0$, the regime of one of the limit cycles and chaos was observed, while for $-4.0 < x < -2.0$, a limit cycle of another family is added. Figure 10.8 shows projections of the co-existing three attractors with their basins of attraction, namely, period-1 and period-2 limit cycles and the chaotic attractor. This regime was chosen to visualize the complexity of a quasi-attractor with the finite number of coexisting attractors[4]. In the case where $\lambda_1 = 0$, the systems (10.13) and (10.14) demonstrate cascades of period doubling bifurcations. The component λ_2 takes different negative values, and the basins of attraction of different attractors change their structure

---

[4]Theoretically, when parameters are varied slightly, a quasi-attractor includes an infinite number of coexisting limit-cycle regimes which undergo an infinite sequence of different bifurcations.

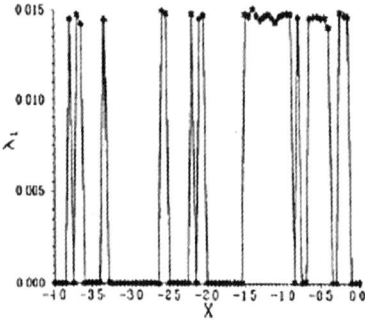

Fig. 10.7 The component λ_1 of the Anishchenko–Astakhov oscillator (10.13) and (10.14) as a function of $-4 \leq x \leq 0$ for the parameters $m = 1.42$ and $g = 0.097$. Reused with permission from Anishchenko, V. and Strelkova, G., Discrete Dynamics in Nature and Society (1998). Copyright 1998, Hindawi Publishing Corporation.

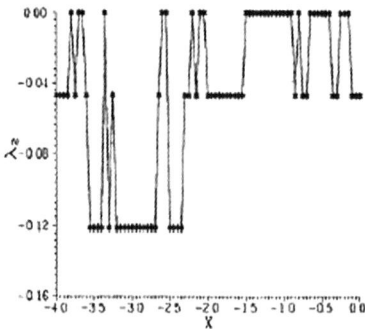

Fig. 10.8 The component λ_2 of the Anishchenko–Astakhov oscillator (10.13) and (10.14) as a function of $-4 \leq x \leq 0$ for the parameters $m = 1.42$ and $g = 0.097$. Reused with permission from Anishchenko, V. and Strelkova, G., Discrete Dynamics in Nature and Society (1998). Copyright 1998, Hindawi Publishing Corporation.

with fractal boundaries. From Figs. 10.7, 10.8, 10.9, and 10.10, it is possible to conclude that system (10.13) and (10.14) is a quasi-attractor. This can be seen from certain periodic components in the ACF and sudden peaks at certain typical characteristic frequencies in the spectrum.

10.5 Chua's circuit

Chua's circuit is one of the famous 3-D dynamical systems because its system of equations is one of the simplest piecewise-linear systems, while

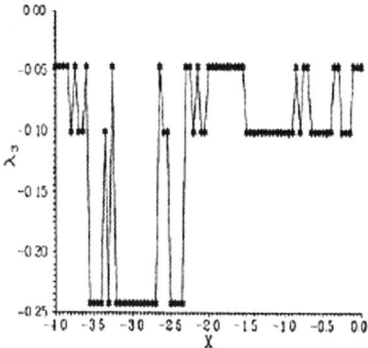

Fig. 10.9 The component λ_3 of the Anishchenko–Astakhov oscillator (10.13) and (10.14) as a function of $-4 \leq x \leq 0$ for the parameters $m = 1.42$ and $g = 0.097$. Reused with permission from Anishchenko, V. and Strelkova, G., Discrete Dynamics in Nature and Society (1998). Copyright 1998, Hindawi Publishing Corporation.

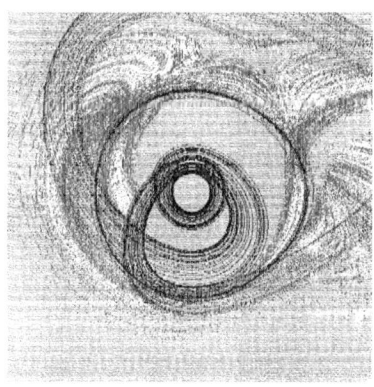

Fig. 10.10 Projections of the co-existing attractors in the Anishchenko–Astakhov oscillator (10.13) and (10.14) for $m = 1.42$ and $g = 0.097$ and the structure of the basins of their attraction in the Poincaré section at $z = 1$. Reused with permission from Anishchenko, V. and Strelkova, G., Discrete Dynamics in Nature and Society (1998). Copyright 1998, Hindawi Publishing Corporation.

dynamically is very complicated and rich in bifurcations and strange attractors. This circuit can display more than 890 chaotic attractors including smooth Chua's systems [Billota et al. (2007)], and the existence of several of them can be proved numerically [Matsumoto (1984), Zhong & Ayrom (1985)], analytically by two independent methods [Chua et al. (1986), Matsumoto et al. (1988)], and experimentally [Matsumoto et al. (1985), Komuro et al. (1991)]. Chua's circuit is the subject of hundreds of papers

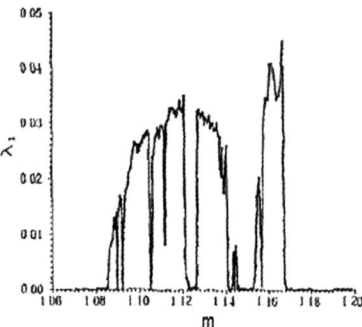

Fig. 10.11 The largest Lyapunov exponent λ_1 as a function of parameter m ($g = 0.2$) for the Anishchenko–Astakhov oscillator (10.13) and (10.14). Reused with permission from Anishchenko, V. and Strelkova, G., Discrete Dynamics in Nature and Society (1998). Copyright 1998, Hindawi Publishing Corporation.

covering all topics related to this system [Madan (1993)]. Chua's circuit is easy to implement for potential applications [Chua (1993), Madan (1993)]. However, the complexity of Chua's circuit led to discovery of a large family of equivalent circuits [Chua & Lin (1990), Altman (1993a), Chua et al. (1993a-b), Pospisil et al. (1995), Chua et al. (1995), Pospisil & Brzobohaty (1996), Pospisil et al. (1999-2000)].

The double scroll system is described by a third-order autonomous differential equation. In particular, we will choose the dimensionless form given by (10.15) and (10.16) below of [Matsumoto et al. (1985)], which we rewrite in the equivalent form

$$\begin{cases} x' = \alpha\,(y - h(x)) \\ y' = x - y + z \\ z' = -\beta y \end{cases} \quad (10.15)$$

where

$$h(x) = m_1 x + \frac{1}{2}(m_0 - m_1)\,(|x+1| - |x-1|) \quad (10.16)$$

is the canonical piecewise-linear equation describing an odd-symmetric, three-segment, piecewise-linear curve having a breakpoint at $x = \pm 1$, a slope equal to $m_0 = a + 1 < 0$ in the inner segment, and a slope $m_1 = b + l > 0$ in the outer segments, respectively, namely,

$$h(x) = \begin{cases} m_1 x + m_0 - m_1, & x \geq 1 \\ m_0 x, & |x| \leq 1 \\ m_1 x - m_0 + m_1, & x \leq -1. \end{cases} \quad (10.17)$$

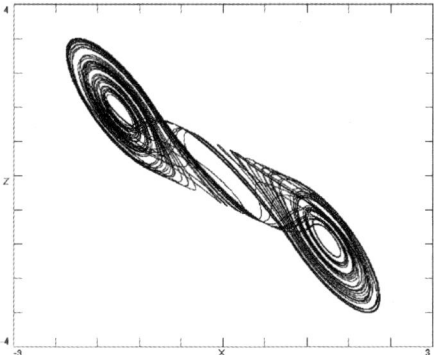

Fig. 10.12 The classic double scroll attractor obtained from (10.15) and (10.16) for $\alpha = 9.35$, $\beta = 14.79$, $m_0 = -\frac{1}{7}$, and $m_1 = \frac{2}{7}$.

The double scroll equation has a chaotic attractor [Chua et al. (1986)] for $\alpha = 9.35$, $\beta = 14.79$, $m_0 = -\frac{1}{7}$, and $m_1 = \frac{2}{7}$ as shown in Fig. 10.12. Chua's system of equations, or specifically the piecewise-linear function h given by (10.17) can be modified to take other forms of smooth or non-smooth functions. This operation is called the *generalization of Chua's circuit*. For example, the original piecewise-linear function is replaced by a discontinuous function in [Mahla & Palhares (1993), Lamarque et al. (1999)], by a C^∞ "*sigmoid function*" in [Brown (1993b)], by a cubic polynomial $h(x) = c_0 x^3 + c_1 x$ in [Altman (1993b), Hartley & Mossayebi (1993), Khibnik et al. (1993)], and by an $x|x|$ function in [Tang et al. (2003)]. These generalizations conserve some properties of the original systems and display other new phenomena [Madan (1992-1993), Chua et al., (1993a-b)]. The number of equilibrium points can be extended enough to generate chaotic attractors with several scrolls [Suykens & Vandewalle (1993), Yalcin (2001-2002), Tang et al. (2001), Aziz-Alaoui (2002)]. For more details see [Billota et al. 2007]. For the rigorous proof of chaos in the system (10.15) and (10.16) see [Zeraoulia & Sprott (2010)] where the Poincaré map was calculated to prove rigorously the existence of chaos in the double scroll family. On the other hand, the Shilnikov criteria introduced for proving chaos in Chua's system (10.15) and (10.16) was applied when there exists a homoclinic or heteroclinic orbit. The existence of these orbits was proved analytically using the piecewise-linear geometry of the vector field associated with Chua's system (10.15) and (10.16). For more explanations, consider the following notations: Vectors in $\mathbb{R}^3$ are denoted by $X = (x, y, z)^T$, σ and ω are the real and imaginary parts of a complex eigenvalue, $\xi(x) : \mathbb{R}^3 \longrightarrow \mathbb{R}^3$ denotes the piecewise-linear vector field evaluated at $x \in \mathbb{R}^3$, and the set $\mathcal{L}$ indicates the generalized family of vector fields ξ, called the *piecewise-linear vector field family*. $E^c(0)$ is a 2-D eigenspace corresponding to complex eigenvalue $\tilde{\sigma}_0 + j\tilde{\omega}_0$ at O. $E^r(0)$ is a 1-D eigenspace corresponding to the real eigenvalue $\tilde{\gamma}_0$ at O. $E^c(P^+)$ is a 2-D eigenspace corresponding to the complex eigenvalue $\tilde{\sigma}_1 + j\tilde{\omega}_1$ at P^+. $E^r(P^+)$ is a 1-D eigenspace corresponding to the real eigenvalue $\tilde{\gamma}_1$ at P^+, and $L_0 = U_1 \cap E^C(0), L_1 = U_1 \cap E^C(P^+), L_2 = \{x \in U_1, \xi(x) \| U_1\}$. Here, $\xi(x) \| U_1$ means that the vector $\xi(x)$ lies on a plane parallel to U_1, and L_2 is a straight line. Second, some geometric properties of the vector fields ξ of the generalized family $\mathcal{L}$ are summarized in the following lemma:

Lemma 10.2. *Any member ξ of the family $\mathcal{L}$ satisfies the following properties:*

(P.0) ξ is a continuous, piecewise-linear vector field.

(P.1) ξ is symmetric with respect to the origin, i.e., $\xi(-x,-y,-z) = -\xi(x,y,z)$ for all $(x,y,z) \in \mathbb{R}^3$.

(P.2) There are two planes U_1, U_{-1} which are symmetric with respect to the origin, and they partition $\mathbb{R}^3$ into three closed regions D_1, D_0, and D_{-1}.

(P.3) In each region D_i, $i = -1, 0, 1$, the vector field ξ is affine, i.e., $D\xi(x,y,z) = M_i$ for all $(x,y,z) \in D_i$, where $D\xi$ denotes the Jacobian matrix of ξ, and M_i denotes a 3×3 real constant matrix.

(P.4) ξ has three equilibrium points, $P^- \in D_{-1}, O \in D_0$, and $P^+ \in D_1$, where $O = (0,0,0) \in \mathbb{R}^3$.

(P.5) Each matrix M_i has a pair of complex conjugate eigenvalues and one real eigenvalue labeled, respectively, by $\tilde{\sigma}_0 \pm j\tilde{\omega}_0$ and $\tilde{\gamma}_0$ for M_0, and $\tilde{\sigma}_1 \pm j\tilde{\omega}_1$ and $\tilde{\gamma}_1$ for M_{-1} and M_1, where $\tilde{\omega}_0 > 0, \tilde{\omega}_1 > 0, \tilde{\gamma}_i \neq 0, i = 0, 1$, and $j^2 = -1$.

(P.6) The eigenspace associated with either the real or the complex eigenvalue at each equilibrium point is not parallel to U_1 or U_{-1}.

For the double scroll Eq. (10.15) and (10.16), there are several types of chaotic attractors which include those cited previously in [Matsumoto (1984), Matsumoto et al. (1986), Parker & Chua (1987), Bartissol & Chua (1988), Silva (1991)], period doubling types, and periodic window types. The first type of chaotic attractor is a Rössler screw-type attractor that is sandwiched between the eigenspace through P^+ and the eigenspace through 0 because it has a screw-like structure first reported by Rössler [Rössler (1979)]. An odd-symmetric image of this attractor has also been observed between the eigenspaces through P^- and O, as expected. These two Rössler screw-type attractors are separated by the eigenspace through O. The second type of chaotic attractor is the *double scroll*, which has already been extensively reported [Matsumoto (1984), Matsumoto et al. (1986)] and which spans all three regions D_{-1}, D_0, and D_1.

Other types of interesting chaotic attractors are discussed here. Indeed, another subfamily of $\mathcal{L}_0$ was studied exactly with the same method as for the double scroll by extending it in a more general fashion than was done in [Chua et al. (1986)] using the *Poincaré half-map technique*[5] allowing one to detect homoclinic and heteroclinic orbits and to locate the region in parameter space for which stable attracting sets exist. In addition, the boundaries between the return/transfer/escape regions and a period-one limit cycle were predicted using these maps [Parker & Chua (1987)].

This family is called the *dual double scroll family*[6], namely, the following definition:

Definition 10.3. The dual double scroll family is the subset of $\mathcal{L}_0$ that satisfy the properties $(P-i)_{i=0,6}$ and that have $\tilde{\gamma}_0 < 0$ and $\tilde{\gamma}_1 > 0$.

The dynamics of the double scroll family [Chua et al. (1986)] and the dynamics of the dual double scroll family [Parker & Chua (1987)] are quite different. For example, in the double scroll family, entry points either transfer or return, while for the dual double scroll family, the opposite stability type of the equilibrium point in region D_1 leads to a third possibility of escape.

The *double-hook family*[7] $\mathcal{F}_s$ [Bartissol & Chua (1988), Silva (1991)] is a derivative of the double scroll family and exhibits chaotic behavior both numerically and experimentally [Bartissol & Chua (1988)] and analytically [Silva (1991)]. The analysis of the double hook is also done using Poincaré maps and searching for homoclinic and heteroclinic orbits, thus proving horseshoes. The double-hook family $\mathcal{F}_s$ is obtained by replacing property (P.5) given in Lemma 10.2 by the following:

(P.5)': M_0 has three real eigenvalues denoted by η_0, μ_0, and ν_0, where $\eta_0 \mu_0 > 0$ and $\eta_0 \nu_0 > 0$, while $M_{\pm 1}$ has a pair of complex conjugate eigenvalues $\sigma_1 \pm j\omega_1$ and one real eigenvalue γ_1 with $\omega_1 > 0$ and $\gamma_1 \neq 0$.

---

[5] These maps are defined with no assumptions on the system dynamics and with the largest possible domain.

[6] The name "dual double scroll family" comes from the fact that this family describes the same circuit as the double scroll equation (10.35) and (10.36) except for the single nonlinear element which is the dual of the double scroll nonlinearity.

[7] Because its basic structure appeared as two "fishhooks" in tandem, connected at their pointed ends.

In this case, $P^{\pm}$ are still unstable saddle-foci, and the origin is a saddle-node instead of a saddle-focus. This implies a dramatic change in the behavior and analysis of the resulting system.

Definition 10.4. The double-hook family $\mathcal{F}_s$, is defined as the subset of $\mathcal{L}_0$ such that

$$\eta_0 < 0, \mu_0 > 0, \nu_0 > 0, \tilde{\sigma}_1 > 0, \text{ and } \tilde{\gamma}_1 < 0. \tag{10.18}$$

Changing the conditions (10.18), one obtains the following definition:

Definition 10.5. The dual double-hook family $\mathcal{F}_s^*$ is the subfamily of $\mathcal{L}_0$ defined by the condition

$$\eta_0 > 0, \mu_0 > 0, \nu_0 < 0, \tilde{\sigma}_1 < 0, \text{ and } \tilde{\gamma}_1 > 0. \tag{10.19}$$

In this case, the stabilities of the various eigenlines and eigenplanes associated with the equilibria O and $P^{\pm}$ are reversed. Thus by some minor sign and flow reversal changes, the results for $\mathcal{F}_s^*$ are found from the analysis of the subfamily $\mathcal{F}_s$. The remaining vector fields in $\mathcal{L}_0$ outside of $\mathcal{F}_s \cup \mathcal{F}_s^*$ cannot exhibit horseshoe chaos through the Shilnikov mechanism because the equilibrium points $P^{\pm}$ are no longer saddle foci.

10.5.1 Homoclinic and heteroclinic orbits

It was shown in [Chua et al. (1986)] that the double scroll family (10.15) and (10.16) is chaotic in the sense of Shilnikov's theorem given in Sec. 5.3.2. In particular, it was shown that there exist parameters such that the trajectory along the unstable real eigenvector $E^r(O)$ from the origin will enter the stable eigenspace $E^c(O)$ and hence will return to the origin. By symmetry, the trajectory along the other unstable real eigenvector would behave in the same way, i.e., there exist homoclinic orbits.

Theorem 10.6. *(Chaos in the double scroll) [Chua et al. (1986)]: The double scroll system (10.15) and (10.16) is chaotic in the sense of Shilnikov's theorem for some parameters m_0, m_1, α, and β. In particular, if $m_0 = -\frac{1}{7}$, $m_1 = \frac{2}{7}$, and $\alpha = 7$, then there exists some β in the range $6.5 \leq \beta \leq 10.5$ such that the hypotheses of Shilnikov's theorem are satisfied.*

The search for a heteroclinic orbit in Chua's circuit (10.15) and (10.16) is done by method similar to the one used for homoclinic orbits introduced in [Mees & Chapman (1987)]. Such a demonstration has been given in [Mees &

Chapman (1985)] where a computer-calculated *hole filling heteroclinic orbit* is obtained for $(\alpha, \beta, m_0, m_1) = (9.439, 13.987, -0.614, -1.256)$. Additional insights and conditions for the appearance of the double scroll attractor are given in [Kahlert & Chua (1987)]. In [Matsumoto et al. (1988)], three key inequalities were studied by giving verifiable error bounds to the quantities involved in the inequalities with the assistance of a computer. This provides another rigorous proof that the double scroll is chaotic in the sense of Shilnikov. Theorem 10.1 implies the existence of a homoclinic orbit at the origin with a horseshoe embedded in a neighborhood of this orbit. Homoclinic orbits through the other two equilibrium points P^+ and P^- can also occur for appropriate parameter values [Mees & Chapman (1987)]. Thus Chua's circuit (10.15) and (10.16) has a positively and negatively invariant Cantor set containing infinitely many saddle-type (unstable) periodic orbits of arbitrarily long periods, uncountably many bounded nonperiodic orbits, and a dense orbit. Moreover, the horseshoe persists under perturbations. Indeed, in [Blazquez & Tuma (1993d)], the chaotic behavior of the solutions of a 3-D dynamical system, in particular Chua's circuit (10.15) and (10.16), was studied in the neighborhood of a homoclinic orbit and for two heteroclinic orbits to an equilibrium point of the saddle-focus type and to saddle-focus points which form a closed contour, respectively. This study suggests that one can find a subfamily of solutions of the considered system such that for each given sequence of natural numbers there exists an orbit of this family such that the *number of turns* around the equilibrium points is given by the sequence in each lap[8]. In particular, there exist infinitely many periodic solutions and homoclinic orbits for those solutions. Consider the equation

$$x' = f(x), x \in \mathbb{R}^3, f \in C^k \ (k \geq 3). \tag{10.20}$$

Assume that there exist two isolated, hyperbolic equilibrium points P_i, $i = 1, 2$ of the saddle-node type with a spectrum $\sigma(L_i) = \{\lambda_i, \rho_i \pm i\omega_i\}$ of the linear part $L_i = Df(P_i)$, $i = 1, 2$. Then one has the following results proved in [Blazquez & Tuma (1993d)]:

Theorem 10.7. *Assume that (1)*

$$\begin{cases} \rho_i < 0, \lambda_i > 0, \omega_i \neq 0, i = 1, 2 \\ \rho_1\omega_2 + \lambda_2\omega_1 > 0, \rho_2\omega_1 + \lambda_1\omega_2 > 0. \end{cases} \tag{10.21}$$

---

[8]And under some conditions, this orbit touches one stable manifold, and it tends to the equilibrium point.

(2) There exist two connections $\Gamma_i = \{p_i(t), t \in \mathbb{R}\}, i = 1, 2$ such that

$$\begin{cases} \lim_{t \to -\infty} p_1(t) = P_1, \lim_{t \to \infty} p_1(t) = P_2 \\ \lim_{t \to -\infty} p_2(t) = P_2, \lim_{t \to \infty} p_2(t) = P_1. \end{cases} \quad (10.22)$$

Then there exists a subsystem of solutions in a one-to-one correspondence with the set

$$\Omega_a = \{(..., j_n, ...) : j_n \in \mathbb{N}^*, j_{n+1} < aj_n\} \quad (10.23)$$

for some real a such that

$$1 < a < \min\left\{-\frac{\lambda_1 \omega_2}{\rho_2 \omega_1}, -\frac{\lambda_2 \omega_1}{\rho_1 \omega_2}\right\}. \quad (10.24)$$

In other words, Theorem 10.2 says that one can find a subfamily of solutions of (10.20) in general, and in Chua's circuit (10.15) and (10.16) in particular, such that for each given sequence of natural numbers there exists an orbit of the family such that the "number of turns" around the equilibrium points is given by the sequence in each lap. In particular, there exists an infinite number of periodic solutions and homoclinic orbits to those periodic solutions.

Theorem 10.8. *Assume that (1)*

$$\begin{cases} \rho_i < 0, \lambda_i > 0, \omega_i > 0, i = 1, 2 \\ \rho_1 \omega_2 + \lambda_2 \omega_1 > 0, \rho_2 \omega_1 + \lambda_1 \omega_2 > 0. \end{cases} \quad (10.25)$$

(2) There exist two connections $\Gamma_i = \{p_i(t), t \in \mathbb{R}\}, i = 1, 2$ such that

$$\begin{cases} \lim_{t \to -\infty} p_1(t) = P_1, \lim_{t \to \infty} p_1(t) = P_2 \\ \lim_{t \to -\infty} p_2(t) = P_2, \lim_{t \to \infty} p_2(t) = P_1. \end{cases} \quad (10.26)$$

Then for every $m \in \mathbb{Z}$, there exists a subsystem of solutions in one-to-one correspondence with the set

$$\Omega_a^m = \{(..., j_m) : j_n \in \mathbb{N}^*, j_{n+1} < aj_n, n < m\}, \quad (10.27)$$

which represents orbits asymptotic to P_1 if $m = 1 \mod(2)$ and to P_2 if $m = 0 \mod(2)$. In other words, Theorem 10.3 says that for each given infinite sequence, there exists a solution with the behavior described in Theorem 10.2, but it touches one stable manifold and it tends to the equilibrium point.

Theorem 10.9. *Assume that (1)*

$$\begin{cases} \rho_i < 0, \lambda_i > 0, \omega_i > 0, i = 1, 2 \\ \lambda_1 \lambda_2 - \rho_1 \rho_2 > 0. \end{cases} \quad (10.28)$$

(2) There exist two connections $\Gamma_i = \{p_i(t), t \in \mathbb{R}\}, i = 1, 2$ such that

$$\begin{cases} \lim_{t \to -\infty} p_1(t) = P_1, \lim_{t \to \infty} p_1(t) = P_2 \\ \lim_{t \to -\infty} p_2(t) = P_2, \lim_{t \to \infty} p_2(t) = P_1. \end{cases} \quad (10.29)$$

Then there exists a subsystem of solutions in a one-to-one correspondence with Ω_a for some real a such that

$$1 < a < \frac{\lambda_1 \lambda_2}{\rho_1 \rho_2}. \quad (10.30)$$

The two Theorems 10.2 and 10.3 can be generalized to a Banach space as follows:

Definition 10.6. *A linear operator A on a Banach space X is said to be sectorial if A is a closed, densely defined operator such that for some $\phi \in \left(0, \frac{\pi}{2}\right)$ and some $\mu \geq 1$ and a real number a, the sector*

$$S_{a,\phi} = \{\lambda : \phi \leq |\arg(\lambda - a)| \leq \pi, \lambda \neq a\} \quad (10.31)$$

is in the solvant set of A and

$$\left\| (\lambda - A)^{-1} \right\| \leq \frac{\mu}{|\lambda - a|}, \text{ for all } \lambda \in S_{a,\phi}. \quad (10.32)$$

Now consider an equation of the form of Chua's system

$$z' + Az = f(z) \quad (10.33)$$

where A is a sectorial operator in a Banach space X and $f \in C^k(X^\alpha, X), 0 \leq \alpha < 1$ such that there exist two equilibrium points P_1 and P_2, and if we linearize about P_i, $i = 1, 2$, we have the operator $L_i = A - Df(P_i)$, $i = 1, 2$. Then one has the following [Blazquez & Tuma (1993d)]:

Theorem 10.10. *Assume that (1) There exists $\mu_1^i, \mu_2^i, ..., \mu_n^i$ eigenvalues of L_i with negative real parts, μ_1^i, μ_2^i complex conjugates and $\mathrm{Re}\left(\mu_j^i\right) < \rho_i = \mathrm{Re}\left(\mu_1^i\right) < 0, i = 1, 2, j = 3, 4, ..., n$.*

(2) The remainder of the spectrum has a positive real part, and there exists a simple eigenvalue $\lambda_i \in \mathbb{R}^+$ such that $\mathrm{Re}(\mu) > \lambda_i > 0$ for $\mu \in \sigma(L) - \{\mu_j^i, \lambda_i\}$.

(3)

$$\begin{cases} \rho_1 \omega_2 + \lambda_2 \omega_1 > 0 \\ \rho_2 \omega_1 + \lambda_i \omega_2 > 0. \end{cases} \quad (10.34)$$

(4) There exist two heteroclinic points Γ^i, $i = 1, 2$. Γ^1 connects P_1 with P_2, and Γ^2 connects P_2 with P_1 such that $\dim\left(W_{c_i}^+ \cap W_{c_i}^-\right) = 1$, where $W_{c_i}^+$ and $W_{c_i}^-$ are the spaces at $c_i \in X$ tangent to the stable and unstable manifolds.

Then there exists a subsystem of solutions in one-to-one correspondence with set

$$\Omega_a = \{(..., j_n, ...) : j_n \in \mathbb{N}^*, j_{n+1} < aj_n\} \qquad (10.35)$$

for some real a such that

$$1 < a < \min\left\{-\frac{\lambda_1 \omega_2}{\rho_2 \omega_1}, -\frac{\lambda_2 \omega_1}{\rho_1 \omega_2}\right\}. \qquad (10.36)$$

Note that the proof of these theorems can be deduced from the proof of Theorem 10.2 based on the return maps defined for the system (10.20). A generalization of these theorems to a Banach space with infinite dimension gives the same results as above in view of the existence of subsystems of solutions in one-to-one correspondence with the sets Ω_a and Ω_a^+ [Blazquez & Tuma (1993a)].

10.5.2 The geometric model

In Sec. 6.3, we discussed the three types of strange attractors, and in this section we continue in a more specific way the presentation of the so-called *geometric model* defined for Chua's Eq. (10.15) and (10.16). Indeed, in [Belykh & Chua (1992)], a type of strange attractor generated by an odd-symmetric three-dimensional vector field with a saddle-focus having two homoclinic orbits at the point O was reported, and it was shown that this type is related to the double scroll, and it is different from the Lorenz-type (see Chapter 8) or quasi-attractors discussed in that chapter. This new type contains unstable points[9] in addition to the Cantor set structure of hyperbolic points. In particular, the corresponding two-dimensional Poincaré map has a strange attractor with no stable orbits. See more details in [Belykh & Chua (1992)]. In conclusion, this geometric model[10] is much more complicated than the Lorenz-type attractors [Afraimovich et al. (1983)].

Thus from the above results discussed in this chapter, it is easy to see that Chua's circuit is very interesting in its mathematical analysis because of the existence of infinitely many stability windows for some parameter values, which implies a sensitive dependence of the structure of the attractor on small variations of parameters [Shilnikov (1993b)].

---

[9]This implies that the points from the stable manifolds of the hyperbolic points must necessarily attract the unstable points.
[10]This model represents the simplest idealization of the double scroll attractor.

(1) Chua's circuit is a universal problem because it exhibits a number of distinct routes to chaos: period doubling bifurcations, breakdown of an invariant torus, intermittency, *etc.*
(2) The *double scroll Chua's attractor* is formed by a pair of nonsymmetric spiral Chua's attractors with three equilibrium states of a saddle-focus type visible in the attractor, which indicates that this attractor is multistructural, making it different from the other existing attractors of 3-D systems discussed in Sec. 6.3.
(3) Chua's Eqs. (10.15) and (10.16) are *close* in the sense that the bifurcation portraits of the equations define a 3-D normal form for bifurcations of an equilibrium point with three zero characteristic exponents for the case with additional symmetry and that of a periodic orbit with three multipliers equal to -1.
(4) The results given in [Ovsyannikov & Shilnikov (1991), Gonchenko *et al.* (1993)] imply that the attractors that occur in Chua's circuit (10.15) and (10.16) are new and essentially more complicated objects in their mathematical nature than it seemed before. This complication is due to the presence of structurally unstable Poincaré homoclinic orbits in either the attractor itself, or in an attractor of a nearby system[11].

Therefore, a *complete description* of the dynamics and bifurcations in Chua's equations is impossible, as it is for many other models because it was shown in [Gonchenko *et al.* (1993)] that systems with infinitely many structurally unstable periodic orbits of any degree of degeneracy are dense in the Newhouse regions. This complicated behavior is a result of the existence of structurally unstable Poincaré homoclinic orbits in the attractor itself, or in an attractor of a nearby system.

10.6 Exercises

Exercise 210. Show by example that quasi-attractors have the following properties:

(a) They separatrix loops of saddle-focuses or homoclinic orbits of saddle cycles occur at the point of tangency of their stable and unstable manifolds.

(b) A map of *Smale's horseshoe-type* appears in the neighborhood of their trajectories.

---

[11] These orbits arise from the tangency of the stable and the unstable manifolds of some saddle periodic orbit (cycles).

(c) The quasi-attractor is the unified limit set of the whole attracting set of trajectories including a subset of both chaotic and stable periodic trajectories which have long periods and weak and narrow attractor basins and stability regions.

(d) The basins of attraction of stable cycles are very narrow.

(e) Some orbits do not ordinarily reveal themselves in numerical simulations except for some quite large stability windows where they are clearly visible.

Exercise 211. Explain the principal causes generating quasi-attractors.

Exercise 212. Give an example of a 2-D map with several coexisting attractors, and discuss the possible structures of their basin boundaries.

Exercise 213. Show that the Hénon map (10.1) has the following properties:

(a) It has two fixed points, it is invertible, it is conjugate to its inverse, and its inverse is also a quadratic map. The parameter b is a measure of the rate of area contraction (dissipation).

(b) It is the most general 2-D quadratic map with the property that the contraction is independent of the variables x and y.

(c) It reduces to the quadratic map when $b = 0$, which is conjugate to the logistic map (6.25).

(d) It has bounded solutions over a range of a and b values, and a portion of this range (about 6%) yields chaotic solutions.

(e) For $a = 1.4$ and $b = 0.3$, the Hénon map (10.1) has a chaotic attractor with a correlation dimension of 1.25 ± 0.02 and a capacity dimension of 1.261 ± 0.003.

Exercise 214. Prove the Lemma 10.1.

Exercise 215. (a) Show and illustrate the fact that all the limit subsets of the Anishchenko–Astakhov oscillator (10.13) and (10.14) undergo bifurcations as the parameters are varied. Hint: Use the nonrobustness of the system, and show numerically the dependence of the LCE spectrum exponents on the parameter m.

(b) Using the autocorrelation function (ACF) introduced in Sec. 3.3.2 and the power spectrum PSD, show the presence of stable and saddle cycles in the quasi-attractor at $m = 1.5$ and $g = 0.2$ together with chaotic limit subsets that manifest themselves in the structure of the Anishchenko–Astakhov oscillator (10.13) and (10.14).

(c) Show that the ACF decreases exponentially on average with time and that the power spectrum is continuous.

(d) Plot the probability distribution density $\rho(x,y)$ for the chaotic regime in the Anishchenko–Astakhov oscillator (10.13) and (10.14) for $m = 1.5$ and $g = 0.2$ in the absence of noise and in the presence of additive noise with intensity $d = 10^{-3}$ introduced to all the systems (10.13) and (10.14).

Exercise 216. Find regions in the parameter space of Chua's circuit (10.15) and (10.16) with multiple attractors.

Exercise 217. Give an example of the dual double scroll attractor. Hint: Show that the parameters $\alpha = 9, \beta = 13.8, m_0 = \frac{2}{7}$, and $m_1 = -\frac{1}{7}$ satisfy the conditions of Definition 10.3.

Exercise 218. Draw the geometry associated with the canonical dynamical vector field ξ for the double-hook family $\mathcal{F}_s$, and give an example of a typical orbit. Hint: Use the parameters $\alpha = -4.9150, \beta = -3.6425$, $m_1 = -1.3475$, and $m_0 = -0.5013$.

Exercise 219. (Open problem) The mathematical description of quasi-attractors is still an open problem because almost all nonhyperbolic attractors are obscured by noise.

Exercise 220. (Open problem) [Palis (2008)]

(a) Are there values of the parameters a and b of the Hénon map (10.1) ($b \neq 0$) and a nondegenerate continuum Λ such that $H(\Lambda) = \Lambda$ and $H|_\Lambda$ is transitive. See [Barge & Kennedy (2007)].

(b) Let $f : M \to M$ be a diffeomorphism of the two-dimensional manifold M with $p \in M$ a hyperbolic fixed saddle ($Df(p)$ having eigenvalues λ_1 and λ_2 with $0 < |\lambda_1| < 1 < |\lambda_2|$). Suppose that there is a point q in the intersection of one branch of the unstable manifold $W^u_+(p)$ and the stable manifold $W^s(p)$ and that the intersection of $W^u_+(p)$ with $W^s(p)$ at q is not topologically transverse. Is $cl(W^u_+(p))$ not chainable? See [Barge (1987)].

(c) Under what conditions (if any) on the continuous map $f : I \to I$ of the compact interval I is there an embedding of the inverse limit space (I, f) into the plane so that the shift map $\tilde{f} : (I, f) \to (I, f)$ extends to a diffeomorphism of the plane?

(d) If M is a nonseparating plane continuum and $f : M \to M$ is a mapping, does f have a periodic point?

(e) Let $\{p_1, p_2, ..., p_n\}$ be a set of $n \geq 2$ distinct points in the sphere $\mathbb{S}^2$. Is there a homeomorphism of $\mathbb{S}^2 - \{p_1, p_2, ..., p_n\}$ such that every orbit of the homeomorphism is dense?

(f) Is there a homeomorphism of $\mathbb{R}^n$, $n \geq 3$ such that every orbit of the homeomorphism is dense?

Exercise 221. (Open problem)

(a) Is there a positive Lyapunov exponent for the standard map, *i.e.*, the map of the two-dimensional torus defined by $(x,y) \to (2x - y + K\sin 2\pi x, x)(\mathrm{mod}\,\mathbb{Z}^2)$? See [Duarte (1994), Baraviera (2000), Benedicks (2002)].

(b) Are there for Lebesgue almost every parameter (a,b) in $\{(a,b) \in R2 : 0 < a < 2, b > 0\}$ at most finitely many coexisting strange attractors and stable periodic orbits for the Hénon map (10.1)? See [Colli (1998), Benedicks (2002)].

(c) For the parameter values for which only finitely many Hénon-like attractors or sinks coexists, do the respective basins cover Lebesgue almost all points of the phase space? See [Benedicks (2002)].

(d) Is the set of parameters for which the Hénon map (10.1) is hyperbolic dense in the parameter space? See [Graczyk & Swiatek (1997), Lyubich (1997), Lyubich (2000), Benedicks (2002)].

Chapter 11

Robust Chaos in One-Dimensional Maps

In this chapter, we give several methods to generate robust chaos in one-dimensional maps. The most mathematical tool used here is the unimodality of real functions in an interval $I \subset \mathbb{R}$ introduced in Sec. 11.1 because the most important topological property of a unimodal map is that it stretches and folds the interval I into itself. The importance of S-unimodal maps is discussed in Sec. 11.1.1. The relation between unimodality and hyperbolicity is presented in Sec. 11.1.2. Classification of unimodal maps of the interval is done in Sec. 11.1.3, with their statistical properties in Sec. 11.1.5. In Sec. 11.1.4, we present the concept of unimodal Collet–Eckmann maps with their properties and their relation with hyperbolicity. In Sec. 11.2, we describe a conjecture of Barreto–Hunt–Grebogi–Yorke claiming that robust chaos cannot appear in smooth systems. Counter examples of this conjecture can be found in Secs. 11.2.1, 11.2, and 11.3. The relation between border-collision bifurcation and robust chaos is discussed in some detail in Sec. 11.3. Some exercises and open problems are given in Sec. 11.4.

11.1 Unimodal maps

In this section, we discuss the concept of unimodal maps with their properties and importance in generating robust chaos. First, we recall some basic tools concerning periodic points of one-dimensional maps. In particular, periodic points and their stability are defined through their Jacobian derivatives.

Definition 11.1. A point x is periodic with period-p if $f^p(x) = x$ and p is the smallest positive integer with this property. The stability of a period-p

orbit is governed by:

$$m = (f_p)'(x) = f'(x_p)f'(x_{p-1})...f'(x_1) \qquad (11.1)$$

where the derivatives are evaluated at each point of the orbit $x_{k+1} = f(x_k)$. Using relation (11.1), it is possible to classify periodic orbits as follows: superstable, if $m = 0$ (this is equivalent to the condition that the orbit contains a critical point); stable, if $0 < |m| < 1$; neutral, if $|m| = 1$; and unstable, if $|m| > 1$. For two-dimensional maps (see Sec. 11.2), the identification of the *spine locus* (an idea in smooth higher-dimensional maps that is an analogue of critical points in one dimension) is more involved. Let $x \in \mathbb{R}^2 \to F(x, a)$, such that F has n parameters so that $a \in \mathbb{R}^n$. Hence the following:

Definition 11.2. A period-p orbit is asymptotically stable if $|\lambda_i| < 1, i = 1, 2$ where the λ'_is are the eigenvalues of M, the Jacobian matrix of the p-times iterated map

$$M = DF^p(x) = DF(x_p)DF(x_{p-1})...DF(x_1). \qquad (11.2)$$

For smooth, one-dimensional maps, a periodic window is constructed around a spine locus, which corresponds to parameter values that display superstable periodic orbits as discussed in [Barreto et al. (1997)]. The concept of a *topologically attracting set* has a crucial role in the classification of unimodal maps given in Sec. 11.1.3.

Definition 11.3. A periodic point x of period-p is *topologically attracting* if its basin of attraction $B(x) = B(\{x, f(x), ..., f^{p-1}(x)\})$ has a nonempty interior.

It is clear that stable and superstable periodic orbits are attracting. A neutral periodic orbit may or may not be attracting. Secondly, we give several definitions and important results about the concept of unimodality and its dynamical consequences, in particular, the essential types of attractors that may appear for unimodal maps, namely, a nonrepelling periodic orbit, or a transitive cycle of intervals, or a Cantor set of the solenoid type.

Definition 11.4. A continuous map $f(x) : I \to I$ is unimodal on the interval $I = [a, b]$ if a is a fixed point with b its other preimage ($f(a) = f(b) = a$), and there is a unique maximum at $c \in (a, b)$ such that $f(x)$ is strictly increasing on $x \in [a, c]$ and strictly decreasing on $x \in (c, b]$.

In this case, the only maximum value of $f(x)$ is $f(c)$, and there are no other local maxima. If c is a periodic point for the map f, then it is a so-called *super attractor*.

Definition 11.5. The critical point c is called nondegenerate if the second derivative at the point is not zero.

Definition 11.6. The map f is regular if either the ω-limit set of its critical point c does not contain neutral periodic points or the ω-limit set of c coincides with the orbit of some neutral periodic point. In other words, a unimodal map f is said to be regular if it is hyperbolic (recall Definition 6.1) and if its critical point is nondegenerate and is not periodic or quasi-periodic.

In this case, the set of regular maps in Definition 11.6 coincides with the set of unimodal maps which are structurally stable. Recall Definition 4.3 and see [Kozlovski (2003)] and Theorem 11.2(b) below, *i.e.*, the class of regular maps is open in the C^2 topology and dense in any smooth, and even analytic, topology. For example, if the map has negative Schwarzian derivative (11.4) below, then this map is regular. Example of unimodal functions include the case of quadratic polynomials of the form $a_1 x^2 + b_1 x + c_1, 2x^2 - y^2$, and $xy + xz + yz$, the logistic map (6.25), and the tent map

$$x_{n+1} = \begin{cases} ax_n, \text{ for } x_n < \tfrac{1}{2} \\ a(1 - x_n), \text{ for } x_n \geq \tfrac{1}{2} \end{cases} \quad (11.3)$$

where $a > 1$. The map (11.3) is continuous but not smooth because of the corner at $x = \tfrac{1}{2}$. The map (11.3) generates robust chaos (in the sense of Definition 2.5) as shown in [Alligood *et al.* (1996)]. In this case, the Barreto–Hunt–Grebogi–Yorke Conjecture 11.2 below cannot be applied because map (11.3) is not a smooth system. From the above two examples, a unimodal function can be defined only at discrete values, can be continuous or discontinuous, can have discontinuous first derivatives, *etc.*

11.1.1 S-unimodal maps

In this section, we present some important results about the subject of S-unimodal maps.

Definition 11.7. A map $f : I = [a, b] \longrightarrow I$ is S-unimodal on the interval I if: (a) The function $f(x)$ is of class C^3. (b) The point a is a fixed point with b its other preimage, i.e., $f(a) = f(b) = a$. (c) There is a unique maximum at $c \in (a, b)$ such that $\varphi(x)$ is strictly increasing on $x \in [a, c)$ and strictly decreasing on $x \in (c, b]$. (d) The function f has a negative Schwarzian derivative, i.e.,

$$S(f, x) = \frac{f'''(x)}{f'(x)} - \frac{3}{2}\left(\frac{f''(x)}{f'(x)}\right)^2 < 0 \tag{11.4}$$

for all $x \in I - \{y, f'(y) = 0\}$.

Since only the sign of $S(f, x)$ is used, one can use the product

$$\hat{S}(f, x) = 2f'(x) f'''(x) - 3(f''(x))^2, \tag{11.5}$$

which has the same sign as $S(f, x)$. For example, consider the logistic map (6.25). Obviously, the logistic map is an S-unimodal map on $[0, 1]$. Since it has a unique maximum at $c = \frac{1}{2}$ and negative Schwarzian derivative, there can be at most one attracting periodic orbit with the critical point in its basin of attraction for any $\mu \in [0, 4]$. When $\mu = 4$, the orbit with initial value $x_0 = c$ maps into two iterates to the fixed point $x = 0$, which is unstable $|h_4'(0)| = 4 > 1$. We apply Theorem 11.1 below and conclude that $h_4(x)$ does not have any stable periodic orbits, and there is a unique chaotic attractor. However, if the parameter $\mu = 4$ is slightly modified, stable periodic orbits appear. Therefore, we say that this map generates fragile chaos. In [Blokh & Lyabich (1991)], some results on one-dimensional measurable dynamics of one-dimensional S-unimodal maps can be found with their proofs. In particular, it was shown that if f is an S-unimodal map of the interval having no limit cycles, then the map f is ergodic with respect to the Lebesgue measure and has a unique attractor Λ in the sense of Milnor and coincides with the conservative kernel of f. Furthermore, there are no strongly wandering sets of positive measure and if f has a finite measure, then it has positive entropy. The importance of S-unimodal maps in chaos theory comes from the theorem given in [Singer (1978)] claiming that each attracting periodic orbit attracts at least one critical point or boundary point. Thus as a result, an S-unimodal map can have at most one periodic attractor which will attract the critical point. This result is used to formulate Theorem 11.1 below with its proof given in [Andrecut & Ali (2001b)].

Theorem 11.1. *Let $f(x) : I \to I$ be an S-unimodal map on the interval $I = [a, b]$, then each attracting periodic orbit attracts at least one critical point or boundary point. Furthermore, each neutral periodic orbit is attracting.*

An important result can be derived from Theorem 11.1 as follows:

Corollary 11.1. *An S-unimodal map can have at most one periodic attractor which will attract the critical point.*

Some examples can be found in Secs. 11.2 and 11.3.

11.1.2 Relation between unimodality and hyperbolicity

In this section, we discuss the possible existing relations between unimodality and hyperbolicity. This fact is very important in our case because hyperbolicity implies structural stability, which is a robustness criterion in the sense of robust chaos used in this book. Indeed, it was shown in [Kozlovski (2003)] that the C^k structural stability Conjecture 11.2 is true for unimodal maps. In other words, it was proved that axiom A maps (recall Definition 6.3) are dense in the space of C^k unimodal maps in the C^k topology (recall Definition 4.2) for $k = 1, 2, ..., \infty, \omega$. The symbol ω indicates that the map is analytic. Let $C^\omega(\Delta)$ denote the space of real analytic functions defined on the interval which can be holomorphically extended to a Δ-neighborhood of this interval in the complex plane. Thus the following results were proved in [Kozlovski (2003)]:

Theorem 11.2. *(a) Axiom A maps (recall Definition 6.3) are dense in the space of $C^\omega(\Delta)$ unimodal maps in the $C^\omega(\Delta)$ topology (Δ is an arbitrary positive number). (b) A C^k unimodal map f is C^k structurally stable (recall Definition 4.3) if and only if the map f satisfies the axiom A conditions and its critical point is nondegenerate and nonperiodic, $k = 2, ..., \infty, \omega$. (If $k = \omega$, then one should consider the space $C^\omega(\Delta)$).*

On one hand, Theorem 11.2(a) gives only global perturbations of a given map, *i.e.*, to perturb a map in a small neighborhood of a particular point and to obtain a nonconjugate map. On the other hand, in Theorem 11.2(b), the number k is greater than one because any unimodal map can be C^1 perturbed to a nonunimodal map, and thus there are no C^1 structurally stable unimodal maps because the topological conjugacy preserves the number of turning points. The same reason implies that the critical point of a structurally stable map should be nondegenerate. The following results proved also in [Kozlovski (2003)] discuss sufficient conditions for *combinatorially equivalence*, which is a weaker version of the concept of topological conjugacy because if two maps are topologically conjugate, then they are combinatorially equivalent.

Theorem 11.3. *Let X be an interval and $f_\lambda : X \to X$ be an analytic family of analytic unimodal regular maps with a nondegenerate critical point, $\lambda \in \Omega \subset \mathbb{R}^N$ where Ω is an open set. If the family f_λ is nontrivial in the sense that there exist two maps in this family which are not combinatorially equivalent, then axiom A maps are dense in this family. Moreover, let T_{λ_0} be a subset of Ω such that the maps f_{λ_0} and $f_{\lambda'}$ are combinatorially equivalent for $\lambda' \in T_{\lambda_0}$, and the iterates of the critical point of f_{λ_0} do not converge to some periodic attractor. Then the set T_{λ_0} is an analytic variety. If $N = 1$, then $T_{\lambda_0} \cap Y$, where the closure of the interval Y is contained in Ω, has finitely many connected components.*

In other words, two unimodal maps f and $\tilde{f}$ are *combinatorially equivalent* if there exists an order-preserving bijection $h : \cup_{n \geq 0} f^n(c) \to \cup_{v \geq 0} \tilde{f}^v(\tilde{c})$ such that $h(f^n(c)) = \tilde{f}^n(\tilde{c})$ for all $n \geq 0$, where c and $\tilde{c}$ are critical points of f and $\tilde{f}$, i.e., the mappings f and $\tilde{f}$ are combinatorially equivalent if the order of their forward critical orbit is the same. The proof of the above results in this section uses the following strategies:

(1) Prove the C^k structural stability conjecture only for analytic maps with nondegenerate critical point, i.e., $k = \omega$, because the analytic maps are dense in the space of C^k maps.
(2) Use the tools described in [Kozlovski (2000)] to avoid the proof of the fact that Schwarzian derivative is negative, i.e., use the concept of nonflat critical point (recall Definition 11.8) instead of the negative Schwarzian derivative condition (11.4).
(3) Use the concept of *Epstein class*, and estimate the sum in some power greater than 1^1 of lengths of intervals from an orbit of some interval. Use Lemma 2.4 in [de Faria & de Melo (2000)] to estimate the shape of *pullbacks* of disks. But in the case of the above results, the estimation of the sum is not needed.
(4) Prove the renormalization theorem, i.e., for a given unimodal analytical map with a nondegenerate critical point, there is an *induced holomorphic polynomial-like map*. The proof is done if one generalizes the notion of polynomial-like maps for this finitely renormalizable map because one can show that the classical definition does not work in this case for all maps. The similar result for infinitely renormalizable maps was proved in [Levin & van Strien (1998)].

---
[1]This sum is small if the last interval in the orbit is small.

(5) Use the *method of quasi-conformal deformations* to construct a perturbation of any given regular analytic map, and show that any analytic map can be included in a nontrivial analytic family of unimodal regular maps.

More general results about the density of hyperbolicity in dimension-one are given in [Kozlovski *et al.* (2007)] as follows:

Theorem 11.4. *(a) (Density of hyperbolicity for real polynomials). Any real polynomial can be approximated by hyperbolic real polynomials of the same degree. (b) (Density of hyperbolicity in the C^k topology). Hyperbolic (i.e., axiom A) maps are dense in the space of C^k maps of the compact interval or the circle, $k = 1, 2, ..., 1, \omega$.*

Theorem 11.4(a) is a solution of the second part of Smale's eleventh problem for the 21st century given in [Smale (2000)]. More details about the proofs and problems related to the topic of density of hyperbolicity can be found in [Jakobson (1971), Lehto & Virtanen (1973), Singer (1978), Mané *et al.* (1983), Douady & Hubbard (1985), Bers & Royden (1986), Martens (1990), Yoccoz (1990), Lyubich (1991-1993-1994), Slodkowski (1991), Sullivan (1991-1997), Martens *et al.* (1992), de Melo & van Strien (1993), Graczyk & Swiatek (1997), Blokh & Misiurewicz (1998a-b), Levin & van Strien (1998), Shishikura (1998), de Faria & de Melo (2000), Kozlovski (2000-2003), Kozlovski *et al.* (2007)]. This list of references contains the best known and basic work in this field.

11.1.3 *Classification of unimodal maps of the interval*

In this section, we give a recent result about the classification of unimodal maps of the interval in terms of the types of their possible attractors. Before that, we need the following definition:

Definition 11.8. *The critical point of f is C^r nonflat of order l if, near c, f can be written as $f(x) = \pm |\phi(x)|^l + f(c)$ where ϕ is a C^r diffeomorphism. The critical point is C^r nonflat if it is C^n nonflat of some order $l > 1$. The set of critical points of f is denoted by $Crit$.*

An example of such a situation is the case of a C^4 unimodal map f with a critical point c such that $f''(c) \neq 0$. In this case, the point c is C^3 nonflat of order two, this last type of points was used in an essential way to exclude the possibility of *wild Cantor attractors* [Bruin *et al.* (1996)].

A recent result for the problem of classification of the measure theoretic attractors of general C^3 unimodal maps with quadratic critical points was given in [Graczyk et al. (2004)] using essentially the concept of the *decay of geometry*. A part of this work is a positive answer of an old question known as *Milnor's problem*, whose subject is whether the metric and topological attractors coincide for a given smooth unimodal map. Before stating this classification, we need the following definitions about *restrictive intervals*, *renormalizations*, *metric attractors*, and *topological attractors*.

Definition 11.9. A compact interval J is restrictive if J contains the critical point of f in its interior, and, for some $n > 0$, $f^n(J) \subseteq J$, and $f^n|_J$ is unimodal. In particular, f^n maps the boundary of J into itself. This restriction of f^n to J is called a renormalization of f. We say that f is infinitely renormalizable if it has infinitely many restrictive intervals.

As mentioned before, it was shown in [Graczyk et al. (2004)] that attractors of unimodal maps are either a nonrepelling periodic orbit, or a transitive cycle of intervals, or a Cantor set of the solenoid type. This classification is based first on the work of Mané [Mané (1985)] and the problem of classification of metric attractors containing the nondegenerate critical point announced in [Graczyk et al. (2001)].

Definition 11.10. (a) A transitive cycle of intervals is a finite union C of compact intervals such that C is invariant under f, C contains the critical point of f in its interior, and the action of f on C is transitive (it has a dense orbit).

(b) We say that f has a Cantor set of the solenoid type if f is infinitely renormalizable, the solenoid then being the ω-limit set of the critical point.

(c) A forward invariant compact set A is called a (minimal) metric attractor for some dynamics if the basin of attraction $B(A) = \{x : \omega(x) \subset A\}$ of A has positive Lebesgue measure and $B(A^c)$ has Lebesgue measure zero for every forward invariant compact set A^c strictly contained in A.

(d) A set is *nowhere dense* if its closure has an empty interior, and *meager* if it is a countable union of nowhere dense sets.

(e) A forward invariant compact set A is called a (minimal) topological attractor if $B(A)$ is not meager while $B(A^c)$ is meager for every forward invariant compact set A^c strictly contained in A.

Milnor's problem mentioned above has a long and turbulent history; see for example [de Melo & van Strien (1993), Lyubich (1994), Graczyk & Swiatek (1999), Graczyk et al. (2005)]. An example of a recent solution of this problem can be found in [Lyubich (1994)] for the class of C^3 unimodal maps with negative Schwarzian derivative and a quadratic critical point, but this solution does not provide a complete proof. A correct solution using different techniques can be found in [Graczyk et al. (2005)]. A negative solution when the critical point has high order is given in [Bruin et al. (1996)]. The solution of Milnor's problem in the generic case can be found in [Kozlovskii (2008)] where it was shown that a generic C^3 unimodal map has finitely many metric attractors which are all attracting cycles using the C^3 stability theorem given in [Kozlovskii (1998)]. In [Graczyk et al. (2004)], Milnor's problem was solved for smooth unimodal maps with a quadratic critical point as done in Theorem 11.5 below. This solution is based on two key developments. The first, given in [Graczyk et al. (2005)], uses decay of geometry for a class of C^3 nonrenormalizable box mappings with finitely many branches and negative Schwarzian derivative everywhere except at the critical point which must be quadratic. The second as given in [Kozlovskii (2000)], recovers a negative Schwarzian derivative for smooth unimodal maps with nonflat critical points, *i.e.*, the first return map to a neighborhood of the critical value has negative Schwarzian derivative. Finally, in [Graczyk et al. (2004)], the full proof was given with an explanation of the structure of metric attractors not containing the critical point based on the following work of Mané [Mané (1985)]:

Theorem 11.5. *Let I be a compact interval and $f : I \to I$ be a C^3 unimodal map with C^3 a nonflat critical point of order two. Then the ω-limit set of Lebesgue almost every point of I is either*
1. *a nonrepelling periodic orbit, or*
2. *a transitive cycle of intervals, or*
3. *a Cantor set of the solenoid type.*

Corollary 11.2. *(a) Every metric attractor of f is either*
1. *a topologically attracting periodic orbit, or*
2. *a transitive cycle of intervals, or*
3. *a Cantor set of the solenoid type.*
There is at most one metric attractor of type other than 1.
(b) The metric and topological attractors of f coincide.

Theorem 11.6. *Let I be a compact interval and $f : I \to I$ be a C^3 unimodal map with a C^3 nonflat and nonperiodic critical point. Then there exists a real analytic diffeomorphism $h : I \to I$ and an (arbitrarily small) open interval U such that, putting $g = h \circ f \circ h^{-1}$, U is a regularly returning* (for g) neighborhood of the critical point of g, and the first return map of g to U has a uniformly negative Schwarzian derivative.

Proposition 11.1. *Let I be a compact interval and $f : I \to I$ be a C^3 unimodal map with a C^3 nonflat critical point of order 2. Then for Lebesgue almost every $x \in I$, either $\omega(x)$ does not contain a critical point or $\omega(x)$ coincides with either*
1. *a super-attracting periodic orbit, or*
2. *a transitive cycle of intervals, or*
3. *a Cantor set of the solenoid type.*

More details about the topic of classification of unimodal maps of the interval can be found in [Mané (1985), Bruin et al. (1996), Graczyk et al. (2001), Graczyk et al. (2004), de Melo & van Strien (1993), Lyubich (1994), Kozlovskii (1998), Graczyk & Swiatek (1999), Kozlovskii (2000), Graczyk et al. (2005), Kozlovskii (2008)].

11.1.4 Collet–Eckmann maps

First, we give the following definitions about the concept of *unimodal Collet–Eckmann maps* with their properties and their relation to hyperbolicity. Indeed, let B^k be a closed unitary ball in $\mathbb{R}^k$.

Definition 11.11. A k-parameter family of unimodal maps is a map $\Gamma : B^k \times I \to I$ such that for $p \in B^k, \gamma_p(x) = \Gamma(p,x)$ is a unimodal map. Such a family is said to be C^n or analytic, according to Γ being C^n or analytic.

The natural topology can be introduced in spaces of smooth families (C^n with $n = 2, ..., \infty$), but it is not necessary to introduce any topology in the space of analytic families.

Definition 11.12. (a) A unimodal map f is called Collet–Eckmann (CE) if there exist constants $C > 0, \lambda > 1$ such that for every $n > 0$,

$$|Df^n(f(0))| > C\lambda^n. \tag{11.6}$$

(b) A unimodal map f is called *backwards Collet–Eckmann* (BCE) if there exist constants $C > 0, \lambda > 1$ such that for any $n > 0$ and any x with $f^n(x) = 0$, we have

$$|Df^n(x)| > C\lambda^n. \tag{11.7}$$

In this case, the unimodal Collet–Eckmann and the backwards Collet–Eckmann maps are strongly hyperbolic along the critical orbit. By a result of Nowicki (see [de Melo & van Strien (1993)]) for S-unimodal maps, Collet–Eckmann maps are also backwards Collet–Eckmann maps. Collet–Eckmann maps have the following properties:

(1) They are characterized by a positive Lyapunov exponent for the critical value.
(2) They have the *best possible* near hyperbolic properties: exponential decay of correlations, validity of central limit and large deviations theorems, good spectral properties, and zeta functions [Keller & Nowicki (1992), Young (1992)].
(3) They have a robust statistical description, with a good understanding of stochastic perturbations: strong stochastic stability [Baladi & Viana (1996)] and rates of convergence to equilibrium ([Baladi et al. (2002)].

There are several ways to estimate how fast is the recurrence of the critical orbit, two of which are polynomial recurrence (P) which holds if there exists an $\alpha > 0$ such that

$$|f^n(0)| > n^{-\alpha} \quad (11.8)$$

for n sufficiently large, and subexponential recurrence (SE) which holds if for all $\alpha > 0$,

$$|f^n(0)| > e^{-\alpha n} \quad (11.9)$$

for n sufficiently large. An example of the Collet–Eckmann map is the map given by

$$f_c : z \to z^d + c \quad (11.10)$$

which was studied in [Avila et al. (2008)] where it was shown that for real parameters, the map (11.10) is either hyperbolic, or Collet–Eckmann, or infinitely renormalizable, and for some complex parameters the map (11.10) is either hyperbolic or infinitely renormalizable. The method of proof is based on controlling the spacing between consecutive elements in the *principal nest* of parapuzzle pieces. Indeed, for $d \geq 2$ and fixed, let $M = M_d = \{c \in \mathbb{C}, \text{ the Julia set of } f_c \text{ is connected}\}$ be the corresponding *multibrot set* (the set of values in the complex plane whose absolute value remains below some finite value throughout iterations of the above map). First, note that when $c \notin M$ or d is odd and $c \in \mathbb{R}$, then the dynamics of the map (11.10) is always trivial. Second, the most important dynamics of map (11.10) was observed when $c \in M$ and d is even because in this case, if $c \in M \cap \mathbb{R}$, then the map f_c is a unimodal map. The following results were proved in [Avila et al. (2008)]:

Theorem 11.7. *(a) For almost every $c \in M_d \cap \mathbb{R}$, the map (11.10) is either regular, or Collet–Eckmann, or infinitely renormalizable. (b) For almost any $c \in \mathbb{C}$, the map (11.10) is either hyperbolic or infinitely renormalizable.*

Corollary 11.3. *For any even criticality d, the set of parameters $c \in M_d \cap \mathbb{R}$ for which the wild attractor exists has zero Lebesgue measure.*

More details about the topic of Collet–Eckmann maps can be found in [Keller & Nowicki (1992), Young (1992), de Melo & van Strien (1993), Baladi & Viana (1996), Baladi et al. (2002-2008)].

11.1.5 Statistical properties of unimodal maps

In this section, we describe some statistical properties of unimodal maps. We begin briefly with the case of the quadratic maps. It was shown by Jakobson and Benedicks-Carleson, that nonregular unimodal maps correspond to a positive measure set of parameters in a large (C^2 open) set of parameterized families. In [Lyubich (2002), Avila & Moreira (2005)] are some results about the dynamics of typical nonregular quadratic maps. In particular, a description of the asymptotic behavior of the critical orbit is done in [Avila & Moreira (2005)]. In Theorem 11.8(a) below, an estimate of hyperbolicity is given using the Collet–Eckmann condition. In Theorem 11.8(b) below, an estimate of the recurrence of the critical point is given[2].

Theorem 11.8. *(a) Almost every nonregular, real quadratic map satisfies the Collet–Eckmann condition,*

$$\liminf_{n \to \infty} \frac{\ln(|Df^n(f(0))|)}{n} > 0. \tag{11.11}$$

(b) Almost every nonregular real quadratic map has polynomial recurrence of the critical orbit with exponent 1,

$$\limsup_{n \to \infty} -\frac{ln(|f^n(0)|)}{\ln(n)} = 1. \tag{11.12}$$

In other words, the set of n such that $|f^n(0)| < n^{-\gamma}$ is finite if $\gamma > 1$ and infinite if $\gamma < 1$.

---
[2]The critical point is nonrecurrent for regular maps, and it actually converges to the periodic attractor.

The proof of Theorem 11.8(b) is an answer to the conjecture of Sinai, and it the first proof of polynomial estimates for the recurrence of the critical orbit which is valid for a positive measure set of nonhyperbolic parameters. The results given in [Lyubich (2002)] and in Theorem 11.8 above were subsequently extended to typical analytic (and even smooth) unimodal maps in [Avila et al. (2003)] and finally in all generality in [Avila & Moreira (2005a)]. In [Avila & Moreira (2003)] the following results were proved:

Theorem 11.9. *(a) Let Γ be a standard k-parameter family of S-unimodal maps. Then for almost every $p \in B^k$, f^p is either regular or satisfies CE and P.*

(b) Let Γ be a standard k-parameter family of S-unimodal maps. Let Γ_n be a sequence of C^2 families such that $\Gamma_n \to \Gamma$ in the C^2 topology. Let X_n be the set of parameters $p \in B^k$ such that Γ_n is either regular or satisfies BCE, CE, SE, and WR. Then $|X_n| \to 1$.

Using the concept of the *phase-parameter relation*, [Avila & Moreira (2005a)] showed the following results:

Theorem 11.10. *(a) Let f_λ be a one-parameter nontrivial analytic family*[3] *of unimodal maps. Then f_λ satisfies the phase-parameter relation at almost every nonregular parameter.*

(b) Let f_λ be a nontrivial analytic family of unimodal maps (any number of parameters). Then almost every parameter is either regular or has a renormalization which is topologically conjugate to a quadratic polynomial.

(c) The set of nonhyperbolic, noninfinitely renormalizable complex quadratic parameters has zero Lebesgue measure.

From Theorem 11.10(a), the phase-parameter relation has many remarkable consequences for the study of the dynamical behavior of typical parameters, and Theorem 11.10(b) allows one to reduce the study of typical unimodal maps to the special case of unimodal maps which are quasi-quadratic[4].

In other words, Theorem 11.10 says that a typical nonregular unimodal map f possess a *unique* nontrivial chaotic attractor A_f. This attractor has the following properties:

(a) The chaotic attractor A_f is a transitive finite union of intervals.

---
[3] An analytic family of unimodal maps is nontrivial if regular parameters are dense (in particular nontrivial analytic families are dense in any topology).
[4] Which is persistently topologically conjugate to a quadratic polynomial.

(b) Periodic orbits are dense in A_f.
(c) The chaotic attractor A_f is the support of an absolutely continuous invariant measure μ_f.
(d) The chaotic attractor A_f has excellent stochastic properties due to the Collet–Eckmann condition.
(e) The measure-theoretical dynamics of f can be described by μ_f and finitely many trivial attractors (hyperbolic periodic orbits).
(f) The chaotic attractor A_f is simultaneously a metric and a topological attractor in the sense of Milnor [Livsic (1972)].

The proof of Theorem 11.10(a) can be divided in four parts:

(1) Describe a complex analogous of the phase-parameter relation for certain families of complex return type maps following the works of [Lyubich (2000), Avila & Moreira (2005b)].
(2) Show that through any given analytic finitely renormalizable with a recurrent critical point unimodal map f, there exists an analytic family f_λ which gives rise to a full family of complex return type maps, and use step (1).
(3) Show that if the phase-parameter relation is valid for one transverse family at f, then it is valid for all transverse families at f [Avila et al. (2003)].
(4) Use a simple generalization of [Avila et al. (2003)] to conclude first that a nontrivial family of unimodal maps is transverse to the topological class of almost every nonregular parameter, and second that typical parameters are finitely renormalizable with a recurrent critical point.

The proof of Theorem 11.10(b) was obtained using a parameter exclusion argument.

Corollary 11.4. *(a) Let f_λ be a nontrivial analytic family of unimodal maps (in any number of parameters). Then almost every nonregular parameter is Collet–Eckmann, and its critical point is polynomially recurrent.*

(b) In generic smooth ($C^k, k = 2, ..., \infty$) families of unimodal maps (any number of parameters), almost every parameter is regular or has a renormalization which is conjugate to a quadratic map, is Collet–Eckmann, and its critical point is subexponentially recurrent.

(c) Let f_λ be a nontrivial analytic family of unimodal maps (any number of parameters). Then almost every parameter is either regular or has a polynomially recurrent critical point with exponent 1.

First, Corollary 11.4(a-b) gives an answer to the Palis conjecture 11.2 in the general unimodal case. Second, the dichotomies in Corollary 11.4(a-b) imply that the dynamics of typical nonregular unimodal maps have the same excellent statistical description as the quasi-quadratic case governed by Theorem 11.10 [Avila & Moreira (2003)].

In [Artur & Gustavo (2005)] are some possible relations (obtained through an analysis of the physical measure) between the geometry of chaotic attractors of typical analytic unimodal maps and the behavior of the critical orbit, *i.e.*, an explicit formula relating the combinatorics of the critical orbit with the exponents of periodic orbits.

Theorem 11.11. *(a) Let f_λ be a nontrivial analytic family of unimodal maps. Then for almost every nonregular parameter λ and for every periodic orbit p in the nontrivial attractor A_{f_λ}, the exponent of p is determined by an explicit combinatorial formula involving the kneading sequence of f and the itinerary of p.*

(b) Let f_λ be a nontrivial analytic family of quasi-quadratic maps. Then for almost every nonregular parameter λ, the critical point belongs to the basin of μ_{f_λ} (the absolutely continuous invariant measure of f_λ).

Corollary 11.5. *In the setting of Theorem 11.11(b), one also has equality between the Lyapunov exponent of the critical value and the Lyapunov exponent of μ_{f_λ}.*

More details about the statistical properties of unimodal maps are contained in [Lyubich (2000-2002), Avila & Moreira (2003–2005a-b), Avila et al. (2003)].

11.2 The Barreto–Hunt–Grebogi–Yorke conjecture

In this section, we describe a conjecture of Barreto–Hunt–Grebogi–Yorke stated in [Barreto et al. (1997)] claiming that robust chaos cannot appear in smooth systems. In particular, they address some possible circumstances for this phenomenon in high-dimensional chaos. The two best known examples of dynamical systems that support this conjecture are, first, the fact that the bifurcation diagram of the smooth logistic map is densely populated by periodic windows as shown in [Collet & Eckmann (1980), Chapter 1], and second, the fact that a continuously chaotic bifurcation diagram is obtained for the tent map (11.3), which is nonsmooth. However,

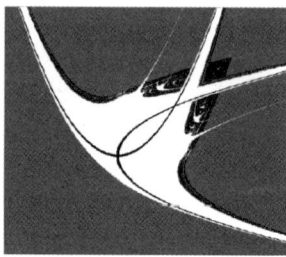

Fig. 11.1 The bifurcation diagram for $x \to \left(x^2 - a\right)^2 - b$. The axes are $a, b \in [-2, 3]$. The areas lead to stable periodic orbits. The light gray (in the original source) points lead to divergent trajectories. Dark gray (in the original source) points display chaos with one positive Lyapunov exponent. The spine locus is superimposed in black which delineates the shape of the window. Reprinted (Figure 1) with permission from [Barreto, E., Hunt, B. R., Grebogi, C., and Yorke, J. A., Phys. Rev. Lett. 78, 4561, 1997. Copyright (1997) by the American Physical Society].

this conjecture is false due to the counterexamples given in [Andrecut & Ali (2001a-b), Gabriel (2004), Jafarizadeh & Behnia (2002)]. These counterexamples are listed in Sec. 11.2.1.

This conjecture is based on following remarks and ideas:

(1) In one-dimensional chaotic dynamical systems, stable behavior (windows[5]) appear as system parameters traverse chaotic regions as shown in Fig. 11.1 for the example of the quadratic map $x \to x^2 - a$. This phenomenon has been proved to be true in [Graczyk & Swiatek (1997)].
(2) Most known chaotic systems display *fragile* chaos (see Chapter 10) in the sense that a slight alteration of a large number N of parameters will destroy the chaos and give a stable periodic orbit.
(3) The idea that a window is constructed around a *spine locus* (see Sec. 11.1), *i.e.*, for the quadratic family $x \to x^2 - a$, the windows are intervals in the one-dimensional space of parameters built around isolated spine points, and the spine locus in this case[6] corresponds to parameter values that give rise to superstable orbits. For the two-parameter quadratic family $x \to \left(x^2 - a\right)^2 - b$, the spine locus consists of the two parabolas (the black curves defined by the condition $m = 0$) shown in Fig. 11.1. More generally, this is the case for maps that contain critical points [Yorke *et al.* (1984)].

---

[5]The concept of windows here means regions in the parameter space of chaotic systems that correspond to stable behavior.
[6]Also for one-dimensional maps.

The dimension of the spine locus has a crucial importance in the formation of this conjecture because it determines the geometry of the window as follows: If the spine is a point, then the window (called *windows limited*) will typically have limited extent. This is to be contrasted with windows that have spines of higher dimension and the window extends (called *windows extended*) along the entire length of the spine, as shown in Fig. 11.1. In particular, the spine locus is of codimension-one in the parameter space because the condition $m = 0$ is a single constraint. In other words, the value of the dimension of the spine locus gives a criterion to expect when one can find windows, *i.e.*, if the map under consideration has n parameters, then varying these parameters defines an n-dimensional accessible parameter manifold in the full parameter space. From this step, the criterion for the appearance of windows is when this accessible parameter manifold intersects or come close to a spine locus for some period-p. In this case, if the codimension of the spine locus is k, then at least $n = k$ is the dimension of the accessible parameter manifold for point intersections to generically occur, and these windows are constructed around isolated spine points, and therefore are limited. Now, if the accessible parameter manifold is of a higher dimension, then typical intersections occur in higher-dimensional sets, and extended windows can be expected in the accessible parameter space. Finally, arbitrarily small perturbations to k parameters can stabilize one of the unstable periodic orbits which are dense in the chaotic attractor, and windows are dense when $n \geq k$.

Now for a two-dimensional map F, the identification of the spine locus is more involved, using only the trace T and the determinant $D = \det M = \det D^p F$ as follows:

(1) In the region of parameter space that exhibits only one positive Lyapunov exponent $\lambda_1 > 0$ such that $\lambda_1 + \lambda_2 < 0$, the map is asymptotically area-contracting, and $|\det DF| < 1$ over a trajectory. If this trajectory is periodic, then $D = \det M(x) = \lambda_1 \lambda_2 \simeq 0$ for sufficiently high p, and the spine loci in this region are of codimension-one.

(2) In the region of parameter space that exhibits two positive Lyapunov exponents (hyperchaos), the stability conditions can be formulated in terms of the trace $T = \lambda_1 + \lambda_2$ and determinant $D = \lambda_1 \lambda_2$ of M. In this case, stability implies that these numbers must fall within a triangular region (called *stability triangle* and every point in the parameter space that leads to a stable orbit maps to a particular point within this triangle) in D versus T space as shown in Fig. 11.2.

Fig. 11.2 The stability triangle. The trace (horizontal) $T \in [-2, 2]$, and the determinant (vertical) $D \in [-1, 1]$. Reprinted (Figure 2) with permission from [Barreto, E., Hunt, B. R., Grebogi, C., and Yorke, J. A., Phys. Rev. Lett. 78, 4561, 1997. Copyright (1997) by the American Physical Society].

(3) $D = T = 0$ (called *nilpotent* points) have a central importance because in this case because the spine locus for windows consists of nilpotent parameter values (points where the colors come together in Fig. 11.2) and is of codimension-two because the restriction of D and T to the stability triangle represents two constraints.

The above ideas were illustrated using a physically motivated map, called the *kicked double rotor* which is a two-dimensional, two-parameter map studied in [Gardini et al. (1994)] and given by

$$(x, y) \to \left(\alpha x (1 - x) + \left(1 - \frac{\alpha}{4}\right) y, \beta y (1 - y) + \left(1 - \frac{\beta}{4}\right) x \right). \quad (11.13)$$

Indeed, Fig. 11.3(a) shows a region of parameter space dominated by area-contracting chaos with one positive Lyapunov exponent and very many extended windows. Figure 11.3(b) shows a region of area-expanding chaos with one positive Lyapunov exponent and many extended windows. In the above two cases, the spines are one-dimensional. For the case of the regions where two positive Lyapunov exponents are found, the spines consist of isolated nilpotent points. In this case, a large number of limited windows[7] are observed in the two-dimensional parameter space as shown in Fig. 11.4(a). Fig. 11.4(b) shows a blowup of a period-five window (constructed around the two isolated nilpotent points at the top and bottom of the central blue, with the same for the other windows) region with the interior colored according to Fig. 11.2. The case of high-dimensional maps, namely d-dimensional maps, can be studied in the same way taking in account that

---

[7]It seems that these windows are dense and limited.

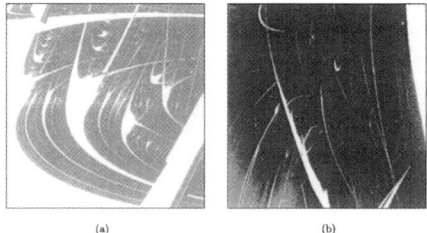

(a)   (b)

Fig. 11.3 (a) A region in parameter space $(\alpha, \beta) \in [3.4722, 3.4857] \times [1.078, 1.316]$ dominated by area-contracting chaos with one positive Lyapunov exponent (light gray). White areas lead to asymptotically stable orbits. A dense set of extended windows is seen. In (b) $(\alpha, \beta) \in [2.876, 3.288] \times [1.932, 2.46]$, and dark gray areas indicate area-expanding chaos with one positive Lyapunov exponent. Again, a dense set of extended windows is seen, as predicted by the conjecture described in Sec. 11.2. Reprinted (Figure 3) with permission from [Barreto, E., Hunt, B. R., Grebogi, C., and Yorke, J. A., Phys. Rev. Lett. 78, 4561, 1997. Copyright (1997) by the American Physical Society].

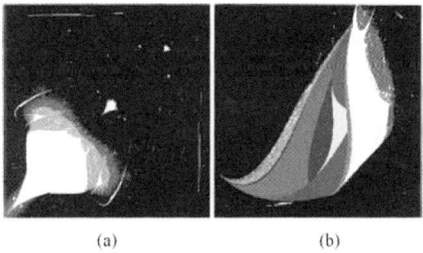

(a)   (b)

Fig. 11.4 (a) A parameter space region $(\alpha, \beta) \in [3, 4 \times [3, 4]$ dominated by chaos with two positive Lyapunov exponents (black). The shading is otherwise as in Fig. 11.3. A dense set of limited windows is seen. In (b) a window from within (a) is magnified, $(\alpha, \beta) \in [3.375, 3.42] \times [2.87, 2.9825]$, and the interior of the period-five region is colored according to Fig. 11.2. Two nilpotent points, forming the spine, are evident at the top and bottom of the central blue region. Reprinted (Figure 4) with permission from [Barreto, E., Hunt, B. R., Grebogi, C., and Yorke, J. A., Phys. Rev. Lett. 78, 4561, 1997. Copyright (1997) by the American Physical Society].

the matrix M has a characteristic polynomial of degree d in λ since it is $d \times d$ and the coefficients c_i can be written as the sum of all possible distinct product combinations of the eigenvalues taken λ_i at a time for $i = 1, 2, ..., d$. The stability conditions are $|\lambda_i| < 1$, $i = 1, 2, ..., d$ that determine a volume in the coefficient space $\{c_1, c_2, ..., c_d\}$, and stability occurs if the numbers $c_1, c_2, ..., c_d$ lie within this volume. In this case, the spines of windows are given by points in parameter space that correspond to the center of this

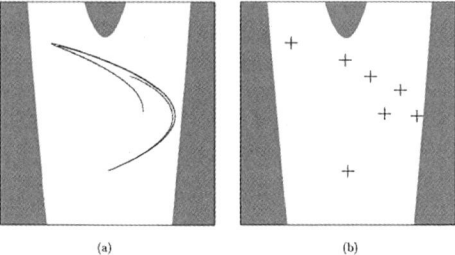

Fig. 11.5 Illustration of a dispelled chaotic attractor, using the Hénon map (6.88) with $a = \rho$ and $b = 0.3$. The figures show $(x,y) \in [22.5, 2.5] \times [2.5, 2.5]$. In (a) $\rho = 1.22$ and the white region is the basin of the chaotic attractor shown in black. The gray region is the basin of infinity. In (b) $\rho = 1.23$ and trajectories originating in the vicinity of the attractor in (a) now converge to the attracting period-seven orbit shown (crosses). We say that the chaotic attractor in (a) is dispelled for the map with $\rho = 1.23$. Reprinted (Figure 5) with permission from [Barreto, E., Hunt, B. R., Grebogi, C., and Yorke, J. A., Phys. Rev. Lett. 78, 4561, 1997. Copyright (1997) by the American Physical Society].

volume, *i.e.*, $c_1 = c_2 = ... = c_d = 0$. Note that these equations may be solved numerically to locate particular windows. Generally, some of these conditions restricting the parameter space are automatically satisfied by the map under consideration. Hence unstable periodic orbits have at most k expanding directions in any attractor on which all invariant measures give k positive Lyapunov exponents. This implies that the transition to stability must satisfy at most k conditions. Thus the spines of windows are of codimension-k in the parameter space.

Now, to state the main conjecture, consider a smooth map f from a region $S \subset \mathbb{R}^d$ to itself that exhibits a chaotic attractor Λ with k positive Lyapunov exponents, with the assumption that all invariant measures supported on Λ yield the same k. Let g be a map close to f[8]. Thus one has the following definition:

Definition 11.13. We say that Λ is *dispelled* for g if almost all points in a neighborhood of Λ belong to basins of attracting periodic orbits of g.

The situation of Definition 11.13 is illustrated in Fig. 11.5 using the Hénon map (6.88) with $a = \rho$ and $b = 0.3$. Now it is possible to define fragile chaos using the concept of a dispelled set as follows:

Definition 11.14. If there exist (possibly rare) functions arbitrarily close to f for which the attractor Λ is dispelled, we say that Λ is fragile.

---

[8]This means that $f(x)$ and $g(x)$ are close and that all first partial derivatives of g are close to those of f.

Finally, let f_a be an n-parameter family of functions with $a \in \mathbb{R}^n$ such that $f_0 = f$ and f_a depends smoothly on a. Let d be the dimension of the state vector and n be the number of accessible parameters. Then the concept of a window can be formulated as follows:

Definition 11.15. The window set W is the set of a values such that Λ is dispelled for f_a.

We now state the Barreto–Hunt–Grebogi–Yorke conjecture announced in [Barreto et al. (1997)]:

Conjecture 11.1. *(Windows conjecture). Let f be a smooth map from a region $S \subset \mathbb{R}^d$ to itself that exhibits a fragile chaotic attractor Λ with $k \geq 1$ positive Lyapunov exponents, where all invariant measures supported on Λ yield the same k. Let W be the window set corresponding to a typical family f_a, where $a \in \mathbb{R}^n$ and $f_0 = f$.*

(1) If $n < k$, there exists a neighborhood of $a = 0$ entirely outside of W.

(2) If $n \geq k$, W is dense in a neighborhood of $a = 0$, and the components of W are limited.

(3) If $n - k$, W is dense in a neighborhood of $u = 0$, and the components of W are extended.

Conjecture 11.2 says that for d-dimensional maps, a slight variation of n parameters can destroy the chaos if $n \geq k$, where k is the number of positive Lyapunov exponents. In particular, as mentioned in the beginning of this section, it follows that robust chaos cannot occur in smooth one-dimensional systems where $d = n = k = 1$. At the end of this section, we make some remarks about Conjecture 11.2 as follows:

(a) It is possible to say that in cases (2) and (3) of Conjecture 11.2 that the window W consists of a union of connected subsets W_i; these are the individual windows.

(b) Intuitively in case (2) of Conjecture 11.2, the subsets W_i get smaller and smaller as they converge to zero because successively smaller neighborhoods of zero implies that the diameters[9] of W_i decrease to zero. While in case (3), this property does not hold (in this case, the components W_i are called *extended*) because the sets W_i may be quite long in the vicinity of zero as shown in Fig. 11.1.

(c) The Conjecture 11.2 is a local property because it describes the local structure of parameter space in the vicinity of a point $a = 0$ that

---
[9]The diameter of a set is the maximum distance between two points of the set.

displays chaotic attractors. But it seems that this conjecture can play a crucial role in the context of the global structure of the dynamics.

(d) It seems that in more general higher-dimensional situations, chaos with several positive Lyapunov exponents has the property that the set of parameter values that display chaos has a nonzero Lebesgue measure as shown in [Jacobson (1981)] for the one-dimensional quadratic family.

11.2.1 Counter-examples to the Barreto–Hunt–Grebogi–Yorke conjecture

The robustness of chaos in the sense of Definition 2.5 is expected to be relevant for any practical applications of chaos, and it was shown to exist in a general family of piecewise-smooth two-dimensional maps, but it was conjectured to be impossible for smooth unimodal maps in Conjecture 11.2 [Barreto et al. (1997), Banerjee (1998)]. However, it is shown in [Andrecut & Ali (2001b-c)] that the following 1-D maps have robust chaotic attractors:

$$f(x, \alpha) = \frac{1 - x^\alpha - (1 - x)^\alpha}{1 - 2^{1-\alpha}}, \alpha \geq 3 \tag{11.14}$$

and

$$f(\phi(x), v) = \frac{1 - v^{\pm\phi(x)}}{1 - v^{\pm\phi(c)}} \tag{11.15}$$

where α and v are bifurcation parameters, $\phi(x)$ is unimodal with a negative Schwarzian derivative (but not necessarily chaotic), and c is the critical point of $\phi(x)$, i.e., $\phi'(c) = 0$ and v obeys

$$\left| \frac{\ln(v) \phi'(0)}{1 - v^{\pm\phi(c)}} \right| > 1, v > 0, v \neq 1. \tag{11.16}$$

As can be seen, these examples of robust chaos in unimodal smooth maps rely on the use of noninteger powers. Note that the map

$$g(x, \alpha) = 1 - x^k - (1 - x)^k, k \geq 3 \tag{11.17}$$

describes the complex dynamics of a *stochastic Boolean network* with an infinite number of connected elements. Such a stochastic network might mimic the random interaction between two-state neurons in a neural network. The parameter of the map is k, and it represents the network's connectivity. The map (11.17) is chaotic, and the route to chaos is period doubling bifurcations. In this section, we consider a slight modification of the map (11.17), namely the map (11.14).

The first counter-example of Conjecture 11.2 can be found in [Andrecut & Ali (2001b)] where the occurrence of robust chaos in a one-dimensional smooth system was investigated by a general theorem with a practical procedure for constructing S-unimodal maps that generate robust chaos.

Theorem 11.12. *Let $\varphi_\nu(x) : I = [a, b] \to I$ be a parametric S-unimodal map with the unique maximum at $c \in (a, b)$ and $\varphi_\nu(c) = b, \forall \nu \in (\nu_{min}, \nu_{max})$. Then $\varphi_\nu(x)$ generates robust chaos for $\nu \in (\nu_{min}, \nu_{max})$.*

Proof. Using Theorem 11.1 for the S-unimodal map $\varphi_\nu(x)$ for all $\nu \in (\nu_{min}, \nu_{max})$, it follows that there is at most one attracting periodic orbit with the critical point in its basin of attraction for all $\nu \in (\nu_{min}, \nu_{max})$. In this case, the orbit with initial value $x_0 = c$ maps in two iterates to the fixed point $x = a$ ($\varphi_\nu(c) = b, \varphi_\nu(b) = a$), which is unstable because

$$|\varphi_\nu'(a)| \geq \frac{\varphi_\nu(c) - \varphi_\nu(a)}{c - a} = \frac{b - a}{c - a} > 1. \qquad (11.18)$$

Hence the map $\varphi_\nu(x)$ does not have any stable periodic orbits, and there is a unique chaotic attractor for all $\nu \in (\nu_{min}, \nu_{max})$. $\square$

Theorem 11.12 gives the general conditions for the occurrence of robust chaos in S-unimodal maps, but it does not give any procedure for the construction of the S-unimodal map $\varphi_\nu(x)$. A procedure for constructing S-unimodal maps that generate robust chaos from the composition of two S-unimodal maps is given in [Andrecut & Ali (2001a-b), Jafarizadeh & Behnia (2002), Gabriel (2004)]. As an example of the above result, consider the map $f_\nu(x)$ given by

$$f_\nu(x) = \frac{1 + \nu - \nu^x - \nu^{1-x}}{1 + \nu - 2\sqrt{\nu}}, \nu > 0, \nu \neq 1, x \in [0, 1]. \qquad (11.19)$$

The map (11.19) has the following properties:

(1) $f_\nu(0) = f_\nu(1) = 0$.
(2) $f_\nu(x)$ is a smooth map, $\forall x \in [0, 1], \forall \nu > 0, \nu = 1$.
(3) $c = \frac{1}{2}, f_\nu'(c) = 0, f_\nu(c) = 1, \forall \nu > 0, \nu = 1$.
(4) $\lim_{\nu \to 0^+} f_\nu(x) = 1, \forall x \in [0, 1]$.
(5) $\lim_{\nu \to 1^\pm} f\nu(x) = h_4(x) = 4x(1 - x), \forall x \in [0, 1]$.
(6) $f_\nu(x)$ has negative Schwarzian derivative ($x \neq c, \nu \neq 1$),

$$S(f_\nu(x), x) = -[\ln(\nu)]^2 \left[1 + \frac{3}{2}\left(\frac{\nu^x + \nu^{1-x}}{\nu^x - \nu^{1-x}}\right)^2\right] < 0. \qquad (11.20)$$

(7) It is possible to restrict the study of the map (11.19) to the interval $\nu \in (0,1)$ because $f_\nu(x) = f_{1/\nu}(x), \forall \nu > 0, \nu \neq 1$.

From the above items (1)–(6), it follows that the map (11.19) is S-unimodal, and in this case, the orbit with initial value $x_0 = c = \frac{1}{2}$ maps in two iterates to the fixed point $x = 0$ which is unstable because $|f'_\nu(0)| > \frac{f_\nu(c)}{c} = 2 > 1$. The application of Theorem 11.12 implies that the map (11.19) does not have any stable periodic orbits and there is a unique chaotic attractor for all $\nu \in (0,1)$. The numerical computation of the Lyapunov exponent for the map (11.19) is given by

$$\lambda(\nu) = \lim \frac{1}{T} \sum_{t=0}^{T} \left| \frac{d}{dx} f_\nu(x_t) \right| \simeq \ln(2), \forall \nu \in (0,1), \quad (11.21)$$

and it confirms the analytical result. A general procedure for constructing S-unimodal maps that generate robust chaos was done based on Theorem 11.12 as follows:

Corollary 11.6. *Let $f(x): I \to I$ and $g(x): I \to I$ be S-unimodal maps on the interval $I = [0,1]$, with $c_f \in (0,1)$ the unique maximum of f and $c_g \in (0,1)$ the unique maximum of g. Then the map*

$$F_\nu(x) = \frac{f\left(\nu \frac{g(x)}{g(c_g)}\right)}{f(\nu)}, x \in I \quad (11.22)$$

is S-unimodal, and it generates robust chaos for $\nu \in (0, c_f)$.

Proof. The critical points of the map $F_\nu(x)$ are given by the solutions of the equation

$$F'_\nu(x) = \frac{\nu g'(x)}{f(\nu) g(c_g)} \left(\frac{df(y)}{dy} \bigg|_{y=\nu \frac{g(x)}{g(c_g)}} \right) = 0. \quad (11.23)$$

Equation (11.23) has the unique solution $c_F = c_g$, $(g'(c_g) = 0)$ if

$$\frac{df(y)}{dy} \bigg|_{y=\nu \frac{g(x)}{g(c_g)}} > 0, \quad (11.24)$$

or equivalently

$$\nu \in (0, c_f). \quad (11.25)$$

Thus one has $F_\nu(0) = F_\nu(1) = 0$, and the map (11.22) has a negative Schwarzian derivative,

$$S(F_\nu(x), x) = (S(f(y), y)|_{y=g(x)}) \left(\nu \frac{g'(x)}{g(c_g)} \right)^2 + S(g(x), x) < 0 \quad (11.26)$$

because

$$\begin{cases} S\left(f(y),y\right) < 0 \\ S\left(g(x),x\right) < 0 \end{cases} \quad (11.27)$$

for all $\nu \in (0, c_f)$. Hence the map $F_\nu(x)$ is S-unimodal for all $\nu \in (0, c_f)$. In this case, the orbit with initial value $c_F = c_g$ maps in two iterates to the fixed point $x = 0$ which is unstable because $F_\nu(x)$ is S-unimodal with the unique maximum $F_\nu(c_F) = 1, c_F = c_g \in (0,1), \forall \nu \in (0, c_f)$, and the inequality

$$|F'_\nu(0)| > \frac{F_\nu(c_F)}{c_F} = \frac{1}{cF} > 1. \quad (11.28)$$

Thus the map $F_\nu(x)$ does not have any stable periodic orbits, and there is a unique chaotic attractor, $\forall \nu \in (0, c_f)$, because all the conditions of Theorem 11.12 are satisfied. □

An example of Theorem 11.12 is given in Exercise 242 where it was shown in [Andrecut & Ali (2001b)] that for $f(x) = \pi^{-1}\sin(\pi x)$ and $g(x) = \sqrt{x}(1 - \sqrt{x})$, the function $F_\nu(x)$ generates robust chaos for $x \in [0,1]$ and $\nu \in (0, \frac{1}{2})$, and numerical computation of the Lyapunov exponent gives $\lambda(\nu) = \ln(2) > 0 \forall \nu \in (0, \frac{1}{2})$.

The second counter example to Conjecture 11.2 can be found in [Andrecut & Ali (2001a)] where a one-dimensional smooth map $x_{n+1} = f(x_n, \alpha)$ is presented, and it was found that it generates robust chaos in a large domain of the parameter space (α). The method of analysis is based on the idea that the map does not have periodic windows and the Lyapunov exponent is positive in a very large domain of the parameter space (α). The invariant measures introduced in Sec. 3.1 and the ergodicity theory introduced in Sec. 3.3 for the map are also investigated analytically and confirmed numerically. Finally, two physical applications were suggested for random number generation and cryptography.

The parameter of the map (11.14) is $\alpha \in [2, \infty)$. The use of graphical representations (see Exercise 244) of the map(11.14) for $\alpha = 2$ and 3 show that the map (11.14) is identical to the logistic map (6.25) with the parameter $a = \mu = 4$. For larger values of α as $\alpha = 9$ and $\alpha = 15$, the map (11.14) looks like a *flat* version of the logistic map (6.25). Thus the map (11.14) is a unimodal map, with similar chaotic behavior to the logistic map (6.25) when $a = \mu = 4$. The map (11.14) is a smooth map with only one critical point at $x = \frac{1}{2}$ $(f'\left(\frac{1}{2}, \alpha\right) = 0, \forall \alpha \geq 2)$ and two fixed points, $x = 0$, and it is unstable because $f'(0, \alpha) = \alpha\left(1 - 2^{1-\alpha}\right) > 1$. The second fixed point is the real solution of the equation

$$f(x^*, \alpha) = x^*, x^* \in (0,1), \forall \alpha \geq 2. \tag{11.29}$$

This fixed point is also unstable because $|f'(x^*, \alpha)| > 1, \forall \alpha \geq 2$. In this case, the Schwarzian derivative of the map (11.14) is negative, and its bifurcation diagram shows that the map (11.14) does not have periodic windows. This can be proved also using the same method as in the case of logistic map (6.25) with $\mu = 4$ because the Schwarzian derivative is negative and the map has only one critical point, at $x = \frac{1}{2}$ for $\alpha \geq 2$. This implies that if an orbit starts with $x = \frac{1}{2}$, then it converges to the attractor, but if the orbit started with $x = \frac{1}{2}$, after two iterates, for all α, then it converges to $x = 0$ which is an unstable fixed point. Thus the map (11.14) does not have attracting periodic orbits for $\alpha \geq 2$.

The Lyapunov exponent of the map (11.14) is given by

$$\lambda(\alpha) = \lim \frac{1}{T} \sum_{t=0}^{T} \left| \frac{d}{dx} f(x_t, \alpha) \right| \simeq \ln(2) \tag{11.30}$$

where the numerical calculation of $\lambda = \lambda(\alpha)$ has been done for very long orbits ($T = 10^6$) and an accuracy of 500 digits using an arbitrary precision mathematical library (Maple). In this case, the Lyapunov exponents are positive, $\lambda(\alpha) > 0$, and almost constant, i.e., the Lyapunov exponents are decreasing very slowly with increasing α with a maximum value at $\lambda(2) = \ln(2)$, and it corresponds to the logistic map (6.25) with $a = 4$. On the other hand, the least squares fit in the range $2 \leq \alpha \leq 102$ shows that the very slow decrease is almost linear with an approximate regression equation given by

$$\lambda(\alpha) = 0.69445258 - 0.0013054\alpha. \tag{11.31}$$

Finally, the above analysis proves that the smooth map (11.14) generates robust chaos in a very large domain of the parameter space, namely $2 \leq \alpha \leq 102$, and hence it is ergodic, i.e., all trajectories approach every point in the interval $(0,1)$ arbitrarily closely over the course of time as shown in [Eckmann & Ruelle (1985)]. Thus the trajectories of the map (11.14) densely fill the interval $x \in (0,1)$ for an arbitrary initial x_0, and the probability density $P(x)$ for finding the trajectory at $x \in (0,1)$ is constant.

Definition 11.16. We say that the probability density $P(x)$ is an invariant distribution (invariant measure) of the mapping $f(x)$ if

$$P(x) = P(f^{-1}(x)) \tag{11.32}$$

and if it satisfies the normalization condition

$$\int P(x)dx = 1. \tag{11.33}$$

In this case, $P(x)$ is singular for a parameter value corresponding to a stable, period-p orbit. Thus $P(x)$ consisting of p δ-*functions* (with an integrated value of $\frac{1}{p}$) at the p stable fixed points of the map. Parameters giving rise to chaotic motion yield a unique (discontinuous in x and nonzero over a finite range of x) equilibrium distribution. For ergodic systems, a unique equilibrium distribution is generated by repeated iteration of the map with the condition that the time average is equal to the space average over the distribution of all initial conditions x_0.

Generally, there is no analytical method to calculate the invariant distribution, but numerically, $P(x)$ can be constructed from (11.32) by considering that the number of trajectories $P(x)dx$ within some small interval dx in x is equal to the number within the corresponding intervals at the preimage points of the mapping under consideration. The fact that there are two preimage points x_1 and x_2 for a unimodal map implies the following functional equation:

$$P(x)dx = P(x_1)dx_1 + P(x_2)dx_2. \tag{11.34}$$

Writing

$$\frac{dx}{dx_i} = \left|\frac{df}{dx}\right|_{x=x_i}, i=1,2, \tag{11.35}$$

one has

$$P(x) = \frac{P(x_1)}{\left|\frac{df}{dx}\right|_{x=x_1}} + \frac{P(x_2)}{\left|\frac{df}{dx}\right|_{x=x_2}}. \tag{11.36}$$

This functional equation cannot be solved analytically, but a numerical solution can be obtained by iterating (11.36) as follows:

(1) Choose an initial distribution $P_i(x) = 1, \forall x \in (0,1)$.
(2) Evaluate the right-hand side of (11.36) using $P_i(x)$ to obtain the next iterate $P_{i+1}(x)$.
(3) Repeat step (2) until convergence is achieved.

For the case of the Eq. (11.14), the interval $(0,1)$ has been discretized using a step $\Delta x = 10^{-4}$, and *calibration* of the algorithm has been done using the logistic map (6.25) which exhibits chaotic motion with a nonzero distribution over a finite range of x for $\mu = 4$, in which the procedure converges rapidly, *i.e.*, typically in five iteration steps for an error less than 10^{-3}% to the following analytical form of the invariant distribution [Eckmann & Ruelle (1985)]:

$$P(x) = \frac{1}{\pi\sqrt{x(1-x)}}. \qquad (11.37)$$

For the map (11.14), the invariant distribution $P(x,\alpha), x \in (0,1), \alpha \geq 2$ calculated using the above algorithm shows that the qualitative aspect of the analytical function (11.37) is conserved, *i.e.*, symmetry (relative to the average $x = \frac{1}{2}$) on the interval $(0,1)$ and the strong delocalization (the delocalization increases upon increasing the parameter α).

The Lyapunov exponent (11.31) can be found using the relation

$$\lambda(\alpha) = \int_0^1 \left|\frac{d}{dx}f(x,\alpha)\right| P(x)\,dx \qquad (11.38)$$

because for ergodic systems, it is possible to replace time averages by spatial averages over the invariant distribution as shown in [Grossman & Thomae (1977), Shaw (1981), Eckmann & Ruelle (1985), Diakonos & Schmelcher (1996), Pingel *et al.* (1999)]. Hence numerical calculation using Eq. (11.38) confirms the result obtained using (11.31).

The third counterexample can be found in [Gabriel (2004)], where it was shown that the following map is robust

$$f_{m,n}(x,a) = 1 - 2(ax^m + bx^n), \qquad (11.39)$$

where $0 \leq a \leq 1, b = 1 - a$, and m,n are even and positive, and without loss of generality one can assume that $m \leq n$. The construction of the map (11.39) is a generalization (in the interval $[-1,1]$) of the results obtained by S. Thomae, (Unpublished, private communication from S. Sinha) showing numerically the linear interpolation between the fully chaotic logistic map (6.25) and quartic map

$$f_{2,4}(x,a) = 1 - 2(ax^2 + bx^4), 0 \leq a \leq 1, b = 1 - a \qquad (11.40)$$

The proof of [Gabriel (2004)] consists of two steps. The first is to show that the critical point is not attracted to any stable periodic orbit, and the second is to show that the Schwarzian derivative of the map (11.39) is negative for the whole $-1 \leq x \leq 1$ range. Indeed, the critical point of the map (11.39) at $x_c = 0$ falls after two iterations to the unstable fixed point at $x^* = -1$. In this case, there are no super-stable periodic orbits for the map (11.39) as would be needed for the spine locus (see Sec. 11.1) around which the periodic windows appear [Barreto *et al.* (1997)].

For proving the second item, the Schwarzian derivative is given by
$$\begin{cases} \hat{S}\left(f_{m,n}\left(x,a\right),x\right) = -P\left(m,a\right)x^{2m-4} - P\left(n,b\right)x^{2n-4} + Qx^{m+n-4} \\ P\left(m,a\right) = 4a^{2}m^{2}\left(m^{2}-1\right) \\ Q = Q\left(m,n,a\right) = 8abmn\left(m^{2}+n^{2}-3mn+1\right) = 8abmn\hat{Q}. \end{cases}$$
(11.41)

Some helpful remarks are the following:

(1) For even $m > 0$ and $a \geq 0$, one gets $P > 0$, and therefore it suffices to discuss only the sign of Q when finding the sign of $\hat{S}\left(f_{m,n}\left(x,a\right),x\right)$.
(2) If $a = 0$ or $a = 1$, all maps of the form $f_m(x) = 1 - 2x^m$ with even $m > 0$ have negative Schwarzian derivative for all $-1 \leq x \leq 1$.
(3) For $a, b, m,$ and n nonnegative, one only needs to check the sign of $\hat{Q}$.

To check the last item (3), define $\delta = m - n > 0$. Thus $\hat{Q} = \delta^2 - m^2 - m\delta + 1 < 0$ for all m. Hence, for each value of δ, one can verify that there are a few extra values of m that leave $\hat{Q}$ negative. Let m_1 be the smallest such m so that the first values of the sequence of pairs are as follows:

$$(\delta, m_1) : (2,2)\,(4,4)\,(6,4)\,(8,6)\,(10,8)\,(12,8)\,\ldots\,etc. \quad (11.42)$$

This enumeration is still too conservative since the condition $\hat{Q} < 0$ is sufficient but not necessary for $\hat{S}\left(f_{m,n}\left(x,a\right),x\right) < 0$. In this case, m_1 is an upper bound for the minimum possible m, because the behavior of $\hat{S}\left(f_{m,n}\left(x,a\right),x\right)$ shows that, for the values of m_1 given above, the minimum value of m (called m_0) may be smaller than m_1. The pairs (δ, m_0) found numerically are as follows:

$$(\delta, m_0) : (2,2), (4,2), (6,4), (8,4), (10,4), (12,6)\ldots\,etc. \quad (11.43)$$

Thus numerical simulations show that $f_{m,m+\delta}$ has a negative Schwarzian derivative for any $m \geq m_0(\delta)$ in the range $0 \leq a \leq 1$ and that $f_{m,n}$ displays robust chaos for at least a small neighborhood $a_{\min} \leq a \leq 1$ for all even $n > m > 0$ integer pairs that have been tested. The above method is still not applicable. For example, $f_{2,8}$, which corresponds to $\delta = 6, m = 2$, has a positive Schwarzian derivative for most considered ranges of the x and a, which implies the coexistence of several attractors. However, replacing the given values of m and n, one finds

$$\hat{S}\left(f_{2,8}\left(x,a\right),x\right) = -48\left(336b^{2}x^{12} - 56abx^{6} + a^{2}\right). \quad (11.44)$$

Thus $\hat{S}\left(f_{2,8}\left(x,a\right),x\right) < 0$ for $a = 1$, and a sign change in $\hat{S}\left(f_{2,8}\left(x,a\right),x\right)$ appears at the first root x satisfying

$$\frac{bx^6}{a} = \frac{1}{12} \pm \frac{1}{6\sqrt{7}} \quad (11.45)$$

of the equation $\hat{S}(f_{2,8}(x,a),x) = 0$. Replacing $b = 1 - a$, then the largest value of a that can be obtained from Eq. (11.45) (which comes from $x = 1$ and the minus sign in the square root) is $0.980066 < a < 1$ in which $f_{2,8}$ displays robust chaos. The invariant distributions of the maps (11.39) are similar to that of the logistic map (6.25). For a few pairs m, n at several values of a, these invariant distributions were calculated numerically, in which they are smooth, concave, and with divergences at $x = \pm 1$. Generally, the divergences at the two extremes are controlled by m, the smallest exponent, and are of the form

$$\rho(x) \propto (1 - |x|)^{-\left(1 - \frac{1}{m}\right)} \tag{11.46}$$

using the expression for the *Frobenius–Perron equation* for the evolution of their instantaneous distribution $\rho_m(x, t)$ under a mapping $f(x)$

$$\begin{cases} \rho_m(x, t+1) = \sum_i \frac{\rho_m(y_i, t)}{\left|\frac{df_m(y_i)}{dy}\right|}, \\ f_m(x) = 1 - 2x^m \end{cases} \tag{11.47}$$

where y_i are the preimages of x under the mapping f_m. More details can be found in [Gabriel (2004)].

11.2.2 Robust chaos without the period-n-tupling scenario

In [Jafarizadeh & Benhia (2002)], a new hierarchy of many-parameter families of maps of the interval $[0, 1]$ defined as ratios of polynomials with an invariant measure along with the Kolmogorov-Sinai (KS) entropy were calculated analytically for arbitrary values of the parameters. This family of maps under consideration differs from the usual maps by the non-occurrence of a period doubling or period-n-tupling cascade bifurcation to chaos. These maps have a single fixed point attractor, and they bifurcate directly to chaos without having a period-n-tupling scenario.

The family of maps studied in [Jafarizadeh & Benhia (2002)] is defined on the interval $[0, 1]$ as the ratio of polynomials of degree N given by

$$\Phi_N(x, \alpha) = \frac{\alpha^2 (1 + (-1)^N F_1(-N, N, \frac{1}{2}, x))}{((\alpha^2 + 1) + (\alpha^2 - 1)(-1)^N F_1\left(-N, N, \frac{1}{2}, x\right))} \tag{11.48}$$

or

$$\Phi_N(x, \alpha) = \frac{\alpha^2 (T_N(\sqrt{x}))^2}{1 + (\alpha^2 - 1)(T_N(\sqrt{x}))^2} \tag{11.49}$$

where N is an integer greater than 1, and the hypergeometric polynomials of degree N, $F_1\left(-N, N, \frac{1}{2}, x\right)$ are given by

$$F_1\left(-N, N, \frac{1}{2}, x\right) = (-1)^N \cos\left(2N \arccos \sqrt{x}\right) = (-1)^N T_{2N}\left(\sqrt{x}\right) \tag{11.50}$$

where $T_N(x)$ are Chebyshev polynomials of type I [Wang & Guo (1989)]. We have (see Exercise 244) that $F_1\left(-N, N, \frac{1}{2}, x\right)$ and $T_N(x)$ map the unit interval $[0, 1]$ into itself. The function $\Phi_N(x, \alpha)$ is called an $(N-1)$-nodal map, and it has $(N-1)$ critical points in $[0, 1]$ since its derivative is proportional to the derivative of the hypergeometric polynomial $F_1\left(-N, N, \frac{1}{2}, x\right)$ which is itself a hypergeometric polynomial of degree $(N-1)$.

The Schwarzian derivative of the map (11.49) is given by

$$S\left(\Phi_N\left(x, \alpha\right)\right) = S\left(F_1\left(-N, N, \frac{1}{2}, x\right)\right) \leq 0, \tag{11.51}$$

and it has at most $(N+1)$ attracting periodic orbits as shown in [Devaney (1982)]. New hierarchies of families of many-parameter chaotic maps with an invariant measure can be obtained from the maps (11.49) simply by the composition of these maps. Indeed, for $n \in \mathbb{N}$, let the functions $\Phi_{N_k}(x, \alpha_k), k = 1, 2, ..., n$ be defined by (11.49), and let $\Phi_{N_1, N_2, ..., N_n}^{\alpha_1, \alpha_2, ..., \alpha_n}(x)$ denote their composition which can be written in terms of them in the following form:

$$\begin{cases} \Phi_{N_1, N_2, ..., N_n}^{\alpha_1, \alpha_2, ..., \alpha_n}(x) = g_1 = \Phi_{N_1}\left(\Phi\left(N_2 \cdots \underbrace{\left(\Phi_{N_n}(x, \alpha_n), \alpha_{(n-1)}\right)}_{n} ..., \alpha_2\right), \alpha_1\right) \\ g_1 = \Phi_{N_1} \circ \Phi_{N_2} \circ ... \circ \Phi_{N_n}(x). \end{cases} \tag{11.52}$$

In this case, the Schwarzian derivative of the map (11.52) is also negative because these maps consist of compositions of $(N_k - 1)$-nodals $(k = 1, 2, ..., n)$ maps with negative Schwarzian derivative. Hence they are $(N_1 N_2 ... N_n - 1)$-nodal maps [Devaney (1982)]. Therefore, maps (11.52) have at most $N_1 N_2 ... N_n + 1$ attracting periodic orbits [Devaney (1982)]. In fact, it was shown in [Jafarizadeh & Benhia (2002)] that these maps have only a single period-one stable fixed point as follows:

Let $\Phi^{(m)}$ be the m-composition of these functions. Then the derivative of $\Phi^{(m)}$ at its possible $m \times n$ periodic points of an m-cycle defined by

$$\begin{cases} x_{\mu, k+1} = \Phi_{N_k}(x_{\mu, k}, \alpha_k), x_{1, \mu+1} = \Phi_{N_n}(x_{n, \mu}, \alpha_N), \\ x_{1,1} = \Phi_{N_n}(x_{m,n}, \alpha_n), \mu = 1, 2..., m, k = 1, 2, ..., n \end{cases} \tag{11.53}$$

are given by

$$\left|\frac{d\Phi^{(m)}}{dx}\right| = \prod_{\mu=1}^{m} \prod_{k=1}^{n} \left|\frac{N_k}{\alpha_k}\left(\alpha_k^2 + \left(1 - \alpha_k^2\right) x_{\mu, k}\right)\right|. \tag{11.54}$$

Since for $x_{\mu,k} \in [0,1]$, one has
$$\min\left(\alpha_k^2 + 1 - \alpha_k^2 x_{\mu,k}\right) = \min\left(1, \alpha_k^2\right). \tag{11.55}$$
Therefore,
$$\min\left|\frac{d\Phi^{(m)}}{dx}\right| = \prod_{k=1}^{n}\left(\frac{N_k}{\alpha_k}\min\left(1,\alpha_k^2\right)\right)^m. \tag{11.56}$$
Hence (11.56) is definitely greater than 1 for
$$\prod_{k=1}^{n}\frac{1}{N_k} < \prod_{k=1}^{n}\alpha_k < \prod_{k=1}^{n}N_k. \tag{11.57}$$
Hence these maps do not have any kind of m-cycle or periodic orbits in the region of the parameter space defined by (11.57), i.e., they are chaotic. Relation (11.54) implies that $\left|\frac{d\Phi^{(m)}}{dx}\right|$ at the $m \times n$ periodic points of the m-cycle belonging to the interval $[0,1]$, and vary between $\prod_{k=1}^{n}(\alpha_k N_k)^m$ and $\prod_{k=1}^{n}\left(\frac{N_k}{\alpha_k}\right)^m$ for $\prod_{k=1}^{n}\alpha_k < \prod_{k=1}^{n}\frac{1}{N_k}$ and between $\prod_{k=1}^{n}\left(\frac{N_k}{\alpha_k}\right)^m$ and $\prod_{k=1}^{n}(\alpha_k N_k)^m$ for $\prod_{k=1}^{n}\alpha_k > \prod_{k=1}^{n}\frac{1}{N_k}$. Note that in special case of odd integer values of $N-1, N_2, ..., N_n$, one has that $x = 1$ and $x = 0$ belong to one of the m-cycles, and $\left|\frac{d\Phi^{(m)}}{dx}\right| < 1$ because $\prod_{k=1}^{n}\alpha_k < \prod_{k=1}^{n}\frac{1}{N_k}$ or $(\prod_{k=1}^{n}\alpha_k > \prod_{k=1}^{n}N_k)$. Therefore, the curve of $\Phi^{(m)}$ starts at $x = 1$ ($x = 0$) beneath the bisector and then crosses it at the previous (next) periodic point with slope greater than one because the formula (11.54) implies that the slope at the fixed points increases with the decreasing (increasing) $|x_{\mu,k}|$. In this case, all periodic points of n-cycles are unstable except for $x = 1$ ($x = 0$) because the slope is greater than one, and this is possible only if $x = 1$ ($x = 0$) is the only period-one fixed point of these maps (11.49). Hence all m-cycles except for possible period-one fixed points $x = 1$ and $x = 0$ are unstable. For $\alpha_k > 0, k = 1, 2, ..., n$ and $\prod_{k=1}^{n}\alpha_k < \prod_{k=1}^{n}\frac{1}{N_k}$ and for odd integer values of $N_1, N_2, ..., N_n$, the fixed point $x = 0$ is a stable fixed point of these maps. If one of the integers $N_k, k = 1, 2, ..., n$ is even, then $x = 0$ is not a stable fixed point, but the fixed point at $x = 1$ is stable for $\prod_{k=1}^{n}\alpha_k > \prod_{k=1}^{n}N_k$ and $\alpha_k < \infty, k = 1, 2, ..., n$ for all integer values of $N_1, N_2, ..., N_n$.

Some examples of the above situation are given by

$$\begin{cases} \Phi_2^\alpha(x) = \frac{\alpha^2(2x-1)^2}{4x(1-x)+\alpha^2(2x-1)^2} \\ \Phi_3^\alpha(x) = \frac{\alpha^2 x(4x-3)^2}{\alpha^2 x(4x-3)^2+(1-x)(4x-1)^2} \\ \Phi_4^\alpha(x) = \frac{\alpha^2(1-8x(1-x))^2}{\alpha^2(1-8x(1-x))^2+16x(1-x)(1-2x)^2} \\ \Phi_5^\alpha(x) = \frac{\alpha^2 x\left(16x^2-20x+5\right)^2}{\alpha^2 x(16x^2-20x+5)^2+(1-x)(16x^2-(2x-1))} \\ \Phi_{2,2}^{\alpha_1,\alpha_2}(x) = \frac{\alpha_1^2 x\left(4x(x-1)+(2x-1)^2\alpha_2^2\right)^2}{\alpha_1^2\left(4x(x-1)+(2x-1)^2\alpha_2^2\right)^2-16x\alpha_2^2(2x-1)^2(x-1)} \\ \Phi_{2,3}^{\alpha_1,\alpha_2}(x) = \frac{\alpha_1^2\left((x-1)(4x-1)^2+x(4x-3)^2\alpha_2^2\right)^2}{\alpha_1^2\left((x-1)(4x-1)^2+x(4x-3)^2\alpha_2^2\right)-4x\alpha_2^2(x-1)(4x-1)^2(4x-3)^2} \\ \Phi_{3,2}^{\alpha_1,\alpha_2}(x) = \frac{\alpha_2^2\left((x-1)(4x-1)^2+x(4x-3)^2\alpha_1^2\right)^2}{\alpha_2^2\left((x-1)(4x-1)^2+x(4x-3)^2\alpha_1^2\right)-4x\alpha_1^2(x-1)(4x-1)^2(4x-3)^2} \\ \Phi_{3,3}^{\alpha_1,\alpha_2}(x) = \frac{\alpha_1^2\alpha_2^2 x(4x-3)^2\left(3(x-1)(4x-1)^2+x(4x-3)^2\alpha_2^2\right)}{\left(-(x-1)^3(4x-1)^6+3x(4x-3)^2\left(3\alpha_1^2-2\right)\right)(x-1)^2(4x-1)^4\alpha_2^2+h} \end{cases}$$

(11.58)

where

$$h = 3x^2\alpha_2^4 - 3 + 2\alpha_1^2(x-1)(4x-1)^2(4x-3)^4 + \alpha_1^2\alpha_2^6 x^3(4x-3)^6. \quad (11.59)$$

The conjugate or isomorphic maps for the maps (11.49) are given by

$$\begin{cases} \tilde{\Phi}_{N_1,N_2,\ldots,N_n}^{\alpha_1,\alpha_2,\ldots,\alpha_n}(x) = \frac{1}{\alpha_1^2}\tan^2\left(N_1 \arctan g_2\right) \\ g_2 = \sqrt{\frac{1}{\alpha_2^2}\tan^2\left(N_2 \arctan \ldots \sqrt{\frac{1}{\alpha_n^2}\tan^2 N_n \arctan \sqrt{x}\ldots}\right)} \end{cases} \quad (11.60)$$

where

$$\tilde{\Phi}_{N_k}(x,\alpha_k) = h \circ \Phi_{N_k}(x,\alpha_k) \circ h^{(-1)} = \frac{1}{\alpha_k^2}\tan^2\left(N_k \arctan \sqrt{x}\right), \quad (11.61)$$

and $h(x) = \frac{1-x}{x}$ maps $I = [0,1]$ into $[0,\infty)$ and transforms the maps $\Phi_{N_k}(x,\alpha_k)$ into $\tilde{\Phi}_{N_k}(x,\alpha_k)$. The conjugate maps (11.60) are very useful in the derivation of the invariant measure and calculation of the KS-entropy of the maps (11.49). See Exercise 244.

11.2.3 The B-exponential map

In this section, we present a one-dimensional generalization of the logistic map (6.25) described in [Mahesh et al. (2006)] called the "*B-exponential map*". This map exhibits robust chaos for all real values of the parameter $B \geq e^{-4}$. Generally, it is still possible to generate a number of maps, all of which exhibit robust chaos by means of topological conjugacy. For example, one can see robust chaos in the so-called "*generalized tent map*" introduced also in [Mahesh et al. (2006)]. As a practical application, a pseudo-random number generator based on this map was proposed by chaotically hopping between different trajectories for different values of B. This generator is called the *BEACH* (*B*-exponential all-chaotic map hopping) *pseudo-random number generator* and successfully passes stringent statistical randomness tests such as ENT, NIST, and Diehard.

The B-exponential map $GL(B, x)$ is defined by

$$\begin{cases} x_{n+1} = GL(B, x_n) \\ GL(B, x) = \frac{B - xB^x - (1-x)B^{1-x}}{B - \sqrt{B}}, 0 \leq x \leq 1 \text{ and } B \in \mathbb{R}^+ \end{cases} \quad (11.62)$$

where B is the adjustable parameter. The map (11.62) has the following properties:

(1) The B-exponential map (11.62) is a generalization of the logistic map (6.25) because of the following property[10]:
$$\lim_{B \to 1} GL(B, x) = 4x(1-x). \quad (11.63)$$

(2) For all x, the B-exponential map (11.62) tends to a constant function (with a value of 1) as B tends to ∞.

(3) In the interval $[0, 1]$, the function $GL(B, x)$ is a linear combination of $f(x) = xB^x$ and $f(1-x) = (1-x)B^{1-x}$ which are both single-hump maps for all B.

(4) The critical point of $f(x)$ is $-\frac{1}{\ln(B)} < 0$, and that of $f(1-x)$ is $1 + \frac{1}{\ln(B)} > 1$ for $B > 1$. Hence we know that $GL(B, x)$ is unimodal for $B > 1$, while for $B < 1$, these two critical points lie within $[0, 1]$ on either sides of $x = 0.5$.

(5) For $0 < B < e^{-4}$, $GL(B, x)$ is no longer unimodal because it loses its surjectivity and $x = 0.5$ is a local minimum.

(6) For all $B \geq e^{-4}$, $GL(B, x)$ is unimodal because $x = 0.5$ is a local maximum[11] and $GL(B, 0) = 0, GL(B, 0.5) = 1$. If there is a critical point in $[0, 0.5)$ and it is a local maximum, then there has to be another critical point in $[0, 0.5]$, and this means that $GL(B, x)$ will have at least five critical points in $[0, 1]$ owing to symmetry around $x = 0.5$. Hence $GL(B, x)$ is unimodal for $B \geq e^{-4}$.

(7) The B-exponential map (11.62) is concave for a wide range of B.

(8) Numerical estimation for $B \geq e^{-4}$ shows that the Lyapunov exponents defined by

$$\lambda(B) = \lim_{T \to \infty} \frac{1}{T} T \sum_{t=0}^{T} \left| \frac{d}{dx} GL(B, x)|_{x=x_t} \right| (3) \quad (11.64)$$

---
[10]This can be done using L'Hospital's rule.
[11]$B = e^{-4}$ is a point of inflexion.

for the B-exponential map (11.62) appear to be constant and equal to $\ln 2$.

(9) The bifurcation diagram of the B-exponential map (11.62) shows that this map is chaotic for a large range of B and that there is full chaos, with surjective mapping, for an infinite range of B. This property is very useful in generating pseudo-random numbers.

The verification of these properties is left to the reader in Exercise 246.

The most interesting property of the B-exponential map (11.62) is that it has robust chaos $B \geq e^{-4}$ as stated in the following results proved in [Mahesh et al. (2006)]:

Theorem 11.13. *(a) The B-exponential map (11.62) is topologically conjugate to the logistic map (6.25) for $e^{-4} \leq B < \infty$.*
(b) The B-exponential map (11.62) is chaotic for all real $B \geq e^{-4}$.
(c) The B-exponential map (11.62) exhibits robust chaos $B \geq e^{-4}$.

Proof. (a) This can be done by the transitivity of topological conjugacy relation. Indeed, since there is only one critical point for $GL(B,x)$ at $x = 0.5$ for $e^{-4} \leq B < \infty$, one can define a diffeomorphism of $GL(B,x)$ to the line $2x$ (standard tent map (11.3)) which is topologically conjugate to the logistic map (6.25) and for $0 \leq x < 0.5$ by associating the appropriate heights or by defining another diffeomorphism to the line $2 - 2x$ for $0.5 \leq x \leq 1$.

(b) Because there is no universal definition of chaos, the authors in [Mahesh et al. (2006)] use the following conditions as necessary and sufficient for chaos: Determinism, surjectivity and boundedness, sensitive dependence on initial conditions, positive topological entropy, and denseness of periodic points. Indeed, the map (11.62) is a deterministic map of $[0,1] \to [0,1]$ for $B \geq e^{-4}$, surjective on $[0,1]$ for $B \geq e^{-4}$, and bounded for all $0 \leq x \leq 1$. The map (11.62) has sensitive dependence on initial conditions (continuously) because it has a positive Lyapunov exponents of $\ln 2 = 0.6931$ (almost everywhere) for every $B \geq e^{-4}$. The map (11.62) has a positive topological entropy ($\simeq \ln 2 > 0$) because its symbolic dynamics is such that all possible transitions (0 to 0, 0 to 1, 1 to 0 and 1 to 1) are achieved and the Markov partitions are zero (if $0 \leq x < 0.5$) and one (if $0.5 \leq x \leq 1$). Successive iterations of the map (11.62) mix the domain, i.e., for every pair of open sets $A, B \subset [0,1]$, there is a $k > 0$ such that $T^{(k)}(A) \cap B \neq \emptyset$. Hence the map (11.62) is topologically transitive. Finally, the fact that the map (11.62) is topologically conjugate to the logistic map

(6.25) implies that periodic points are dense in $[0,1]$ for the map (11.62).
(c) The function $GL(B \geq e^{-4}, x)$ has only one critical point at $x = \frac{1}{2}$, and numerically it is possible to find that the Schwarzian derivative of the map (11.62) is negative. In this case, using the result in [Alligood et al. (1997)], it is possible to say that there can be at most one attracting periodic orbit with a critical point in its basin of attraction. This is a result of the fact that the function $GL(B, x)$ is smooth and unimodal for $B \geq e^{-4}$ and has one unstable fixed point at $x = 0.5$ because $|GL'(B, 0)| > 1$. This fixed point ends in a value zero after two iterations. Hence the map (11.62) does not have any attracting periodic orbits for $B \geq e^{-4}$. □

It was shown in [Aguirregabiria (2008)] that all examples of one-dimensional maps (expect (11.39) and (11.49)) displaying robust chaos are conjugate to the logistic map (6.25), which give a new easy way of generating smooth one-dimensional maps displaying robust chaos in the whole range of its parameter. The main characterizations of these maps are that the Lyapunov exponent varies widely with the parameter because they are conjugated through nonsmooth homeomorphisms similar to *Minkowski's question mark function* [Minkowski (1904)][12], *i.e.*, Lyapunov exponents take rather different values depending on the value of the parameter, and that the condition of negative Schwarzian derivative is not necessary for robust chaos. Indeed, the Schwarzian derivative (11.4) is invariant under linear fractional transformations [Jackson (1991), Appendix D], *i.e.*, that for constants a, b, c and d one has

$$S\frac{af(x)+b}{cf(x)+d} = Sf(x). \qquad (11.65)$$

Let $f : [0,1] \to [0,1]$ be any S-unimodal map of class C^3. Then the map

$$f_r(x) \equiv \frac{(1+r)f(x)}{f(c)+rf(x)}, (-1 < r < \infty) \qquad (11.66)$$

is S-unimodal for all $r > -1$ using (11.65) and the formula

$$f'_r(x) = \frac{(1+r)f(c)f'(x)}{[f(c)+rf(x)]^2}. \qquad (11.67)$$

Now, for

$$r > r_0 = \frac{f(c)}{f'(0)} - 1, \qquad (11.68)$$

---

[12] Minkowski's function is continuous with a continuous strictly increasing inverse, and its derivative is zero almost everywhere and infinite or undefined otherwise [Viader et al. (1998), Paradis et al. (2001)].

the origin is an unstable fixed point. Thus the map (11.67) has no stable periodic orbit because $f_r^2(c) = 0$. The Lyapunov exponent is positive for all $r > r_0$ as shown by numerical calculations using the formula

$$\lambda(\nu) = \lim_{N \to \infty} \sum_{n=1}^{T} \ln |f_r'(x_n)|. \qquad (11.69)$$

For $f(x) = x(1-x)$ in (11.66), the Lyapunov exponent is negative for $r < r_0 = -\frac{3}{4}$ and becomes positive at $r > r_0$ for which the map (11.66) displays a chaotic attractor. For the counterexample maps given above, the Lyapunov exponent is always $\ln 2$ or very close, but for the map (11.66) this exponent varies continuously with r. Indeed, the maximum value of this exponent is $\lambda_{\max} = \ln 2$, reached at $r = 0$, which corresponds to the case of the logistic map (6.25), which in turn is conjugate to the tent map (11.3). Similar results were obtained for the asymmetric map $f(x) = x(1-x^2)$ and for $f(x) = \sin \pi x$ with different values of r_0 and the location of the maximum. In conclusion, the natural invariant density is $\rho_0(x) = 2x(1-x) - \frac{1}{2}$. Generally, the robustness of chaos is guaranteed by Singer's Theorem 11.1 as shown in the previous examples. Now, let $f(x)$ be a S-unimodal map. Then the Schwarzian derivative $S(f_r)$ of the map f, defined by

$$f_r(x) = \frac{1 + r(x-c)^2}{f(c)} f(x) \qquad (11.70)$$

has a complicated formula which implies the impossibility of sign analysis. For $f(x) = x(1-x)$, map (11.70) is S-unimodal for $-4 < r < 4$ and displays robust chaos for $-3 < r < 4$. Similar results can be obtained with $f(x) = x$. All the smooth maps discussed above have negative Schwarzian derivative. This condition is restrictive and can be destroyed by a smooth change of the x coordinate [de Melo & van Strien (1988)] or by small perturbations [Jackson (1991), Appendix D]. In fact, this condition is not necessary to have robust chaos in one-dimensional smooth maps because the map (11.66) with the Singer's function [Jackson (1991), Appendix D] given by

$$f(x) = 7.86x - 23.31x^2 + 28.75x^3 - 13.3x^4 \qquad (11.71)$$

has a positive Schwarzian derivative in a subinterval of $[0,1]$ and robust chaos after the origin becomes unstable at $r \approx -0.88156$ by numerical calculation of the Lyapunov exponent with a maximum value of $\lambda_{\max} \approx 0.62$, which is smaller than the maximum value of $\ln 2$ obtained in all previous examples. The same things can be seen for the one-parameter family of maps given by

$$f_r(x) = \left(\frac{f(x)}{f(c)} \right)^r, (r > 0) \qquad (11.72)$$

and $f(x) = x(1-x)$. Indeed, the map (11.72) is S-unimodal only when $r = 1$. For $r > 1$, the function has a minimum at the origin, and $x = 0$ is a stable fixed point with a tiny basin of attraction. This point attracts the generic orbit, after a chaotic transient, which may be very long for values of r just above 1. For $0 < r < 1$, the Schwarzian derivative of (11.72) is positive near the origin and near $x = 1$ because

$$Sf_r(x) \sim \frac{1-r^2}{2x^2}, \text{ as } x \to 0 \quad (11.73)$$

implies that the map (11.72) is not even C^3. Hence robust chaos arises in map (11.72) after the fixed point located in the interval $\left(\frac{1}{2}, 1\right)$ becomes unstable at $r \simeq 0.1759$. Again, the maximum Lyapunov exponent is $\ln 2$ and is reached at $r = 1$. Similar results can be obtained with $f(x) = x\left(1 - x^2\right)$ and $f(x) = \sin \pi x$.

From the above study, it seems that all the above maps are conjugate because they have qualitatively similar dynamics, *i.e.*, the graphs of all the maps $g = f_r$ are similar; they start from $g(0) = 0$, increase monotonically until $g(c) = 1$, and then decrease monotonically until $g(1) = 0$. Namely, let g and $\tilde{g}$ be two of these maps. Then there exists a homeomorphism ϕ on $[0,1]$ such that $\tilde{g} = \phi \circ g \circ \phi^{-1}$, or there exists a continuous change of variables $x \to \tilde{x} = \phi(x)$, with continuous inverse, such that

$$\phi[g(x)] = \tilde{g}[\phi(x)], \forall x \in [0,1] \quad (11.74)$$

in which the dynamical systems $x_{n+1} = g(x_n)$ and $\tilde{x}_{n+1} = \tilde{g}(x_n)$ have essentially equivalent dynamics. In particular, if the unstable periodic orbits are dense for g, then they are dense for $\tilde{g}$, and if ϕ and ϕ^{-1} are smooth, then g and $\tilde{g}$ have the same Lyapunov exponent [Ott (2002)]. This a simple method of constructing families of maps with robust chaos, *i.e.*, take a map $f(x)$ that displays robust chaos and a smooth homeomorphism $\phi(x)$ on $[0,1]$ depending continuously on a parameter r. Then the map $f_r = \phi \circ f^{-1}$ also displays robust chaos for all values of r. As an example, take $f(x) = 4x(1-x)$ and $\phi(x) = x^r$ for $r > 0$ to obtain

$$f_r(x) = 4^r x \left(1 - x^{\frac{1}{r}}\right)^r, (r > 0) \quad (11.75)$$

whose Lyapunov exponent will be $\lambda = \ln 2$ for all $r > 0$. However, map (11.75) is not S-unimodal except for $r = 1$ because its Schwarzian derivative becomes positive near $x = 0$ for $r > 1$ and near $x = 1$ for $0 < r < 1$. In this case, the solution of the map (11.75) is $x_n = \sin^{2r}\left(2^n \arcsin x_0^{\frac{1}{2r}}\right)$, and its natural invariant density is $\rho(x) = \pi^2 r^2 x^2 \left(x^{-\frac{1}{r}} - 1\right)^{-\frac{1}{2}}$.

Another way to generate robust chaos is to use a method of successive approximations for two maps g and $\tilde{g}$ as follows:

$$\phi_0(x) = x \qquad (11.76)$$

$$\phi_{n+1}(x) = \tilde{g}^{-1}[\phi_n(g(x))], (n = 0, 1, 2, ...). \qquad (11.77)$$

The inverse function $\tilde{g}^{-1}$ is two-valued in this kind of map, but the right preimage is given by the condition that if c and $\tilde{c}$ are the critical points of g and $\tilde{g}$, then $\phi(x) > \tilde{c}$ when $x > c$. The method given by (11.76) and (11.77) converges quickly when $\tilde{g}(x) = 4x(1-x)$ and g is one of the maps (11.14) or (11.15). The accuracy of this method was verified using the fact that the *question mark function* is the homeomorphism conjugating the tent map (11.3) and the Farey map [Panti (2008)] defined as

$$g(x) = \begin{cases} \frac{x}{1-x}, 0 \leq x \leq \frac{1}{2} \\ \frac{1-x}{x}, \frac{1}{2} \leq x \leq 1. \end{cases} \qquad (11.78)$$

An alternate method to construct the functions g and $\tilde{g}$ uses the fact that Minkowski's question mark function can be recursively constructed using the Farey sequence and continuity [Girgensohn (1996)]. Indeed, one starts from the critical point $x_0 = c$ since $y_0 = \phi(x_0) = \phi(c) = \tilde{c}$. Then for each pair $(x_n, y_n \equiv \phi(x_n))$ already computed, one can calculate two new pairs

$$(x_{n+1}, y_{n+1} = \phi(x_{n+1})) = \left(g_\pm^{-1}(x_n), \tilde{g}_\pm^{-1}(y_n)\right) \qquad (11.79)$$

where $g_-^{-1}(x)$ is the value of y satisfying $g(y) = x$ and $y \leq c$, while $y = g_+^{-1}(x)$ is given by the conditions $g(y) = x$ and $y > c$. Analogous definitions are used for $\tilde{g}_\pm^{-1}$. This method gives the same results as the successive approximations, and it works for other pairs of maps constructed by means of (11.66), (11.70), or (11.72). From the above analysis, it is clear that all the maps considered in this section are thus conjugate to $f(x) = 4x(1-x)$, and, in consequence [Ott (2002)], to the tent map. Since the function ϕ and its inverse are smooth, all these maps share the Lyapunov exponent $\lambda = \ln 2$. Numerical analysis shows (in all the above cases) that the corresponding Lyapunov exponents are different because they are conjugated by nonsmooth homeomorphisms. The graphs of these functions looks very similar to the graph of Minkowski's question mark function [Minkowski (1904)].

11.3 Border-collision bifurcation and robust chaos

In this section, we discuss the relations between border-collision bifurcations and robust chaos in one-dimensional piecewise-smooth maps. The importance of these maps can be seen from the fact that the dynamics of a number of switching circuits can be represented by one-dimensional piecewise-smooth maps under discrete modeling, in particular in power electronic circuits [Wolf et al. (1994), Deane & Hamill (1996), Tse & Chan (1997), Banerjee & Chakrabarty (1998), Banerjee et al. (2000), Robert & Robert (2002)].

First, recall that the analysis of this relation is based on ingredients, the first of which is the affinity of the corresponding normal forms for fixed points on borders, and second is the behavior of fixed points and periodic points depending on the bifurcation parameter for the scenarios associated with the various cases.

Consider the one-dimensional piecewise-smooth system

$$x_{k+1} = f(x_k, \mu) = \begin{cases} g(x, \mu), x \leq x_b \\ h(x, \mu), x \geq x_b \end{cases} \quad (11.80)$$

where μ is the bifurcation parameter and x_b is the border. The map $f : \mathbb{R} \times \mathbb{R} \to \mathbb{R}$ is assumed to be piecewise-smooth, *i.e.*, f depends smoothly on x everywhere except at x_b, where it is continuous in x. It is also assumed that f depends smoothly on μ everywhere. Let R_L and R_R be the two regions in state space separated by the border

$$\begin{cases} R_L = \{x : x \leq x_b\} \\ R_R = \{x : x \geq x_b\}. \end{cases} \quad (11.81)$$

Let $x_0(\mu)$ be a possible path of fixed points of f, with this path depending continuously on μ. Suppose also that the fixed point hits the boundary at a critical parameter value μ_b: $x_0(\mu_b) = x_b$.

11.3.1 *Normal form for piecewise-smooth one-dimensional maps*

The normal form of the piecewise-smooth one-dimensional map (11.80) is given by [Yuan (1997), Banerjee et al. (2000)]

$$G_1(x, \mu) = \begin{cases} ax + \mu, x \leq 0 \\ bx + \mu, x \geq 0. \end{cases} \quad (11.82)$$

The normal form (11.82) at a fixed point on the border is a piecewise affine approximation of the map in the neighborhood of the border point x_b. The method of derivation of such a form is as follows: Let $\bar{x} = x - x_b$ and $\bar{\mu} = \mu - \mu_b$. Then Eq.(11.80) becomes

$$\bar{f}(\bar{x}, \bar{\mu}) = \begin{cases} g(\bar{x} + x_b, \bar{\mu} + \mu_b), \bar{x} \leq 0 \\ h(\bar{x} + x_b, \bar{\mu} + \mu_b), \bar{x} \geq 0. \end{cases} \quad (11.83)$$

Hence for map (11.83), the border is at $\bar{x} = 0$, and the state space is divided into two halves, $\mathbb{R}_- = (-\infty, 0]$ and $\mathbb{R}_+ = [0, \infty)$, and the fixed point of (11.80) is at the border for the parameter value $\bar{\mu} = 0$. Expanding $\bar{f}$ to first order about $(0,0)$ gives

$$\begin{cases} \bar{f}(\bar{x}, \bar{\mu}) = \begin{cases} a\bar{x} + \bar{\mu}v + O(\bar{x}, \bar{\mu}), \bar{x} \leq 0 \\ b\bar{x} + \bar{\mu}v + O(\bar{x}, \bar{\mu}), \bar{x} \geq 0 \end{cases} \\ a = \lim_{x \to 0^-} \frac{\partial}{\partial x} \bar{f}(\bar{x}, 0) \\ b = \lim_{x \to 0^+} \frac{\partial}{\partial x} \bar{f}(\bar{x}, 0) \\ v = \lim_{x \to 0} \frac{\partial}{\partial \mu} \bar{f}(\bar{x}, 0). \end{cases} \quad (11.84)$$

Note that the last limit in (11.84) does not depend on the direction of approach to zero by x due to the smoothness of f in μ. Assume that $v \neq 0$, $|a| \neq 1$, and $|b| \neq 1$ so that the nonlinear terms are negligible close to the border. Define a new parameter $\mu'' = \bar{\mu}v$ and neglect the higher-order terms as in [Yuan (1997), Banerjee et al. (2000)]. Then the 1-D normal form is given by

$$G_1(\bar{x}, \bar{\mu}) = \begin{cases} a\bar{x} + \mu'', \bar{x} \leq 0 \\ b\bar{x} + \mu'', \bar{x} \geq 0, \end{cases} \quad (11.85)$$

which has the same form as (11.82).

11.3.2 *Border-collision bifurcation scenarios*

In this section, we recall the occurrence of various types of border-collision bifurcations from x_b for μ near μ_b. Let x_R^* and x_L^* be the possible fixed points of the system near the border to the right $(x > x_b)$ and left $(x < x_b)$ of the border, respectively. Then in the normal form of (11.82), one has $x_R^* = \frac{\mu}{1-b} \geq 0$ if $b < 1$, and $x_L^* = \frac{\mu}{1-a} \leq 0$ if $a < 1$.

The partitioning of the parameter space into regions with the same qualitative bifurcation phenomena is shown in Fig. 11.6. The details are discussed below.

11.3.2.1 *Scenario A. Persistent fixed point (nonbifurcation)*

Border-collision bifurcation scenarios can be obtained by various combinations of the parameters $a \geq b$ as μ is varied because the normal form (11.82)

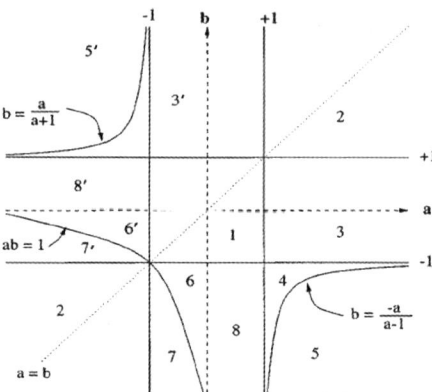

Fig. 11.6 The partitioning of the parameter space into regions with the same qualitative bifurcation phenomena. Numbering of the cases are as discussed in the text. 1) Period-1 to Period-1; 2) No attractor to no attractor; 3) No fixed point to Period-1; 4) No fixed point to chaotic attractor; 5) No fixed point to unstable chaotic orbit (no attractor); 6) Period-1 to Period-2; 7) Period-1 to no attractor; and 8) Period-1 to periodic or chaotic attractor. The regions shown with primed numbers have the same bifurcation behavior as the unprimed ones when μ is varied in the opposite direction. Reprinted (Figure 2) with permission from [Banerjee, S., Karthik, M. S., Yuan, G. and Yorke, J. A., IEEE Trans, Circuits and Systems, 47(3), 389-394, 2000. Copyright (2000) by IEEE].

is invariant under the transformation $x \to -x$, $\mu \to -\mu$, and $a \rightleftarrows b$. The following two situations lead to this scenario:

Scenario A1: Scenario A1 (Persistence of Stable Fixed Point) or Period-1 $\to$ Period-1 in [Yuan (1997), Banerjee et al. (2000)]. If

$$-1 < b \leq a < 1, \tag{11.86}$$

then a stable fixed point for $\mu < 0$ persists and remains stable for $\mu > 0$. Here, the fixed point changes continuously as a function of the bifurcation parameter, and the eigenvalue associated with the system linearization at the fixed point changes discontinuously from a to b at $\mu = 0$. Figure 11.7 shows the dependence of the map (11.80) and its fixed points on μ near the border with a single eigenvalue (the slopes of the map on the two sides of the vertical axis), which changes discontinuously at the border.

Scenario A2: (Persistence of Unstable Fixed Point) or No Attractor $\to$ No Attractor in [Yuan (1997), Banerjee et al. (2000)]. If

$$1 < b \leq a \tag{11.87}$$

or

$$b \leq a < -1, \tag{11.88}$$

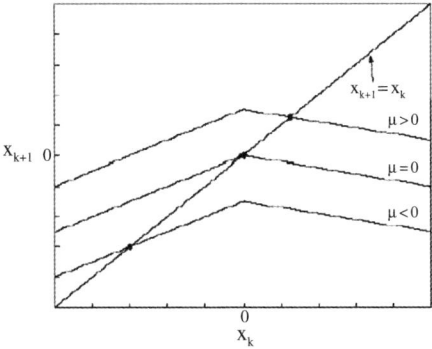

Fig. 11.7 Dependence of the first return map and its fixed point on μ for Scenario A ($-1 < b < a < 1$). Intersections of the map with the line $x_{k+1} = x_k$ are the fixed points [Hassouneh & Abed, 2002].

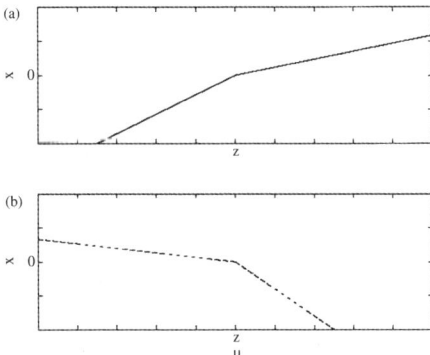

Fig. 11.8 Bifurcation diagrams for Scenario A. A solid line represents a stable fixed point, whereas a dashed line represents an unstable fixed point. (a) A typical bifurcation diagram for Scenario A1 (b) A typical bifurcation diagram for Scenario A2 (i) and (ii) [Hassouneh & Abed, 2002].

then an unstable fixed point for $\mu < 0$ persists and remains unstable for $\mu > 0$. Here, there is one unstable fixed point (that depends continuously on μ) for both positive and negative values of μ, no local attractors exist, and the system trajectory diverges for all initial conditions. Typical bifurcation diagrams for Scenarios A1 and A2 are shown in Fig. 11.8.

11.3.2.2 Scenario B. Border-collision pair bifurcation

For other values of the parameters a and b given in (11.86), (11.87), and (11.88), there are two main kinds of border-collision bifurcations, namely, *border-collision pair bifurcations* which are similar to saddle-node bifurcations (or tangent bifurcations) in smooth systems, and *border-crossing bifurcations*, which have some similarities with period doubling bifurcations in smooth maps (supercritical period doubling bifurcations in smooth maps with one distinction). In border-collision pair bifurcations, the map (11.80) has two fixed points (one side of the border and the other fixed point is on the opposite side) for positive (respectively, negative) values of μ, and no fixed points for negative (respectively, positive) values of μ. Hence the border-collision pair bifurcation occurs if

$$b < 1 < a. \tag{11.89}$$

There are three situations that lead to this scenario as follows:

Scenario B1: (Merging and Annihilation of Stable and Unstable Fixed Points) or No Fixed Point → Period-1 in [Yuan (1997), Banerjee et al. (2000)]. If

$$-1 < b < 1 < a, \tag{11.90}$$

then there is a bifurcation from no fixed point to two period-1 fixed points. Here, there is no fixed point for $\mu < 0$, while there are two fixed points x_L^* (unstable) and x_R^* (stable) for $\mu \geq 0$ as shown in Fig. 11.9(a).

Scenario B2: (Merging and Annihilation of Two Unstable Fixed Points, Plus Chaos) or No Fixed Point → Chaos in [Yuan (1997), Banerjee et al. (2000)]. If

$$\begin{cases} a > 1 \\ -\frac{a}{a-1} < b < -1, \end{cases} \tag{11.91}$$

then there is a bifurcation from no fixed point to two unstable fixed points plus a growing chaotic attractor as μ is increased through zero as shown in Fig. 11.9(b).

Scenario B3: (Merging and Annihilation of Two Unstable Fixed Points) or No Fixed Point → No attractor in [Yuan (1997), Banerjee et al. (2000)]. If

$$\begin{cases} a > 1 \\ b < -\frac{a}{a-1}, \end{cases} \tag{11.92}$$

then there is a bifurcation from no fixed point to two unstable fixed points as μ is increased through zero as shown in Fig. 11.9(c), and the system trajectory diverges for all initial conditions.

The three typical bifurcation diagrams for Scenario B are shown in Fig. 11.9.

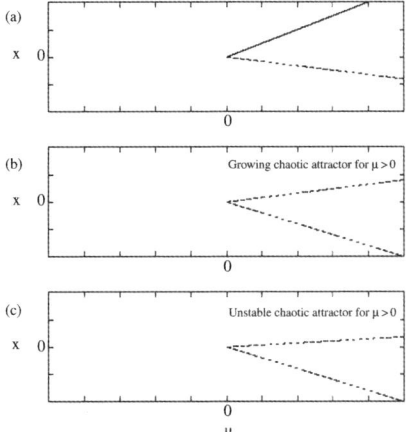

Fig. 11.9 Bifurcation diagrams for Scenario B. A solid line represents a stable fixed point, whereas a dashed line represents an unstable fixed point. (a) A typical bifurcation diagram for Scenario B1 (b) A typical bifurcation diagram for Scenario B2 (c) A typical bifurcation diagram for Scenario B3 [Hassouneh and Abed, 2002].

11.3.2.3 Scenario C. border-crossing bifurcation

In a border-crossing bifurcation, the fixed point persists and crosses the border as μ is varied through zero, and other attractors or repellers appear or disappear as a result of the bifurcation. Indeed, a border-crossing bifurcation occurs if
$$\begin{cases} a > -1 \\ b < -1. \end{cases} \quad (11.93)$$
There are three situations that lead to this scenario.

Scenario C1: (Supercritical Border-Collision Period Doubling) or Period-1 → Period-2 in [Yuan (1997), Banerjee et al. (2000)]. If
$$\begin{cases} b < -1 < a < 0 \\ ab < 1, \end{cases} \quad (11.94)$$
then there is a bifurcation from a stable fixed point to an unstable fixed point plus a stable period-2 orbit. The first condition of (11.94) implies that there is a bifurcation from a stable period-1 fixed point to an unstable period-1 fixed point, and the second implies the emergence of a stable period-2 solution for $\mu > 0$.

Scenario C2: (Subcritical Border-Collision Period Doubling) or Period-1 → No Attractor in [Yuan (1997), Banerjee et al. (2000)]. If
$$\begin{cases} b < -1 < a < 0 \\ ab > 1, \end{cases} \quad (11.95)$$

then there is a bifurcation from a stable fixed point to an unstable fixed point. In this case, there is a period-1 attractor and an unstable period-2 orbit for $\mu < 0$ and an unstable fixed point for $\mu > 0$, and the system trajectory diverges to infinity for $\mu > 0$.

Scenario C3: (Emergence of a Periodic or Chaotic Attractor from Stable Fixed Point) or Period-1 $\to$ Periodic or Chaotic Attractor in [Yuan (1997), Banerjee et al. (2000)]. If

$$\begin{cases} 0 < a < 1 \\ b < -1, \end{cases} \quad (11.96)$$

then there is a bifurcation from a stable fixed point to an unstable fixed point plus a period-n attractor with $n \geq 2$ or a chaotic attractor as μ is increased through zero. The specific scenario depends on the pair (a, b) as shown in Fig. 11.11.

For more details, the reader can see [Nusse & Yorke (1995)].

To show the bifurcation of a subcritical period-2 orbit for $\mu < 0$ in Scenario C2, consider the first and second return maps (for $\mu < 0$) given respectively by

$$x_{k+1} = \begin{cases} ax_k + \mu, x_k \leq 0 \\ bx_k + \mu, x_k \geq 0 \end{cases} \quad (11.97)$$

and

$$x_{k+2} = \begin{cases} abx_k + \mu(1+b), x_k \leq -\frac{\mu}{a} \\ a^2 x_k + \mu(1+a), -\frac{\mu}{a} \leq x_k \leq 0 \\ abx_k + \mu(1+a), x_k \geq 0. \end{cases} \quad (11.98)$$

The first return map (11.97) has a stable fixed point, $x_L^* = \frac{\mu}{1-a}$, and the second return map (11.98) has three fixed points, one of which coincides with x_L^* and two unstable ones $x_1^* = \frac{\mu(1+b)}{1-ab}$ and $x_2^* = \frac{\mu(1+a)}{1-ab}$, which form a period-2 orbit for the first return map (11.97). This result is shown in Fig. 11.10. From the above analysis, it is possible to determine regions in the ab-plane in which robust chaos can occurs, namely, Scenario B2 or Scenario C3, i.e., (11.91) or (11.96) that give the interval for robust chaos denoted by

$$RC = (1, \infty) \times \left(-\frac{a}{a-1}, -1\right) \cup (0, 1) \times (-\infty, -1) \quad (11.99)$$

as shown in Fig. 11.11. In fact, for $\mu > 0$, the behavior is robust chaotic for the whole region of (11.91) and a significant portion of (11.96), i.e., the chaotic attractor of the 1-D normal form (11.82) is robust in the contiguous region of the parameter space where no periodic windows exist [Banerjee et al. (2000)].

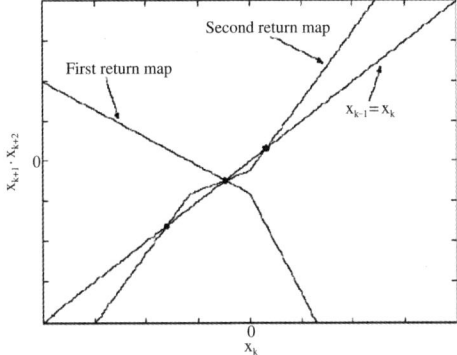

Fig. 11.10 First and second return maps for Scenario C2 showing a subcritical border-collision period doubling $((a, b, \mu) = (-0.7, -2.5, -0.04))$. Intersections of the maps with the line $x_{k+1} = x_k$ are the fixed points. [Hassouneh and Abed, 2002].

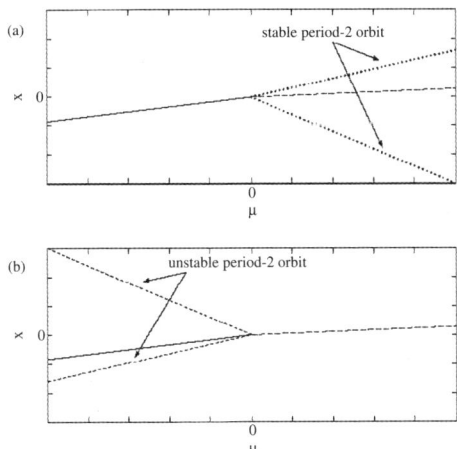

Fig. 11.11 Typical bifurcation diagrams for Scenario C1 and Scenario C2. A solid line represents a stable fixed point, whereas a dashed line represents an unstable fixed point. Dotted lines represent period-2 orbits: stable (dark) and unstable (light). (a) Supercritical border-collision period doubling (Scenario C1, $b < -1 < a < 0$ and $ab < 1$) (b) Subcritical border-collision period doubling (Scenario C2, $b < -1 < a < 0$ and $ab > 1$) [Hassouneh and Abed, 2002].

11.3.3 *Robust chaos in one-dimensional singular maps*

These types of 1-D singular maps play a key role in the theory of up-embedability of graphs [Chen & Liu (2007)]. In [Alvarez-Llamoza *et al.* (2008)], the critical behavior of the Lyapunov exponent of a 1-D singular

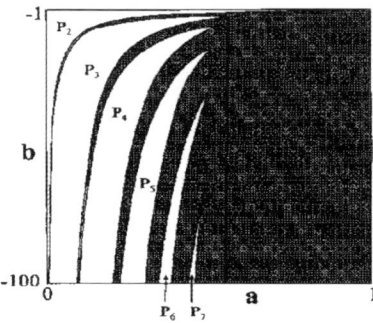

Fig. 11.12 Schematic drawing of the parameter region $0 < a < 1, b < -1$, showing the type of attractor for $\mu > 0$. The white regions correspond to period-n attractors (P_n : $n = 2, 3, ..., 7$), and the shaded regions have chaotic attractors. Reprinted (Figure 4) with permission from [Banerjee, S., Karthik, M. S., Yuan, G., and Yorke, J. A., IEEE Trans, Circuits and Systems, 47(3), 389-394, 2000. Copyright (2000) by IEEE].

map (which has only one face on a surface) near the transition to robust chaos by type-III intermittency was determined for a family of one-dimensional singular maps. The calculation of critical boundaries separating the region of robust chaos from the region of stable fixed points was given and discussed.

First, intermittent chaos is characterized by irregular switchings between long periodic-like signals, called the *laminar phase*, and comparatively short *chaotic bursts*. The statistical signature of intermittency is given by the scaling relations describing the dependence of the average length $\langle l \rangle$ of the laminar phases with a control parameter ϵ that measures the distance from the bifurcation point. For chaotic systems, this sign is a positive largest Lyapunov exponent λ[13]. For type-III intermittency, one has $\langle l \rangle \sim \epsilon^{-1}$ and $\lambda \sim \epsilon^{\frac{1}{2}}$ when $\epsilon \to 0$ [Pomeau & Manneville (1980)]. In [Khan et al. (1992), Fukushima & Yazaki (1995)], it was shown that $\langle l \rangle \sim \epsilon^v$ with $\frac{1}{2} < v < 1$.

The following classification of different types of intermittencies was given in [Pomeau & Manneville (1980)] according to the local Poincaré map associated with the system under consideration:

Definition 11.17. (a) Type-I intermittency occurs by a tangent bifurcation when the Floquet's multiplier for the Poincaré map crosses the unit circle in the complex plane through $+1$.

---

[13]The Lyapunov exponent can be considered as an order parameter that characterizes the transition to chaos by type-III intermittency.

(b) Type-II intermittency is due to a Hopf's bifurcation which appears as two complex eigenvalues of the Floquet's matrix cross the unit circle off the real axis.

(c) Type-III intermittency is associated to an inverse period doubling bifurcation whose Floquet's multiplier is -1.

Theoretical and numerical studies of type-III intermittency can be found in [Pomeau & Manneville (1980), Mayer-Kress & Haken (1984), Kodama et al. (1991), Khan et al. (1992), Pellegrini et al. (1993), Fukushima & Yazaki (1995), Russo et al. (2006)].

In [Alvarez-Llamoza et al. (2008)], the following singular map

$$\begin{cases} x_{n+1} = f(x_n) = \mu - |x_n|^z \\ \frac{d}{dx}\left(\mu - |x|^z\right) = -z|x|^{z-1} \end{cases} \quad (11.100)$$

was given as a simple model of a dynamical system displaying robust chaos associated with type-III intermittency[14], where $n \in \mathbb{Z}$, μ is a real parameter, and $|z| < 1$ describes the order of the singularity at the origin. Maps (11.100) are unbounded and display robust chaos on a single interval of the parameter μ. Robust chaos is also found in *logarithmic maps* [Kawabe & Kondo (1991), Cosenza & Gonzalez (1998)].

The Schwarzian derivative of (11.100) is always positive

$$S(f,x) = \frac{1-z^2}{2x^2} > 0, \text{ for all } |z| < 1. \quad (11.101)$$

Thus map (11.100) is not a unimodal map, and it does not exhibit a sequence of period doubling bifurcations. If a stable fixed point loses its stability at some critical value of the parameter to yield robust chaos, then condition (11.101) implies the occurrence of an inverse period doubling bifurcation. The map (11.100) has two stable fixed points for each value of z, $x_-^* < 0$ and $x_+^* \geq 0$, satisfying $f(x^*) = x^*$ and $|f'(x^*)| < 1$, where the fixed point x_-^* becomes unstable at the parameter value

$$\mu_-(z) = |z|^{\frac{z}{1-z}} - |z|^{\frac{1}{1-z}} \quad (11.102)$$

through an inverse period doubling bifurcation that gives robust chaos by type-III intermittency, while the fixed point x_+^* originates from a tangent bifurcation at the value

$$\mu_+(z) = |z|^{\frac{z}{1-z}} + |z|^{\frac{1}{1-z}} \quad (11.103)$$

---

[14]Robust chaos not associated with type-III intermittency has also been discovered in smooth, continuous, one-dimensional maps given is this chapter.

with a type-I intermittency transition to chaos. The boundaries $\mu_+(z)$ and $\mu_-(z)$ correspond to the Lyapunov exponent values $\lambda = 0$. For $z \in (0,1)$, the behaviors of the fixed points are interchanged; x_-^* displays a tangent bifurcation at (11.102), and a type-I intermittent transition to chaos occurs, while the fixed point x_+^* undergoes an inverse period doubling bifurcation at (11.103), giving the scenario for a type-III intermittent transition to chaos, i.e., the transition to chaos by type-III intermittency occurs at the critical parameter values $\mu_m(z) = \mu_-(z)$ for $z \in (-1,0)$, and $\mu_m(z) = \mu_+(z)$ for $z \in (0,1)$. Hence, near the critical boundary $\mu_m(z)$ within the chaotic region, the map (11.100) shows type-III intermittency. In this case, type-I or type-III intermittencies appear at the boundaries of the robust chaos intervals, i.e., $z = -0.5$ or $z = 0.5$. The width of the interval for robust chaos in the parameter μ for a given $|z| < 1$ is

$$\Delta\mu(z) = \mu_+(z) - \mu_-(z) = 2|z|^{\frac{1}{1-z}}. \qquad (11.104)$$

The Lyapunov exponent is positive in the robust chaos interval $\Delta\mu(z)$, and the transition to chaos through type-I intermittency is smooth, while the transition to chaos by type-III intermittency has a discontinuity (due to the sudden loss of stability of the fixed point associated with the inverse period doubling bifurcation) in the derivative of the Lyapunov exponent at the parameter values corresponding to the critical curve $\mu_m(z)$. The behavior of the Lyapunov exponent near the boundary $\mu_m(z)$ within the chaotic region is characterized by a scaling relation

$$\lambda \sim \epsilon^{\beta(z)} \qquad (11.105)$$

where $\epsilon = |\mu - \mu_m|$ and $\beta(z)$ is a critical exponent expressing the order of the transition. The exponent $\beta(z)$ is calculated from the slopes of each curve in the log-log plot of the Lyapunov exponent versus ϵ. Thus the scaling behavior of the Lyapunov exponent at the transition to chaos is characterized by the values of the critical exponent β in the interval $0 \leq \beta < 1/2$ as the singularity z of the map varies in $(-1,1)$. For more clarity, see Exercise 248.

11.4 Exercises

Exercise 222. (a) Show that stable and superstable periodic orbits are attracting.

(b) Find sufficient conditions in which neutral periodic orbit are attracting.

Exercise 223. Show that a map with a negative Schwarzian derivative is regular.

Exercise 224. Show that the quadratic map $p_a(x) = a - x^2$, $-\frac{1}{4} \leq a \leq 2$ is unimodal, and check its regularity in the sense of Definition 11.6.

Exercise 225. Show that the uniqueness of c in Definition 11.4 in the interval $[0, 1]$ implies the following:
(a) $c_1 = f(c)$ is the maximum of f on $[0, 1]$.
(b) If $x \neq c$, then $f(x) < c_1$.
(c) $c_1 > 0$.
(d) $c \in (0, 1)$.
(e) $Df > 0$ on $[0, c)$, and $Df < 0$ on $(c, 1]$.

Exercise 226. Show that the quadratic polynomials of the form $a_1 x^2 + b_1 x + c_1$, $2x^2 - y^2$, and $xy + xz + yz$ are unimodal functions.

Exercise 227. (a) Show the conjugacy between the logistic map (6.25) and the tent map (11.3).
(b) Show that the tent map (11.3) is continuous but not smooth.
(c) Show that the tent map (11.3) generates robust chaos (in the sense of Definition 2.5).

Exercise 228. Show that the logistic map (6.25) is an S-unimodal map on $[0, 1]$, and show that this map displays fragile chaos.

Exercise 229. Prove Theorem 11.1.

Exercise 230. Show that if a map f is an S-unimodal map of an interval having no limit cycles, then the map f is ergodic with respect to the Lebesgue measure and has a unique attractor Λ in the sense of Milnor and coincides with the conservative kernel of f. Furthermore, show that if there are no strongly wandering sets of positive measure and if f has a finite measure, then it has positive entropy. See [Blokh & M.Lyabich (1991)].

Exercise 231. Prove Corollary 11.1.

Exercise 232. Show that if two maps are topologically conjugate, then they are combinatorially equivalent.

Exercise 233. Show that a critical point c of a C^4 unimodal map f such that $f''(c) \neq 0$ is C^3 nonflat of order 2.

Exercise 234. (Open problem) (*Milnor's problem*) The subject is whether the metric and topological attractors coincide for a given smooth unimodal map. See [de Melo & van Strien (1993), Lyubich (1994), Graczyk & Swiatek (1999), Graczyk *et al.* (2005)].

Exercise 235. (a) Show that the unimodal Collet–Eckmann map and the backwards Collet–Eckmann map are strongly hyperbolic (see Chapter 6) along the critical orbit.

(b) Show that the Collet–Eckmann maps are also backwards Collet–Eckmann maps. See [de Melo & van Strien (1993)].

(c) Show that the Collet–Eckmann maps have a positive Lyapunov exponent at the critical value.

(d) Show that the Collet–Eckmann maps have the *best possible* near hyperbolic properties: exponential decay of correlations, validity of central limit and large deviations theorems, good spectral properties, and zeta functions. See [Keller & Nowicki (1992), Young (1992)].

(e) Show that the Collet–Eckmann maps have a robust statistical description, with a good understanding of stochastic perturbations: strong stochastic stability [Baladi & Viana (1996)] and rates of convergence to equilibrium [Baladi *et al.* (2002)].

(f) Show that the map (11.10) is a Collet–Eckmann map for real parameters.

Exercise 236. Prove Theorem 11.8. See [Avila & Moreira 2005(b)].

Exercise 237. Prove Theorem 11.9. See [Avila & Moreira (2003)].

Exercise 238. Prove Theorem 11.10. See [Avila & Moreira 2005(a)].

Exercise 239. (a) Show that the bifurcation diagram of the smooth logistic map (6.25) is densely populated by periodic windows. See [Collet & Eckmann (1980), Chap.1].

(b) Show that a continuously chaotic bifurcation diagram is obtained for the tent map (11.3) (11.3), which is nonsmooth.

(c) Show that most known chaotic systems display *fragile* chaos (see Chapter 10) in the sense that a slight alteration of a large number N of parameters will destroy the chaos and give a stable periodic orbit.

(d) Show that for the quadratic family $x \to x^2 - a$, the windows are intervals in the one-dimensional space of parameters built around isolated spine points, and the spine locus in this case corresponds to parameter values that give rise to superstable orbits.

(e) Show that for the two-parameter quadratic family $x \to (x^2 - a)^2 - b$, the spine locus consists of two parabolas (the black curves defined by the condition $m = 0$) in Fig. 11.1.

(f) Explain the importance of the dimension of the spine locus.

(g) Explain the method of identification of the spine locus for two-dimensional maps.

(h) Explain the method of identification of the spine locus for a d-dimensional (high-dimensional) map.

Exercise 240. Show that the map (11.17) is chaotic, and the route to chaos is period doubling bifurcations.

Exercise 241. Show that the map (11.19) has the following properties:
(a) $f_\nu(0) = f_\nu(1) = 0$.
(b) $f_\nu(x)$ is a smooth map, $\forall x \in [0, 1], \forall \nu > 0, \nu = 1$.
(c) $c = \frac{1}{2},\,, f'_\nu(c) = 0, f_\nu(c) = 1, \forall \nu > 0, \nu = 1$.
(d) $\lim_{\nu \to 0^+} f_\nu(x) = 1, \forall x \in [0, 1]$.
(e) $\lim_{\nu \to 1^\pm} f\nu(x) = h_4(x) = 4x(1-x), \forall x \in [0, 1]$.
(f) $f_\nu(x)$ has a negative Schwarzian derivative $(x \neq c, \nu \neq 1)$.
(g) It is possible to restrict the study of the map (11.19) to the interval $\nu \in (0, 1)$ because $f_\nu(x) = f_{1/\nu}(x), \forall \nu > 0, \nu \neq 1$.

Exercise 242. Consider $f(x) = \pi^{-1} \sin(\pi x)$ and $g(x) = \sqrt{x}(1 - \sqrt{x})$.
(a) Show that both maps are S-unimodal on the interval $x \in [0, 1]$, and they have the following critical points: $c_f = \frac{1}{2}$ and $c_g = \frac{1}{4}$.
(b) Verify that all the conditions in Theorem 11.12 are satisfied.
(c) Deduce that the map $F_\nu(x) = \frac{\sin(4\pi\nu\sqrt{x}(1-\sqrt{x}))}{\sin(\pi\nu)}$ generates robust chaos for $x \in [0, 1]$ and $\nu \in (0, \frac{1}{2})$.
(d) Compute numerically the Lyapunov exponent $\lambda(\nu)$ of $F_\nu(x)$ for $\nu \in (0, \frac{1}{2})$.

Exercise 243. The purpose of this exercise is to show that the map (11.14) has similar chaotic behavior to the logistic map (6.25) when $a = \mu = 4$. To do that, the reader must solve the following problems:
(a) Plot in the same plane x-$f(x, \alpha)$ the corresponding phases of the map (11.14) for $\alpha = 2, 5,$ and 9.
(b) Plot the fixed points x_α of the map (11.14) as a function of $\alpha \in [0, 100]$.
(c) Plot the graph of $f'(x_\alpha, \alpha)$ as a function of $\alpha \in [0, 100]$.

(d) Plot in the same plane the Schwarzian derivative $S(f(x,\alpha),x)$ of the map (11.14) as function of x for $\alpha = 2, 9$, and 15.
(e) Plot the bifurcation diagram of the map (11.14) for $\alpha \in [0, 40]$.
(f) Plot the Lyapunov exponent of the map (11.14) for $\alpha \in [0, 40]$.

Exercise 244. (a) Prove that the maps $F_1\left(-N, N, \frac{1}{2}, x\right)$ given by (11.50) and $T_N(x)$ map the unit interval $[0, 1]$ into itself.
(b) Show that $\Phi_N(x,\alpha)$ has $(N-1)$ critical points in unit interval $[0, 1]$.
(c) Derive the formula (11.51).
(d) Show that the maps (11.52) have negative Schwarzian derivative.
(e) Show that the maps (11.52) have at most $N_1 N_2 ... N_n + 1$ attracting periodic orbits. See [Devaney (1982)].
(f) Show that the maps (11.52) have only a single period-one stable fixed point.
(g) Calculate the SRB measure and the Kolmogorov–Sinai entropy for the map (11.49) using the conjugate maps (11.60).

Exercise 245. (a) Plot the map $GL(B, x)$ versus x for several values of $0 < B < e^{-4}$.
(b) Show that the function $GL(B, x)$ is unimodal for $e^{-4} \leq B < \infty$.
(c) Show all the properties of the map (11.62) cited after equation (11.62).
(d) Plot the Lyapunov exponents using relation (11.64) with 10^4 iterations of the B-exponential map (11.62) for $B \geq e^{-4}$.
(e) Plot the bifurcation diagram of the B-exponential map (11.62) for $0 \leq B \leq 10^4$.
(f) Show numerically that the Schwarzian derivative of the map (11.62) is negative.
(g) Show that the numerator of the B-exponential map (11.62) is topologically conjugate to the logistic map (6.25) at $B = \phi^2$ where ϕ is the golden mean ($\phi = \frac{1+\sqrt{5}}{2}$). Determine the range for chaos in this case, and draw the bifurcation diagram for $0 \leq B \leq 3$ and for $2 \leq B \leq 2.8$.

Exercise 246. Show that the chaotic attractor of the one-dimensional normal form (11.82) is robustly chaotic for the whole region of (11.91) and a significant portion of (11.96) where no periodic windows exist. See [Banerjee et al. (2000)].

Exercise 247. (a) Draw the bifurcation diagrams of the map (11.100) for two different values of the singularity exponent $z = -0.5$ and $z = 0.5$.

(b) Plot the critical boundaries $\mu_-(z)$ and $\mu_+(z)$ of the robust chaotic region for the singular maps on the space of parameters (μ, z), and find the regions of robust chaos.

(c) Plot the Lyapunov exponent λ as a function of the parameter μ for (11.100) for two values of z, $z = -0.5$ and $z = 0.5$.

(d) Make a log-log plot of the Lyapunov exponent as a function of $\epsilon = |\mu - \mu_m|$ for several values of z, $z = -0.65$ (circles), $z = -0.55$ (crosses), $z = -0.45$ (squares), and $z = -0.25$ (diamonds).

(e) Plot the critical exponent β as a function of z for the transition to chaos by type-III intermittency, and indicate the error bars.

Exercise 248. (Open problems) (Discussion with Prof. Soumitro Banerjee)

(a) The invariant density function (the fixed point of the Frobenius–Perron operator) can be easily obtained if the map is Markov. Can one obtain the number of partitions and the densities in those partitions when the map is non-Markov (without recourse to numerics)?

(b) The occurrence of high-period orbits has not been extensively studied. The conditions of existence and stability for high-period orbits can be obtained for "maximal" periodic orbits (with one point in L and all others in R, and *vice versa*). But nonmaximal orbits have not been studied.

(c) No formulation for the calculation of invariant density is available.

(d) The dynamics of piecewise-linear discontinuous maps is practically unexplored.

Chapter 12

Robust Chaos in 2-D Piecewise-Smooth Maps

In this chapter, we discuss robust chaos in two-dimensional piecewise-smooth maps. In Sec. 12.1.1, we give the normal form of these maps to facilitate the study of bifurcations and chaos. Section 12.1.2 provides details about bifurcations and chaos obtained using such a normal form, including general classifications and different routes to chaos. Section 12.1.3 deals with regions in the parameter space of the normal form with nonrobust chaos, and Sec. 12.1.4 describes regions for robust chaos. In Sec. 12.1.5, we give a proof of unicity of the orbits for the normal form, and in Sec. 12.1.6 a proof of the robustness of chaos in two-dimensional piecewise-smooth maps is given along with two examples. In Sec. 12.2, the same strategy as above is applied to the two-dimensional noninvertible piecewise-linear maps with their normal form as given in Sec. 12.2.1, and a proof of robust chaos is given in Sec. 12.2.2 with some examples. In Sec. 12.3, some exercises and open problems are provided to fix the ideas and experiences.

12.1 Robust chaos in 2-D piecewise-smooth maps

Power electronics is an area with wide practical application [Robert & Robert (2002), Banerjee & Grebogi (1999), Banerjee *et al.* (1998), Hassouneh *et al.* (2002), Banerjee & Verghese (2000), Kowalczyk (2005), Banerjee *et al.* (1999)]. It is concerned with the problem of the efficient conversion of electrical power from one form to another. *Power converters* [Banerjee *et al.* (1999)] exhibit several nonlinear phenomena such as border-collision bifurcations, coexisting attractors (alternative stable operating modes or fragile chaos), and chaos (apparently random behavior). These phenomena are created by switching elements [Banerjee *et al.* (1999)]. Recently, several researchers have studied border-collision bifurcations in piecewise-smooth systems [Robert & Robert (2002), Banerjee & Grebogi (1999), Banerjee et al. (1998), Hassouneh et al. (2002), Banerjee & Verghese (2000), Kowalczyk (2005), Banerjee et al. (1999)]. Piecewise-smooth systems can exhibit classical smooth bifurcations, but if the bifurcation occurs when the fixed point is on the border, there is a discontinuous change in the elements of the Jacobian matrix as the bifurcation parameter is varied. A variety of such border-collision bifurcations (see Sec. 12.1.2) have been reported in [Robert & Robert (2002), Banerjee & Grebogi (1999)], and under certain conditions, border-collision bifurcations produce robust chaos (see Sec. 12.1.4).

12.1.1 *Normal form for 2-D piecewise-smooth maps*

In this section, we discuss the derivation of the normal form for 2-D piecewise-smooth maps and its importance in studying bifurcations and chaos in this type of system. Indeed, consider the following 2-D piecewise-smooth system given by

$$g(x,y;\rho) = \begin{pmatrix} g_1 = \begin{pmatrix} f_1(x,y;\rho) \\ f_2(x,y;\rho) \end{pmatrix}, \text{ if } x < S(y,\rho) \\ \\ g_2 = \begin{pmatrix} f_1(x,y;\rho) \\ f_2(x,y;\rho) \end{pmatrix}, \text{ if } x \geq S(y,\rho) \end{pmatrix} \quad (12.1)$$

where the smooth curve $x = S(y,\rho)$ divides the phase plane into two regions R_1 and R_2, given by

$$R_1 = \{(x,y) \in \mathbb{R}^2, x < S(y,\rho)\} \quad (12.2)$$

$$R_2 = \{(x,y) \in \mathbb{R}^2, x \geq S(y,\rho)\}, \quad (12.3)$$

and the boundary between them as

$$\Sigma = \{(x,y) \in \mathbb{R}^2, x = S(y,\rho)\}. \quad (12.4)$$

It is assumed that the functions g_1 and g_2 are both continuous and have continuous derivatives. Then the map g is continuous, but its derivative is discontinuous at the borderline $x = S(y,\rho)$. It is further assumed that the one-sided partial derivatives at the border are finite, and in each subregion R_1 and R_2, the map (12.1) has one fixed point in R_1 and one fixed point in R_2 at a value ρ_* of the parameter ρ.

It is shown in [Banerjee & Grebogi (1999)] that the normal form of the map (12.1)[1] is given by

$$N(x,y) = \begin{cases} \begin{pmatrix} \tau_L & 1 \\ -\delta_L & 0 \end{pmatrix} \begin{pmatrix} x \\ y \end{pmatrix} + \begin{pmatrix} 1 \\ 0 \end{pmatrix} \mu, & \text{if } x < 0 \\ \begin{pmatrix} \tau_R & 1 \\ -\delta_R & 0 \end{pmatrix} \begin{pmatrix} x \\ y \end{pmatrix} + \begin{pmatrix} 1 \\ 0 \end{pmatrix} \mu, & \text{if } x > 0, \end{cases} \quad (12.5)$$

where μ is a parameter and $\tau_{L,R}$ and $\delta_{L,R}$ are the traces and determinants of the corresponding matrices of the linearized map in the two subregion R_L and R_R given by

$$\begin{cases} R_L = \{(x,y) \in \mathbb{R}^2, x \leq 0, y \in \mathbb{R}\} \\ R_R = \{(x,y) \in \mathbb{R}^2, x > 0, y \in \mathbb{R}\} \end{cases} \quad (12.6)$$

evaluated at the fixed point

$$P_L = \left(\frac{\mu}{1 - \tau_L + \delta_L}, -\frac{\delta_L \mu}{1 - \tau_L + \delta_L} \right) \in R_L \quad (12.7)$$

(with eigenvalues $\lambda_{L1,2}$) and at the fixed point

$$P_R = \left(\frac{\mu}{1 - \tau_R + \delta_R}, -\frac{\delta_R \mu}{1 - \tau_R + \delta_R} \right) \in R_R \quad (12.8)$$

(with eigenvalues $\lambda_{R1,2}$), respectively. The stability of the fixed points is determined by the eigenvalues of the corresponding Jacobian matrix, *i.e.*,

$$\lambda = \frac{1}{2} \left(\tau \pm \sqrt{\tau^2 - 4\delta} \right) \quad (12.9)$$

in the regions R_L and R_R the map (12.5) is smooth, and the boundary between them is given by

$$\Sigma = \{(x,y) \in \mathbb{R}^2, x = 0, y \in \mathbb{R}\}. \quad (12.10)$$

12.1.2 Border-collision bifurcations and robust chaos

In this section, we summarize the known sufficient conditions for the possible bifurcation phenomena in the normal form (12.5). In particular, we concentrate our attention on the cases leading to robust chaos.

---

[1] Since the nature of border-collision bifurcations depends on the local character of the map in the neighborhood of the fixed point.

12.1.2.1 General classification

The dynamics of piecewise-smooth maps is the first reason for some investigations into border-collision bifurcation phenomena [Feigin (1970), Nusse & Yorke (1992)]. Later the development is based on the observation that most power electronic circuits yield piecewise-smooth maps under discrete-time modeling, and nonsmooth bifurcations are quite common in them [Yuan et al. (1998)]. In a two-dimensional piecewise-linear map, the stability of the fixed points is determined by the eigenvalues of the corresponding Jacobian matrices [Feigin (1970), di Bernardo et al. (1999)] where the existence of period-1 and period-2 orbits before and after the border collision was studied with a classification of border-collision bifurcations in n-dimensional piecewise-smooth systems based on the various cases depending on the number of real eigenvalues that are greater than 1 or less than -1. Other methods of classification can be found for example in [Banerjee & Grebogi (1999), Banerjee et al. (2000)] for one- and two-dimensional maps, where they consider the asymptotically stable orbits (including chaotic orbits) before and after border collision, along several proofs of the existence of various types of border-collision bifurcations using essentially the trace and the determinant of the Jacobian matrix on the two sides of the border.

The results on border-collision bifurcations in two-dimensional systems available in the literature require that $|\delta_L| < 1$ and $|\delta_R| < 1$, but it was proved that attractors can exist if the determinant on one side is greater than unity in magnitude provided that the determinant in the other side is smaller than unity. The situation $|\delta_L| > 1$ and $\delta_R = 0$ (which occurs in some classes of power electronic systems) has been studied in [Parui & Banerjee (2002)].

12.1.2.2 The possible types of fixed points of the normal form of map (12.5)

The possible types of fixed points of the normal form of map (12.5) are given by the following:

(1) For a positive determinant:

(1-a) For $2\sqrt{\delta} < \tau < (1 + \delta)$, then the Jacobian matrix has two real eigenvalues $0 < \lambda_{1L}, \lambda_{2L} < 1$, and the fixed point is a regular attractor.

(1-b) For $\tau > (1+\delta)$, then the Jacobian matrix has two real eigenvalues $0 < \lambda_{1L} < 1, \lambda_{2L} > 1$, and the fixed point is a regular saddle.

(1-c) For $-(1 + \delta) < \tau < -2\sqrt{\delta}$, then the Jacobian matrix has two real eigenvalues $-1 < \lambda_{1L} < 0, -1 < \lambda_{2L} < 0$, and the fixed point is a flip attractor.

(1-d) For $\tau < -(1+\delta)$, then the Jacobian matrix has two real eigenvalues $-1 < \lambda_{1L} < 0, \lambda_{2L} < -1$, and the fixed point is a flip saddle.

(1-e) For $0 < \tau < 2\sqrt{\delta}$, then the Jacobian matrix has two complex eigenvalues $|\lambda_{1L}|, |\lambda_{2L}| < 1$, and the fixed point is a clockwise spiral.

(1-g) For $-2\sqrt{\delta} < \tau < 0$, then the Jacobian matrix has two complex eigenvalues $|\lambda_{1L}|, |\lambda_{2L}| < 1$, and the fixed point is a counter-clockwise spiral.

(2) For a negative determinant

(2-a) For $-(1+\delta) < \tau < (1+\delta)$, then the Jacobian matrix has two real eigenvalues $-1 < \lambda_{1L} < 0, 0 < \lambda_{2L} < 1$, and the fixed point is a flip attractor.

(2-b) For $\tau > 1 + \delta$, then the Jacobian matrix has two real eigenvalues $\lambda_{1L} > 1, -1 < \lambda_{2L} < 0$, and the fixed point is a flip saddle.

(2-c) For $\tau < -(1+\delta)$, then the Jacobian matrix has two real eigenvalues $0 < \lambda_{1L} < 1, \lambda_{2L} < -1$, and the fixed point is a flip saddle.

12.1.3 Regions for nonrobust chaos

The known cases in which a locally unique fixed point before the border yields, after the border, either a new locally unique fixed point or a locally unique period-2 attractor, *i.e.*, no robust chaos, are as follows depending on the sign of the system determinants on both sides of the normal form (12.5):

12.1.3.1 The case of positive determinants on both sides of the border

Scenario A: (*Locally unique stable fixed point on both sides of the border*). If

$$\begin{cases} \delta_L > 0, \delta_R > 0 \\ -(1+\delta_L) < \tau_L < (1+\delta_L) \\ -(1+\delta_R) < \tau_R < (1+\delta_R), \end{cases} \quad (12.11)$$

then a stable fixed point persists as the bifurcation parameter μ is increased (or decreased) through zero.

For the parameter range given by (12.11), a stable fixed point yields a stable fixed point after the border crossing with or without extraneous periodic orbits emerging from the critical point. These cases are as follows:

(1) If $2\sqrt{\delta_L} < \tau_L < (1+\delta_L), -(1+\delta_R) < \tau_R < -2\sqrt{\delta_R}$, then a regular attractor yields a flip attractor.

(2) If $2\sqrt{\delta_L} < \tau_L < (1+\delta_L), 2\sqrt{\delta_R} < \tau_R < (1+\delta_R)$, then a regular attractor yields a regular attractor.

(3) If $-(1+\delta_L) < \tau_L < -2\sqrt{\delta_L}, -(1+\delta_L) < \tau_L < -2\sqrt{\delta_L}$, then a flip attractor yields a regular attractor.

(4) If $-(1+\delta_L) < \tau_L < -2\sqrt{\delta_L}, -(1+\delta_R) < \tau_R < -2\sqrt{\delta_R}$, then a flip attractor yields a flip attractor.

(5) If $2\sqrt{\delta_L} < \tau_L < (1+\delta_L), -2\sqrt{\delta_R} < \tau_R < 2\sqrt{\delta_R}$, then a regular attractor yields a spiral attractor.

(6) If $-2\sqrt{\delta_L} < \tau_L < 2\sqrt{\delta_L}, 2\sqrt{\delta_R} < \tau_R < (1+\delta_R)$, then a spiral attractor yields a regular attractor.

(7) If $-(1+\delta_L) < \tau_L < -2\sqrt{\delta_L}, -2\sqrt{\delta_R} < \tau_R < 2\sqrt{\delta_R}$, then a flip attractor yields a spiral attractor.

(8) If $-2\sqrt{\delta_R} < \tau_R < 2\sqrt{\delta_R}, -(1+\delta_R) < \tau_R < -2\sqrt{\delta_R}$, then a spiral attractor yields a flip attractor.

(9) If $0 < \tau_L < 2\sqrt{\delta_L}, -2\sqrt{\delta_R} < \tau_R < 0$, then a clockwise spiral attractor yields a counterclockwise spiral attractor.

(10) If $-2\sqrt{\delta_L} < \tau_L < 0, 0 < \tau_R < 2\sqrt{\delta_R}$, then a counterclockwise spiral attractor yields a clockwise spiral attractor.

(11) If $0 < \tau_L < 2\sqrt{\delta_L}, 0 < \tau_R < 2\sqrt{\delta_R}$, then a clockwise spiral attractor yields a clockwise spiral attractor.

(12) If $-2\sqrt{\delta_L} < \tau_L < 0, -2\sqrt{\delta_R} < \tau_R < 0$, then a counterclockwise spiral attractor yields a counterclockwise spiral attractor.

Note that extraneous periodic orbits occur[2] at the bifurcations in the cases (9) and (10). For the cases (5), (6), (11), and (12), there are no known examples, but also no proof has been reported that extraneous periodic orbits do not occur. For the cases (1) to (4), the extraneous periodic orbits cannot appear, *i.e.*, the fixed points on both sides of the border are locally unique and stable (see Theorem 12.1). This happens when the fixed point changes from (1) a regular attractor to a flip attractor, (2) a regular attractor to a regular attractor, (3) a flip attractor to a regular attractor, and (4) a flip attractor to a flip attractor, as μ is varied through its critical value.

Scenario B: (*Supercritical period doubling border-collision bifurcation*). In this case, there are two regions in the parameter space where period doubling border-collision bifurcation occurs, *i.e.*, a locally unique stable

---
[2]*i.e.*, either before, after, or on both sides of the border.

fixed point leads to an unstable fixed point plus a locally unique attracting period-two orbit. This scenario is divided into Scenario B1 and Scenario B2 as follows:

Scenario B1:
$$\begin{cases} \delta_L > 0, \delta_R > 0 \\ -(1+\delta_L) < \tau_L < -2\delta_L \\ \tau_R < -(1+\delta_R) \\ \tau_R \tau_L < (1+\delta_R)(1+\delta_L). \end{cases} \quad (12.12)$$

Scenario B2:
$$\begin{cases} \delta_L > 0, \delta_R > 0 \\ 2\delta_L < \tau_L < (1+\delta_L) \\ \tau_R < -(1+\delta_R) \\ \tau_R \tau_L > -(1-\delta_R)(1-\delta_L). \end{cases} \quad (12.13)$$

12.1.3.2 The case of negative determinants on both sides of the border

It is shown in Theorem 12.1 that if the determinants on both sides of the border are negative, then if the fixed point is stable, it is locally unique. The condition for a locally unique stable fixed point on both sides of the border is given in Scenario C below.

Scenario C: (*Locally unique stable fixed point on both sides of the border*). If
$$\begin{cases} \delta_L < 0, \delta_R < 0 \\ -(1+\delta_L) < \tau_L < (1+\delta_L) \\ -(1+\delta_R) < \tau_R < (1+\delta_R), \end{cases} \quad (12.14)$$
then a locally unique stable fixed point leads to a locally unique stable fixed point as μ is increased through zero.

Scenario D: (*Supercritical border-collision period doubling*). If
$$\begin{cases} \delta_L < 0, \delta_R < 0 \\ -(1+\delta_L) < \tau_L < (1+\delta_L) \\ \tau_R < -(1+\delta_R) \\ \tau_R \tau_L < (1+\delta_R)(1+\delta_L) \\ \tau_R \tau_L > -(1-\delta_R)(1-\delta_L), \end{cases} \quad (12.15)$$
then a locally unique stable fixed point to the left of the border for $\mu < 0$ crosses the border and becomes unstable, and a locally unique period-two orbit is born as μ is increased through zero, *i.e.*, this is a condition for a supercritical period doubling border collision with no extraneous periodic orbits.

12.1.3.3 The case of negative determinant to the left of the border and positive determinant to the right of the border

If the determinant is negative to the left of the border, i.e., $\delta_L < 0$, then the eigenvalues are real. If the determinant is positive to the right of the border, i.e., $\delta_R > 0$, then the eigenvalues are real if $\tau_R^2 > 4\delta_R$. Thus a sufficient condition for having a locally unique fixed point leading to a locally unique fixed point as μ is varied through the critical value is given as follows:

Scenario E: (*Locally unique stable fixed point on both sides of the border*). This occurs if

$$\begin{cases} \delta_L < 0, \delta_R > 0 \\ -(1+\delta_L) < \tau_L < (1+\delta_L) \\ -(1+\delta_R) < \tau_R < (1+\delta_R) \\ \tau_R^2 > 4\delta_R. \end{cases} \quad (12.16)$$

The conditions (12.16) can be divided into the following two cases:

Scenario E1: This occurs if

$$\begin{cases} \delta_L < 0, \delta_R > 0 \\ -(1+\delta_L) < \tau_L < (1+\delta_L) \\ -(1+\delta_R) < \tau_R < -2\delta_R. \end{cases} \quad (12.17)$$

Scenario E2: This occurs if

$$\begin{cases} \delta_L < 0, \delta_R > 0 \\ -(1+\delta_L) < \tau_L < (1+\delta_L) \\ 2\delta_R < \tau_R < (1+\delta_R). \end{cases} \quad (12.18)$$

12.1.3.4 The case of positive determinant to the left of the border and negative determinant to the right of the border

If the determinant is negative to the right of the border, i.e., $\delta_R < 0$, then the eigenvalues are real. If the determinant is positive to the left of the border, i.e., $\delta_L > 0$, then the eigenvalues are real if $\tau_L^2 > 4\delta_L$. Thus a sufficient condition for having a locally unique fixed point leading to a locally unique fixed point as μ is varied through the critical value is given as follows:

Scenario F: (*Locally unique stable fixed point on both sides of the border*). This occurs if

$$\begin{cases} \delta_L > 0, \delta_R < 0 \\ -(1+\delta_L) < \tau_L < (1+\delta_L) \\ \tau_L^2 > 4\delta_L \\ -(1+\delta_R) < \tau_R < (1+\delta_R). \end{cases} \quad (12.19)$$

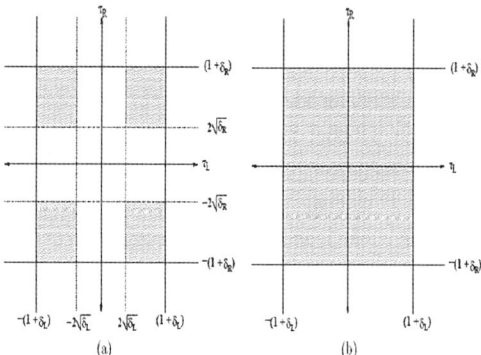

Fig. 12.1 No bifurcation occurs as μ is increased (decreased) through zero in the shaded regions. Only the path of the fixed point changes at $\mu = 0$. (a) $0 < \delta_L < 1$ and $0 < \delta_R < 1$. (b) $-1 < \delta_L < 0$ and $-1 < \delta_R < 0$.

The conditions (12.19) can be divided into the following two cases:

Scenario F1: This occurs if

$$\begin{cases} \delta_L > 0, \delta_R < 0 \\ -(1+\delta_L) < \tau_L < -2\delta_L \\ -(1+\delta_R) < \tau_R < (1+\delta_R). \end{cases} \quad (12.20)$$

Scenario F2: This occurs if

$$\begin{cases} \delta_L > 0, \delta_R < 0 \\ 2\delta_L < \tau_L < (1+\delta_L) \\ -(1+\delta_R) < \tau_R < (1+\delta_R). \end{cases} \quad (12.21)$$

In fact, there are no known conditions for supercritical border-collision period doublings that occur without EBOs when the determinants on both sides of the border are of opposite signs.

The Scenarios A–F are shown in Figs. 12.1 and 12.2, *i.e.*, the parameter ranges in the space $(\tau_L, \delta_L, \tau_R, \delta_R)$ such that the fixed points are unique attractors on both sides of the border (in these Figures, δ_L and δ_R are fixed, whereas τ_L and τ_R are variables).

12.1.3.5 *Stable fixed point leading to stable fixed point plus extraneous periodic orbits*

Furthermore, there are certain border-collision bifurcations that, while not causing a *catastrophic collapse* of the system, may lead to undesirable system behavior [Banerjee & Grebogi (1999), Banerjee *et al.* (2000)], *i.e.*, the case of a stable fixed point leading to a stable fixed point plus extraneous periodic orbits, that can display multiple attractor bifurcations on either

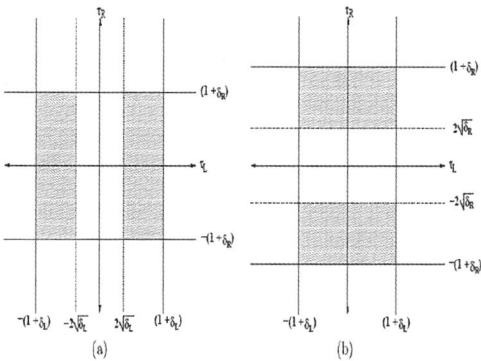

Fig. 12.2 No bifurcation occurs as μ is increased (decreased) through zero in the shaded regions. Only the path of the fixed point changes at $\mu = 0$. (a) $0 < \delta_L < 1$ and $-1 < \delta_R < 0$. (b) $-1 < \delta_L < 0$ and $0 < \delta_R < 1$.

side of the border or both sides of the border in addition to the stable fixed points. Conditions for this case are not currently available. A first example of this situation is the case of a stable fixed point plus a period-4 attractor to a stable fixed point plus a period-3 attractor as μ is increased through zero. This situation is displayed by the map

$$l(x,y) = \begin{cases} \begin{pmatrix} 0.50 & 1 \\ -0.90 & 0 \end{pmatrix} \begin{pmatrix} x \\ y \end{pmatrix} + \begin{pmatrix} 1 \\ 0 \end{pmatrix} \mu, & \text{if } x < 0 \\ \begin{pmatrix} -1.22 & 1 \\ -0.36 & 0 \end{pmatrix} \begin{pmatrix} x \\ y \end{pmatrix} + \begin{pmatrix} 1 \\ 0 \end{pmatrix} \mu, & \text{if } x > 0, \end{cases} \quad (12.22)$$

and shown in Fig. 12.3(a). In this case, the fixed point for $\mu < 0$ is spirally attracting and for $\mu > 0$ is a flip attractor.

The second example is given by the case of a stable fixed point to a stable fixed point plus a period-7 attractor displayed by the map

$$m(x,y) = \begin{cases} \begin{pmatrix} 1.6 & 1 \\ -0.8 & 0 \end{pmatrix} \begin{pmatrix} x \\ y \end{pmatrix} + \begin{pmatrix} 1 \\ 0 \end{pmatrix} \mu, & \text{if } x < 0 \\ \begin{pmatrix} -1.4 & 1 \\ -0.6 & 0 \end{pmatrix} \begin{pmatrix} x \\ y \end{pmatrix} + \begin{pmatrix} 1 \\ 0 \end{pmatrix} \mu, & \text{if } x > 0, \end{cases} \quad (12.23)$$

and shown in Fig. 12.3(b). In this case, the fixed point for $\mu < 0$ is spirally attracting and for $\mu > 0$ is also spirally attracting with an opposite sense of rotation.

12.1.4 Regions for robust chaos: undesirable and dangerous bifurcations

First, we give a definition of the so-called *dangerous bifurcations:*

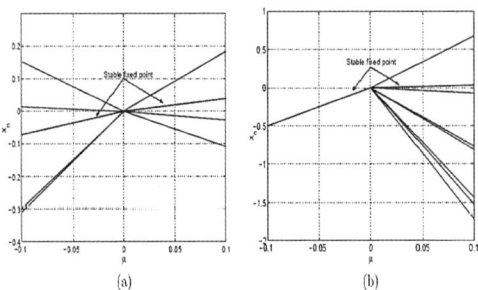

Fig. 12.3 (a) Bifurcation diagram of (12.22). (b) Bifurcation diagram of (12.23).

Definition 12.1. The dangerous bifurcations considered begin with a system operating at a stable fixed point on one side of the border, say the left side.

The main dangerous bifurcations that can result from a border collision are the following:

12.1.4.1 Border-collision pair bifurcation

This occurs if

$$\begin{cases} -(1+\delta_L) < \tau_L < (1+\delta_L) \\ \tau_R > (1+\delta_R), \end{cases} \quad (12.24)$$

where a stable fixed point and an unstable fixed point merge and disappear as μ is increased through zero. This is analogous to saddle-node bifurcations in smooth maps. In this case, the system trajectory diverges for positive values of μ since no local attractors exist.

12.1.4.2 Subcritical border-collision period doubling

This occurs if

$$\begin{cases} -(1+\delta_L) < \tau_L < (1+\delta_L), \\ \tau_R < -(1+\delta_R), \\ \tau_R \tau_L > (1+\delta_R)(1+\delta_L), \end{cases} \quad (12.25)$$

where a bifurcation from a stable fixed point and an unstable period-2 orbit to the left of border to an unstable fixed point to the right of the border occurs as μ is increased through zero.

12.1.4.3 Supercritical border-collision period doubling

This type of bifurcation is not classified as a *dangerous bifurcation*, but it is undesirable in some applications. An example of this situation is the one called the *cardiac conduction model* proposed in [Sun et al. (1995)] as a two-dimensional piecewise-smooth map in which the atrial His interval A is that between cardiac impulse excitation of the lower interatrial septum to the bundle of His as

$$\begin{pmatrix} A_{n+1} \\ B_{n+1} \end{pmatrix} = f(A_n, R_n, H_n) \qquad (12.26)$$

where

$$f(A_n, R_n, H_n) = \begin{cases} \begin{cases} A_{\min} + D_n, & \text{if } A_n < 130 \\ A_{\min} + C_n, & \text{if } A_n > 130 \end{cases} \\ R_n = \exp\left(-\frac{(H_n + A_n)}{\tau_{fat}}\right) + \gamma \exp\left(-\frac{H_n}{\tau_{fat}}\right) \\ D_n = R_{n+1} + (201 - 0.7 A_n) \exp\left(-\frac{H_n}{\tau_{rec}}\right) \\ C_n = R_{n+1} + (500 - 3.0 A_n) \exp\left(-\frac{H_n}{\tau_{rec}}\right), \end{cases} \qquad (12.27)$$

where the variables and constants in map (12.27) are as follows: $R_0 = \gamma \exp\left(-\frac{H_n}{\tau_{fat}}\right)$, H_0 is the initial H interval, and the parameters $A_{\min}$, τ_{fat}, γ, and τ_{rec} are positive constants. The variable H_n (which is usually the bifurcation parameter) is the interval between bundle of His activation and the subsequent activation (the atrioventricular nodal recovery time). The variable R_n (which is usually the bifurcation parameter) is a drift in the nodal conduction time.

Model (12.27) incorporates physiological concepts of recovery, facilitation, and fatigue. In this case, two factors determine the atrioventricular (AV) nodal conduction time, the time interval from the atrial activation to the activation of the bundle of His and the history of activation of the node. This model confirms and predicts several experiments concerning complex rhythms of nodal conduction. Especially, alternans rhythms, with an alternation in conduction time from beat to beat, display period doubling bifurcations in the theoretical model in which the map f is piecewise-smooth and is continuous at the border $A_b = 130$ ms.

For $\tau_{rec} = 70$ ms, $\tau_{fat} = 30000$ ms, $A_{\min} = 33$ ms, and $\gamma = 0.3$ ms, a stable period-2 orbit is born after the border collision as seen in Fig. 12.4.

12.1.4.4 Stable fixed point leading to chaos

This situation is also called *instant chaos* where chaotic behavior develops following a border collision. In this case, a strange bifurcation route to

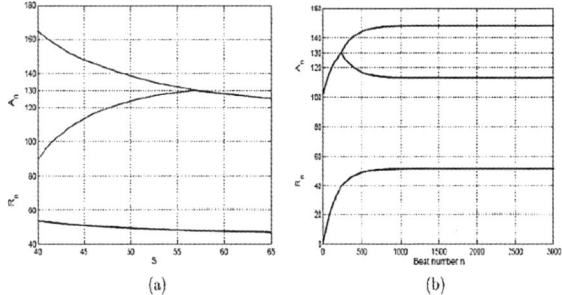

Fig. 12.4 (a) Bifurcation diagram for A_n and for R_n for (12.27) with H_n as a bifurcation parameter and $\tau_{rec} = 70$ ms, $\tau_{fat} = 30000$ ms, $A_{\min} = 33$ ms, and $\gamma = 0.3$ ms. (b) Iterations of map (12.27) showing the alternation in A_n as a result of a period doubling bifurcation. The parameter values are the same as in (a) with $H_n = 45$ ms.

chaos is found in a piecewise-linear, second-order, nonautonomous differential equation derived from a simple electronic circuit, *i.e.*, the behavior of the system changes directly to chaos (without a period doubling bifurcation or intermittency) when a limit cycle loses its stability. See [Ohnishi & Inaba (1994)] for more details.

12.1.5 *Proof of unicity of orbits*

It was proved in [Hassouneh *et al.* (2002)] that coexisting attractors cannot occur under some conditions on the eigenvalues of the fixed points of the map (12.5). Namely, the following result was shown:

Theorem 12.1. *When the eigenvalues on both sides of the border are real, if an attracting orbit exists, it is unique (i.e., coexisting attractors cannot occur).*

Proof. When the determinant is positive, there are six types of fixed points in a linearized, dissipative, two-dimensional discrete system as discussed in Sec. 12.1.3. Hence we have the following cases:

(a) If the fixed points on both sides are saddles, then period-2 or chaos occurs on the unstable manifolds of the saddles. Thus a heteroclinic intersection must exist because the stable eigenvector at P_R has a positive slope $m_1 = (-\frac{\delta_R}{\lambda_{1R}})$, and the unstable eigenvector at P_L has a negative slope given by $(\frac{-\delta_L}{\lambda_{1L}})$. Using the lambda lemma[3] [Alligood *et al.* (1996)], it

---
[3] If a curve C crosses a stable manifold transversely, then each point of the unstable manifold of the same saddle fixed point is a limit point of $\bigcup_{n>0} f^n(C)$.

follows that the two unstable manifolds come arbitrarily close to each other because the unstable manifold of P_L has transverse intersections with the stable manifold of P_R. Finally, the attractor must be unique.

(b) If the fixed points on both sides are stable with real eigenvalues, then for all $\mu < 0$, initial conditions in R_L converge on the stable manifold associated with the larger eigenvalue, and then converge onto P_L along that eigenvector. Similarly, all initial conditions in R_R see the virtual fixed point P_R which is in R_L, and move to R_L along a stable manifold, which folds at the intersection with the x-axis [Banerjee et al. (2000)] which implies the existence of a transverse heteroclinic intersection with a stable manifold of P_L. Hence the stable manifolds of P_L and P_R come arbitrarily close to each other, and therefore all initial conditions in R_R also converge onto P_L. A similar proof can be done for $\mu > 0$, and trajectories starting from all initial conditions converge on the stable fixed point P_L.

(c) If the fixed point in R_L is stable and that in R_R is unstable, then for $\mu < 0$, a mechanism like case (b) operates, and for $\mu > 0$, a mechanism like case (a) operates, preventing the occurrence of coexisting attractors. □

12.1.6 *Proof of robust chaos*

It is shown in [Banerjee et al. (1998-1999)] that the resulting chaos from the 2-D map (12.5) is robust in the following cases:

12.1.6.1 *Case 1*

$$\begin{cases} \tau_L > 1 + \delta_L, \text{ and } \tau_R < -(1+\delta_R) \\ 0 < \delta_L < 1, \text{ and } 0 < \delta_R < 1, \end{cases} \quad (12.28)$$

where the parameter range for boundary crisis is given by

$$\delta_L \tau_L \lambda_{1L} - \delta_L \lambda_{1L} \lambda_{2L} + \delta_R \lambda_{2L} - \delta_L \tau_R + \delta_L \tau_L - \delta_L - \lambda_{1L}\delta_L > 0, \quad (12.29)$$

where the inequality (12.29) determines the condition for stability of the chaotic attractor. The robust chaotic orbit continues to exist as τ_L is reduced below $1 + \delta_L$.

12.1.6.2 *Case 2*

$$\begin{cases} \tau_L > 1 + \delta_L, \text{ and } \tau_R < -(1+\delta_R) \\ \delta_L < 0, \text{ and } -1 < \delta_R < 0 \\ \frac{\lambda_{1L}-1}{\tau_L - 1 - \delta_L} > \frac{\lambda_{2R}-1}{\tau_R - 1 - \delta_R} \end{cases} \quad (12.30)$$

The condition for stability of the chaotic attractor is also determined by (12.29). However, if the third condition of (12.30) is not satisfied, then the condition for existence of the chaotic attractor changes to

$$\frac{\lambda_{2R} - 1}{\tau_R - 1 - \delta_L} < \frac{(\tau_L - \delta_L - \lambda_{2L})}{(\tau_L - 1 - \delta_L)(\lambda_{2L} - \tau_R)}. \quad (12.31)$$

12.1.6.3 Case 3

The remaining ranges for the quantities $\tau_{L,R}$ and $\delta_{L,R}$ can be determined in some cases using the same logic as in the above two cases, or there is no analytic condition for a boundary crisis, and it has to be determined numerically. Due to the symmetric region of the parameter space with the roles of R_L and R_R interchanged, where the same phenomena are observed for $\mu < 0$, we give here the proof of the above statements for only $\mu > 0$ as follows: First, if

$$(1 + \delta_R) > \tau_R > -(1 + \delta_R), \quad (12.32)$$

then the map (12.5) always has a periodic attractor for $\mu > 0$ because for $\mu > 0$, the fixed point in P_L is a regular saddle and P_R is an attractor[4], and the condition $\tau_R = -(1 + \delta_R)$ results in a nongeneric situation where all points on the line joining the points $\left(\frac{\mu}{1+\delta_R}, 0\right)$ and $\left(0, -\frac{\delta_R \mu}{1+\delta_R}\right)$ are fixed points of the second iterate. Secondly, for very low values of the determinant, the main attractor is not chaotic even for $\tau_L > 2\sqrt{\delta_L}$, as periodic orbits become stable [Nusse & Yorke (1995), Maistrenko et al. (1993)], and for one-dimensional systems, the parameter range for robust chaos is bounded by $\tau_R = 1$, $\tau_R > -\frac{\tau_L}{\tau_L - 1}$, and the lower limit of τ_L is given by the conditions of existence of various periodic windows. Thus if the determinants on the two sides are unity, the area in $\tau_L - \tau_R$ space for robust chaos shrinks to zero. Third, robust chaos occurs for $\mu > 0$ on the region of parameter space given by

$$\begin{cases} \tau_L > 1 + \delta_L \\ \tau_R < -(1 + \delta_R). \end{cases} \quad (12.33)$$

If $0 \leq \delta_L < 1$ and $0 \leq \delta_R < 1$ for (12.33), P_L is a regular saddle, and P_R is a flip saddle. Let $\mathcal{U}_L$ and $\mathcal{S}_L$ be the unstable and stable manifolds of P_L and $\mathcal{U}_R$ and $\mathcal{S}_R$ be the unstable and stable manifold of P_R, respectively. For (12.5), all intersections of the unstable manifolds with $x = 0$ map to the line $y = 0$, and the sets $\mathcal{U}_L$ and $\mathcal{U}_R$ experience folds along the x-axis, and

---
[4]This is like a saddle-node bifurcation occurring on the border.

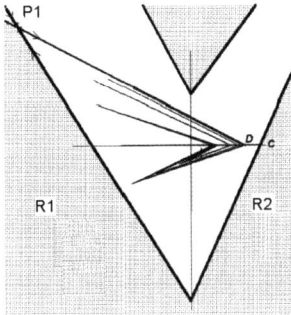

Fig. 12.5 The stable and unstable manifolds of P_1 for $\tau_1 = 1.7$, $\delta_1 = 0.5$, $\tau_2 = -1.7$, and $\delta_2 = 0.5$. Reprinted (Figure 3) with permission from Banerjee, S., Yorke, J. A., and Grebogi, C., Phys. Rev. Lett. 80(14) 3049–3052, 1998. Copyright (1998) by the American Physical Society.

all images of fold points will be fold points and the $\mathcal{S}_L$ and $\mathcal{S}_R$ fold along the y-axis, and all preimages of fold points are fold points. The reason is that one linear map changes to another linear map across the $x = 0$ line.

Now, let $\lambda_{1,2L}$ be the eigenvalues on side R_L and $\lambda_{1,2R}$ be the eigenvalues at side R_R. For condition (12.33), one has $\lambda_{1L} > \lambda_{2L} > 0$ and $\lambda_{2R} < \lambda_{1R} < 0$. The slope of the stable eigenvector at P_R is $m_1 = -\frac{\delta_R}{\lambda_{1R}} > 0$, and the slope of the unstable eigenvector is $m_2 = \frac{-\delta_R}{\lambda_{2R}} > 0$. Thus the points of $\mathcal{U}_R$ to the left of the y-axis map to points above the x-axis because points on an eigenvector map to points on the same eigenvector and since points on the y-axis map to the x-axis. This implies that the set $\mathcal{U}_R$ has an angle $m_3 = \frac{\delta_L \lambda_{2R}}{\delta_R - \tau_L \lambda_{2R}} < 0$ after the first fold. Hence there must be a transversal homoclinic intersection in R_R and therefore an infinity of homoclinic intersections and the existence of a chaotic orbit. Note that the basin boundary of the resulting chaotic attractor is formed by the set $\mathcal{S}_L$ which is folds at the y-axis and intersects the x-axis at point C shown in Fig. 12.5. The portion of the set $\mathcal{U}_L$ to the left of P_L goes to infinity, and the portion to the right of P_L leads to the chaotic orbit. Thus the set $\mathcal{U}_L$ meets the x-axis at point D shown in Fig. 12.5 and then undergoes repeated foldings, leading to an intricately folded compact structure as shown in Fig. 12.5. This description helps when looking at the stability of this chaotic orbit.

The slope of the unstable eigenvector at P_L is given by $-\frac{\delta_L}{\lambda_{1L}} < 0$, which implies the existence of a heteroclinic intersection with the set $\mathcal{S}_R$. Using the lambda lemma [Alligood et al. (1996)], it follows that for each point

q on $\mathcal{U}_R$ and for each ϵ neighborhood $N_\epsilon(q)$, there exist points of $\mathcal{U}_L$ in $N_\epsilon(q)$ because both $\mathcal{U}_L$ and $\mathcal{U}_R$ have transverse intersections with $\mathcal{S}_R$. In this case, the attractor must span $\mathcal{U}_R$ on one side of the heteroclinic point since $\mathcal{U}_L$ comes arbitrarily close to $\mathcal{U}_R$. Thus the attractor is unique because all initial conditions in R_L converge on $\mathcal{U}_L$ and all initial conditions in R_R converge on $\mathcal{U}_R$ and since there are points of $\mathcal{U}_L$ in every neighborhood of $\mathcal{U}_R$. Furthermore, this attractor is robust and cannot be destroyed by small changes in the parameters because small changes in the parameters can only cause small changes in the Lyapunov exponents when the chaotic attractor is stable. This orbit continues to exist as τ_L is reduced below $(1 + \delta_L)$ because in this case, there is no fixed point in R_L for $\mu > 0$, but the invariant manifolds suffer only a slight change. The robustness of this chaotic attractor can be seen from the fact that the invariant manifold of R_L associated with λ_{1L} still forms the attractor and cannot exists for $\tau_L < 2\sqrt{\delta_L}$ since the eigenvalues become complex. Thus multiple attractors cannot exist because for $\tau_L < 2\sqrt{\delta_L}$, there is a sudden reduction in the size of the attractor as it spans only $\mathcal{U}_R$.

Figure 12.5 shows that no point of the attractor can be to the right of point D. Hence, if D lies towards the left of C, the chaotic orbit is stable, while if D falls outside the basin of attraction, it is an unstable chaotic orbit or chaotic saddle. Thus the condition for stability of the chaotic attractor is obtained as (12.29).

Theorem 12.2. *For $1 > \delta_L > 0$, and $1 > \delta_R > 0$, the normal form (12.5) exhibits robust chaos in a portion of parameter space bounded by the conditions $\tau_R = -(1 + \delta_R)$, $\tau_L > 2\sqrt{\delta_L}$, and (12.29) as shown in Fig. 12.6.*

The same logic applies for the cases with a negative determinant, *i.e.*, for $-1 < \tau_R < 0$, or for $\delta_L < 0$ and $\delta_R < 0$, or for $\delta_L < 0$, and the eigenvalues are real for all τ_L, then the set $\mathcal{U}_L$ converges on $\mathcal{U}_R$ from one side. If the conditions of **Case 2** above hold, then the intersection of $\mathcal{U}_L$ with the x-axis is the rightmost point of the attractor, and (12.29) still gives the parameter range for boundary crisis. But, if (12.29) is not satisfied, the intersection of $\mathcal{U}_R$ with the x-axis becomes the rightmost point of the attractor, and the condition of existence of the chaotic attractor changes to (12.30). But, if $\delta_L < 0$ and $\delta_R > 0$, the set $\mathcal{U}_L$ does not approach $\mathcal{U}_R$ from one side. Thus if (12.31) is not satisfied, then numerical calculations are required because there is no analytic condition for boundary crisis.

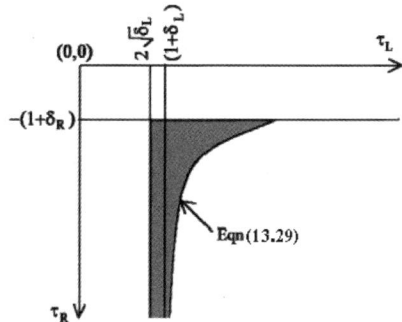

Fig. 12.6 Schematic diagram of the parameter space region of the normal form (13.5) where robust chaos is observed for $0 < \delta_{1,2} < 1$ and $\mu > 0$. Here, $\delta_{1,2} = \delta_{L,R}$. Reprinted (Figure 4) with permission from Banerjee, S., Yorke, J. A., and Grebogi, C., Phys. Rev. Lett. 80(14) 3049–3052, 1998. Copyright (1998) by the American Physical Society.

In what follows, we give two examples, the first of which is the boost converter, and the second is a purely mathematical construction of 2-D maps displaying robust chaos.

12.1.6.4 Example 1

Several examples of robust chaos in 2-D piecewise-smooth systems can be found in [Robert & Robert (2002), Banerjee & Grebogi (1999), Banerjee et al. (1998), Hassouneh et al. (2002b), Banerjee & Verghese (2001), Kowalczyk (2005)].

The first example of 2-D maps exhibiting robust chaos is the one given in [Banerjee et al. (1998)], which is a practical example from electrical engineering known as the *boost converter* whose circuit as shown in Fig. 12.7 is widely used in regulated dc switch-mode power supplies. This circuit consists of a controlled switch S, an uncontrolled switch D, an inductor L, a capacitor C, and a load resistor R. The circuit works as follows: When S is turned on, the current in L increases, and energy is stored in it, and when S is turned off, the stored energy in L drops, and the polarity of the inductor voltage changes so that it adds to the input voltage. The voltage across L and the input voltage together *boost* the output voltage to a value higher than the input voltage. Controlling the switching by current feedback, known as *current-mode control*, regulates the output current. In this case, the switch is turned on by clock pulses that are spaced T seconds apart, and when the switch is closed, then the inductor current increases until it reaches the specified reference value I_{ref}. The switch opens when

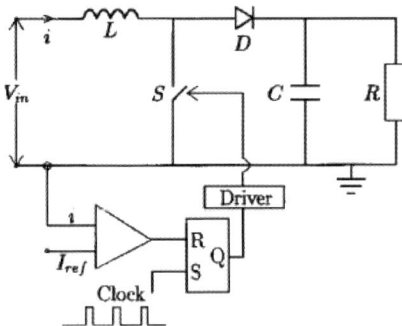

Fig. 12.7 The current mode controlled boost converter. Reprinted (Figure 1) with permission from Banerjee, S., Yorke, J. A., and Grebogi, C., Phys. Rev. Lett. 80(14) 3049–3052, 1998. Copyright (1998) by the American Physical Society.

$i = I_{ref}$ with the assumption that any clock pulse arriving during the on-period is ignored. Hence, if the switch has opened, the next clock pulse causes it to close.

The discrete-time modeling was obtained by observing the state variables at every clock instant. A state can evolve from one clock instant to the next by two ways. First, if the on-time $T_{on} = \frac{L(I_{ref}-i_n)}{V_{in}} < T$, then the evolution between observation instants includes one on-period and one off-period. The derivation of the two-dimensional map for $T_{on} < T$ is done by assuming that the waveforms are linear between clock instants and neglecting the higher-order Taylor terms, that is

$$\begin{cases} i_{n+1} = I_{ref} + \frac{1}{L}\left(V_{in} - v_n + \frac{v_n T_{on}}{CR}\right)(T - T_{on}) \\ v_{n+1} = v_n - \frac{v_n T_{on}}{CR} + \left(\frac{I_{ref}}{C} - \frac{v_n}{CR} + \frac{v_n T_{on}}{C^2 R^2}\right)(T - T_{on}). \end{cases} \quad (12.34)$$

Now, if the clock pulse arrives while $i < I_{ref}$, then the switch remains on between the observation instants, and if $T_{on} \geq T$, then the map is given by

$$\begin{cases} i_{n+1} = i_n + \frac{V_{in}}{L}T \\ v_{n+1} = v_n - \frac{v_n T}{CR}. \end{cases} \quad (12.35)$$

The borderline between the two cases described by (12.34) and (12.35) is given by the situation where the current reaches I_{ref} exactly at the arrival of the next clock pulse, i.e., $I_{border} = I_{ref} - \frac{V_{in}T}{L}$. From Fig. 12.8, it is clear that the bifurcation diagram and the Lyapunov exponent of the boost converter governed by Eqs. (12.34) and (12.35) confirm that there is no periodic window or coexisting attractor in the parameter range $R = 241$ to 500 V.

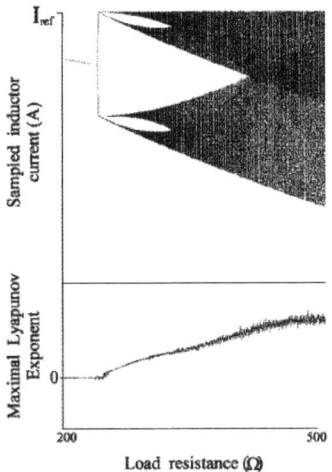

Fig. 12.8 Bifurcation diagram and Lyapunov exponent of the boost converter. The parameter values are $c = 220$ mF, $I_{ref} = 0.5$ A, $V_{in} = 30$ V, $T = 400$ ms, and $L = 0.1$ H. Reprinted (Figure 2) with permission from Banerjee, S., Yorke, J. A., and Grebogi, C., Phys. Rev. Lett. 80(14) 3049–3052, 1998. Copyright (1998) by the American Physical Society.

12.1.6.5 *Example 2*

The second example of 2-D piecewise-linear mappings displaying robust chaos is the one given in [Zeraoulia & Sprott (2011)]. Indeed, consider the unified piecewise-smooth chaotic mapping that contains the Hénon [Hénon (1976)] (6.88) and the Lozi systems (8.88) [Lozi (1978)] defined by

$$U(x,y) = \begin{pmatrix} 1 - 1.4 f_\alpha(x) + y \\ 0.3x \end{pmatrix}, \qquad (12.36)$$

where $0 \leq \alpha \leq 1$ is the bifurcation parameter and the function f_α is given by

$$f_\alpha(x) = \alpha |x| + (1 - \alpha) x^2. \qquad (12.37)$$

It is easy to see that for $\alpha = 0$, one has the original Hénon map (6.88), and for $\alpha = 1$, one has the original Lozi map (8.88), and for $0 < \alpha < 1$, the unified chaotic map (12.36) is chaotic with different kinds of attractors. In this case, it was shown rigorously that the unified system (12.36) has robust chaotic attractors for $0.493122734 \leq \alpha < 1$ as shown in Fig. 12.9. Some corresponding robust chaotic attractors are shown in Fig. 12.11. These

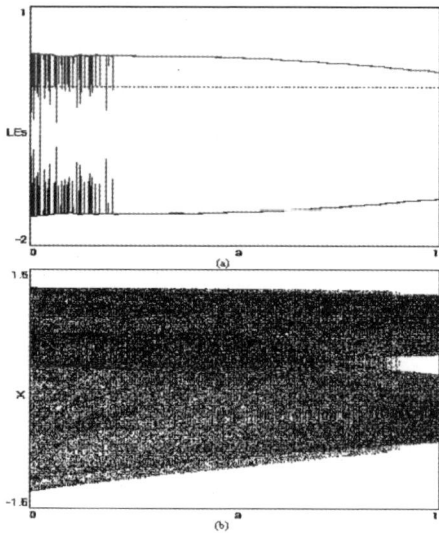

Fig. 12.9 (a) Variation of the Lyapunov exponents of the unified map (12.36) for $0 \leq \alpha \leq 1$. (b) Bifurcation diagram for the unified chaotic map (12.36) for $0 \leq \alpha \leq 1$.

chaotic attractors cannot be destroyed by small changes in the parameters since small changes in the parameters can only cause small changes in the Lyapunov exponents. Hence the range for the parameter $0 \leq \alpha < 1$, in which the map (12.36) converges to a robust chaotic attractor, is approximately 50.6 88 percent. This result was also verified numerically by computing Lyapunov exponents and bifurcation diagrams as shown in Fig. 12.9. For $\alpha < 0.493122734$, the chaos is not robust in some ranges of the variable α because there are numerous small periodic windows as shown in Fig. 12.10(b) such as the period-8 window at $\alpha = 0.025$. Furthermore, for $\alpha = 0.114$, there are some periodic windows. We note also the existence of some regions in α where the largest Lyapunov exponent is positive, but this does not guarantee the unicity of the attractor, contrary to the case of $\alpha \in [0.493122734, 1[$ where the attractor is guaranteed to be unique due to the analytical expressions.

12.2 Robust chaos in noninvertible piecewise-linear maps

In [Kowalczyk (2005)], the author investigates border-collision bifurcations in 2-D piecewise-linear maps that are noninvertible in one region. This

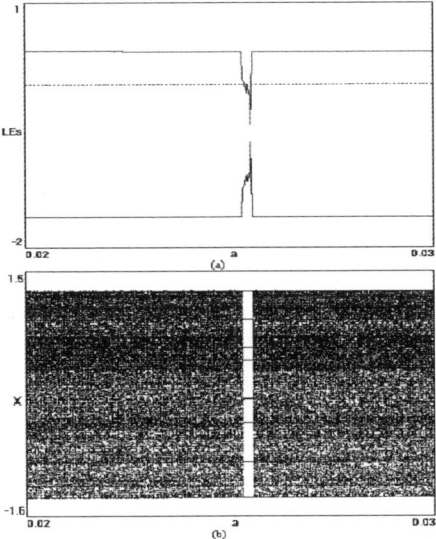

Fig. 12.10 (a) Variation of the Lyapunov exponents for the unified chaotic map (13.36) for $0.02 \leq \alpha \leq 0.03$. (b) Bifurcation diagram for the unified chaotic map (13.36) for $0.02 \leq \alpha \leq 0.03$ showing a period-8 attractor obtained for $\alpha = 0.025$.

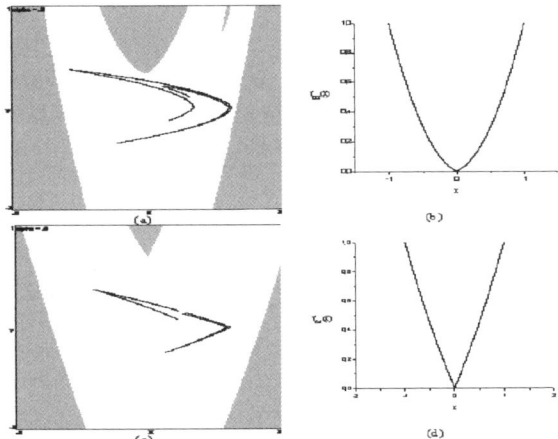

Fig. 12.11 (a) The transition Hénon-like chaotic attractor obtained for the unified chaotic map (13.36) with its basin of attraction (white) for $\alpha = 0.2$. (b) Graph of the function $f_{0.2}$. (c) The transition Lozi-like chaotic attractor obtained for the unified chaotic map (13.36) with its basin of attraction (white) for $\alpha = 0.8$. (d) Graph of the function $f_{0.8}$.

type of map is the normal form for *grazing–sliding bifurcations* in three-dimensional Filippov type systems. Robust chaos was proved to be an essential feature for these maps. The main differences with respect to the above analysis are the following: First, the authors of [Banerjee & Grebogi (1999)] consider only the case where the determinants of the piecewise-linear normal form on both sides of the bifurcation boundary are less then unity, *i.e.*, $|\delta_L| < 1$ and $|\delta_R| < 1$. Secondly, if the fixed points are purely real and are of the saddle type, then the corresponding eigenvectors exhibit nondifferentiable folds (corners), *i.e.*, the stable manifold (the unstable manifolds) folds at every intersection with the y-axis (at every intersection with the x-axis), and every preimage of the fold point is a fold for both cases. Third, the occurrence of a homoclinic intersection of the unstable and stable manifolds is no longer obvious if the map is noninvertible in one of its regions. Compare with [Banerjee & Grebogi (1999)] and [Parui & Banerjee (2002)] where it was implicitly assumed (without any justification, *i.e.*, no proofs of border-collision bifurcations leading to the onset of chaos in map (12.1)) that for piecewise-linear noninvertible maps of type (12.38) below, the same argument can be used to prove the onset of chaos due to the existence of border-collisions.

12.2.1 *Normal forms for two-dimensional noninvertible maps*

Following [di Bernardo *et al.* (2002)], it was shown in [Kowalczyk (2005)] that a normal form map for grazing-sliding bifurcations can be obtained in which the discontinuity on one side of the boundary has a corank of -1:

$$\Pi(x,y,\mu) = \bar{x} = \begin{cases} \begin{pmatrix} \tau_L & 1 \\ -\delta_L & 0 \end{pmatrix} \begin{pmatrix} x \\ y \end{pmatrix} + \begin{pmatrix} 1 \\ 0 \end{pmatrix} \mu, & \text{if } x \leq 0 \\ \begin{pmatrix} \tau_R & 1 \\ 0 & 0 \end{pmatrix} \begin{pmatrix} x \\ y \end{pmatrix} + \begin{pmatrix} 1 \\ 0 \end{pmatrix} \mu, & \text{if } x > 0. \end{cases} \quad (12.38)$$

Let Π_L and Π_R be the two submappings defined by

$$\Pi_L = M\bar{x} + C\mu, \text{ if } x \leq 0 \quad (12.39)$$

$$\Pi_R = N\bar{x} + C\mu, \text{ if } x > 0 \quad (12.40)$$

where

$$\begin{cases} M = \begin{pmatrix} \tau_L & 1 \\ -\delta_L & 0 \end{pmatrix}, N = \begin{pmatrix} \tau_R & 1 \\ 0 & 0 \end{pmatrix} \\ C = \begin{pmatrix} 1 \\ 0 \end{pmatrix}, \bar{x} = \begin{pmatrix} x \\ y \end{pmatrix}. \end{cases} \quad (12.41)$$

Obviously, $\Pi_L : R_L \to R_L \cup R_R$ and is smooth and invertible while $R : R_L \to \{x \in \mathbb{R}, y = 0\}$ is smooth but noninvertible.

The analysis of map (12.38) is based on the distinction between admissible fixed points and virtual fixed points of map (12.38) defined by [Kowalczyk (2005)] as follows:

Definition 12.2. (a) Admissible fixed points of map (12.38) are fixed points of Π_L and Π_R that lie in the domain of definition of these two submappings.

(b) [Banerjee & Grebogi (1999)] Virtual fixed points of map (12.38) are fixed points of Π_L and Π_R but existing outside the corresponding domain of definition.

Thus it follows from Definition 12.2 that admissible fixed points of Π_L lie in R_L and of Π_R in R_R. By contrast, the virtual fixed point of Π_L belongs to R_R and of Π_R to R_L. In this case, the admissible fixed points of Π_L and Π_R are denoted by the letters A, a and B, b, respectively (see [Feigin (1994)]), and the virtual fixed points of Π_L are denoted by $\bar{A}, \bar{a}$ and of Π_R by $\bar{B}, \bar{b}$, i.e., the capital letters refer to the stable fixed points and lower case to unstable ones, where

$$A, a = \left(\frac{\mu}{1 - \tau_L + \delta_L}, -\frac{\delta_L \mu}{1 - \tau_L + \delta_L} \right) \text{ for } \frac{\mu}{1 - \tau_L + \delta_L} \leq 0 \qquad (12.42)$$

$$B, b = \left(\frac{\mu}{1 - \tau_R}, 0 \right), \text{ for } \frac{\mu}{1 - \tau_R} > 0. \qquad (12.43)$$

By definition, $\frac{\mu}{1-\tau_L+\delta_L} > 0$ for a virtual fixed point of Π_L, and $\frac{\mu}{1-\tau_R} \leq 0$ for a virtual fixed point of Π_R. Furthermore, the fixed point of Π_R needs to lie on the x-axis since every point which belongs to R_R is mapped onto the x-axis. The corresponding eigenvalues of $A, a, \bar{A}, \bar{a}$ are given by

$$\lambda_{1,2} = \frac{\tau_L \pm \sqrt{\tau_L - 4\delta_L}}{2}, \qquad (12.44)$$

and for the points $B, b, \bar{B}, \bar{b}$ as

$$\lambda_{1L} = \tau_R, \lambda_{2L} = 0. \qquad (12.45)$$

Thus the border-collision bifurcation can be defined for the map (12.38) as follows [Kowalczyk (2005)]:

Definition 12.3. Consider piecewise-linear map (12.38), and suppose that the fixed points of this map, which are fixed points of Π_L and Π_R, depend

smoothly on the parameter μ in some small neighborhood $\varepsilon > 0$ of the
origin. Suppose that a fixed point $(x, y) = (x^*, y^*)$

(1) for $-\varepsilon < \mu < 0$ belongs to R_L,
(2) for $\mu = 0$ belongs to the boundary Σ, and
(3) for $0 < \mu < \varepsilon$ belongs to R_R.

Then we say that the fixed point of (12.38) undergoes a border-collision bifurcation, and the fixed point is called *the border-crossing fixed point*.

Definition 12.3 implies that under the variation of μ in map (12.38), a stable admissible border-crossing fixed point of Π_L (which undergoes a border-collision) will move to another region. Thus it will become a virtual fixed point of Π_L.

12.2.2 Onset of chaos: Proof of robust chaos

In this section, we present four cases for the occurrence of robust chaos in the map (12.38), *i.e.*, border-collision from stable fixed point to flip saddle, two-piece invariant sets, three-piece invariant sets, and N-piece invariant sets.

12.2.2.1 Border-collision from stable fixed point to flip saddle

For this case, the border-collision from a stable admissible fixed point A existing for $\mu < 0$, is characterized by real eigenvalues, to an admissible saddle. This scenario is possible in the parameter regions labeled as '1' and '2' in Fig. 12.12. The conditions for this case are

$$\delta_L > \tau_L - 1 \qquad (12.46)$$

$$\tau_R < -1 \qquad (12.47)$$

which represents a border-collision of the stable admissible fixed point to an admissible flip saddle with a period-2 point involved in the border-collision bifurcations with the variation of the bifurcation parameter μ through the origin. Let $\bar{A}$ be the virtual fixed point obtained from the transformation of A under border-collision (see Fig. 12.13(a)). In the same way, let b be an admissible flip saddle of (12.38) existing for $\mu > 0$ (see Fig. 12.13(b)), which is a virtual flip for $\mu < 0$ (see Fig. 12.13(a)). Note that A and $\bar{A}$ can lie either above or below the x-axis, whereas b and $\bar{b}$ need to lie on the x-axis itself.

With these assumptions, the dynamics of the map (12.38) can be summarized as follows:

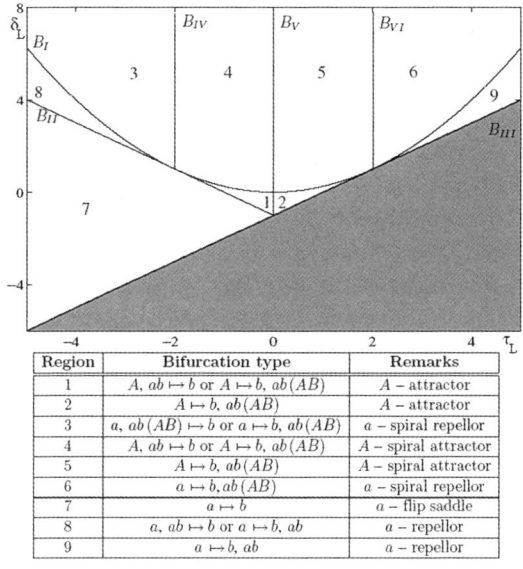

Fig. 12.12 Partition of the parameter space τ_L, δ_L into regions characterized by different border collision bifurcation scenarios as μ increases for $\tau_R < -1$. The behavior of the simple period-1 and period-2 points is given in the table. *Flip saddle* refers to a saddle type fixed point with one eigenvalue less than -1. The six borderlines indicated by B_I to B_{VI} are described in the text. Reused with permission from Kowalczyk, P., Nonlinearity (2005). Copyright 2005, IOP Publishing Ltd.

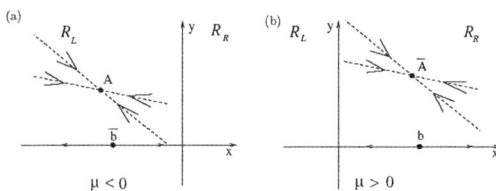

Fig. 12.13 Schematic representation of the fixed points of (12.38) before (a) and after (b) the border-collision bifurcation from an admissible stable fixed point to an admissible saddle. Lines with arrows schematically depict eigendirections of both fixed points. Reused with permission from Kowalczyk, P., Nonlinearity (2005). Copyright 2005, IOP Publishing Ltd.

(1) For $\mu < 0$, the fixed point A is attracting for both regions R_L and R_R. Indeed, consider some small ε-neighborhood near the origin of the piecewise-linear map (12.38). All the points in this neighborhood which lie in region R_L will be attracted to the stable fixed point A, and

points from region R_R will be injected to region R_L after at most two iterations. The reason is that in the first iteration, any point in R_R is mapped onto the x-axis, *i.e.*, if the first iteration maps onto some point x such that $x > 0$ and the virtual fixed point $\bar{b}$ lies on the semi-axis $x \leq 0$, then in the next iteration, the point x must get mapped onto the $x \leq 0$ semi-axis.

(2) For $\mu = 0$, the fixed point $A = 0$ is an attracting point for the map (12.38). Indeed, the point $x = 0 \in R_L$ will attract some set of points in the ε-neighborhood of the origin from the R_L region, and some set of points in the neighborhood of the origin lying in R_R will be injected into an ε-neighborhood of the origin of the R_L region. The reason is due to the existence of the virtual saddle.

(3) For $\mu > 0$, there is a possibility of some other attractors, higher-periodic points, quasi-periodic behavior, or a chaotic orbit because the stable admissible fixed point A becomes the virtual fixed point $\bar{A}$ of (12.38), and the unstable virtual saddle $\bar{b}$ becomes the unstable admissible fixed point b. Thus by continuity, $\bar{A}$ attracts some set of points which belong to R_L from an ε-neighborhood of the origin. Hence these points will be mapped onto R_R but not into $\bar{A}$. Furthermore, the fixed point b of R_R is unstable, which implies that these points and points lying in R_R will not be attracted to b but will be mapped onto a segment of the $x \leq 0$-axis after a finite number of iterations of (12.38) because b is of a flip type.

12.2.2.2 Two-piece invariant sets

For this case, some points in the neighborhood of the origin within R_L are mapped into R_R, and all the points in R_R are mapped onto the x-axis. A segment of the x-axis might form an Ω-limit set for these points. Thus the following result was proved in [Kowalczyk (2005)]:

Proposition 12.1. *If the border-collision bifurcation from an admissible fixed point attractor to an admissible flip saddle is exhibited by map (12.38) under the variation of μ and the conditions*

$$\tau_L < -\frac{1}{1+\tau_R} \tag{12.48}$$

$$\tau_L(1+\tau_R) - \delta_L\left(1 + \frac{1}{\tau_R}\right) < 0 \tag{12.49}$$

$$\tau_L(1+\tau_R) - \delta_L\left(1 + \frac{1}{\tau_R}\right) + 1 > 0 \tag{12.50}$$

are satisfied, then there exists an attractor born in the border-collision bifurcation which must necessarily lie within the piecewise-linear continuous invariant segment KLC such that

$$\begin{cases} K = ((\tau_R + 1)\mu, 0) \\ L = (\mu, 0) \\ C = ((\tau_L \tau_R + \tau_L + 1)\mu, -\delta_L(\tau_R + 1)\mu) \end{cases} \quad (12.51)$$

(see Fig. 12.14).

Proof. For $\mu > 0$ there exists a set of points in the neighborhood of the origin of both regions R_L, R_R that is mapped onto the negative part of the x-axis denoted by KO in Fig. 12.14. The origin O is mapped onto the point $L = (\mu, 0)$, and by continuity, the image of KO joins the x-axis at this point, but since it cannot lie on the x-axis itself, it must exhibit a nondifferentiable kink here. The image of KO is the segment LC in Fig. 12.14. Now, if (12.48) holds, then the segment LC does not cross the y-axis. If (12.48) holds, then the image of L is the point $K = ((\tau_R + 1)\mu, 0)$. If (12.49) and (12.50) are satisfied, then C is mapped onto $x \in KL$ under the action of (12.40). It must be considered the image of LC under the action of (12.40). In this case, K is a preimage of C, and it follows that any attractor born in a border-collision bifurcation must necessarily lie within the piecewise-linear continuous invariant segment KLC denoted by Ξ in Fig. 12.4. The size of Ξ was defined by finding points $K, L,$ and C, i.e., $C = ((\tau_L \tau_R + \tau_L + 1)\mu, -\delta_L(\tau_R + 1)\mu) \in R_R$ since $L = (\mu, 0)$, $K = ((\tau_R + 1)\mu, 0)$, and C is an image of K (see Fig. 12.4). The existence of the attractor within Ξ is a result of the fact that KO attracts some points from the neighborhood of the origin from both regions R_L and R_R, and from the additional condition on C that it is mapped into the segment KL. The latter condition is sufficient for the existence of this attractor. □

The other case of this situation comes when C is mapped outside the segment KL, i.e., there is a possibility of the existence of higher-periodic points or chaotic attractors living on Ξ. The parameter regions for this case, i.e., $A \to b, ab$ or $A \to b, AB$ are the regions labeled '1' and '2' in Fig. 12.12. The geometry of the set Ξ implies the existence of a switching between R_L and R_R, i.e., one of the period-2 points must lie on KO and the other on LC. Depending on the position of $\tau_L \tau_R - \delta_L$ with respect to the unit circle, the period-2 points can be either stable or unstable. If this period-2 point is stable, i.e., $|\tau_L \tau_R - \delta_L| < 1$, the attractor lies within the set Ξ, is stable, and is the only attractor born in the bifurcation. The uniqueness of this

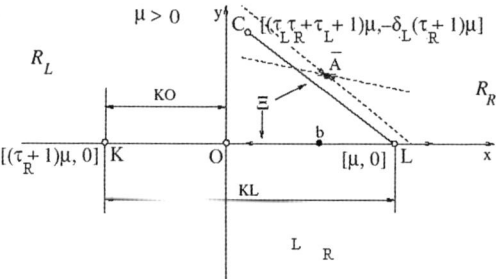

Fig. 12.14 An example of a limit set of (12.38) after the border-collision bifurcation from a stable fixed point to a flip saddle. Reused with permission from Kowalczyk, P., Nonlinearity (2005). Copyright 2005, IOP Publishing Ltd.

attractor is a result that comes from the fact that any attractor born in the border-collision must lie on the set Ξ, and there is no other possible attracting set born in the bifurcations. Now if $|\tau_L \tau_R - \delta_L| > 1$, i.e., the period-2 point is unstable, then there are not any stable periodic orbits within Ξ, but Ξ itself is attracting as shown in the following result proved in [Kowalczyk (2005)]:

Proposition 12.2. 2. *If the border-collision bifurcation from an admissible fixed point attractor to an admissible flip saddle accompanied by the birth of unstable period-2 points is exhibited by map (12.38) and conditions (12.48)–(12.50) hold, then there is an attractor born in the border-collision bifurcation which is a four-, two-, or one-piece chaotic attractor limiting on Ξ.*

Proof. It suffices to show that there are no stable higher-periodic points within Ξ. If Π_L denotes iteration of the map (Eq. (12.39)) governing the dynamics in the R_L region, and correspondingly Π_R denotes iteration of the map (Eq. (12.40)) governing the dynamics in the R_R region, then any periodic point must be composed of any finite string of $\Pi_L \Pi_R$ iterations and any number of Π_R iterations. Because the period-2 point is unstable, any string of $\Pi_R \Pi_L$ iterations cannot produce a stable periodic point nor can additional iterations Π_R since both the eigenvalues of (12.40) are purely real and one of them is less than -1 ($\tau_R < -1$). Furthermore, it is impossible to have a sequence containing $\Pi_L \Pi_L$ since this would imply that the periodic point does not belong to Ξ. Hence there are no stable periodic points in Ξ, and the only possible attractor (quasi-periodic behavior can be excluded since Ξ cannot form an invariant circle) is a chaotic attractor limiting on Ξ or on subsets of Ξ.

On the other hand, the border-collision bifurcations, *i.e.*, from a fixed point attractor to a four-, two-, or one-piece chaotic attractor depends on the quantities τ_L, τ_R and δ_L. Indeed, if $\tau_L\tau_R - \delta_L < -1$, then four-, two-, or one-piece chaotic attractors can be born in the border-collision bifurcations, but close to it, the chaotic attractor is organized around period-2 points. If $\tau_L\tau_R - \delta_L \ll -1$, then the chaotic attractor is robust in the sense of Definition 2.5, *i.e.*, it might fill greater parts of the set Ξ until the limit set of the chaotic attractor is dense on Ξ. The method of analysis is as follows: The chaotic attractor limits on the set Ξ which depends continuously on the parameters τ_L, τ_R, δ_L, and μ. The destruction (due to a small variation of μ away from the border-collision bifurcation point) of this chaotic attractor under the variation of any of these parameters is due to the destruction of the two-piece set Ξ or due to the stabilization of the period-2 point. Thus the set Ξ scales linearly with the variation of μ, and the stability of a fixed or higher-periodic point of (12.38) is μ-invariant.

The effect of the variation of the remaining parameters on the set Ξ and on the unstable period-2 point can be seen as follows: If the image of the point K crosses the y-axis under the variation of some parameter, then the structure of the two-piece invariant set Ξ can be changed. If the image of K is bounded away from the y-axis, *i.e.*, conditions (12.48)–(12.50) hold, this must still be true for sufficiently nearby parameter values since the dependence of the map on the parameters is smooth away from the y-axis. Now, if the eigenvalues describing the stability of the period-2 point lie away from the unit circle, then continuous parameter variations, within a certain range, cannot stabilize the period-2 point. (It was proven earlier that even if there exist other unstable higher-periodic points living on Ξ, provided that the period-2 point is unstable, they are also unstable.) In conclusion, the chaotic attractor is indeed robust due to the fact that any invariant set born in the border-collision bifurcation from an admissible fixed point attractor to an admissible flip saddle must limit on the set Ξ. □

An example of a robust chaotic attractor with two-pieces is the one generated by the map describing the grazing-sliding bifurcation scenario leading to the onset of chaos discussed in [di Bernardo *et al.* (2003)] in which the matrices M and N are given by

$$M = \begin{pmatrix} 0.85401 & 1 \\ -0.009 & 0 \end{pmatrix}, N = \begin{pmatrix} \tau_R & 1 \\ 0 & 0 \end{pmatrix}. \tag{12.52}$$

Several values of τ_R show the occurrence of a border-collision to a stable period-2 point and then to a four-, two-, and finally a one-piece chaotic attractor under a variation of the bifurcation parameter μ. Indeed, for

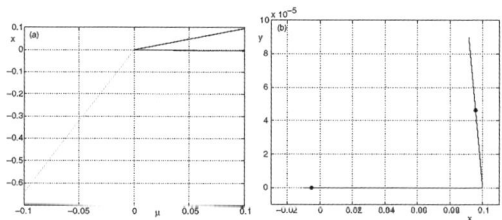

Fig. 12.15 (a) Bifurcation diagram presenting period doubling due to a border-collision bifurcation scenario for (12.52) with $\tau_R = -1.1$ and (b) period-2 points (•) for $\mu = 0.1$. Note that as predicted, they lie within the Ξ-black piecewise-linear line. Reused with permission from Kowalczyk, P., Nonlinearity (2005). Copyright 2005, IOP Publishing Ltd.

$\tau_R < -1$, the map (12.38) with (12.52) is characterized by a stable fixed point $A/\bar{A} = \left(\frac{\mu}{0.1550}, 0.0581\mu\right)$ with eigenvalues 0.8433 and 0.0107, and a saddle point $b/\bar{b} = \left(\frac{\mu}{1-\tau_R}, 0\right)$ with eigenvalues τ_R and 0 ($\bar{b}$ is a saddle fixed point of a flip type (eigenvalues of N are 0 and $\tau_R < -1$)). For $\mu < 0$, the only admissible fixed point is A, and there is a virtual saddle $\bar{b}$. For $\mu > 0$ the only admissible fixed point is a saddle point b, and A becomes a virtual fixed point $\bar{A}$. There are no other fixed points of (12.52). In this case, the period-2 point is born in the border-collision as shown by region '2' of Fig. 12.12. To see a transition to a stable period-2 point, we need the eigenvalues of MN to be within the unit circle. In our case the eigenvalues of MN are $\lambda_{1L} = 0$, $\lambda_{2L} = 0.854\tau_R - 0.009$. If $\tau_R = -1.1$, then $\lambda_{2L} = -0.9484$ and $|\lambda_{2L}| < 1$. Thus period-2 points can be found using some appropriate substitutions. If $\mu = 0.1$, then one has

$$x^* = (0.095617, 0.000046), \, x^{**} = (-0.005132, 0) \quad (12.53)$$

that lie within the set Ξ, i.e., the point x^{**} lies on segment KO since $K = ((\tau_R + 1)\mu = -0.01, 0)$ and x^* on LC segment since $L = (0.1, 0)$ and $C = (0.091460, 0.000090)$ as shown in Fig. 12.15(b) with the bifurcation diagram depicted in Fig. 12.15(a). In this case, conditions (12.48) and (12.50) are satisfied for the numerical values of τ_L, τ_R, and δ_L given in (12.52). For $\tau_R = -1.3$, one has $\tau_L \tau_R - \delta_L = -1.1192 < -1$. In this case for $\mu = 0.1$, one has $L = (0.1, 0)$ and $K = (-0.03, 0)$, and the image of C (which lies in R_R) belongs to the segment KL. Thus the structure and properties of Ξ remain unchanged, and there are unstable periodic points within Ξ which confirms the robustness of the chaotic attractor born in a border-collision bifurcation as shown in Fig. 12.16(a) where if the value of τ_R is close to -1 the chaotic attractor is organized near the two periodic points, i.e., flipping on either side of each one as shown in Fig. 12.16(b). Further

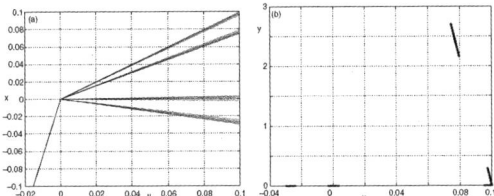

Fig. 12.16 Similar to Fig. 12.15 but for $\tau_R = -1.3$. (a) Bifurcation diagram. (b) Ω-limit set. Reused with permission from Kowalczyk, P., Nonlinearity (2005). Copyright 2005, IOP Publishing Ltd.

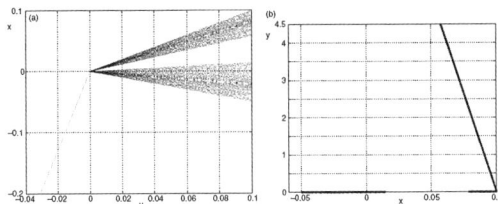

Fig. 12.17 Similar to Fig. 12.15 but for $\tau_R = -1.5$. (a) Bifurcation diagram. (b) Ω-limit set. Reused with permission from Kowalczyk, P., Nonlinearity (2005). Copyright 2005, IOP Publishing Ltd.

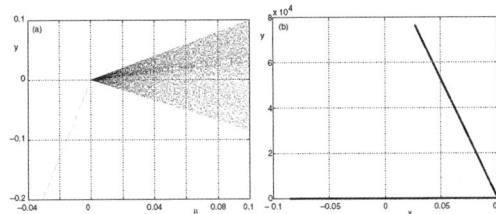

Fig. 12.18 Similar to Fig. 12.15 but for $\tau_R = -1.85$. (a) Bifurcation diagram. (b) Ω-limit set. Reused with permission from Kowalczyk, P., Nonlinearity (2005). Copyright 2005, IOP Publishing Ltd.

decrease of τ_R show that the limit set cover a bigger part of the set Ξ. Thus for $\tau_R = -1.5$, a two-piece chaotic attractor shown in Figs. 12.17(a) and 12.17(b) is obtained. A decrease of τ_R even further (namely, $\tau_R = -1.85$) shows a sudden jump to a one-piece chaotic attractor as shown in Fig. 12.18. This is the case of the *dry-friction oscillator model* studied in [Yoshitake & Sueoka (2000)].

12.2.2.3 Three-piece invariant sets

For this case, the image of the point C (see Fig. 12.14) is not mapped onto the segment KL but outside KL in which it is impossible to have any stable higher-periodic points within the set Ξ formed as in Fig. 12.14. In this case, the segment KO must be regarded as expanding under forward iteration of (12.38). In this case, the image of KO must cross the y-axis, and there is a possibility of the existence of period-3 stable periodic points or of the birth of a chaotic attractor living on a higher number of piecewise-linear segments. Thus the following result was proved in [Kowalczyk (2005)]:

Proposition 12.3. *If the border-collision bifurcation from an admissible fixed point attractor to an admissible flip saddle is exhibited by the normal form map (12.38) under the variation of μ and conditions*

$$\tau_L > -\frac{1}{1+\tau_R} \tag{12.54}$$

$$\tau_L^2(1+\tau_R) + \tau_L + 1 - \delta_L\left(1+\frac{2}{\tau_R}\right) > 0 \tag{12.55}$$

$$\tau_L^3(1+\tau_R) + \tau_L^2(1+\tau_L\tau_R) - \delta_L\tau_L\left(1+\frac{1}{\tau_R}\right) - \frac{\delta_L}{\tau_R} + 1 > 0 \tag{12.56}$$

hold, then there exists an attractor born in the border-collision bifurcation which must necessarily lie within the piecewise-linear continuous invariant set built from segments GEI and LH such that

$$\begin{cases} G = \left((\tau_R(\frac{\delta_L}{\tau_L}) + \tau_R + 1)\mu, 0\right) \\ E = \left(((\frac{\delta_L}{\tau_L}) + 1)\mu, 0\right) \\ I = (v, -\delta_L(\tau_R\delta_L + \tau_L\tau_R + \tau_L + 1)\mu) \\ \nu = (\tau_L(\tau_R\delta_L + \tau_L\tau_R + \tau_L + 1) + 1)\mu + w \\ L = (\mu, 0) \\ H = \left((\tau_R\delta_L + \tau_L\tau_R + \tau_L + 1)\mu, -\delta_L((\frac{\tau_R}{\tau_L})\delta_L + \tau_R + 1)\mu\right) \\ w_1 = -\delta_L\left(\left(\frac{\tau_R}{\tau_L}\right)\delta_L + \tau_R + 1\right)\mu \end{cases} \tag{12.57}$$

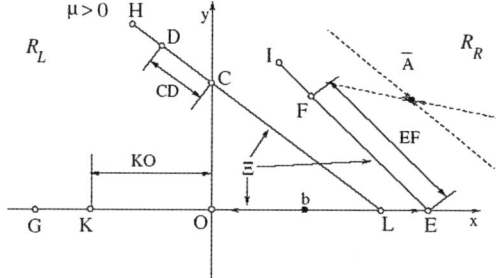

Fig. 12.19 An example of a limit set of (12.38) after the border-collision bifurcation consisting of three piecewise-linear segments. Reused with permission from Kowalczyk, P., Nonlinearity (2005). Copyright 2005, IOP Publishing Ltd.

(see Fig. 12.19). This attractor is either a period-3 point or a six-, three- or one-piece chaotic attractor.

Proof. Consider the segment KO shown in Fig. 12.19 such that its image LD crosses the y-axis, which implies that (12.54) must hold. The segment LD contains a segment denoted by CD, which requires additional iteration of (12.39) to get mapped into the region R_R. Segment LC is mapped in one iteration onto the x-axis, and by continuity its image EF must join the image of LC (joining the x-axis at the point E). In this case, using the fact that C is mapped outside KL, it follows that the point E must lie outside the segment KL but in the region R_R, and the maximum value of x at E is greater than the maximum value of x of the preimage of KO. From (12.54) it also follows that $\tau_L > 0$, and then if (12.55) holds in the forward iteration, the piecewise-linear segment OEF is mapped within the segment GE (condition (12.55) confirms that the value of x in the image of F is less than that of x at E (see Fig. 12.19). Since OE is mapped onto GO, then it is sufficient to consider forward iterations of GO only in which its image LH will cross the y-axis. Finally, the image of LH will form a piecewise-linear continuous segment KEI lying partly on the x-axis (the part KE), where at point E it is joined by segment EI (see Fig. 12.19). Thus the image of EI is the main objective of this proof. If condition (12.56) is satisfied, the image of EI lies within the segment GE, and a bounded set Ξ that must contain the Ω-limit set is obtained. The set Ξ will be formed by three linear segments, GE, EI, and LH. The coordinates of points K, L, C, D, E, F, G, H, and I, which can be obtained from forward iterations of KO, are given by

$$\begin{cases} K = ((\tau_R + 1)\mu, 0), L = (\mu, 0), C = \left(0, \frac{\delta_L}{\tau_L}\mu\right) \\ D = ((\tau_L\tau_R + \tau_L + 1)\mu, -\delta_L(\tau_R + 1)\mu) \\ E = \left((\frac{\delta_L}{\tau_L} + 1)\mu, 0\right) \\ F = ((\tau_L^2\tau_R + \tau_L^2 + \tau_L + 1)\mu - (\delta_L\tau_R + \delta_L)\mu, w_2) \\ w_2 = -\delta_L(\tau_L\tau_R + \tau_L + 1)\mu \\ G = \left((\tau_R(\frac{\delta_L}{\tau_L}) + \tau_R + 1)\mu, 0\right) \\ H = \left((\tau_R\delta_L + \tau_L\tau_R + \tau_L + 1)\mu, -\delta_L((\frac{\tau_R}{\tau_L})\delta_L + \tau_R + 1)\mu\right) \\ I = (v, -\delta_L(\tau_R\delta_L + \tau_L\tau_R + \tau_L + 1)\mu) \\ v = (\tau_L(\tau_R\delta_L + \tau_L\tau_R + \tau_L + 1) + 1)\mu + w_3 \\ w_3 = -\delta_L((\tau_R/\tau_L)\delta_L + \tau_R + 1)\mu. \end{cases} \quad (12.58)$$

The possible Ω-limit sets on Ξ are higher-periodic points or a chaotic attractor. To distinguish between these different scenarios, a stable higher-periodic point was considered. The simplest scenario is a birth of stable period-2 points in which the set Ξ has the geometry described above. More complex geometry (see Fig. 12.19) is the case where the period-2 points born in the bifurcation are unstable. This geometry plays a decisive role in determining possible attractors, *i.e.*, there is a possibility of the existence of a period-3 point whose periodic points lie on the segments GK, HD, and IF which is stable if the eigenvalues of the composed matrix MMN lie within the unit circle. If these period-3 points are unstable, then there are not any stable higher-periodic points on Ξ, but Ξ itself is attracting. The fact that neither Π_R, $\Pi_L \Pi_R$, nor $\Pi_L \Pi_L \Pi_R$ iterations correspond to stable periodic points (Π_R reflects stability of b, $\Pi_L \Pi_R$ of ab, and $\Pi_L \Pi_L \Pi_R$ of aab periodic points), their composition cannot produce any stable periodic point (the eigenvalues of the matrices M and N are purely real) which implies the nonexistence of stable periodic points within Ξ, *i.e.*, any higher-periodic point must contain any sequence of Π_R, $\Pi_L \Pi_R$, and/or $\Pi_L \Pi_L \Pi_R$. Furthermore, there is no any periodic point living on Ξ containing $\Pi_L \Pi_L \Pi_L$ sequences since the maximum number of subsequent iterations of (12.38) in the region R_L is two. This is a result of the character of the set Ξ that any point which belongs to segment GO shown in Fig. 12.19 in a maximum of two iterations is mapped into the region R_R, which implies that any $\Pi_L \Pi_L$ sequence must be followed by at least one Π_R iteration. Since maps (12.39) and (12.40) are linear, there cannot exist any points of period-2 or higher which are confined only to R_L or R_R. Thus a sudden jump to a one-, three-, or six-piece chaotic attractor can be observed. The existence of this type of chaotic attractor comes from the fact that if one of the eigenvalues of the mapping describing the period-3 point is close to -1, the Ω-limit set is organized around the period-3 points, and the chaotic attractor become dense on the entire set Ξ with the variation of some parameter such as τ_R. The robustness of this attractor can be confirmed using similar arguments to the one presented in the case of a two-piece invariant set. Finally, since (12.54) implies positive τ_L, birth of the described three-piece invariant sets can be observed only in the parameter space '2' in Fig. 12.12. □

An example of a robust chaotic attractor with three pieces is the one generated by the map defined by the following formulas of the matrices M and N:

$$M = \begin{pmatrix} 0.5 & 1 \\ -0.06 & 0 \end{pmatrix}, N = \begin{pmatrix} \tau_R & 1 \\ 0 & 0 \end{pmatrix}. \quad (12.59)$$

In this case, and due to the border-collision bifurcation scenario, one can observe a transition from a period-1 point to a period-3 point, or to a six-, three-, or one-piece chaotic attractor. Indeed, with a variation of τ_R from $\tau_R = 3.5$, there are different cases of border-collision bifurcation scenarios under variation of the bifurcation parameter μ, all linked to the existence of the three-piece piecewise-linear continuous set Ξ as shown in Fig. 12.19. For $\tau_R = -3.5$, there exists a stable admissible fixed point and a virtual saddle of the flip type, which under the variation of μ, due to the border collision, becomes a virtual stable fixed point and an admissible flip saddle. The values of points K, L, and C to I for $\mu = 0.1$ are given by

$$\begin{cases} K = (-0.25, 0), L = (0.1, 0), C = (0, 0.012) \\ D = (-0.025, 0.0150), E = (0.112, 0), F = (0.1025, 0.0015) \\ , G = (-0.292, 0), H = (-0.0460, 0.01752) \\ I = (0.09452, 0.00276). \end{cases} \quad (12.60)$$

In this case, there exists a three-piece set Ξ, and since the eigenvalues of MN for $\tau_R = -3.5$ are 0 and -1.81, the period-2 point born in the border collision is unstable, thus the lowest possible stable periodic point is of period-3 iterated twice in R_L and once in R_R because the eigenvalues of MMN lie within the unit circle. Thus border-collision bifurcations from a fixed point to a period-3 attractor are possible under a variation of μ. This result is shown in Fig. 12.20(a). In Fig. 12.20(b), the period-3 points obtained from numerical simulations of map (12.59) are shown to lie within the set Ξ on segments GK, HD, and IF, exactly as predicted

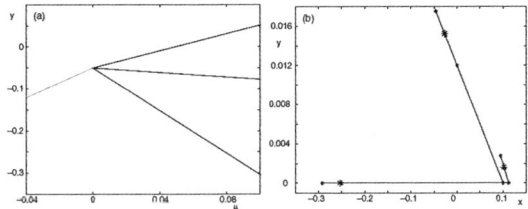

Fig. 12.20 (a) Bifurcation diagram representing the birth of a stable period-3 point due to the border-collision bifurcation scenario for (12.59) with $\tau_R = -3.5$ and (b) period-3 points for $\mu = 0.1$. Note that as predicted, they lie within Ξ, which is denoted in the figure by three black line segments. Small asterisks denote the points $K, L,$ and C to I forming the set Ξ. Reused with permission from Kowalczyk, P., Nonlinearity (2005). Copyright 2005, IOP Publishing Ltd.

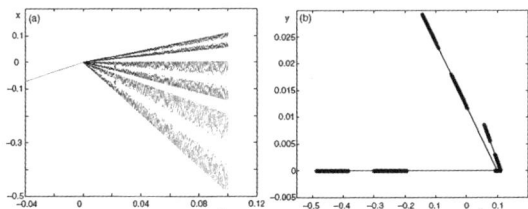

Fig. 12.21 Similar to Fig. 12.20 but for $\tau_R = -5.25$. (a) Bifurcation diagram. (b) Ω-limit set. Reused with permission from Kowalczyk, P., Nonlinearity (2005). Copyright 2005, IOP Publishing Ltd.

by the analysis. Because the eigenvalues of MMN for $\tau_R \approx -5.105$ are 0 and -1, then the birth of a chaotic attractor in the border-collision bifurcations for τ_R less than -5.105 is possible. Thus for $\tau_R = -5.25$, there is a border-collision bifurcation from a fixed point attractor to a six-piece chaotic attractor filling parts of the set Ξ as shown in Fig. 12.21(a). The corresponding Ω-limit set is depicted in Fig. 12.21(b). The six-piece structure of this chaotic attractor is a result of the fact that for $\tau_R = -5.25$, one of the eigenvalues of MMN is close to -1, and the chaotic attractor is organized around the period-3 points. Fig. 12.22(a) shows a bifurcation diagram for the variation of the bifurcation parameter μ with $\tau_R = -5.5$, in which the border-collision bifurcation leads to the birth of the three-piece chaotic attractor shown in Fig. 12.22(b) where thin black lines denote the set Ξ. For $\tau_R = -6.5$, a border collision from a stable fixed point attractor to a one-piece chaotic attractor is observed. The bifurcation diagram and Ω-limit set of the attractor are depicted in Fig. 12.23, where the points

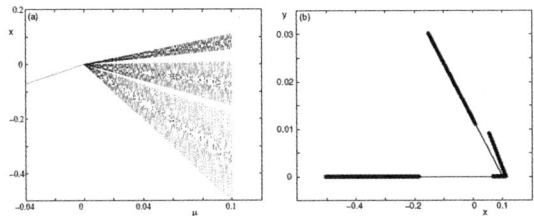

Fig. 12.22 Similar to Fig. 12.20 but for $\tau_R = -5.5$. (a) Bifurcation diagram. (b) Ω-limit set. Reused with permission from Kowalczyk, P., Nonlinearity (2005). Copyright 2005, IOP Publishing Ltd.

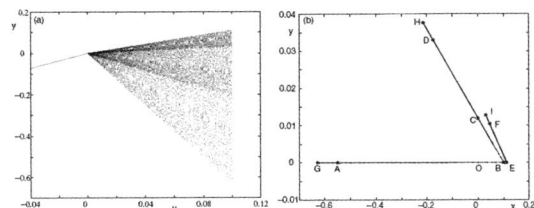

Fig. 12.23 Similar to Fig. 12.20 but for $\tau_R = -6.5$. (a) Bifurcation diagram. (b) Ω-limit set. Reused with permission from Kowalczyk, P., Nonlinearity (2005). Copyright 2005, IOP Publishing Ltd.

K, L, and C to I are denoted by asterisks in Fig. 12.23(b). They define the set Ξ which coincides with the chaotic attractor obtained numerically from (12.59) for $\tau_R = -6.5$.

12.2.2.2.4 N-piece invariant sets

In this case, there is the possible existence of a set Ξ formed by a higher number of piecewise-linear segments, four, five, *etc.* For example, if EI shown in Fig. 12.19 is not mapped within GE (violation of (12.56)) or if the maximum value of x in the image of OEF is greater than at E (violation of (12.55)), then any bounded set must necessarily contain more than three piecewise-linear segments. The analysis of this case can be done using the above techniques.

12.3 Exercises

Exercise 249. Explain the method for the creation of border-collision bifurcations in power converters. See [Banerjee *et al.* (1999)].

Exercise 250. Find an example of a piecewise-smooth system that exhibits classical smooth bifurcations.

Exercise 251. (a) Show that the normal form of the map (12.1) is given by (12.5). See [Banerjee & Grebogi (1999)].
  (b) Show that in the regions R_L and R_R the map (12.5) is smooth.
  (c) Show that the fixed points of map (12.5) are given by (12.7) and (12.8) with eigenvalues $\lambda_{L1,2}$ and $\lambda_{R1,2}$ given by (12.9).
  (d) Find regions for stable, saddle, and unstable fixed points.
  (e) Plot all the regions in the space $(\tau_L, \delta_L, \tau_R, \delta_R)$ of map (12.5) yielding nonrobust chaos.

Exercise 252. Find regions (if possible) of robust chaos in maps (12.22) and (12.23).

Exercise 253. (a) Find fixed points of the cardiac conduction model described by the map (12.27), and study their stability.
  (b) Find regions of robust chaos in map (12.27). See [Sun et al. (1995)].

Exercise 254. Find an example of a piecewise-linear map that displays instant chaos.

Exercise 255. (a) Give four cases and a detailed study for the occurrence of robust chaos in the map (12.38).
  (b) Find regions in the parameters space of map (12.38) in which a set Ξ formed by a higher number of piecewise-linear segments, four, five, seven, and eight can be observed.

Exercise 256. Consider the two-dimensional map given by [Zeraoulia & Sprott (2008d)]

$$\begin{cases} x_{n+1} = ax_n(1-x_n) \\ y_{n+1} = (b+cx_n)y_n(1-y_n). \end{cases} \quad (12.61)$$

(a) Show that the Lyapunov exponents of the map (12.61) are given by

$$\begin{cases} \lambda_1 = \lim_{N \to +\infty} \frac{1}{N} \sum_{n=1}^{n=N} \ln a\,|2x_n - 1| \\ \lambda_2 = \lim_{N \to +\infty} \frac{1}{N} \sum_{n=1}^{n=N} \ln |b+cx_n|\,|2y_n - 1|. \end{cases} \quad (12.62)$$

(b) Show that for all $n \in \mathbb{N}$, and if $0 \leq a \leq 4, 0 \leq x_0 \leq \frac{a}{4}$, then $0 \leq x_n \leq \frac{a}{4}$.

(c) Show that for all $n \in \mathbb{N}$, and if $0 \leq a \leq 4, 0 \leq x_0 \leq \frac{a}{4}, 0 \leq y_0 \leq \frac{4b+ac}{16}, 0 < b \leq 4 - \frac{ac}{4}, 0 < c < \frac{16}{a}$, then $0 \leq y_n \leq \frac{4b+ac}{16}$.

(d) Show that for all $n \in \mathbb{N}$, and if $0 \leq a \leq 4, 0 \leq x_0 \leq \frac{a}{4}, 0 \leq y_0 \leq \frac{4b+ac}{16}, 0 < b \leq 4 - \frac{ac}{4}, 0 < c < \frac{16}{a}$, then $\|(x_n, y_n)\|_2 \leq \frac{8abc+16a^2+16b^2+a^2c^2}{256}$.

(e) Show that if $0 \leq x_0 \leq \frac{a}{4}, 0 \leq y_0 \leq \frac{4b+ac}{16}, 0 \leq a \leq 4, 0 < c < \frac{16}{a}, 0 < b < 4 - \frac{ac}{4}$, then $\lambda_2 \leq \ln \frac{(4b+ac+8)(4b+ac)}{32}$.

(f) Show that the map (12.61) is asymptotically stable if $0 \leq x_0 \leq \frac{a}{4}$, $0 \leq y_0 \leq \frac{4b+ac}{16}$, $0 < a < 3$, $0 < b < \sqrt{3} - \frac{1}{4}ac - 1$, and $0 < c < \frac{4(\sqrt{3}-1)}{a}$.

(g) Show that the map (12.61) is almost periodic if $0 \leq x_0 \leq \frac{a}{4}, 0 \leq y_0 \leq \frac{4b+ac}{16}$, $3 \leq a \leq 3.5699457$, $0 < b < \sqrt{3} - \frac{1}{4}ac - 1$, and $0 < c < \frac{4(\sqrt{3}-1)}{a}$.

(h) Show that the map (12.61) is almost chaotic if $0 \leq x_0 \leq \frac{a}{4}, 0 \leq y_0 \leq \frac{4b+ac}{16}$, $3.5699457 < a < 4$, $0 < b < \sqrt{3} - \frac{1}{4}ac - 1$, and $0 < c < \frac{4(\sqrt{3}-1)}{a}$.

(i) Show that if $(b,c) \in \,]0, \sqrt{3} - \frac{1}{4}ac - 1[\, \times \,]0, \frac{4(\sqrt{3}-1)}{a}[\, = J$, then the onset of chaos in the modulated logistic map (12.61) is the same as in the logistic map (6.25), i.e., the critical value of a is seen to be 3.5699457, where periodicity just ends (the accumulation point).

(j) Plot the bifurcation diagram of the logistic map (6.25) for $0 \leq a \leq 4$ and the bifurcation diagrams of the modulated logistic map (12.61) for $0 \leq a \leq 4$ and (i) $b = 3.5, c = 0.5$, (ii) $b = 4, c = -1$, (iii) $b = -1, c = 4$. (iv) $b = -2, c = 2$. Deduce a relation between the dynamics of the logistic map (6.25) and the map (12.61).

(k) Plot (with the basin of attraction) and show the hyperchaoticity of the orbit obtained for the modulated map (12.61) for $a = 3.6$, $b = 3.5$, and $c = 0.5$, and for $a = 3.6, b = 4$, and $c = -1$.

(l) Find regions of ab-space with $c = 0.5$ for multiple attractors, the regions of unbounded attractors, and the regions of a single attractor for the modulated map (12.61).

(m) Deduce that the map (12.61) is robust outside the intervals $(a,b) \in [3, 3.7] \times [-2, -1.6]$ and $(a,b) \in [3, 3.7] \times [2.5, 3.1]$.

(n) Find regions of robust chaos for the two maps:

$$\begin{cases} x_{n+1} = |ax_n(1-x_n)| \\ y_{n+1} = (b+cx_n)y_n(1-y_n) \end{cases}$$

$$\begin{cases} x_{n+1} = ax_n(1-x_n) \\ y_{n+1} = |(b+cx_n)y_n(1-y_n)|. \end{cases}$$

(12.63)

Exercise 257. Consider the following modified Hénon map [Zeraoulia & Sprott (2008a)]:

$$f(x_n, y_n) = \begin{pmatrix} x_{n+1} \\ y_{n+1} \end{pmatrix} = \begin{pmatrix} 1 - a \sin x_n + b y_n \\ x_n \end{pmatrix}, \quad (12.64)$$

or equivalently,

$$x_{n+1} = 1 - a \sin x_n + b x_{n-1}. \quad (12.65)$$

(a) Show that *for all values of a and b, the sequence $(x_n)_n$ given in (12.65) satisfies the following inequality:* $|1 - x_n + b x_{n-2}| \leq |a|$.

(b) Show that *for every $n > 1$ and for all values of a and b, and for all values of the initial conditions $(x_0, x_1) \in \mathbb{R}^2$, the sequence $(x_n)_n$ satisfies the following equalities:*

(1) If $b \neq 1$, then

$$x_n = \begin{cases} \frac{b^{\frac{n-1}{2}}-1}{b-1} + b^{\frac{n-1}{2}} x_1 - a \sum_{p=1}^{p=\frac{n-1}{2}} b^{p-1} \sin x_{n-(2p-1)}, & \text{if } n \text{ is odd} \\ \frac{b^{\frac{n}{2}}-1}{b-1} + b^{\frac{n}{2}} x_0 - a \sum_{p=1}^{p=\frac{n}{2}} b^{p-1} \sin x_{n-(2p-1)}, & \text{if } n \text{ is even.} \end{cases}$$

(12.66)

(2) If $b = 1$, then

$$x_n = \begin{cases} \frac{n-1}{2} + x_1 - a \sum_{p=1}^{p=\frac{n-1}{2}} \sin x_{n-(2p-1)}, & \text{if } n \text{ is odd} \\ \frac{n}{2} + x_0 - a \sum_{p=1}^{p=\frac{n}{2}} \sin x_{n-(2p-1)}, & \text{if } n \text{ is even.} \end{cases} \quad (12.67)$$

(c) Show that the *fixed points (l, l) of the map (12.65) exist if one of the following conditions holds:*

(i) If $a \neq 0$ and $b \neq 1$, then l satisfies the following conditions:

$$\begin{cases} 1 - a \sin l + (b-1) l = 0, \text{ and } l \leq \frac{1+|a|}{1-b}, & \text{if } b > 1 \\ \frac{1+|a|}{1-b} \leq l, & \text{if } b < 1. \end{cases} \quad (12.68)$$

(ii) If $b = 1$ and $|a| \geq 1$, then l is given by $l = \arcsin\left(\frac{1}{a}\right)$.
(iii) If $b \neq 1$ and $a = 0$, then l is given by $l = \frac{1}{1-b}$.
(iv) If $a = 0$ and $b = 1$, there are no fixed points for the map (12.65).

(d) Show that the *orbits of the map (12.65) are bounded for all $a \in \mathbb{R}$, $|b| < 1$, and all initial conditions $(x_0, x_1) \in \mathbb{R}^2$.*

(e) Show that the map (12.65) possesses unbounded orbits in the following subregions of $\mathbb{R}^4$:

$$\Omega_2 = \left\{(a, b, x_0, x_1) \in \mathbb{R}^4 / \ |b| > 1, |x_0|, |x_1| > \frac{|a|+1}{|b|-1}\right\} \quad (12.69)$$

and

$$\Omega_3 = \left\{(a, b, x_0, x_1) \in \mathbb{R}^4 / \ |b| = 1, \text{ and } |a| < 1\right\}. \quad (12.70)$$

(f) Plot the chaotic multifold attractors of the map (12.65) obtained for (i) $a = 2.4, b = -0.5$. (ii) $a = 2, b = 0.2$. (iii) $a = 2.8, b = 0.3$. (iv) $a = 2.7, b = 0.6$. (v) $a = 3.4, b = -0.8$. (vi) $a = 3.6, b = -0.8$. (vii) $a = 4, b = 0.5$. (viii) $a = 4, b = 0.9$.

(g) Find regions of dynamical behaviors in ab-space for the map (12.65).

(h) Plot the bifurcation diagram and the variation of the Lyapunov exponents for the map (12.65) for $b = 0.3$ and $-1 \le a \le 4$ and for $-150 \le a \le 200$ with $b = 0.3$ for different initial conditions.

(i) Find regions of ab-space with multiple attractors. Hint: Use the interval $(a, b) \in [-1, 4] \times [-2, 2]$. For example, show that with $a = 2$ and $b = -0.6$, a two-cycle $(1.314326, -0.584114)$ coexists with a period-3 strange attractor. Similarly, for $a = 2.2$ and $b = -0.36$, there is a strange attractor surrounded by a second period-3 strange attractor. Deduce some regions in ab-space that support robust chaos.

(j) Find regions of robust chaos for the two maps

$$\begin{cases} x_{n+1} = |1 - a\sin x_n + by_n| \\ y_{n+1} = x_n \end{cases} \text{ and } \begin{cases} x_{n+1} = 1 - a\sin x_n + by_n \\ y_{n+1} = |x_n|. \end{cases} \quad (12.71)$$

Exercise 258. (Open problems) (Discussion with Soumitro Banerjee)

(a) The occurrence of high-period orbits has not been extensively studied. The conditions for existence and stability of high-period orbits can be obtained for "maximal" periodic orbits (those with one point in L and all others in R, and *vice versa*). However, nonmaximal orbits have not been studied.

(b) No formulation for the calculation of invariant density is available.

(c) The dynamics of piecewise-linear discontinuous maps is practically unexplored.

(d) The piecewise-linear map is a valid approximation of a piecewise-smooth map in the neighborhood of the borderline. However, many new phenomena can occur if one considers the possible nonlinear character of the map on either side of the border. Such maps have been probed only for the square-root singularity case. All other possibilities are unexplored.

Exercise 259. (Open problem) Find conditions for supercritical border-collision period doubling that occurs without EBOs when the determinants on both sides of the border are of opposite signs.

Exercise 260. (Open problem) Find regions in the space $\tau_L, \delta_L, \tau_R, \delta_R$ of map (12.5) in which a stable fixed point leads to a stable fixed point plus extraneous periodic orbits.

Bibliography

Aarts, J. M. and Fokkink, R. J., The classification of solenoids. Proceedings Amer. Math. Soc. **111** (1991) 1161–1163.

Abraham, R., *Introduction to Morphology*. Publ. Dept. Math. Lyon. (1972) 38–114.

Abraham, R. and Marsden, J. E., *Foundations of Mechanics*. Benjamin/Cummings Publishing, Reading Mass (1978).

Abraham, R. and Robbin, J., *Transversal Mappings and Flows*. Benjamin (1967).

Abraham, R. and Smale, S., Nongenericity of Ω-stability. *Global analysis I*. Proceedings Symp. Pure Math. AMS **14** (1970) 5–8.

Abdenur, F., Bonatti, C., Crovisier, S., Diaz, L., and Wen, L., Periodic points and homoclinic classes. Ergod. Theor. Dyn. Syst. **27** (2007) 1–22.

Abdenur, F., Bonatti, C., and Crovisier, S., Non-uniform hyperbolicity for C^1-generic diffeomorphisms. Preprint arXiv:0809.3309 (2008).

Adler, R. and Weiss, B., Similarity of automorphisms of the torus. Memoirs of the American Mathematical Society **98**. American Mathematical Society, Providence, RI (1970).

Adrangi, B. and Chatrath, A., Non-linear dynamics in futures prices evidence from the coffee, sugar and cocoa exchange. Applied Financial Economics **13**(4) (2003) 245–256.

Afraimovich, V. S. and Hsu, S. B., *Lectures on Chaotic Dynamical Systems*. Am. Math. Soc., Providence, RI (2003) AMS/IP Studies in Advanced Mathematics, Vol. **28**.

Afraimovich, V. S. and Shilnikov, L. P., Strange attractors and quasiattractors in *Nonlinear Dynamics and Turbulence* (G. I. Barenblatt, G. Iooss, and D. D. Joseph, eds.) Pitman, New York (1983) 1–28.

Afraimovich, V. S., Bykov, V. V., and Shilnikov, L. P., On the appearance and structure of Lorenz attractor. DAN SSSR **234** (1977) 336–339.

Afraimovich, V. S., Bykov, V. V., and Shilnikov, L. P., On structurally unstable attracting limit set of the type of Lorenz attractor. Trans. Moscow. Math. Soc. **44** (1982) 153–216.

Afraimovich, V. S., Bykov, V. V., and Shilnikov, L. P., On the structurally unstable attracting limit sets of Lorenz attractor type. Tran. Moscow. Math. Soc. **2** (1983) 153–215.

Afraimovich, V. S., Chernov, N. I., and Sataev, E. A., Statistical properties of two-dimensional generalized hyperbolic attractors. Chaos **5**(1) (1995) 238–252.
Agrachev, A., The curvature and hyperbolicity of Hamiltonian systems. Proceedings of the Steklov Institute of Mathematics **256**(1) (2007) 26–46.
Aguirregabiria, J. M., Robust chaos with variable Lyapunov exponent in smooth one-dimensional maps. Preprint arXiv:0810.3781 (2008).
Aharonov, D., Devaney, R. L., and Elias, U., The dynamics of a piecewise linear map and its smooth approximation. Int. J. Bifurcation and Chaos **7**(2) (1997) 351–372.
Albers, D. J. and Sprott, J. C., Structural stability and hyperbolicity violation in high-dimensional dynamical systems. Nonlinearity **19**(8) (2006) 1801–1849.
Albers, D. J., Sprott, J. C., and Dechert, W. D., Routes to chaos in neural networks with random weights. Int. J. Bifurcation and Chaos **8** (1998) 1463–1478.
Albers, D. J., Sprott, J. C., and Crutchfield, J. P., Persistent chaos in high dimensions. Phys. Rev. E **74** (2006) 057201.
Alefeld, G. and Herzberger, J., *Introduction to Interval Computations*. Academic Press, New York (1983).
Aleksandrov, P. S., *Introduction to Set Theory and General Topology*. Nauka, Moscow (1977); (German translation: VEB Deutscher Verlag der Wissenschaften, Berlin) (1984).
Alexander, J. C., York, J. A., You, Z. P., and Kan, I., Riddled basins. Int. J. Bifurcation and Chaos **2** (1992) 795–813.
Algaba, A., Freire, E., Gamero, E., and Rodriguez-Luis, A. J., On the Takens–Bogdanov bifurcation in the Chua's equation. IEICE Trans. Fund. Electronics Comm. Comput. Sci. **E82-A**(9) (1999) 1722–1728.
Algaba, A., Merino, M., Freire, E., Gamero, E., and Rodriguez-Luis, A. J., On the Hopf-pitchfork bifurcation in the Chua's equation. Int. J. Bifurcation and Chaos **10**(2) (2000) 291–305.
Alligood, K. T. and Sauer, T., Rotation numbers of periodic orbits in the Hénon map. Comm. Math. Phys. **120**(1) (1988) 105–119.
Alligood, K. T., Sander, E., and Yorke, J. A., Crossing bifurcations and unstable dimension variability. Phys. Rev. Lett. **96** (2006) 244103.
Altman, E. J., Bifurcation analysis of Chua's circuit with applications for low-level visual sensing. J. Circuits. Syst. Comput. **3**(1) (1993a) 63–92.
Altman, E. J., Normal form analysis of Chua's circuit with applications for trajectory recognition. IEEE Trans. Circ. Syst. Part II: Analog and Digital Signal Processing **40**(10) (1993b) 675–682.
Alvarez-Llamoza, O., Cosenza, M. G., and Ponce, G. A., Critical behavior of the Lyapunov exponent in type-III intermittency. Chaos, Solitons & Fractals **36**(1) (2008) 150–156.
Alves, F. and Araujo, V., Random perturbations of nonuniformly expanding maps. Astérisque **286** (2003) 25–62.
Alves, J. F., Araújo, V., Pacifico, M. J., and Pinheiro, V., On the volume of singular-hyperbolic sets. Dynamical Systems, an International Journal **22**(3) (2007) 249–267.

Amaricci, A., Bonetto, F., and Falco, P., Analyticity of the Sinai–Ruelle–Bowen measure for a class of simple Anosov flows. J. Math. Phys. **48**(7) (2007) 072701-072701-15.

Amit, D. J., *Modeling Brain Function: The World of Attractor Neural Networks*. Cambridge University Press (1989).

Andrea, S. A., On homeomorphisms of the plane, and their embeddings in flows. Bull. AMS **71** (1965) 381–383.

Andrecut, M. and Ali, M. K., Robust chaos in a smooth system. Int. J. Modern Physics B **15**(2) (2001a) 177–189.

Andrecut, M. and Ali, M. K., On the occurrence of robust chaos in a smooth system. Modern Physics Letters B **15**(12–13) (2001b) 391–395.

Andrecut, M. and Ali, M. K., Example of robust chaos in a smooth map. Europhys. Lett. **54** (2001c) 300–305.

Andrecut, M. and Ali, M. K., Robust chaos in smooth unimodal maps. Phys. Rev. E **64** (2001d) 025203(R) 1–3.

Andrianov, I. V. and Manevitch, L. I., *Asymptotology: Ideas, Methods, and Applications*. Springer, Berlin (2002).

Andronov, A. and Pontryagin, L., Systèmes grossiers. Dokl. Akad. Nauk USSR **14** (1937) 247–251.

Anishchenko, V. S., *Complex Oscillations in Simple Systems*. Nauka, Moscow (1990).

Anishchenko, V. S., *Dynamical Chaos Models and Experiments*. World Scientific, Singapore (1995).

Anishchenko, V. S., Neiman, A. B., Safonova, M. A., and Khovanov, I. A., Multifrequency stochastic resonance. In *Proceedings of Euromech Colloquium on Chaos and Nonlinear Mechanics* (T. Kapitaniak and J. Brindley, eds.) World Scientific, Singapore (1995) 41.

Anishchenko, V. S. and Strelkova, G. I., Attractors of dynamical systems. *Control of Oscillations and Chaos. Proceedings (1997) 1st Int. Conference* **3** (1997) 498–503.

Anishchenko, V. S. and Strelkova, G., Irregular attractors. Discrete Dynamics in Nature and Society **2**(1) (1998) 53–72.

Anishchenko, V. S., Vadivasova, T. E., Strelkova, G. I., and Kopeikin, A. S., Chaotic attractors of two-dimensional invertible maps. Discrete Dynamics in Nature and Society **2** (1998) 249–256.

Anishchenko, V. S., Kopeikin, A. S., and Kurths, J., Studying hyperbolicity in chaotic systems. Phys. Lett. A **270** (2000) 301–307.

Anishchenko, V. S., Astakhov, V. V., Neiman, A. B., Vadivasova, T. E., and Schimansky-Geier, L., *Nonlinear Dynamics of Chaotic and Stochastic Systems: Tutorial and Modern Development*. Springer, Berlin (2002).

Anishchenko, V. S., Luchinsky, D. G, McClintock, P. V. E., Khovanov, I. A., and Khovanova, N. A., Fluctuational escape from a quasi-hyperbolic attractor in the Lorenz system, J. Experimental Theoretical Physics **94**(4) (2002) 821–833.

Anishchenko, V. S., Vadivasova, T. E., Okrokvertskhov, G. A., and Strelkova, G. I., Correlation analysis of dynamical chaos. *Physica* A **325** (2003a) 199–212.

Anishchenko, V. S., Vadivasova, T. E., Kopeikin, A. S., Kurths, J., and Strelkova, G. I., Spectral and correlation analysis of spiral chaos. Fluct. Noise. Lett. **3** (2003b) L213–L221.

Anishchenko, V. S., Vadivasova, T. E., Okrokvertskhov, G. A., and Strelkova, G. I., Correlation analysis of the regimes of deterministic and noisy chaos. J. Comm. Techn. Electr. **48** (2003c) 750–760.

Anishchenko, V. S., Astakhov, V. V., Vadivasova, T. E., Neoeman, A. B., Strelkova, G. I., and Schimansky-Geier, L., *Nonlinear Effects of Chaotic and Stochastic Systems*. Inst. Komp'yut. Issled., Moscow (2003d) (in Russian).

Anishchenko, V. S., Vadivasova, T. E., Strelkova, G. I., and Okrokvertskhov, G. A., Statistical properties of dynamical chaos. Math. Biosciences. Engineering **1**(1) (2004) 161–184.

Anishchenko, V. S., Neiman, A. B., Vadiavasova, T. E., Astakhov, V.V., and Schimansky-Geier, L., *Nonlinear Dynamics of Chaotic and Stochastic Systems* (2nd edn.) Springer Series in Synergetics (2007).

Anosov, D. V., Geodesic flows on closed Riemannian manifolds of negative curvature. Proceedings Steklov Math. Inst. **90** (1967) 1–235.

Anosov, D. V., *Ergodic Theory*. In Hazewinkel, Michiel, Encyclopedia of Mathematics, Kluwer Academic Publishers (2001).

Anosov, D. V., Aranson, S. K., Grines, V. Z., and Plykin, R. V., *Dynamical Systems IX: Dynamical Systems with Hyperbolic Behaviour*. Encyclopedia Mat. Sci. **9** (1995).

Anosov, D. V., Klimenko, A. V., and Kolutsky, G., On the hyperbolic automorphisms of the 2-torus and their Markov partitions. Preprint of Max-Plank Institute for Mathematics (2008).

Arai, Z., On hyperbolic plateaus of the Hénon map. Experiment. Math. **16**(2) (2007a) 181–188.

Arai. Z., On loops in the hyperbolic locus of the complex Hénon map. Preprint (2007b).

Arai. Z. and Mischaikow, K., Rigorous Computations of Homoclinic Tangencies. SIAM J. Appl. Dyn. Syst. **5** (2006) 280–292.

Araujo, V., *Existencia de Atratores Hiperbolicos para Dieomorfismos de Superficies*, Ph.D. Thesis, IMPA (1987).

Araujo, V. and Bessa, M., Dominated splitting and zero volume for incompressible three-flows. Nonlinearity **21**(7) (2008) 1637–1653.

Araujo, V. and Pacifico, M. J., Three dimensional flows. XXV Brazillian Mathematical Colloquium. IMPA, Rio de Janeiro (2007).

Araujo, V. and Pacifico, M. J., What is new on Lorenz-like attractors. Preprint arXiv:0804.3617 (2008).

Araujo, V., Pacifico, M. J., Pujals, E. R., and Viana, M., Singular-hyperbolic attractors are chaotic. Trans. Amer. Math. Soc. **361** (2009) 2431–2485.

Archibald, T., Tension and potential from Ohm to Kirchhoff, Centaurus **31**(2) (1988) 141–163.

Arnéodo, A., Coullet, P., and Tresser, C., Possible new strange attractors with spiral structure. Comm. Math. Phys. **79** (1981) 673–679.

Arnold, V. I., *Geometrical Methods in the Theory of Ordinary Differential Equations*. Springer-Verlag, Berlin (1988).

Arnold, V. I., *Ordinary Differential Equations*. Springer-Verlag, Berlin (1992).

Arnold, V. I. and Avez, A., *Ergodic Problems of Classical Mechanics*. Benjamin, New York (1968).

Aronson, D. G., Chory, M. A., Hall, G. R., and McGehee, R. P., Bifurcations from an invariant circle for two-parameter families of maps of the plane A computer-assisted study. Comm. Math. Phys. **83** (1982) 303–354.

Arov, D. Z., Topological similarity of automorphisms and translations of compact commutative groups. Uspekhi Mat. Nauk **185** (1963) 133–138 (in Russian).

Arrowsmith, D. K. and Place, C. M., *An Introduction to Dynamical Systems*. Cambridge University Press (1990).

Arroyo, A., Singular hyperbolicity for transitive attractors with singular points of 3-dimensional C^2-flows. Bull. Braz. Math. Soc., New Series **38**(3) (2007) 455–465.

Arroyo, A. and Hertz, F. R., Homoclinic bifurcations and uniform hyperbolicity for three-dimensional flows. Ann. Inst. H. Poincaré **20**(5) (2003) 805–841.

Arroyo, A. and Pujals, E., Dynamical properties of singular hyperbolic attractors. Discrete and Continuous Dynamical Systems **19**(1) (2007) 67–87.

Artur, A. and Gustavo, M. C., Statistical properties of unimodal maps: Physical measures, periodic orbits and pathological laminations, Publications Mathématiques **101** (2005) 1–67.

Asaoka, M., A simple construction of C^1-Newhouse domain for higher dimensions. Preprint (2008).

Ashwin, P. and Rucklidge A. M., Cycling chaos its creation, persistence and loss of stability in a model of nonlinear magnetoconvection. Physica D **122**(1) (1998) 134–154.

Aubry, S. J., Anti-integrability in dynamical and variational problems. Physica D **86** (1995) 284–296.

Aulbach, B. and Flockerzi, D., The past in short hypercycles. J. Math. Biol. **27** (1989) 223–231.

Avila, A. and Bochi, J., A generic C^1 map has no absolutely continuous invariant probability measure. Nonlinearity **19**(11) (2006) 2717–2725.

Avila, A. and Moreira, C. G., Statistical properties of unimodal maps smooth families with negative Schwarzian derivative. Astérisque **286** (2003) 81–118.

Avila, A. and Moreira, C. G., Phase-parameter relation and sharp statistical properties for general families of unimodal maps. Contemporary Mathematics **389** (2005a) 1–42.

Avila, A. and Moreira, C. G., Statistical properties of unimodal maps the quadratic family. Annals Math. **161** (2005b) 831–881.

Avila, A., Lyubich, M., and de Melo, W., Regular or stochastic dynamics in real analytic families of unimodal maps. Invent. Math. **154** (2003) 451–550.

Avila, A., Lyubich, M., and Shen, W., Parapuzzle of the multibrot set and typical dynamics of unimodal maps. Preprint arXiv:0804.2197 (2008).

Avrutin, V. and Schanz, M., Crises cascades within robust chaos in piecewise-smooth maps. ENOC-2008 St.-Petersburg, Russia (2008).

Aziz-Alaoui, M. A., Multispiral chaos. *(2000) 2nd Inter. Conference. Control of Oscillations and Chaos. Proceedings (Cat. No.00TH8521) IEEE. Part 1* (2000) 88–91.

Baladi, V. and Viana, M., Strong stochastic stability and rate of mixing for unimodal maps. Ann. Scient. Ec. Norm. Sup. **29** (1996) 483–517.

Baladi, V., Benedicks, M., and Maume-Deschamps, V., Almost sure rates of mixing for i.i.d. unimodal maps. Annales Scientifiques de l'Ecole Normale Superieure **35**(1) (2002) 77–126.

Balint. P. and Gouezel. S., Limit theorems in the stadium billiard. Comm. Math. Phys. **263** (2006) 461–512.

Ballmann, W. and Wojtkowski, M. P., An estimate for the measure theoretic entropy of geodesic flows. Ergod. Theor. Dyn. Syst. **9** (1989) 271–279.

Balmforth, N. J., Solitary waves and homoclinic orbits. Annual review of fluid mechanics. **27** (1995) 335–373 Annual Reviews, Palo Alto, CA.

Bamon, R., Labarca, R., Mané, R., and Pacifico, M. J., The explosion of singular cycles. Publ. Math. IHES **78** (1993) 207–232.

Banerjee, S. and Chakrabarty, K., Nonlinear modeling and bifurcations in the boost converter. IEEE Trans. Power Electron. **13** (1998) 252–260.

Banerjee, S. and Grebogi, C., Border collision bifurcations in two-dimensional piecewise smooth maps. Phy. Rev. E **59**(4) (1999) 4052–4061.

Banerjee, S. and Verghese, G. C., *Nonlinear Phenomena in Power Electronics: Attractors Bifurcations, Chaos, and Nonlinear Control.* IEEE Press, New York (2001).

Banergee, S., York, J. A, and Grebogi, C., Robust chaos. Phys. Rev. Lett. **80**(14) (1998) 3049–3052.

Banerjee, S., Kastha, D., Das, S., Vivek, G., and Grebogi, C., Robust chaos – The theoretical formulation and experimental evidence **ISCAS (5)** (1999) 293–296.

Banerjee, S. Karthik, Guohui, M. S., and Yorke, J. A., Bifurcations in one-dimensional piecewise smooth maps – Theory and applications in switching systems. IEEE Trans. Circ. Syst. Fund. Theor. Appl. **47**(3) (2000) 389–394.

Banks, J. and Dragan, V., Smale's horseshoe map via ternary numbers. SIAM Review **36**(2) (1994) 265–271.

Baraviera, A. T., *Robust Nonuniform Hyperbolicity for Volume Preserving Maps.* Ph.D. Thesis, IMPA (2000).

Barge, M., Homoclinic intersections and indecomposability. Proceedings Amer. Math. Soc. **101** (1987) 541–544.

Barge, M. and Kennedy, J., Continuum theory and topological dynamics. In *Open Problems in Topology II* (Elliott Pearl, ed.) Elsevier, Amsterdam (2007).

Barreto, E., Hunt, B., Grebogi, C., and Yorke, J. A., From high dimensional chaos to stable periodic orbits – The structure of parameter space. Phys. Rev. Lett. **78** (1997) 4561.

Bartissol, P. and Chua, L. O., The double hook (nonlinear chaotic circuits). IEEE Trans. Circ. Syst. **35**(12) (1988) 1512–1522.

Bass, H., Conjecture Jacobienne et opérateurs différentiels. *Mém. Soc. Math. France* **38** (1989) 39–50.

Bass, H., Cornell, E. H., and Wright, D., The Jacobian conjecture – Reduction of degree and formal expansion of inverse. *Bull. Amer. Math. Soc.* **7** (1982) 287–330.

Bauer, M. and Martienssen, W., Quasi-periodicity route to chaos in neural networks. *Europhys. Lett.* **10** (1989) 427–431.

Bautista, S., *Sobre conjuntos hiperbólicos-singulares* (in Portuguese) Thesis, Universidade Federal do Rio de Janeiro (2005).

Bautista, S. and Morales, C., Existence of periodic orbits for singular-hyperbolic sets. *Moscow Mathematical Journal* **6**(2) (2006) 265–297.

Bautista, S., Morales, C., and Pacifico, M. J., There are singular hyperbolic flows without spectral decomposition. Preprint IMPA. Série **A278** (2004).

Bautista, S., Morales, C., and Pacifico, M. J., Intersecting invariant manifolds on singular-hyperbolic sets. Preprint (2005).

Becker, T., Weispfenning, V., and Gröbner, B., *Computational Approach to Commutative Algebra*. Springer-Verlag, New York (1993a).

Bedford, E. and Smillie, J., Polynomial diffeomorphisms of $\mathbb{C}^2$ currents, equilibrium measure and hyperbolicity. *Invent. Math.* **103** (1991) 69–99.

Bedford, E. and Smillie, J., The Hénon family – The complex horseshoe locus and real parameter values. *Contemp. Math.* **396** (2006a) 21–36.

Bedford, E. and Smillie, J., Real Polynomial diffeomorphisms with maximal entropy: Tangencies. *Ergod. Theor. Dyn. Syst.* **26**(5) (2006b) 1259–1283.

Bedford, E. and Smillie, J., Real polynomial diffeomorphisms with maximal entropy II – Small Jacobian. *Ergod. Theor. Dyn. Syst.* **26**(5) (2006c) 1259–1283.

Bedford, T. and Keane, M., and Series, C., (eds.), *Ergodic Theory, Symbolic Dynamics and Hyperbolic Spaces*. Oxford University Press (1991).

Béguin, F. and Bonatti, C., Flots de Smale en dimension 3 – Présentations finies de voisinages invariants d'ensembles selles (in French, English, French summary) (Smale flows in dimension 3 – Finite presentations of invariant neighborhoods of saddle sets). *Topology* **41**(1) (2002) 119–162.

Belykh, V., Models of discrete systems of phase locking. In *Phase Locking Systems* (L. N. Belyustina and V. V. Shakhgil'dyan, eds.) Radio Svyaz, Moscow. (1982) 161–176 (in Russian).

Belykh, V., Chaotic and strange attractors of two-dimensional map. *Math. Sbornik* **186**(3) (1995) (in Russian).

Belykh, V. N. and Chua, L. O., New type of strange attractor from a geometric model of Chua's circuit. *Int. J. Bifurcation and Chaos* **2**(3) (1992) 697–704.

Belykh, V., Belykh, I., and Mosekilde, E., Hyperbolic Plykin attractor can exist in neuron models. *Int. J. Bifurcation and Chaos* **15**(11) (2005) 3567–3578.

Benedicks, M., Non-uniformly Hyperbolic Dynamics – Hénon Maps and Related Dynamical Systems. *ICM 2002* **III** (2002) 1–3.

Benedicks, M. and Carleson, L., On iterations of $1-ax^2$ on $(-1,1)$. *Annals Math.* **122** (1985) 1–25.

Benedicks, M. and Carleson, L., The dynamics of the Hénon maps. Annals Math. **133** (1991) 1–25.

Benedicks, M. and Young, L. S., SBR-measures for certain Hénon maps. Invent. Math. **112** (1993) 541–576.

Benedicks, M. and Young, L. S., Markov extensions and decay of correlations for certain Hénon maps. Asterisque **26** (2000) 113–56.

Benettin, G., Galgani, L., Giorgilli, A., and Strelcyn, J. M., Lyapunov characteristic exponents for smooth dynamical systems and for Hamiltonian systems; a method for computing all of them, Part 2: Numerical applications. Meccanica **15**(9) (1980) 21–30.

Benoist, Y. and Labourie, F., Sur les difféomorphismes d'Anosov à feuilletages stable et instable différentiables. Invent. Math. **111**(2) (1993) 285–308.

Berger, A., *Chaos and Chance: An Introduction to Stochastic Aspects of Dynamics*. Walter de Gruyter (2001).

Berns, D. W., Moiola, J. L., and Chen, G., Quasi-analytical method for period-doubling bifurcation, ISCAS (2001). *The (2001) IEEE Inter. Symposium on Circuits and Syst. (Cat. No.01CH37196). IEEE. Part* **2** (2001a) 739–742.

Bers, L. and Royden, H. L., Holomorphic families of injections. Acta Math. **157** (1986) 259–286.

Bertau, M., Mosekilde, E., and Westerhoff, H. V., *Biosimulation in Drug Development*. Wiley and Sons (2007).

Bertsekas, D. P. and Tsitsiklis, J. N., *Introduction to Probability*. Athena Scientific, Massachusetts (2002).

Bessa, M., The Lyapunov exponents of generic zero divergence 3-dimensional vector fields. Ergod. Theor. Dyn. Syst. **27**(5) (2007) 1445–1472.

Bessa, M. and Duarte, P., Abundance of elliptic dynamics on conservative 3-flows. Preprint arXiv:0709.0700v (2007).

Bhalla, U. S. and Lyengar, R., Robustness of the bistable behavior of a biological signaling feedback loop. Chaos **11** (2001) 221–226.

Bhattacharya, J. and Kanjilal, P. P., On the detection of determinism in a time series. Physica D. **132**(1) (1999) 100–110.

Biham, O. and Wenzel, W., Characterization of unstable periodic orbits in chaotic attractors and repellers. Phys. Rev. Lett. **63** (1989) 819–822.

Bilotta, E., Pantano, P., and Stranges, S., A gallery of Chua attractors. Part I. Int. J. Bifurcation and Chaos **17** (2007) 1–60.

Birkhoff, G. D., Proof of the ergodic theorem. Proceedings Natl. Acad. Sci. USA **17** (1931) 656–660.

Birkhoff, G. D., What is the ergodic theorem? Amer. Math. Monthly **49**(4) (1942) 222–226.

Bischi, G. I. and Gardini, L., Role of invariant and minimal absorbing areas in chaos synchronization. Phys. Rev. E **58** (1998) 5710–5719.

Blazquez, C. M. and Tuma, E., Dynamics of Chua's circuit in a Banach space. J. Circuits Syst. Comput. **3**(2) (1993a) 613–626.

Blazquez, M. and Tuma, E., Strange attractors of the Shilnikov type in Chua's circuit. Int. J. Bifurcation and Chaos **3**(5) (1993b) 1293–1298.

Blazquez, M. and Tuma, E., Strange attractors of the Shilnikov type in Chua's circuit. Int. J. Bifurcation and Chaos **3**(5) (1993c) 1293–1298.
Blazquez, M. and Tuma, E., Chaotic behavior of orbits close to a heteroclinic contour. Int. J. Bifurcation and Chaos **6**(1) (1993d) 69–79.
Bloch, W. L., Extending flows from isolated invariant sets. Ergod. Theor. Dyn. Syst. **15**(6) (1995) 1031–1043.
Blokh, A. and Lyabich, M., Measurable of S-unimodal maps of the interval. Annales Scientifique de l'E.N.S, 4e série **24**(5) (1991) 545–573.
Blokh, A. and Misiurewicz, M., Dense set of negative Schwarzian maps whose critical points have minimal limit sets. Discrete Contin. Dynam. Systems **4** (1998a) 141–158.
Blokh, A. and Misiurewicz, M., Collet–Eckmann maps are unstable. Comm. Math. Phys. **191** (1998b) 61–70.
Bochi, J., Genericity of zero Lyapunov exponents. Ergod. Theor. Dyn. Syst. **22**(6) (2002) 1667–1696.
Bonani, F. and Gilli, M., A harmonic balance approach to bifurcation analysis of limit cycles. ISCAS'99. *Proc.(1999) IEEE Inter. Symp. Circuits Syst. VLSI (Cat. No.99CH36349). IEEE* Part **6** (1999a) 298–301.
Bonani, F. and Gilli, M., A harmonic-balance based method for computing Floquet's multipliers in Lur'e systems. *Proceedings of the 7th Inter. Specialist Workshop on Nonlinear Dynamics of Electronic Systems*, Tech. University (1999b) 13–16.
Bonasera, A., Bucolo, M., Fortuna, L., and Rizzo, A., The d_∞ parameter to characterise chaotic dynamics. *Proceedings of the IEEE-INNS-ENNS Inter. Joint Conference on Neural Networks. IJCNN (2000). Neural Computing – New Challenges and Perspectives for the New Millennium. IEEE Comput. Soc.* Part **5** (2000) 565–570.
Bonatti, C., A local mechanism for robust transitivity. Conference at IMPA, Seminar of Dynamical systems, August (1996).
Bonatti, C. and Diaz, L. J., Persistent nonhyperbolic transitive diffeomorphisms. Annals Math. **143** (1996) 357–396.
Bonatti, C. and Diaz, L. J., Connexions hétérocliniques et généricite d'une infinité de puits ou de sources. Annales Scientifiques de l'école Normal Suprieure de Paris **32**(4) (1999) 135–150.
Bonatti, C. and Viana, M., SRB measures for partially hyperbolic systems whose central direction is mostly contracting. Israel J. Math. **115** (2000) 157–193.
Bonatti, C., Diaz, L. J., and Pujals, E. R., A C^1-generic dichotomy for diffeomorphisms: Weak forms of hyperbolicity or infinitely many sinks or sources. Annals Math. **158** (2003) 355–418.
Bonatti, C., Boyle, M., and Downarowicz, T., The entropy theory of symbolic extensions. Invent. Math. **156** (2004) 119–161.
Bonatti, C., Diaz, L., and Viana, M., *Dynamics Beyond Uniform Hyperbolicity. A Global Geometric and Probabilistic Perspective.* Encyclopedia Mat. Sci. **102**, Mathematical Physics, III. Springer-Verlag, Berlin (2005).
Bonatti, C., Gan, S., and Wen, L., On the existence of non-trivial homoclinic classes. Ergod. Theor. Dyn. Syst. **27** (2007) 1473–1508.

Bonatti, C., Crovisier, S., and Wilkinson, A., Centralizers of C^1 generic diffeomorphisms. Preprint (2008a).
Bonatti, C., Diaz, L., and Fisher, T., Supergrowth of the number of periodic orbits for non-hyperbolic homoclinic classes. Preprint (2008b).
Bonetto, F., Falco, P., and Giuliani, A. Analyticity of the SRB measure of a lattice of coupled Anosov diffeomorphisms of the torus. J. Math. Phys. **45** (2004) 3282–3300.
Bonetto, F., Kupiainen, A., and Lebowitz, J., Absolute continuity of projected SRB measures of coupled Arnold cat map lattices. Ergod. Theor. Dyn. Syst. **25** (2005) 59–88.
Bornholdt, S. and Schuster, H. G., *Handbook of Graphs and Networks – From the Genome to the Internet*. Wiley-VCH, Weinheim (2003).
Bothe, H. G., Strange attractors with topologically simple basins. Topology Appl. **114**(1) (2001) 1–25.
Bourbaki, N., *Topologie Generale Actualités*. Sci. Ind. (2nd edn.) Hermann, Paris 1942–1947 nos. 916 1029; Russian translation of Topologie Generale (Gos. Izdat. Fiz. Mat. Lit., Moscow) Chaps. IV, VI, VII, (1959).
Bowen, R., Topological entropy and axiom A. Proceedings Sympos. Pure Math., Vol. **XIV**, Berkeley, Calif. (1968) 23–41 Amer. Math. Soc., Providence, RI (1970a).
Bowen, R., Markov partitions and minimal sets for axiom A diffeomorphisms. Amer. J. Math. **92** (1970b) 907–918.
Bowen, R., Periodic points and measures for axiom A diffeomorphisms. Trans. Amer. Math. Soc. **15**(4) (1971) 377–397.
Bowen, R., Symbolic dynamics for hyperbolic flows. Amer. J. Math. **95** (1973) 429–460.
Bowen. R., *Equilibrium States and the Ergodic Theory of Anosov Diffeomorphisms*. Lect. Notes Math. **470** Springer-Verlag, Berlin (1975a).
Bowen, R., ω-limit sets for axiom A diffeomorphisms. J. Differential Equations **18**(2) (1975b) 333–339.
Bowen, R. and Ruelle, D., The ergodic theory of axiom A flows. Invent. Math. **29** (1975) 181–202.
Box, G. E. P. and Jenkins, G., *Time Series Analysis: Forecasting and Control*. Holden-Day (1976).
Boyle, M. and Sullivan, M., Equivariant flow equivalence for shifts of finite type. Proceedings of the London Mathematical Society **91**(1) (2005) 184–214.
Boyle, M., Fiebig, D., and Fiebig, U., Residual entropy, conditional entropy and subshift covers. Forum Math. **14** (2002) 713–757.
Breiman, L., *Probability*. Original edition published by Addison-Wesley (1968) reprinted by Society for Industrial and Applied Mathematics (1992).
Brin, M., Bernoulli diffeomorphisms with nonzero exponents. Ergod. Theor. Dyn. Syst. **1** (1981) 1–7.
Brin, M. and Pesin, Y., Partially hyperbolic dynamical systems. Proceedings Sov. Acad. Sci. Ser. Math. (Izvestia) **38** (1974) 170–212.
Brouwer, L., Beweis des ebenen translationssatzes. Annals Math. **72** (1912) 37–54.

Brown, R., Generalizations of the Chua equations, Int. J. Bifurcation and Chaos **2**(4) (1992) 889–909.

Brown, R., From the Chua circuit to the generalized Chua map. J. Circuits Systems & Computers **3**(1) (1993a) 11–32.

Brown, R., Generalizations of the Chua equations. IEEE Trans. Circ. Syst. Fund. Theor. Appl. **40**(11) (1993b) 878–884.

Brown, R., Horseshoes in the measure-preserving Hénon map. Ergod. Theor. Dyn. Syst. **15** (1995) 1045–1059.

Brucks, K. and Bruin, H., *Topics from One-Dimensional Dynamics*, London Mathematical Society Student Texts, Cambridge University Press (2004).

Brucks, K. and Buczolich, Z., Trajectory of the turning point is dense for a co-σ-porouous set of tent maps. Fund. Math. **165** (2000) 95–123.

Brucks, K. M., Misiurewicz, M., and Tresser, C., Monotonicity properties of the family of trapezoidals maps. Comm. Math. Phys. **137** (1991a) 1–12.

Brucks. K. M., Diamond. B., Otero-Espinar. M. V., and Tresser. C., Dense orbits of critical points for the tent map. Contemp. Math. **117** (1991b) 57–61.

Brucks, K. and Misiurewicz. M., Trajectory of the turning point is dense for almost all tent maps. Ergod. Theor. Dyn. Syst. **16** (1996) 1173–1183.

Bruin, H., For almost every tent map, the turning point is typical. Fund. Math. **155** (1998) 215–235.

Bruin, H., Keller, G., Nowicki, T., and van Strien, S., Wild Cantor attractors exist. Annals Math. **143** (1996) 97–130.

Bunimovich, L. A., Statistical properties of Lorenz attractors. In *Nonlinear Dynamics and Turbulence* (G. I. Barenblatt, ed.) Pitman, Boston (1983) 71–92.

Bunimovich, L. A., *Dynamical Systems, Ergodic Theory and Applications.* Encyclopedia of Mathematical Sciences **100** Springer, New York (2000).

Bunimovich, L. A. and Sinai, Y., In *Nonlinear Waves*, edited by Gaponov-Grekhov, A. V. Nauka, Moscow (1980) 212 (in Russian).

Bunimovich, L. A., Sinai, Y., and Chernov, N. I., Markov partitions for two-dimensional hyperbolic billiards. Russ. Math. Surv. **45**(3) (1990) 105–152.

Bunimovich, L. A., Sinai, Y., and Chernov, N. I., Decay of correlations and the central limit theorem for two-dimensional billiards. Russ. Math. Surv. **46**(4) (1991) 47–106.

Burns K., Pugh, C., Shub, M., and Wilkinson, A., Recent results about stable ergodicity. In *Proceedings of Symposia in Pure Mathematics* Vol **69** "*Smooth Ergodic Theory and Its Applications*" (A. Katok, R. de la Llave, Y. Pesin, and H. Weiss, eds.), AMS, Providence, RI (2001) 327–366.

Burns, K., Dolgopyat, D., and Pesin, Y., Partial hyperbolicity, Lyapunov exponents and stable ergodicity. J. Stat. Phys. **108**(5–6) (2002) 927–942.

Butler, L. T. and Gelfreich, V., Positive-entropy geodesic flows on nilmanifolds. Nonlinearity **21**(7) (2008) 1423–1434.

Bykov, V. and Shilnikov, L. P., On the boundaries of the domain of existence of the Lorenz attractor. In *Methods of Qualitative Theory and Theory of Bifurcations.* Gorky State University, Gorky (1989) 151–159 (in Russian).

Caladrini, G., Berns, D., Paolini, E., and Moiola, J., On cyclic fold bifurcations in nonlinear systems. *2000 IEEE Inter. Symposium on Circuits and Systems – Emerging Technologies for the 21st Century. Proceedings (IEEE Cat No.00CH36353)*. Presses Polytech. University Romandes. Part **2** (2000) 485–488.

Campbell, D. K., Galeeva, R., Tresser, C., and Uherka, D. J., Piecewise linear models for the quasiperiodic transition to chaos. Chaos **6**(2) (1996) 121–154.

Cao, Y., A note about Milnor attractor and riddled basin. Chaos, Solitons & Fractals **19** (2004) 759–764.

Cao, Y. and Kiriki, S., The basin of the strange attractors of some Hénon maps. Chaos, Solitons & Fractals **11**(5) (2000a) 729–734.

Cao, Y. and Liu, Z., Strange attractors in the orientation-preserving Lozi map. Chaos, Solitons & Fractals **9** (1998) 1857–1863.

Cao, Y. and Mao, M. J., The non-wandering set of some Hénon maps. Chaos, Solitons & Fractals **11**(13) (2000b) 2045–2053.

Cao, Y., Luzzatto, S., and Rios, I., The boundary of hyperbolicity for Hénon-like families. Ergod. Theor. Dyn. Syst. **28** (2008) 1049–1080.

Carballo, C. M., Morales, C., and Pacifico, M. J., Maximal transitive sets with singularities for generic C^1 vector fields. Bol. Soc. Brasil. Mat. (N.S.) **31**(3) (2000) 287–303.

Carballo, C. M., Morales, C. A., and Pacifico, M. J., Homoclinic classes for generic C^1 vector fields. Ergod. Theor. Dyn. Syst. Preprint (2007).

Casselman, B., Picturing the horseshoe map. Notices Amer. Math. Soc. **52**(5) (2005) 518–519.

Cassels, J. W. S., *Introduction to the Theory of Diophantine Approximation*. Cambridge University Press (1957), Russian translation: Inostr. Lit., Moscow (1961).

Cessac, B., Does the complex susceptibility of the Hénon map have a pole in the upper-half plane? A numerical investigation. Nonlinearity **20**(12) (2007) 2883–2895.

Cessac, B., Doyon, B., Quoy, M., and Samuelides, M., Mean-field equations, bifurcation map and route to chaos in discrete time neural networks. Physica D **74** (1994) 24–44.

Changming, D., The omega limit sets of subsets in a metric space. Czechoslovak Mathematical Journal **55**(1) (2005) 87–96.

Charles, M., Gray, P. K., Andreas, K. E., and Wolf, S., Oscillatory responses in cat visual cortex exhibit inter-columnar synchronization which reflects global stimulus properties. Nature **338** (1989) 334–337.

Chen, G., *Controlling Chaos and Bifurcations in Engineering Systems*. CRC Press, Boca Raton, FL (1999).

Chen, G. and Dong, X., *From Chaos to Order – Methodologies, Perspectives and Applications*. World Scientific, Singapore (1998).

Chen, Y., Liu, Y., Up-embeddability of a graph by order and girth. Graphs and Combinatorics **23**(5) (2007) 521–527.

Chernov, N. I., Ergodic and statistical properties of piecewise linear hyperbolic automorphisms of the 2-torus J. Stat. Phys. **69**(1–2) (1992) 111–134.

Chernov, N. I., Statistical properties of the periodic Lorentz gas. Multidimensional case. J. Stat. Phys. **74**(1-2) (1994) 11–53.

Chernov, N. I., Decay of correlations and dispersing billiards. J. Stat. Phys. **94** (1999) 513–556.

Chernov, N. I., Eyink, G. L., Lebowitz, J. L., and Sinai, Y., Steady-state electrical conduction in the periodic Lorentz gas. Comm. Math. Phys. **154**(3) (1993) 569–601.

Chittaro, F., An Estimate for the entropy of Hamiltonian flows. J. Dynamical and Control Systems **13**(1) (2007) 55–67.

Chow, S. N. and Palmer, K. J., On the numerical computation of orbits of dynamical systems – The higher dimensional case. J. Complexity **8** (1992) 398–423.

Christiansen, F. and Rugh, H. H., Computing Lyapunov spectra with continuous Gram–Schmidt orthonormalization. Nonlinearity **10**(5) (1997) 1063–1073.

Christy, J., Branched surfaces and attractors I – Dynamic branched surfaces. Trans. Amer. Math. Soc. **336** (1993) 759–784.

Chua, L. O., Global unfolding of Chua's circuit. IEICE Trans. Fund. Electronics Comm. Comput. Sci. **E76-A**(5) (1993) 704–734.

Chua, L. O. and Lin, G., Canonical realization of Chua's circuit family. IEEE Trans. Circ. Syst. **37**(7) (1990) 885–902.

Chua, L.O. and Tichonicky, I., 1-D map for the double scroll. IEEE Trans. Circ. Syst. Fund. Theor. Appl. **38**(3) (1991) 233–243.

Chua, L. O., Komuro, M., and Matsumoto, T., The double scroll family. IEEE Trans. Circ. Syst. **CAS-33**(11) (1986) 1073–1118.

Chunyan, Z. and Wang, X., Attractors and quasi-attractors of a flow. J. Applied Mathematics and Computing **23**(1-2) (2007) 411–417.

Cohen, S. D. and Hindmarsh, A. C., CVODE, A stiff/nonstiff ODE solver in C. Comput. Phys. **10** (1996) 138–141.

Collet, P. and Eckmann, J. P., On the abundance of aperiodic behaviour for maps of the interval. Comm. Math. Phys. **73** (1980a) 115–160.

Collet, P. and Eckmann, J. P., *Iterated Maps on the Interval as Dynamical Systems*. Birkhauser, Boston (1980b).

Collet, P. and Levy, Y., Ergodic properties of the Lozi mappings. Comm. Math. Phys. **93** (1984) 461–482.

Colli, E., Infinitely many coexisting strange attractors. Ann. de l'Inst. H. Poincaré. Anal. Non Linéaire **15**(5) (1998) 539–579.

Colmenarez, W., *Algumas Propriedades de Atratores para Fluxos em Dimensao Trés*. Thesis (2002) Universidade Federal do Rio de Janeiro (in Portuguese).

Cook, H., Ingram, W. T., Kuperberg, K., and Lelek, A., *Continua*. Lecture Notes in Pure and Applied Mathematics (1995).

Coomes, B. A., Kocak, H., and Palmer, K. J., Computation of long periodic orbits in chaotic dynamical systems. The Australian Math. Soc. Gazette **24** (1997) 183–190.

Coomes, B. A., Koçak, H., and Palmer, K. J., Homoclinic shadowing. J. Dynamics and Differential Equations **17**(1) (2005) 175–215.

Cooper, R., Winter, A. L., Crow, H. J., and Grey, W., Electroencephalogr. Clin. Neurophysiol. **18** (1965) 217.

Cornfeld, I., Fomin, S. V., and Sinai, Y., *Ergodic Theory*. Springer-Verlag, Berlin (1982).

Cosenza, M. G. and Gonzalez, J., Synchronization and collective behavior in globally coupled logarithmic maps. Prog. Theor. Phys. **100** (1998) 21–38.

Crovisier, S., Nombre de rotation et dynamique faiblement hyperbolique. Thèse de doctorat, Université Paris-Sud (Décembre 2001).

Crovisier, S., Birth of homoclinic intersections – A model for the central dynamics of partially hyperbolic systems. Preprint arXiv:math.DS/0605387 (2006).

Crovisier, S., Partial hyperbolicity far from homoclinic bifurcations. Preprint (2008).

Cvitanović, P., Gunaratne, G., and Procaccia. I., Topological and metric properties of Hénon-type strange attractors. Phys. Rev. A **38** (1988) 1503–1520.

Dafilis, M. P., Liley, D. T. J., and Cadusch, P. J., Robust chaos in model of the electroencephalogram – Implications for brain dynamics. Chaos **11**(3) (2001) 474–478.

Dan, T., Turing instability leads oscillatory systems to spatiotemporal chaos. Prog. Theor. Phys. Suppl. **61** (2006) 119–126.

Danca, M-F. and Codreanu, S., On a possible approximation of discontinuous dynamical systems. Chaos Solitons & Fractals **13**(4) (2002) 681–691.

Davidchack, R., Lai, Y. C., Klebanoff, A., and Bollt, E., Towards complete detection of unstable periodic orbits in chaotic systems. Phys. Lett. A **287** (2001) 99–104.

Davies, H. G. and Rangavajhula, K., Noisy parametric sweep through a period-doubling bifurcation of the Henon map. Chaos, Solitons & Fractals **14**(2) (2002) 293–299.

Davis, M. J., MacKay, R. S., and Sannami, A., Markov shifts in the Hénon family. Physica D **52**(2–3) (1991) 171–178.

Dawson, S., Grebogi, C., Sauer, T., and Yorke, J. A., Obstructions to shadowing when a Lyapunov exponent fluctuates about zero. Phys. Rev. Lett. **73**(14) (1994) 1927–1930.

Deane, J. H. B. and Hamill, D. C., Improvement of power supply EMC by chaos. Electron. Lett. **32**(12) (1996) 1045–1049.

Dedieu, H. and Ogorzalek, M. J., Identifiability and identification of chaotic systems based on adaptive synchronization. IEEE Trans. Circ. Syst. Fund. Theor. Appl. **44**(10) (1997) 948–962.

Dedieu, J. P. and Shub, M., On random and mean exponents for unitarily invariant probability measures on $GL(n, \mathbb{C})$. In *Geometric Methods in Dynamical Systems (II) – Volume in Honor of Jacob Palis*. Asterisque Soc. Math. de France **287** (2003) 1–18.

de Faria, E. and de Melo, W., Rigidity of critical circle mappings II. JAMS **13** (2000) 343–370.

Dellnitz, M. and Junge, O., Set oriented numerical methods for dynamical systems. In *Handbook of Dynamical Systems* **2** (2002) 221–264 North-Holland, Amsterdam.

de Melo, W., Structural stability of diffeomorphisms on two-manifolds. Invent. Math. **21** (1973) 233–246.
de Melo, W. and Palis, J., *Geometric Theory of Dynamical Systems – An Introduction.* Springer-Verlag, New York (1982).
de Melo, W. and van Strien, S., One-dimensional dynamics: The Schwarzian derivative and beyond. Bull. Am. Math. Soc. **18** (1988) 159–162.
de Melo, W. and van Strien, S., *One-Dimensional Dynamics.* Springer-Verlag, Berlin (1993).
Denker, M., The central limit theorem for dynamical systems. Dyn. Syst. Ergod. Th. Banach. Center Publ. **23** (1989) Warsaw PWN Polish Sci. Publ.
de Oliveira, K. A., Vannucci, A., and da Silva, E. C., Using artificial neural networks to forecast chaotic time series. Physica A **284**(1) (2000) 393–404.
Derbyshire, J., *Prime Obsession – Bernhard Riemann and the Greatest Unsolved Problem in Mathematics.* Penguin, New York (2004).
Devaney, R. L., Reversibility, homoclinic points, and the Hénon Map. In *Dynamical Systems, Approaches to Nonlinear Problems in Systems and Circuits.* SIAM, Philadelphia (1988) 3–14.
Devaney, R. L., *An Introduction to Chaotic Dynamical Systems.* Addison-Wesley, New York (1989).
Devaney, R. L. and Nitecki, Z., Shift automorphism in the Hénon mapping. Comm. Math. Phys. **67** (1979) 137–146.
Diakonos, F. K. and Schmelcher, P., On the construction of one-dimensional iterative maps from the invariant density – The dynamical route to the beta function. Phys. Lett. A **211** (1996) 199–203.
Diamond, P., Kloeden, P. E., Kozyakin, V. S., and Pokrovskii, A. V., Semi-hyperbolic mappings. J. Nonlinear Sci. **5** (1995) 419–431.
Diamond, P., Kloeden, P. E., Kozyakin, V. S., and Pokrovskii, A. V., Semi-hyperbolic mappings. Manuscript (2008).
Diaz, L. J., Pujals, E., and Ures, R., Partial hyperbolicity and robust transitivity. Acta Math. **183** (1999) 1–43.
di Bernardo, M., Feigin, M. I., Hogan, S. J., and Homer, M. E., Local analysis of C-bifurcations in n-dimensional piecewise smooth dynamical systems. Chaos, Solitons & Fractals **10**(11) (1999) 1881–1908.
di Bernardo, M., Kowalczyk, P., and Nordmark, A. B., Bifurcations of dynamical systems with sliding: Derivation of normal form mappings. Physica D **170** (2002) 175–205.
di Bernardo, M., Kowalczyk, P., and Nordmark, A. B., Sliding bifurcations: A novel mechanism for the sudden onset of chaos in dry-friction oscillators. Int. J. Bifurcation and Chaos **13** (2003) 2935–2948.
Dieudonné, J., *Foundations of Modern Analysis.* Academic Press (1960).
Ding, C., On the intertwined basins of attraction for planar flows. Applied Mathematics and Computation **148**(3) (2004) 801–805.
Ding, C., The omega limit sets of subsets in a metric space. Czechoslovak Mathematical Journal **55**(1) (2005) 87–96.
Ding, M. and Yang, W., Observation of intermingled basins in coupled oscillators exhibiting synchronized chaos. Phy. Rev. E **54**(3) (1996) 2489–2494.

Djellit, I. and Boukemara, I., Dynamics of a three parameters family of piecewise maps. Facta Universitatis. Nis, Ser. Elec. Energ. **20**(1) (2007) 85–92.

Doering, C. I., Persistently transitive vector fields on three-dimensional manifolds. In *Dynamical Systems and Bifurcation Theory*, Pitman Research Notes in Mathematics Series **160** (1987) 59–89.

Doerner, R., Hubinger, B., and Martienssen, W., Advanced chaos forecasting. Phys. Rev. E **50**(1) (1994) R12–15.

Dogaru, R., Murgan, A. T., Ortmann, S., and Glesner, M., Searching for robust chaos in discrete time neural networks using weight space exploration. 1996 Proceedings **ICNN'96** Washington D.C. 2–6 June (1996).

Dold, A., *Lectures on Algebraic Topology*. Springer-Verlag, Berlin (1980).

Dolgopyat, D., On dynamics of mostly contracting diffeomorphisms. Comm. Math. Phys. **213** (2000) 181–201.

Dolgopyat, D., On mixing properties of compact group extensions of hyperbolic systems. Israel J. Math. **130** (2002) 157–205.

Dolgopyat, D., On differentiability of SRB states for partially hyperbolic systems. Invent. Math. **155** (2004) 389–449.

Dolgopyat, D. and Pesin, Y., On the existence of Bernoulli diffeomorphisms with nonzero Lyapunov exponents on compact smooth manifolds. Preprint (2002).

Dolgopyat, D., Hu, H., and Pesin, Y., An example of a smooth hyperbolic measure with countably many ergodic components. In *Smooth Ergodic Theory and its Applications* by Katok, A. and de la Llave, R., Proceedings of Symposia in Pure Mathematics **69** (2001) 95–106.

Dolgopyat, D. and Pesin, Y., Every compact manifold carries a completely hyperbolic diffeomorphism. Ergod. Theor. Dyn. Syst. **22**(2) (2002) 409–435.

Douady, A. and Hubbard, J. H., On the dynamics of polynomial-like mappings. Ann. Sci. Ecole Norm. Sup. **18** (1985) 287–343.

Driebe, D. J., *Fully Chaotic Maps and Broken Time Symmetry*. Kluwer Academic Publishers (1999).

Drutarovsky, M. and Galajda, P. A., Robust chaos-based true random number generator embedded in reconfigurable switched capacitor hardware. Radioengineering **16**(3) (2007) 120–127.

Duarte, P., Plenty of elliptic islands for the standard family of area preserving maps. Ann. Inst. H. Poincaré Anal. Nonlinéaire **11** (1994) 359–409.

Duchesne, L., Using characteristic multiplier loci to predict bifurcation phenomena and chaos – A tutorial. IEEE Trans. Circ. Syst. Fund. Theor. Appl. **40**(10) (1993) 683–688.

Dullin, H. R., Sterling, D., and Meiss, J. D., Self-rotation number using the turning angle. Physica D **145** (2000) 25–46.

Easton, R. W., *Geometric Methods for Discrete Dynamical Systems*. Oxford University Press (1998).

Eckhorn, R., Bauer, R., Jordan, W., Brosch, M., Kruse, W., Munk, M., and Reitboeck, H. J., Coherent oscillations. Biol. Cybern. **60** (1988) 121–130.

Eckmann, J. P. and Ruelle, D., Ergodic theory of chaos and strange attractors. Rev. Mod. Phys. **57** (1985) 617–656.

El Hamouly, H. and Mira, C., Lien entre les propriétés d'un endomorphisme de dimension unet celles d'un difféomorphisme de dimension deux. C. R. Acad. Sci. Paris Sér. I Math. **293** (1981) 525–528.

Endler, A. and Gallas, J. A. C., Period four stability and multistability domains for the Hénon map. Physica A **295**(1) (2001) 285–290.

Farmer, J. D., Ott, E., and Yorke, J. A., The dimension of chaotic attractors. Physica D **7** (1983) 153–180.

Farrell, F. T. and Jones, L. E., Anosov diffeomorphisms constructed from $\pi_1(Diff(S_n))$. Topology **17**(3) (1978) 273–282.

Feigenbaum, M. J., Universal behavior in nonlinear systems. Los Alamos Sci. **1** (1980) 4-27 (reprinted in *Universality in Chaos* (P. Cvitanovid, ed.) Adam Hilger (1984).

Feigin, M. I., Doubling of the oscillation period with C-bifurcations in piecewise continuous systems. Prikladnaya Matematika i Mechanika **34** (1970) 861–869.

Feigin, M. I., *Forced Oscillations in Systems with Discontinuous Nonlinearities.* Nauka, Moscow (1994) (in Russian).

Feit, S., Characteristic exponents and strange attractors. Comm. Math. Phys. **61** (1978) 249–260.

Feldman, J. and Katok, A., Bernoulli diffeomorphisms and group extensions of dynamical systems with nonzero characteristic exponents. Annals Math. **113** (1981) 159–179.

Feng, B. Y., The heteroclinic cycle in the model of competition between n-species and its stability. Acta Math. Appl. Sinica **14** (1998) 404–413.

Fisher, T., Hyperbolic sets that are not locally maximal. Ergod. Theor. Dyn. Syst. **26**(05) (2006a) 1491–1509.

Fisher, T., Hyperbolic sets with nonempty interior. Discrete and Contin. Dynam. Systems **15**(2) (2006b) 433–446.

Fisher, T., The topology of hyperbolic attractors on compact surfaces. Ergod. Theor. Dyn. Syst. **26**(5) (2006c) 1511–1520.

Fisher, T., Hyperbolic chain recurrent classes for commuting diffeomorphisms. Preprint (2009).

Fisher, T. and Rodriguez-Hertz, J., Quasi-Anosov diffeomorphisms of 3-manifolds. Preprint (2008).

Flaminio, L. and Katok, A., Rigidity of symplectic Anosov diffeomorphisms on low-dimensional tori. Ergod. Theor. Dyn. Syst. **1**(3) (1991) 427–441.

Fokkink, R. J., *The Structure of Trajectories.* Ph.D. Thesis, Technical University of Delft (1991).

Fomin, S. V. and Gelfand, I. M., Geodesic flows on manifolds of constant negative curvature. Uspehi. Mat. Nauk. **7**(1) (1952) 118–137.

Fomin, S. V., Kornfeld, I. P., Sinaj, J. G., *Ergodic Theory.* Nauka, Moscow (1980) (in Russian).

Fontich, E., Transversal homoclinic points of a class of conservative diffeomorphisms. J. Differ. Equations **87** (1990) 1–27.

Formanek, E., Observations about the Jacobian conjecture. Houston J. Math. **20** (1994) 369–380.

Fornæss, J. E. and Gavosto, E. A., Existence of generic homoclinic tangencies for Hénon mappings. J. Geom. Anal. **2** (1992) 429–444.

Fornæss, J. E. and Gavosto, E. A., Tangencies for real and complex Hénon maps: Analytic method. Experiment. Math. **8** (1999) 253–260.

Forni, G., Lyubich, M., Pugh, C., and Shub, M., Partially hyperbolicsynamics, laminations, and Teichmuller flow. Fields Institute Communications (2007).

Franceschini, V., Giberti, C., and Zheng, Z., Characterization of the Lorenz attractor by unstable periodic orbits. Nonlinearity **6** (1993) 251–258.

Franks, J., Anosov diffeomorphisms. Proceedings Sympos. Pure Math. **14** (1970) 61–93.

Franks, J., Necessary conditions for the stability of diffeomorphisms. Trans. AMS **158** (1971) 301–308.

Franks, J. and Williams, R., Anomalous Anosov flows. Global theory of dynamical systems In *Proceedings Internat. Conf., Northwestern University, Evanston, IL, 1979.* Lecture Notes in Math. **819** Springer, Berlin (1980) 158–174.

Freeman, W. J., *Societies of Brains.* Lawrence Erlbaum Associates, Mahwah (1995).

Freire, E., Rodriguez-Luis, A. J., Gamero. E., and Ponce, E., A case study for homoclinic chaos in an autonomous electronic circuit – A trip from Takens–Bogdanov to Hopf–Shilnikov. Physica D **62** (1993) 230–253.

Friedland, S. and Milnor, J., Dynamical properties of plane polynomial automorphisms. Erg. Th. and Dyn. Syst. **9** (1989) 67–99.

Fujii, H., Aihara, K., and Tsuda, I., Corticopetal acetylcholine: A role in attentional state transitions and the genesis of quasi-attractors during perception. Advances in Cognitive Neurodynamics ICCN (2007a).

Fujii, H., Aihara, K., and Tsuda, I., Corticopetal acetylcholine: Possible scenarios on the role for dynamic organization of quasi-attractors. Lecture Notes in Computer Science, Neural Information Processing, 14th International Conference **ICONIP** (2007) Kitakyushu, Japan, November 13–16 (2007) Revised Selected Papers, Part I (2007b) 170–178.

Fujisaka, H. and Sato, C., Computing the number, location and stability of fixed points of Poincaré maps. IEEE Trans. Circ. Syst. I **44**(4) (1997) 303–311.

Funahashi, K. and Nakamura, Y., Approximation of dynamical systems by continuous time recurrent neural networks. Neural Networks **6** (1993) 801–806.

Fu-Yan, S., Shu-Tang, L., and Zong-Wang, L., Image encryption using high-dimension chaotic system. Chinese Physics **16**(12) (2007) 3616–3623.

Gabriel, P., Robust chaos in polynomial unimodal maps. Int. J. Bifurcation and Chaos **14**(7) (2004) 2431–2437.

Galeeva, R., Martens, M., and Tresser, C., Inducing, slopes and conjugacy classes. Israel J. Math. **99** (1997) 123–147.

Galias, Z., Numerical studies of the Hénon map. In Proceedings *Int. Symposium on Scientific Computing, Computed Arithmetic and Validated Numerics, SCAN'97* **XIV5-6** Lyon (1997a).

Galias, Z., Positive topological entropy of Chua's circuit: A computer assisted proof. Int. J. Bifurcation and Chaos **7**(2) (1997b) 331–349.

Galias, Z., Rigorous numerical studies of the existence of periodic orbits for the Hénon Map. J. Universal Computer Science **4**(2) (1998a) 114–125.

Galias, Z., Existence and uniqueness of low-period cycles and estimation of topological entropy for the Hénon map. In *Proceedings Int. Symposium on Nonlinear Theory and its Applications, NOLTA'98* **1** (1998b) 187–190, Crans-Montana.

Galias, Z., All periodic orbits with period $n \leq 26$ for the Hénon map. In *Proceedings European Conference on Circuit Theory and Design, ECCTD'99* **1** (1999) 361–364, Stresa.

Galias, Z., Interval methods for rigorous investigation of periodic orbits. Int. J. Bifurcation and Chaos **11**(9) (2001) 2427–2450.

Galias, Z., Rigorous investigations of piecewise linear circuits. In Proceedings *Int. Conference on Signals and Electronic Systems* **ICSES'02** (2002a) 141–146, Swieradów Zdrój.

Galias, Z., Study of Poincaré map associated with the Chua'a circuit using interval arithmetic. In Proceedings *Int. Symposium on Nonlinear Theory and its Applications* **NOLTA'02 Xi'an, PRC** (2002b) 779–782.

Galias, Z., Mean value form for evaluation of Poincaré map in piecewise linear systems. In *Proceedings European Conference on Circuit Theory and Design, ECCTD'03* **I** (2003) 283–286, Kraków.

Galias, Z., Towards full characterization of continuous systems in terms of periodic orbits. In Proceedings *IEEE Int. Symposium on Circuits and Systems* **ISCAS'04 IV** (2004) 716-719, Vancouver.

Galias, Z., Short periodic orbits and topological entropy for the Chua's circuit. In Proceedings *IEEE Int. Symposium on Circuits and Systems* **ISCAS'06** (2006a) 1667–1670, Kos, Greece.

Galias, Z., Counting low-period cycles for flows. Int. J. Bifurcation and Chaos **16**(10) (2006b) 2873–2886.

Gallas, A. and Jason, C., Structure of the parameter space of the Hénon map. Phys. Rev. Lett. **70**(18) (1993) 2714–2717.

Galias, Z. and Tucker, W., Rigorous study of short periodic orbits for the Lorenz system. IEEE International Symposium on Circuits and Systems **18-21** (2008) 764–767.

Galias, Z. and Zgliczynski, P., Computer assisted proof of chaos in the Lorenz equations. Physica D **115** (1998) 165–188.

Galias, Z. and Zgliczynski, P., Abundance of homoclinic and heteroclinic orbits and rigorous bounds for the topological entropy for the Hénon map. Nonlinearity **14**(5) (2001) 909–932.

Gallavotti, G., Bonetto, F., and Gentile, G., *Aspects of Ergodic, Qualitative and Statistical Theory of Motion*. Springer-Verlag, Berlin (2004).

Gambaudo, J. M. and Tresser, C., Dynamique régulière ou chaotique – Applications du cercle ou l'intervalle ayant une discontinuité. C. R. Acad. Soc. Paris, Sér. **I 300**(10) (1985) 311–313.

Gan, S., Yang, D., and Wen, L., Minimal non-hyperbolicity and index completeness. Preprint Beijing University (2007).

Gardini, L., Abraham, R., Record, R. J., and Fournier-Prunaret, D., A double logistic map. Int. J. Bifurcation and Chaos **4**(1) (1994) 145–176.

Gaspard, P., r-adic one-dimensional maps and the Euler summation formula. J. Physics A **25** (1992) L483–L485.

Gaspard, P., Maps. In *Encyclopedia of Nonlinear Science*. Routledge, New York (2005) 548–553.

Gavrilov, N. K. and Shilnikov, L. P., On three dimensional dynamical system close to systems with a structurally stable homoclinic curve. Math. USSR Sbornik **19** (1973) 139–156.

Gelfreich, V. G., Separatrices splitting for polynomial area-preserving maps. In *Topics in Math. Phys.* (M. Sh. Birman, ed.) **13** (1991) 108–116, Leningrad State University.

Gelfreich, V. G. and Sauzin, D., Borel summation and splitting of separatrices for the Hénon map. Ann. Inst. Fourier **51** (2001) 513–567.

Ghrist, R. W., Holmes, P. J., and Sullivan, M. C., *Knots and Links in Three-Dimensional Flows*. Lecture Notes in Mathematics **1654**, Springer-Verlag, Berlin (1997).

Ghrist, R. W. and Zhirov, A. Yu., Combinatorics of one-dimensional hyperbolic attractors of diffeomorphisms of surfaces (in Russian, Russian summary) Trudy Matematicheskogo Instituta Imeni V. A. Steklova. Rossi kaya Akademiya Nauk. **244** Din. Sist. i Smezhnye Vopr. Geom. (2004) 143–215.

Ghys, E., Codimension one Anosov flows and suspensions. Dynamical systems, Val-paraiso (1986) 59–72 Lecture Notes in Math. **1331** Springer, Berlin (1988).

Ghys, E., Holomorphic Anosov systems. Invent. Math. **119** (1995) 585–614.

Girgensohn, R., Constructing singular functions via Farey fractions. J. Math. Anal. Appl. **203** (1996) 127–141.

Glendinning, P., *Stability, Instability and Chaos*. Cambridge Texts in Applied Mathematics (1994).

Glendinning, P., Hyperbolicity of the invariant set for the logistic map with $\mu > 4$. MIMS EPrint(2006) 99 (2001).

Glendinning, P. and Sparrow, C., Local and global behavior near homoclinic orbits. J. Stat. Phys. **35** (1984) 645–697.

Glendinning, P. and Sparrow, C., Prime and renormalizable kneading invariants and the dynamics of expanding Lorenz maps. Physica D **62** (1993) 22–50.

Golub, G., VanLoan, C. F., *Matrix Computations*. Johns Hopkins University Press, Baltimore (1989).

Gomez, A. and Meiss, J. D., Reversible polynomial automorphisms of the plane – The involutory case. Phys. Lett. A **312**(1) (2003) 49–58.

Gomez, A. and Meiss, J. D., Reversors and symmetries for polynomial automorphisms of the plane. Nonlinearity **17**(3) (2004) 975–1000.

Gomez, G. and Simó, C., Homoclinic and heteroclinic points in the Hénon map. Lect. Notes in Physics **179** (1983) 245–247.

Gomes, J. B. and Ruggiero, R. O. Uniqueness of central foliations of geodesic flows for compact surfaces without conjugate points. Nonlinearity **20**(2) (2007) 497–515.

Goncalves, J. M., Megretski, A., and Dahleh, M. A., Global analysis of piecewise linear systems using impact maps and surface Lyapunov functions. IEEE. Trans. Automatic Control **48**(12) (2003) 2089–2106.

Gonchenko, S. V., Shilnikov, L. P., Turaev, D. V., On models with a structurally unstable homoclinic Poincaré curve. Sov. Math. Dokl. **44**(2) (1992) 422–426.

Gonchenko, S. V., Shilnikov, L. P., and Turaev, D. V., On models with nonrough Poincaré homoclinic curves. Physica D **62** (1993) 1–14.

Gonchenko, S. V., Turaev, D. V., and Shilnikov, L. P., Dynamical phenomena in multi-dimensional systems with a structurally unstable homoclinic Poincaré curve. Russian Acad. Sci. Dokl. Math. **47**(3) (1993) 410–415.

Gonchenko, S. V., Ovsyannikov, I. I., and Simo, C., Three dimensional Hénon-like maps and wild Lorenz-like attractors. Int. J. Bifurcation and Chaos **15** (2005a) 3493–3508.

Gonchenko, S. V., Turaev, D. V., and Shilnikov, L. P., On dynamic properties of diffeomorphisms with homoclinic tangency. J. Math. Sci. **126**(4) (2005b) 1317–1343.

Goodman, S., *Dehn Surgery on Anosov Flows, Geometric Dynamics* (Rio de Janeiro (1981) 300–307 Lecture Notes in Mathematics **1007**, Springer, Berlin (1983).

Gorodnik, A., Open problems in dynamics and related fields. J. Modern Dynamics **1**(1) (2007) 1–35.

Gorodetski, A. S. and YIlyashenko, U. S., Minimal and strange attractors. Int. J. Bifurcation and Chaos **6** (1996) 1177–1183.

Gourmelon, N., Generation of homoclinic tangencies by C^1-perturbations. Preprint Université de Bourgogne (2007).

Graczyk, J. and Swiatek, G., Generic hyperbolicity in the logistic family. Annals Math. **146** (1997) 1–52.

Graczyk, J. and Swiatek, G. S., Survey: Smooth unimodal maps in the (1990)s, Ergod. Theor. Dyn. Syst. **19** (1999) 263–287.

Graczyk, J., Sands, D., and Swiatek, G., La dérivée Schwarzienne en dynamique unimodale. C. R. Acad. Sci. Paris **332** (2001) 329–332.

Graczyk, J., Sands, D. and Swiatek, G., Metric attractors for smooth unimodal maps. Annals Math. **159** (2004) 725–740.

Graczyk, J. Sands, D., and Swiatek, G., Decay of geometry for unimodal maps: negative Schwarzian case. Annals Math. **161**(2) (2005) 613–677.

Grassberger, P. and Procaccia, I., Measuring the strangeness of strange attractors. Physica D **9** (1983) 189–208.

Grassberger, P., Kantz, H., and Moenig, U., On the symbolic dynamics of the Hénon map. J. Phys. A Math. Gen. **22** (1989) 5217–5230.

Grassi, G. and Mascolo, S., A system theory approach for designing crytosystems based on hyperchaos. IEEE Trans. Circ. Syst. Fund. Theor. Appl. **46**(9) (1999) 1135–1138.

Greblicki, W., Nonparametric identification of Wiener systems by orthogonal series. IEEE Trans. on Automatic Control **39** (1994) 2077–2086.

Grebogi, C., Ott, E., and Yorke, J. A., Crises, sudden changes in chaotic attractors, and transient chaos. Physica D **70** (1983) 191–200.

Grebogi, C., Ott, E., and Yorke, J. A., Basin boundary metamorphoses: Changes in sccessible boundary orbits. Physica D **24** (1987a) 243.

Grebogi, C., Ott, E., and Yorke, J. A., Critical exponent of chaotic transients in nonlinear dynamical systems. Phys. Rev. Lett. **57**(11) (1987b) 1284–1287.

Grebogi, C., Ott, E., Romeiras, F., and Yorke, J. A., Critical exponents for crisis-induced intermittency. Phy. Rev. A **36** (1987c) 5365–5380.

Grebogi, C., Ott, E., and Yorke, J. A., Unstable periodic orbits and the dimension of multifractal chaotic attractors. Phys. Rev. A **37** (1988a) 1711–1724.

Grebogi, C., Hammel, S., and Yorke, J. A., Numerical orbits of chaotic processes represent true orbits. Bull. Am. Math. Soc. **19** (1988b) 465–469.

Grebogi, C., Hammel, S. M., Yorke, J. A and Sauer, T., Shadowing of physical trajectories in chaotic dynamics – Containment and refinement. Phys. Rev. Lett. **65** (1990) 1527–1530.

Gribov, A. F. and Krishchenko, A. P., Analytical conditions for the existence of a homoclinic loop in Chua circuits. Comp. Math. Modeling **13**(1) (2002) 75–80.

Grossman, S. and Thomae, S., Invariant distributions and stationary correlation functions of one-dimensional discrete processes. Z. Naturforsch **32** (1977) 1353–1363.

Guckenheimer, J., A strange strange attractor. In *The Hopf Bifurcation and its Spplications* (J. E. Marsden and M. McCracken, eds.) Applied Mathematical Series **19** Springer, Berlin (1976) 368–381.

Guckenheimer, J., On bifurcation of maps of the interval. Inv. Math. **39**(2) (1977) 165–178.

Guckenheimer, J., Sensitive dependence to initial conditions for one dimensional maps. Comm. Math. Phys. **70** (1979) 133–160.

Guckenheimer, J. and Holmes, P., *Nonlinear Oscillations, Dynamical Systems, and Bifurcations of Vector Fields*. Springer-Verlag, New York (1983).

Guckenheimer, J. and Williams, R. F., Structural stability of Lorenz attractors. Publ. Math. IHES **50** (1979) 307–320.

Gumowski, I. and Mira, C., *Dynamique Chaotique*. Toulouse Cepadues Editions (1980).

Hahn, W., *Stability of Motion*. Springer-Verlag, Berlin (1967).

Hammel, S. M., A noise reduction method for chaotic systems. Phys. Lett. A **148** (1990) 421–428.

Hammel, S. M., Yorke, J. A., and Grebogi, C., Do numerical orbits of chaotic dynamical processes represent true orbits? Complexity **3** (1987) 136–145.

Han, P., Perturbed basins of attraction. Mathematische Annalen **337**(1) (2007) 1–13.

Hansen, K. T. and Cvitanovic, P., Bifurcation structures in maps of Hénon type. Nonlinearity **11**(5) (1998) 1233–1261.

Hardy, G. H., *Ramanujan: Twelve Lectures on Subjects Suggested by His Life and Work* (3rd edn.) New York, Chelsea (1999).

Hartley, T. T., The Duffing double scroll. *Proc. Amer. Control Conference* **1** (1989) 419–423.

Hartley, T. T. and Mossayebi, F., Control of Chua's system. J. Circuits, Systems and Computers **3**(1) (1993) 173–194.

Hassard, B., Zhang, J., Hastings, S. P., and Troy, W. C., A computer proof that the Lorenz equations have chaotic solutions. Appl. Math. Lett. **7**(1) (1994) 79–83.

Hasselblatt, B., *Hyperbolic Dynamical Systems.* Handbook of Dynamical Systems **1A**, Elsevier, Amsterdam (2002) 239–319.

Hasselblatt, B., *Dynamics, Ergodic Theory and Geometry.* Mathematical Sciences Research Institute Publications (2007).

Hasselblatt, B. and Pesin,Y., *Partially Hyperbolic Dynamical Systems.* Handbook of Dynamical Systems **1B**, Elsevier, Amsterdam (2005) 1–55.

Hassouneh, M. A. and Abed, E. H., Feedback control of border collision bifurcations in piecewise smooth systems. ISR Technical Report, TR 26 (2002).

Hassouneh, M. A. and Abed, E. H., and Banerjee, S., Feedback control of border collision bifurcations in two-dimensional discrete-time systems. ISR Technical Research Report (2002).

Hasting, S. P. and Troy, W. C., A Shooting Approach to the Lorenz equations. Bull. Amer. Math. Soc. **27** (1992) 298–303.

Hayashi, S., Diffeomorphisms in $C^1(M)$ satisfy axiom A. Ergod. Theor. Dyn. Syst. **12** (1992) 233–253.

Hayashi, S., Connecting invariant manifolds and the solution of the C^1 stability and Ω-stability conjectures for flows. Annals Math. **145** (1997) 81–137.

Haykin, S. *Neural Networks – A Comprehensive Foundation* (2nd edn.) Prentice Hall, Upper Saddle River, NJ (1999).

Hegger, R., Kantz, H., and Schreiber, T., Practical implementation of nonlinear time series methods: The TISEAN package. Chaos **9** (1999) 413–424.

Helstrom, C. W., *Probability and Stochastic Processes for Engineers.* Macmillan Publishing Company, New York (1984).

Hempel, J., 3-Manifolds. Annals of Mathematics Studies **86** Chelsea Publishing, London (2004).

Henk, B., Konstantinos, E., and Easwar, S., Robustness of unstable attractors in arbitrarily sized pulse-coupled networks with delay. Nonlinearity **21**(1) (2008) 13–49.

Hénon, M., Numerical study of quadratic area preserving mappings. Q. Appl. Math. **27** (1969) 291–312.

Hénon, M., A two dimensional mapping with a strange attractor. Comm. Math. Phys. **50** (1976) 69–77.

Hénon, M., On the numerical computation of Poincaré maps. Physica D **5**(2–3) (1982) 412–414.

Hewitt, E. and Ross, K. A., *Abstract Harmonic Analysis.* **I**. Structure of topological groups. Integration theory group representations, Academic Press, New York (1963); Russian translation: Abstract harmonic analysis **I** Structure of topological groups. Integration theory, Nauka, Moscow (1975).

Hirsch, M., *On Invariant Subsets of Hyperbolic Sets. Essays on Topology and Related Topics.* Mémoires dédiés à Georges de Rham (1970) 126–135.

Hirsch, M. W., *Differential Topology.* Number **33** in Graduate Texts in Mathematics. Springer-Verlag, New York (1976).

Hirsch, M. and Pugh, C., Stable manifolds and hyperbolic sets. Proc. Sympos. Pure Math. **XIV**, Berkeley, Calif. (1968) Amer. Math. Soc., Providence, RI (1970) 133–163.

Hirsch, M. W. and Smale, S., *Differential Equations, Dynamical Systems, and Linear Algebra.* Academic Press, New York (1974).

Hirsch, M., Palis, J., Pugh, C., and Shub, M., Neighborhoods of hyperbolic sets. Invent. Math. **9** (1969)/(1970) 121–134.

Hirsch, M., Pugh, C., and Shub, M., *Invariant Manifolds.* Lecture Notes in Math. Springer, Berlin (1977).

Hirsch, M. W., Smale, S., and Devaney, R. L., *Differential Equations, Dynamical Systems and an Introduction to Chaos.* Elsevier, Amsterdam (2004).

Hitzl, D. H. and Zele, F., An exploration of the Hénon quadratic map. Physica D **14** (1985) 305–326.

Hobson, P. R. and Lansbury, A. N., A simple electronic circuit to demonstrate bifurcation and chaos. Phys. Edu. **31**(1) (1996) 39–43.

Hochster, M., *Lectures on Jacobian Conjecture.* sci. math. research post forwarded by I. Algol. Nov. 11 (2004).

Hoensch, U. A., Some hyperbolicity results for Hénon-like diffeomorphisms. Nonlinearity **21** (2008) 587–611.

Hofbauer, F. and Keller G., Ergodic properties of invariant measures for piecewise monotonic transformations. Math. Z. **180** (1982) 119–140.

Holmes, P. J. and Whitley, D.C., Bifurcation of one- and two-dimensional maps. Phil. Trans. Roy. Soc. Lond. **A311** (1984) 43–102.

Hopf, E., Statistik der geodätischen Linien in Mannigfaltigkeiten negativer Krümmung. Leipzig Ber. Verhandl. Sächs. Akad. Wiss. **91** (1939) 261–304.

Horita, V. and Tahzibi, A., Partial hyperbolicity for symplectic diffeomorphisms. Ann. Inst. H. Poincaré – AN **23** (2006) 641–661.

Hornik, K., Stinchocombe, M., and White, H., Universal approximation of an unknown mapping and its derivatives using multilayer feedforward networks. Neural Networks **3**(5) (1990) 551–560.

Hruska, S. L., A numerical method for constructing the hyperbolic structure of complex Hénon mappings. Foundations of Computational Mathematics **6** (2006a) 427–455.

Hruska, S. L., Rigorous numerical studies of the dynamics of polynomial skew products of $\mathbb{C}^2$. Contemp. Math. **396** (2006b) 85–100.

Hsu, G., Ott, E., and Grebogi, C., Strange saddles and the dimension of their invariant manifolds, Phys. Lett. A **127** (1988) 199–204.

Hu, H. and Young, L. S., Nonexistence of SBR measure for some diffeomorphisms that are "almost Anosov". Ergod. Theor. Dyn. Syst. **15** (1995) 67–76.

Hu, H., Pesin, Y., and Talitskaya, A., Every compact manifold carries a hyperbolic ergodic flow. Preprint (2009).

Huang, A., Pivka, L., Wu, C-W., and Franz, M., Chua's equation with cubic nonlinearity. Int. J. Bifurcation and Chaos **6**(12A) (1996) 2175–2222.

Hubbard, J. H. and Sparrow, C., The classification of topologically expansive Lorenz maps. Comm. Pure Appl. Math. **XLIII** (1990) 431–443.

Hunt, T. J., Low dimensional dynamics, bifurcations of cantori and realisations of uniform hyperbolicity. Ph.D. Thesis, University of Cambridge (2000).

Hunter, W. I. and Korenberg, M. J., Identification of nonlinear bilogical systems – Wiener and Hammerstein models. Biological Cybernitics **55** (1986) 135–144.

Hurewicz, W. and Wallman, H., *Dimension Theory*. Princeton University Press (1984).

Hurley, M., Attractors – Persistence and density of their basins. Trans. Amer. Math. Soc. **269** (1982) 247–271.

Isaeva, O. V., Jalnine, A. Yu., and Kuznetsov, S. P., Arnold's cat map dynamics in a system of coupled non-autonomous van der Pol oscillators. Phys. Rev. E **74** (2006) 046207.

Ishii, Y., Towards a kneading theory for Lozi mappings I – A solution of the pruning front conjecture and the first tangency problem. Nonlinearity **10** (1997a) 731–747.

Ishii, Y., Towards a kneading theory for Lozi mappings II – Monotonicity of the topological entropy and Hausdorff dimension of attractors. Comm. Math. Phys. **190** (1997b) 375–394.

Ishii, Y. and Sands, D., Monotonicity of the Lozi family near the tent-maps. Comm. Math. Phys. **198** (1998) 397–406.

Jackson, E. A., *Perspectives of Nonlinear Dynamics* **Vol. 1**. Cambridge University Press (1991).

Jakobson, M. V., Smooth mappings of the circle into itself. Mat. Sbornik **85**(127) (1971) 163–188.

Jakobson, M. V., Absolutely continuous invariant measures for one-parameter families of one-dimensional maps. Comm. Math. Phys. **81** (1981) 39–88.

Jafarizadeh. M. A. and Behnia, S., Hierarchy of chaotic maps with an invariant measure and their compositions. J. Nonlinear. Math. Phys. **9**(1) (2002) 26–41.

Jarvenpaa, E. and Jarvenpaa, M., On the definition of SRB measures for coupled map lattices. Comm. Math. Phys. **220** (2001) 109–143.

Jiang, M., SRB measures for lattice dynamical systems. J. Stat. Phys. **111**(3–4) (2003) 863–902.

Jing-ling, S., Hua-wei, Y., Jian-hua, D., and Hong-jun, Z., Riddled basin of laser cooled-ions in a Paul trap. Chinese Phys. Lett. **13** (1996) 81–84.

Kahan, S. and Sicardi-Schifino, A. C., Homoclinic bifurcations in Chua's circuit. Physica A **262**(1–2) (1999) 144–152.

Kahlert, C., The chaos producing mechanism in Chua's circuit and related piecewise-linear dynamical systems. In *Circuit Theory and Design 87. Proceedings of the European Conference on Circuits Theory and Design* **87** (1987) 269–274.

Kalinin, B. and Sadovskaya, V., On local and global rigidity of quasi-conformal Anosov diffeomorphisms. J. Inst. Math. Jussieu **2**(4) (2003) 567–582.

Kan, I., Kocak, H., and Yorke, J. A., Persistent homoclinic tangencies in the Hénon family. Physica D **83**(4) (1995) 313–325.

Kanjilal, P. P., *Adaptive Prediction and Predictive Control*. Peter Peregrinus Ltd, Stevenage, UK (1995).

Kantz, H., Grebogi, C., Prasad, A., Ying-Cheng, L., and Sinde, E., Unexpected robustness against noise of a class of nonhyperbolic chaotic attractors. Phys. Rev. E **65** (2002) 026209.

Kapitaniak, T., Maistrenko, Y., and Grebogi, C., Bubbling and riddling of higher-dimensional attractors. Chaos, Solitons & Fractals **17**(1) (2003) 61–66.

Kaplan, J. and Yorke, J. A., Chaotic behavior of multidimensional difference equations. In *Functional Differential Equations and Approximation of Fixed Points* (H. O. Peitgen and H. O. Walther, eds.) Springer, New York (1987).

Kathryn, E. L., Lomel, H. E., and Meiss, J. D., Quadratic volume preserving maps – An extension of a result of Moser. Regular and Chaotic Dynamics **33** (1998) 122–131.

Katok, A., Bernoulli diffeomorphism on surfaces. Annals Math. **110** (1979) 529–547.

Katok, A., Lyapunov exponents, entropy and periodic orbits for diffeomorphisms. Inst. Hautes Etudes Sci. Publ. Math. **51** (1980) 137–173.

Katok, A. and Hasselblatt, B., *Introduction to the Modern Theory of Dynamical Systems*, Cambridge University Press (1995).

Kawabe, T. and Kondo, Y., Intermittent chaos generated by logarithmic map. Prog. Theor. Phys. **86** (1991) 581–600.

Kei, E., Takahiro, I., and Akio, T., Design of a digital chaos circuit with nonlinear mapping function learning ability. IEICE Trans. Fundamentals of Electronics, Communications and Computer Sciences **E81-A**(6) (1998) 1223–1230.

Keller, G. and Nowicki, T., Spectral theory, zeta functions and the distribution of periodic points for Collet–Eckmann maps. Comm. Math. Phys. **149** (1992) 31–69.

Kennedy, M. P., Three steps to chaos. I. Evolution. IEEE Trans. Circ. Syst. Fund. Theor. Appl. **40**(10) (1993a) 640–656.

Kennedy, J. and York, J. A., Topological horseshoes. Trans. Amer. Math. Soc. **353** (2001) 2513–2530.

Kennedy, J., Kocak, S., and Yorke, J. A., A chaos lemma. Amer. Math. Monthly **108** (2001) 411–423.

Kennel, M. B. and Isabelle, S., Method to distinguish possible chaos from colored noise and to determine embedding parameters. Phys. Rev. A **46** (1992) 3111–3118.

Keynes, H. B. and Sears, M., Real-expansive flows and topological dimension. Ergod. Theor. Dyn. Syst. **1** (1981) 179–195.

Khan, A. M., Mar, D. J., and Westervelt, R. M., Spatial measurements near the instability threshold in ultrapure G_e. Phys. Rev. B **45** (1992) 8342–8347.

Khibnik, A. I., Roose, D., and Chua, L. O., On periodic orbits and homoclinic bifurcations in Chua's circuit with a smooth nonlinearity. Int. J. Bifurcation and Chaos **3**(2) (1993) 363–384.

Kifer, Y., Some theorems on small random perturbations of dynamical systems. Uspekhi Math. Nauk. **29**(3) (1974) 205–206 (in Russian).

Kirchgraber, U. and Stoffer, D., Possible chaotic motion of comets in the Sun–Jupiter system – A computer-assisted approach based on shadowing. Nonlinearity **17** (2004) 281–300.

Kirchgraber, U. and Stoffer, D., Transversal homoclinic points of the Hénon map. Annali di Matematica Pura ed Applicata **18**(5) (2006) 187–204.

Kiriki, S., Forward limit sets singularities for the Lozi family. Hokkaido Mathematical Journal **33** (2004) 491–510.

Kiriki, S. and Soma, T., Parameter-shifted shadowing property for Lorenz attractors. Trans. Amer. Math. Soc. **357** (2005) 1325–1339.

Kiriki, S. and Soma, T., Parameter-shifted shadowing property of Lozi maps. Dynamical Systems: An International Journal **22**(3) (2007) 351–363.

Kiriki, S., Li, M. C., and Soma, T., Coexistence of homoclinic sets with/without SRB measures in Hénon maps. Preprint (2008).

Klinshpont, N. E., On the problem of topological classification of Lorenz-type attractors. Math. Sbornik **197**(4) (2006) 75–122.

Klinshpont, N. E., Sataev, E. A., and Plykin, R. V., Geometrical and dynamical properties of Lorenz type system. Journal of Physics Conference Series **23** (2005) 96–104.

Kodama, H., Sato, S., and Honda, K., Renormalization-group theory on intermittent chaos in relation to its universality. Prog. Theor. Phys. **86** (1991) 309–314.

Koiran, P., The topological entropy of iterated piecewise affine maps is uncomputable. Discrete Mathematics and Theoretical Computer Science **4** (2001) 351–356.

Komuro, M., Expansive properties of Lorenz attractors. In *The Theory of Dynamical Systems and its Applications to Nonlinear Problems*. World Scientific, Kyoto (1984) 4–26..

Komuro, M., Tokunaga, R., Matsumoto,T., Chua, L. O., and Hotta, A., Global bifurcation analysis of the double scroll circuit. Int. J. Bifurcation and Chaos **1** (1991) 139–182.

Kowalczyk, P., Robust chaos and border-collision bifurcations in non-invertible piecewise-linear maps. Nonlinearity **18** (2005) 485–504.

Kozlovskii, O., *Structural Stability in One-dimensional Dynamics*. Ph.D. Thesis, University of Amsterdam (1998).

Kozlovskii, O., Getting rid of the negative Schwarzian derivative condition. Annals Math. **152** (2000) 743–762.

Kozlovskii, O. S., Axiom A maps are dense in the space of unimodal maps in the C^k topology. Annals Math. **157** (2003) 1–44.

Kozlovskii, O., Stability conjecture for unimodal maps. Manuscript (2008).

Kozlovski, O. S., Shen, W., and van Strien, S., Density of hyperbolicity in dimension one. Annals Math. **166**(1) (2007) 145–182.

Krishchenko, A., Localization of limit cycles. Differentsia'nye Uranvneniya **31**(11) (1995) 1858–1865 (in Russian).

Krishchenko, A., Estimations of domain with cycles. Comput. Math. Appl. **34** (2–4) (1997) 325–332.

Kubo, G. T., Viana, R. L., Lopes, S. R., and Grebogi, C., Crisis-induced unstable dimension variability in a dynamical system. Phys. Lett. A **372** (2008) 5569–5574.

Kupka, I., Contribution à la théorie des champs génériques, Contributions to Differential Equations **2** (1963) 457–484 and **3** (1964) 411–420.

Kuptsov, P. V., Kuznetsov, S. P., Sataev, I. R., Hyperbolic attractor of Smale–Williams type in a system of two coupled non-autonomous amplitude equations. Preprint (2008).

Kuramitsu, M., A classfication of the 3rd order oscillators with respect to chaos. *Proceedings of 1995 International Symposium on Nonlinear Theory and its Applications* **1** (1995) 599–602.

Kuznetsov, S. P., *Dynamical Chaos*. Fizmatlit, Moscow (2001) (in Russian).

Kuznetsov, Y. A., *Elements of Applied Bifurcation Theory* (3rd edn.) Springer, New York (2004).

Kuznetsov, S. P., Example of a physical system with a hyperbolic attractor of the Smale–Williams type. Phys. Rev. Lett. **95** (2005) 144101.

Kuznetsov, S. P., On the Feasibility of a parametric generator of hyperbolic chaos. J. Experimental and Theoretical Physics **106**(2) (2008) 380–387.

Kuznetsov, S. P. and Sataev, I. R., Hyperbolic attractor in a system of coupled non-autonomous van der Pol oscillators – Numerical test for expanding and contracting cones. Phys. Lett. A **365** (2007) 97–104.

Kuznetsov, S. and Seleznev, E., A strange attractor of the Smale–Williams type in the chaotic dynamics of a physical system. J. Exper. Theor. Physics **102**(2) (2006) 355–364.

Kuznetsov, S. P. and Pikovsky, A., Autonomous coupled oscillators with hyperbolic strange attractors. Physica D **232** (2007) 87–102.

Kuznetsov, S. P. and Pikovsky, A., Hyperbolic chaos in the phase dynamics of a Q-switched oscillator with delayed nonlinear feedbacks. Eur. Phys. Lett. **84** (2008) 10013.

Kuznetsov, S. P. and Ponomarenko, V. I., Realization of a strange attractor of the Smale–Williams type in a radiotechnical delay-fedback oscillator. Technical Physics Letters **34**(9) (2008) 771–773.

Labarca, R. and Moreira, C. G., Bifurcations of the essential dynamics of Lorenz maps and the application to Lorenz like flows – Contributions to the study of the expanding case. Bulletin of the Brazilian Mathematical Society **32**(2) (2001) 107–144.

Labarca, R. and Moreira, C. G., Bifurcations of the essential dynamics of Lorenz maps and the application to Lorenz like flows – Contributions to the study of the contracting case. Preprint (2003).

Labarca, R. and Moreira, C. G., Essential dynamics for Lorenz maps on the real line and the Lexicographical World. Ann. Inst. H. Poincaré – AN **23** (2006) 683–694.

Labarca, R. and Pacifico, M. J., Stability of singular horsshoes. Toplogy **25**(3) (1986) 337–352.

Lai, Y., Grebogi, C., and Yorke, J., How often are chaotic saddles nonhyperbolic?. Nonlinearity **6** (1993) 779–797.

Lai, Y.C., Grebogi, C., Yorke, J. A., and Venkataramani, S. C., Riddling bifurcation in chaotic dynamical systems. Phys. Rev. Lett. **77**(1) (1996) 55–58.

Lamarque, C-H., Janin, O., and Awrejcewicz, J., Chua systems with discontinuities. Int. J. Bifurcation and Chaos **9**(4) (1999) 591–616.

Laughton, S. N. and Coolen, A. C. C., Quasi-periodicity and bifurcation phenomena in Ising spin neural networks with asymmetric interactions. J. Phys. A **27** (1994) 8011–8028.

Ledrappier, F., Some properties of absolutely continuous invariant measures on an interval. Ergod. Theor. Dyn. Syst. **1** (1981) 77–93.

Ledrappier, F., Shub, M., Simo, C., and Wilkinson, A., Random versus deterministic exponents in a rich family of diffeomorphisms. J. Stat. Phys. **113** (2003) 85–149.

Lehto, O. and Virtanen, K. I., *Quasiconformal Mappings in the Plane*. Springer-Verlag, New York (1973).

Levin, G. and van Strien, S., Local connectivity of the Julia set of real polynomials. Annals Math. **147** (1998) 471–541.

Li, C. and Chen, G., Estimating the Lyapunov exponents of discrete systems. Chaos **14**(2) (2004) 343–346.

Li, T. Y. and York, J. A., Periodic three implies chaos. Am. Math. Monthly **82** (1975) 985–989.

Li, C. and Xia, X., On the bound of the Lyapunov exponents for continuous systems. Chaos **14**(3) (2004) 557–661.

Liao, S. T., On the stability conjecture. Chinese Annals Math. **1** (1980) 9–30.

Liao, S. T., Hyperbolicity properties of the non-wandering sets of certain 3-dimensional systems. Acta Math. Sci. **3** (1983) 361–368.

Lichtenberg, A. and Lieberman, M., *Regular and Stochastic Motion*. Springer-Verlag, New York (1983).

Liley, D. T. J., Cadusch, P. J., and Wright, J. J., A continuum theory of electrocortical activity. Neurocomputing **26-27** (1999) 795–800.

Lipschutz, S. and Lipson, M., *Schaum's Outlines – Linear Algebra*. Tata McGraw-Hill, Delhi (2001) 69–80.

Liverani. C., Decay of correlations. Annals Math. **142** (1995) 239–301.

Livsic, A. N., The homology of dynamical systems. Uspehi Mat. Nauk. **273**(165) (1972) 203–204.

Lizana, C. and Mora, L., Lower bounds for the Hausdorff dimension of the geometric Lorenz attractor – The homoclinic case. Discrete and Continuous Dynamical Systems **22**(3) (2008) 699–709.

Lohner, R., Computation of guaranteed enclosures for the solutions of ordinary initial and boundary value problems. In *Computational Ordinary Differential Equations* (J. R. Cash and I. Gladwell, eds.) Clarendon Press, Oxford (1992).

Lopshits, A. M., *Computation of Areas of Oriented Figures*. DC Heath and Company, Boston (1963).

Lorenz, E. N., Deterministic non-periodic flow. J. Atmos. Sci. **20** (1963) 130–141.

Lozi, R. and Ushiki, S., Organized confinors and anti-confinors and their bifurcations in constrained Lorenz system. Ann. Télé Commununications **43**(3–4) (1988) 187–208.

Lu, Y-Y., Xue, L-P., Zhu, M-C., and Qiu, S-S., Frequency band estimate and change for chaos systems. J. Shenzhen University Sci. Eng. **20**(2) (2003) 35–41.

Luchinski, D. G. and Khovanov, I. A., Fluctuation-induced escape from the basin of attraction of a quasiattractor. JETP Letters **69**(11) (1999) 825–830.

Luzzatto, S., In *The Mandelbrot Set – Themes and Variations* (Tan Lei, ed.) LMS Lecture Notes **274**, Cambridge University Press (2000).

Luzzatto, S. and Viana, M., Positive Lyapunov exponents for Lorenz-like families with criticalities. Astérisque **261** (2000) 201–237.

Luzzatto, S., Melbourne, I., and Paccaut, F., The Lorenz attractor is mixing. Comm. Math. Phys. **260**(2) (2005) 393–401.

Lyubich, M., On the Lebesgue measure of the Julia set of a quadratic polynomial. Stony Brook IMS preprint 10 (1991).

Lyubich, M., Geometry of quadratic polynomials – Moduli, rigidity and local connectivity. Stony Brook IMS preprint **93-9** (1993).

Lyubich, M., Combinatorics, geometry and attractors of quasi-quadratic maps. Annals Math. **140** (1994) 347–404.

Lyubich, M., Dynamics of quadratic polynomials I–II. Acta Math. **178** (1997) 185–297.

Lyubich, M., Dynamics of quadratic polynomials III – Parapuzzle and SBR measure. Astérisque **261** (2000) 173–200.

Lyubich, M., Almost every real quadratic map is either regular or stochastic. Annals Math. **156** (2002) 1–78.

MacKay, R. S. and Meiss, J. D., Cantori for symplectic maps near the anti-integrable limit. Nonlinearity **5** (1992) 49–160.

MacKay, R. S. and Percival, I. C., Converse KAM: theory and practice. Comm. Math. Phys. **98** (1985) 469–512.

MacKay, R. S. and van Zeijts, J. B. J., Period doubling for bimodal maps – A horseshoe for a renormalization operator. Nonlinearity **1** (1988) 253–277.

Madan, R. N., Observing and learning chaotic phenomena from Chua's circuit. *Proceedings 35th Midwest Symp. Circuits and Systems* **1** (1992) 736–745.

Madan, R. N., Chua's Circuit: A Paradigm for Chaos. World Scientific, Singapore (1993).

Mahesh, C. S., Nithin, N., Prabhakar, G. V., The B-exponential map: A generalization of the logistic map, and its applications in generating pseudorandom numbers. Preprint (2006).

Mahla, A. I. and Badan Palhares, A. G., Chua's circuit with a discontinuous nonlinearity. J. Circuits. Syst. Comput. **3**(1) (1993) 231–237.

Maistrenko, Y. L., Maistrenko,V. L., Popovich, A., and Mosekilde, E., Transverse instability and riddled basins in a system of two coupled logistic maps. Phys. Rev. E **57** (1998) 2713–2724.

Majumdar, M. and Mitra, T., Robust ergodic chaos in discounted dynamic optimization models. Economic Theory **4**(5) (1994) 677–688.

Malykhin, V. I., Connected Space. In Hazewinkel, Michiel, *Encyclopedia of Mathematics*, Kluwer Academic Publishers, Dordrecht (2001).

Mané, R., Contributions to the C^1-stability conjecture. Topology **17** (1978) 386–396.

Mané, R., Expansive homeomorphisms and topological dimension. Trans. Amer. Math. Soc. **252** (1979) 313–319.

Mañé, R., An ergodic closing lemma. Annals Math. **116** (1982) 503–540.

Mané, R., Oseledec's theorem from the generic viewpoint. Proceedings of the 1983 International Congress of Mathematicians **12** Warsaw (1984) 1269–1276.

Mané, R., Hyperbolicity, sinks and measure in one-dimensional dynamics. Comm. Math. Phys. **100**(4) (1985) 495–524.

Mané, R., *Ergodic Theory and Differentiable Dynamics*. Springer-Verlag, Berlin (1987).

Mañé, R., A proof of the C^1 stability cConjecture. Publ. Math. IHES **66** (1988) 161–210.

Mané, R., The Lyapunov exponents of generic area preserving diffeomorphisms. In *International Conference on Dynamical Systems (Montevideo, 1995)*. Pitman Res. Notes Math. Ser., Longman, Harlow **362** (1996) 110–119.

Mané, R., Sad, P., and Sullivan, D., On the dynamics of rational maps. Ann. Sci. Ecole Norm. Sup. **16** (1983) 193–217.

Manning, A., There are no new Anosov diffeomorphisms on tori. Amer. J. Math. **96** (1974) 422–429.

Markarian, R., Billiards with polynomial decay of correlations. Ergod. Theor. Dyn. Syst. **24** (2004) 177–197.

Marotto, F. R., Snap-back repellers imply chaos in $\mathbb{R}^n$. J. Math. Anal. Appl. **3** (1978) 199–223.

Marotto, F. R., Perturbation of stable and chaotic difference equation. J. Math. Anal. Appl. **72**(2) (1979a) 716–729.

Marotto, F. R., Chaotic behavior in the Hénon mapping. Comm. Math. Phys. **68** (1979b) 187–194.

Marsden, J. and McCracken, M., *The Hopf Bifurcation and its Applications*. Appl. Math. Sciences **19** Springer-Verlag, New York (1976).

Martens, M., *Interval Dynamics*. Ph.D. Thesis, Delft (1990).

Martens, M., de Melo, W., and van Strien, S., Julia–Fatou–Sullivan theory for real one-dimensional dynamics. Acta. Math. **168** (1992) 273–318.

Mascagni, M., In *Algorithms for Parallel Processing* (M. T. Heath, A. Ranade, and R. S. Schreiber, eds.) Springer-Verlag, New York (1999).

Mather, J. N., Stability of C^∞ mappings I – The division theorem. Annals Math. **87**(2) (1968a) 89–104.

Mather, J. N., Stability of C^∞ mappings III – Finitely determined map-germs. Inst. Hautes Etudes Sci., Publ. Math. (35) (1968b) 279–308.

Mather, J. N., Stability of C^∞ mappings II – Infinitesimal stability implies stability. Annals Math. **89**(2) (1969) 254–291.

Matsumoto, T., A chaotic attractor from Chua's circuit. IEEE Trans. Circ. Syst. **CAS-31** (1984) 1055–1058.

Matsumoto, T., Chua, L. O., and Tokumasu, K., Double scroll via a two-transistor circuit. IEEE Trans. Circ. Syst. **CAS-33** (1986) 828–835.
Matthews, P. C., Transcritical bifurcation with $O(3)$ symmetry. Nonlinearity **16**(4) (2003) 1499–1473.
Mautner, F. I., Geodesic flows on symmetric Riemann spaces. Annals Math. **65** (1957) 416–431.
Mayer-Kress, G. and Haken, H., Attractors of convex maps with positive Schwarzian derivative in the presence of noise. Physica D **10** (1984) 329–339.
Mazur, M., On some useful conditions for hyperbolicity. 2008 International Workshop on Dynamical Systems and Related Topics. Trends in Mathematics – New Series **10**(2) (2008) 57–64.
Mazur, M., Tabor, J., and Koscielniak, P., Semi-hyperbolicity and hyperbolicity. Discrete Contin. Dynam. Syst. **20** (2008) 1029–1038.
Mazur, M. and Tabor, J., Computational hyperbolicity. Preprint (2009).
McDonald, S. W., Grebogi, C., Ott, E., and Yorke, J. A., Fractal basin boundaries. Physica D **17** (1985) 125–135.
Medvedev, V. and Zhuzhoma, E., There are no structurally stable diffeomorphisms of odd-dimensional manifolds with codimension one non-orientable expanding attractors. Preprint arXiv:math.DS/0404416 (2004).
Mees, A. I. and Chapman, P. B., Homoclinic and heteroclinic orbits in the double scroll attractor. IEEE Trans. Circ. Syst. **34**(9) (1987) 1115–1120.
Meiss, J. D., Average exit times in volume preserving maps. Chaos **7** (1997) 139–147.
Michelitsch, M. and Rössler, O. E., A new feature in Hénon's Map. Comput. & Graphics **13** (1989) 263–275. Reprinted in Chaos and Fractals, A Computer Graphical Journey – Ten Year Compilation of Advanced Research (C. A. Pickover, ed.) Elsevier, Amsterdam (1998) 69–71.
Miller, D. A. and Grassi, G., A discrete generalized hyperchaotic Hénon map circuit. In *Proceedings of the 44th IEEE (2001) Midwest Symposium on Circuits and Systems* **1** (2001) 328–331.
Milnor, J., On the concept of attractor. Comm. Math. Phys. **99** (1985a) 177–195.
Milnor, J., On the concept of attractor: Correction and remarks. Comm. Math. Phys. **102** (1985b) 517–519.
Milnor, J. and Thurston, R. On iterated maps of the interval I and II. Unpublished notes. Princeton University Press (1977).
Minkowski, H., Verhandlungen des III. Internationalen Mathematiker-Kongresses in Heidelberg, Berlin (1904).
Mira, C., Chua's circuit and the qualitative theory of dynamical systems. Int. J. Bifurcation and Chaos **7**(9) (1997) 1911–1916.
Mischaikow, K., Topological techniques for efficient rigorous computations in dynamics. Acta Numerica **11** (2002) 435–477.
Mischaikow, K. and Mrozek, M., Chaos in the Lorenz equations: A computer-assisted proof. Bull. Amer. Math. Soc. **32** (1995) 66–72.
Mischaikow, K. and Mrozek, M., Chaos in the Lorenz equations: A computer assisted proof. Part II, Detail. Math. Comp. **67**(223) (1998) 1023–1046.

Misiurewicz, M., *Nonlinear Dynamics*, New York Acad. Sci., New York (1980).
Misiurewicz, M., Absolutely continuous measures for certain maps of an interval. Publ. Math. IHES **53** (1981) 17–51.
Misiurewicz, M., Unimodal interval maps obtained from the modified Chua's equations. Int. J. Bifurcation and Chaos **3** (1993) 323–332.
Misiurewicz, M. and Szewc, B., Existence of a homoclinic point for the Hénon map. Comm. Math. Phys. **75**(3) (1980) 285–291.
Moerdijk, I. and Mrčun, J., *Introduction to Foliations and Lie Groupoids*. Cambridge University Press (2003).
Molgedey, L., Schuchhardt, J., and Schuster, H. G., Suppressing chaos in neural networks by noise. Phys. Rev. Lett. **69** (1992) 3717–3719.
Moore, C. C., Ergodicity of flows on homogeneous spaces. Amer. J. Math. **88** (1966a) 154–178.
Moore, R. E., *Interval Analysis*. Prentice Hall, Englewood Cliffs, NJ (1966b).
Moore, R. E., *Methods and Applications of Interval Analysis*. SIAM, Philadelphia (1979).
Mora, L. and Vianna, M., Abundance of strange attractors. Acta. Math. **17**(1) (1993) 1–71.
Morales, C. A. Singular-hyperbolic sets and topological dimension. Dynamical Systems **18**(2) (2003) 181–189.
Morales, C., The explosion of singular hyperbolic attractors. Ergod. Theor. Dyn. Syst. **24**(2) (2004a) 577–592.
Morales, C., A note on periodic orbits for singular-hyperbolic flows. Discrete Contin. Dyn. Syst. **11**(2–3) (2004b) 615–619.
Morales, C., Poincaré–Hopf index and singular-hyperbolic sets on 3-balls. Preprint (2006).
Morales, C., Singular-hyperbolic attractors with handlebody basins. J. Dynamical and Control Systems **13**(1) (2007) 15–24.
Morales, C., Topological dimension of singular-hyperbolic attractors. Preprint published at IMPA (2008).
Morales, C., Poincaré–Hopf index and partial hyperbolicity. Ann. Fac. Sci. Toulouse Math. **XVII**(1) (2008) 193–206.
Morales, C. A. and Pacifico, M. J., Attractors and singularities robustly accumulated by periodic orbits. International Conference on Differential Equations **1, 2**. World Scientific, Berlin (1998) 64–67.
Morales, C. and Pacifico, M. J., Mixing attractors for 3-flows. Nonlinearity **14** (2001) 359–378.
Morales, C. and Pacifico, M. J., A dichotomy for three-dimensional vector fields. Ergod. Theor. Dyn. Syst. **23** (2003a) 1575–1600.
Morales, C. and Pacifico, M. J., Transitivity and homoclinic classes for singular-hyperbolic systems. Preprint Série A 208/2003 (2003b).
Morales, C. A. and Pacifico, M. J., Sufficient conditions for robustness of attractors. Pacific J. Mathematics **216**(2) (2004) 327–342.
Morales, C. and Pujals, E., Singular strange attractors on the boundary of Morse–Smale systems. Ann. Sci. École Norm. Sup. **30** (1997) 693–717.

Morales, C., Pacifico, M. J., and Pujals, E., On C^1 robust singular transitive sets for three-dimensional flows. C. R. Acad. Sci. Paris, Série I **326** (1998) 81–86.

Morales, C., Pacifico, M. J., and Pujals, E. R., Singular hyperbolic systems. Proceedings Amer. Math. Soc. **127**(11) (1999) 3393–3401.

Morales, C., Pacifico, M. J., and Pujals, E., Strange attractors across the boundary of hyperbolic systems. Comm. Math. Phys. **211**(3) (2000) 527–558.

Morales, C. A., Pacifico, M. J., and Pujalls, E. R., Robust transitive singular sets for 3-flows are partially hyperbolic attractors or repellers. Annals Math. **160**(2) (2004) 375–432.

Morales, C. A., Pacifico, M. J., and San Martin, B., Expanding Lorenz attractors through resonant double homoclinic loops. SIAM J. Math. Anal. **36**(6) (2005) 1836–1861.

Morales, C. A., Pacifico, M. J. P., and San Martin, B., Contracting Lorenz attractors through resonant double homoclinic loops. SIAM J. Mathematical Analysis **38** (2006) 309–332.

Morosawa, S., Nishimura, Y., Taniguchi, M., and T. Ueda., *Holomorphic Dynamics*. Cambridge University Press (2000).

Morse, M., A one-to-one representation of geodesics on a surface of negative curvature. Amer. J. Math. **43**(1) (1921) 33–51.

Mosekilde, E., Zhusubaliyev, Z. T., Rudakov, V. N., and Soukhterin, E. A., Bifurcation analysis of the Hénon map. Discrete Dynamics in Nature and Society **53** (2000) 203–221.

Moser, J., On the integrability of area preserving Cremona mappings near an elliptic fixed point. Bol. Soc. Mat. Mexicana **2**(5) (1960) 176–180.

Moser, J., On invariant curves of area-preserving mappings of an annulus. Nachr. Akad. Wiss. Göttingen II. Math. Phys. Kl. (1962) 1–20.

Moser, J., On a theorem of Anosov. Differential Equations **5** (1969) 411–440.

Moser, J., *Stable and Random Motions in Dynamical Systems*. Annals Math. Studies, Princeton University Press (1973).

Moser, J., On quadratic symplectic mappings. Math. Zeitschrift (1994) 216–417.

Mukul, M. and Mitra, T., Robust chaos in dynamic optimization models. Ricerche Economiche **48** (1994) 225–240.

Muller, R., Lippert, K., Kuhnel, A., Behn, U., First-order nonequilibrium phase transition in a spatially extended system. Phys. Rev. E **56**(3) (1997) 2658–2662.

Murakami, C., Murakami, W., and Hirose, K., Sequence of global period doubling bifurcation in the Hénon maps. Chaos, Solitons & Fractals **14**(1) (2002) 1–17.

Naudot, V., Strange attractor in the unfolding of an inclination-flip homoclinic orbit. Ergod. Theor. Dyn. Syst. **16**(5) (1996) 1071–1086.

Neimark, Y. and Landa, P., *Stochastic and Chaotic Oscillations*. Nauka, Moscow (1989).

Nemytskii, V. V. and Stepanov, V. V., *Qualitative Theory of Differential Equations*. Princeton University Press (1960).

Nepomuceno, E. G., Takahashi, R. H. C., Amaral, G. F. V., and Aguirre, L. A., Nonlinear identification using prior knowledge of fixed points: A multiobjective approach. Int. J. Bifurcation and Chaos **13**(5) (2003) 1229–1246.

Neumaier, A., *Interval Methods for Systems of Equations*. Cambridge University Press (1990).

Newcomb, R. W. and Sathyan, S., An RC op amp chaos generator. IEEE Trans. Circ. Syst. **CAS-30** (1993) 54–56.

Newhouse, S., On codimension one Anosov diffeomorphisms. Amer. J. Math. **92** (1970a) 761–770.

Newhouse, S., Non-density of Axiom $A(a)$ on $\mathbb{S}^2$. Proceedings AMS Symp. Pure Math. **14** (1970b) 191–202, 335–347.

Newhouse, S. E., Hyperbolic limit sets. Trans. Amer. Math. Soc. **167** (1972a) 125–150.

Newhouse, S. E., The abundance of wild hyperbolic sets and non-smoth stable sets for diffeomorphisms. Publ. Math. IHES **50** (1972b) 101–151.

Newhouse, S., Diffeomorphisms with infinitely many sinks. Topology **13** (1974) 9–18.

Newhouse, S., On simple arcs between structurally stable flows. In *Dynamical Systems*, Proceedings Sympos. Appl. Topology and Dynamical Systems, University of Warwick, Coventry (1973)/(1974). Lect. Notes Math. **468** Springer-Verlag, Berlin (1975) 209–233.

Newhouse, S., Quasi-elliptic periodic points in conservative dynamical systems. Amer. J. Math. **99**(5) (1977) 1061–1087.

Newhouse, S., The abundance of wild hyperbolic sets and non-smooth stable sets for diffeomorphisms. Publ. Math. IHES **50** (1979) 101–51.

Newhouse, S., Asymptotic behavior and homoclinic points in nonlinear systems. Ann. of New York Acad. Sci. **357** (1980) 292–299.

Newhouse, S., New phenomena associated with homoclinic tangencies. Ergod. Theor. Dyn. Syst. **24**(5) (2004a) 1725–1738.

Newhouse, S., Cone-fields, domination, and hyperbolicity. In *Modern Dynamical Systems and Applications*. Cambridge University Press (2004b) 419–432.

Newhouse, S. and Palis, J., Bifurcations of Morse–Smale dynamical systems. In *Dynamical Systems* (M. M. Peixoto, ed.) Proceedings Symp. Bahia, Brazil, July 26–Aug. 14, 1971. Academic Press (1973) 303–366.

Newhouse, S., Ruelle, D., and Takens, F., Occurrence of strange axiom A attractors near quasi periodic flows on $\mathbb{T}^m$, $m \geq 3$. Comm. Math. Phys. **64** (1978) 35–40.

Newhouse, S., Palis, J., and Takens, F., Bifurcations and stability of families of diffeomorphisms. Publ. Math. IHES **57** (1983) 5–71.

Nikolaev, I., *Foliations on Surfaces*. Ergebnisse der Mathematik und ihrer Grenzgebiete. 3. Folge / A Series of Modern Surveys in Mathematics (2001).

Nithin, N. V., Prabhakar, G., Bhat, K. G., Joint entropy coding and encryption using robust chaos. Preprint, nlin/0608051 (2006).

Nossek, J. A., Experimental verification of horseshoes from electronic circuits. Phil. Trans. Royal Soc. London **353**(1701) (1995) 59–64.

Novikov, S. P., The topology of foliations. Trudy Moskov. Mat. Obshch. **14** (1965) 248–278, English translation: Trans. Moscow Math. Soc. **14** (1965) 268–304.

Nusse, H. E. and Yorke, J. A., A procedure for finding numerical trajectories on chaotic saddles. Physica D **36** (1989) 137–156.

Nusse, H. E. and Tedeschini-Lalli, L., Wild hyperbolic sets, yet no chance for the coexistence of infinitely many KLUS-simple newhouse attracting sets. Comm. Math. Phys. **144** (1992) 429–442.

Nusse, H. E. and Yorke, J. A., Border-collision bifurcations for piecewise smooth one-dimensional maps. Int. J. Bifurcation and Chaos **5** (1995) 189–207.

Nusse, H. E. and Yorke, J. A., Basins of attraction. Science **27**(1) (1996) 1376–1380.

Nunez, P. L., *Electric Fields of the Brain*. Oxford University Press (1981).

Nunez, P. L., Toward a quantitative description of large-scale neocortical dynamic function and EEG. Behav. Brain Sci. **23**(3) (2000) 371–437.

Ogorzalek, M., Chaotic regions from double scroll. IEEE Trans. Circ. Syst. **CAS-34**(2) (1987) 201–203.

Ohnishi, M. and Inaba, N., A singular bifurcation into instant chaos in piecewise-linear circuit. IEEE Trans. Circ. Syst. Communications and Computer Sciences **41**(6) (1994) 433–442.

Ott, E., *Chaos in Dynamical Systems*. Cambridge University Press (1993).

Ottino, J. M., *The Kinematics of Mixing: Stretching, Chaos, and Transport*. Cambridge University Press (1989).

Ottino, J. M., Muzzion, F. J., Tjahjadi, M., Franjione, J. G., Jana, S. C., and Kusch, H. A., Chaos, symmetry, and self-similarity: Exploring order and disorder in mixing processes. Science **257** (1992) 754–760.

Ovsyannikov, I. M. and Shilnikov, L. P., On systems with a saddle-focus homoclinic curve, Mat. Sbornik **58** (1986) 557–574; English translation: Math. USSR Sbornik **58** (1987) 557–574.

Ovsyannikov, I. M. and Shilnikov, L. P., Systems with a homoclinic curve of multidimensional saddle-focus type, and spiral chaos. Mat. Sbornik **182** (1991) 1043–1073; English translation: Math. USSR Sbornik **73** (1992) 415–443.

Paar, V. and Pavin, N., Intermingled fractal arnold tongues. Phy. Rev. E **57**(2) (1998) 1544–1549.

Pacifico, M. J., Pujals, E. R., and Viana, M., Sensitiveness and SRB measure for singular hyperbolic attractors. Preprint (2002).

Palacios, A., Cycling chaos in one-dimensional coupled iterated maps. Int. J. Bifurcation and Chaos **12**(8) (2002) 1859–1868.

Palis, J., On the structure of hyperbolic points in Banach spaces. Anais. Acad. Bras. Ciecias **40** (1968).

Palis, J., On Morse–Smale dynamical systems. Topology **8** (1969) 385–405.

Palis, J., A global view of dynamics and a conjecture on the denseness of finitude of attractors. Asterisque **261** (2000) 339–351.

Palis, J., A global perspective for non-conservative dynamics. Ann. Inst. H. Poincaré AN **22** (2005) 485–507.

Palis, J., Open questions leading to a global perspective in dynamics. Nonlinearity **21** (2008) T37–T43.

Palis, J. and Smale, S., Structural stability theorems. In *Global Analysis*, Berkeley 1968 Proceedings Sympos. Pure Math. **XIV**, Amer. Math. Soc. (1970) 223–232.

Palis, J. and Takens, F., *Hyperbolicity and Sensitive Chaotic Dynamic at Homoclinic Bifurcation*. Cambridge University Press (1993).

Palis, J. and Viana, M., High dimension diffeomorphisms displaying infinitely many sinks. Annals Math. **140** (1994) 1–71.

Palmer, K. J., Exponential dichotomies, the shadowing lemma and transversal homoclinic points. In *Dynamics Reported* (U. Kirchgraber and H. O. Walther, eds.) **1**. Wiley and Teubner, Stuttgart (1988).

Panti, G., Multidimensional continued fractions and a Minkowski function. Monatsh. Math. **154** (2008) 247–264.

Paradıs, J., Viader, P., and Bibiloni, L., The derivative of Minkowski $s?(x)$ function. J. Math. Anal. Appl. **253**(1) (2001) 107–125.

Parker, T. and Chua, L., Chaos: A tutorial for engineers. In Proceedings of the IEEE **75** (1987) 982–1008.

Parry, W., *The Lorenz Attractor and a Related Population Model*. Lecture Notes in Math. **729** Springer-Verlag, New York (1979) 169–187.

Parry, W. and Pollicott, M., An analogue of the prime number theorem for closed orbits of axiom A flows. Annals Math. **118** (1983) 573–591.

Parui, S. and Banerjee, S., Border collision bifurcations at the change of state-space dimension. Chaos **12** (2002) 1054–1069.

Pastor-Satorras, R. and Riedi, R. H., Numerical estimates of the generalized dimensions of the Hénon attractor for negative q. J. Physics A **29**(15) (1996) L391–L398.

Pastore, S., Detection of all the equilibrium points of dynamic 1-D autonomous rings. In *Proceedings of the European Conference on Circuit Theory and Design. ECCTD'99. Politecnico di Torino. Part* **2** (1999) 699–702.

Pei-Min, X. and Bang-Chun, W., A new type of global bifurcation in Hénon map. Chinese Phys. **13**(5) (2004) 618–624.

Peixoto, M. Structural stability on two-dimensional manifolds. Topology **1** (1962) 101–120.

Pellegrini, L., Tablino, C., Albertoni, S., and Biardi, G., Different scenarios in a controlled tubular reactor with a countercurrent coolant. Chaos, Solitons & Fractals **3**(5) (1993) 3537–3549.

Pereira, R. F., Pinto, S. E., Viana, R. L., Lopes, S. R., and Grebogi, C., Periodic orbit analysis at the onset of the unstable dimension variability and at the blowout bifurcation. Chaos **17** (2007) 023131.

Perov, A. I. and Egle, I. Yu., On the Poincaré–Denjoy theory of multidimensional differential equations, Differentsial'nye Uravneniya **8** (1972) 801–810; English translation: Differential Equations **8** (1972) 608–615.

Pesin, Y., Geodesic flows on closed Riemannian manifolds without focal points. Proceedings Sov. Acad. Sci, Ser. Math. (Izvestia) **41** (1977a) 1252–1288.

Pesin, Y., Characteristic Lyapunov exponents and smooth ergodic theory. Russian Math. Surveys **32**(4) (1977b) 55–114.

Pesin, Y., Dynamical systems with generalized hyperbolic attractors: Hyperbolic, ergodic and topological properties. Ergod. Theor. Dyn. Syst. **12** (1992) 123–151.

Pesin, Y., *Lectures on Partial Hyperbolicity and Stable Ergodicity.* Zürich Lectures in Advanced Mathematics, EMS (2004).

Pesin, Y. and Weiss, H. (eds.), Proceedings Symp. Pure. Math., Amer. Math. Soc. (2001).

Peter, G., On the determination of the basin of attraction of discrete dynamical systems. J. Difference Equations and Applications **13**(6) (2007) 523–546.

Peter, G. and Heiko, W., Lyapunov function and the basin of attraction for a single-joint muscle-skeletal model. J. Mathematical Biology **54**(4) (2007) 453–464.

Petersen, K., *Ergodic Theory.* Cambridge Studies in Advanced Mathematics. Cambridge University Press (1990).

Petrisor, E., Entry and exit sets in the dynamics of area preserving Hénon map. Chaos, Solitons & Fractals **17**(4) (2003) 651–658.

Phillipson, P. E. and Schuster, P., Bifurcation dynamics of three-dimensional systems. Int. J. Bifurcation and Chaos **10**(8) (2000) 1787–1804.

Pilyugin, S. Y., *Shadowing in Dynamical Systems.* Lect. Notes Math. **1706** Springer, Berlin (1999).

Pingel, D., Schmelcher, P., and Diakonos, F. K., Theory and examples of the inverse Frobenious–Perron problem for complete chaotic maps. Chaos **9** (1999) 357–366.

Pinto, A. A., Rand, D. A., and Ferreira, F., *Fine Structures of Hyperbolic Diffeomorphisms.* Springer Monographs in Mathematics (2008).

Pivka, L., Wu, C -W., and Anshan, H., Lorenz equation and Chua's equation. Int. J. Bifurcation and Chaos **6**(12B) (1996) 2443–2489.

Plykin, R. V., Sources and sinks for A-diffeomorphisms of surfaces. Math. USSR Sbornik **23** (1974) 233–253.

Plykin, R. V., The existence of attracting (repelling) periodic points of axiom A diffeomorphisms of the projective plane and of the Klein bottle. Uspekhi Mat. Nauk. **323** (1977) 179 (in Russian).

Plykin, R. V., Hyperbolic attractors of diffeomorphisms. Internat. Topology. Conf. (Moscow State University, Moscow) (1979); Uspekhi Mat. Nauk. **353** (1980) 94–104; English translation: Russian Math. Surveys **353** (1980) 109–121.

Plykin, R. V., On the geometry of hyperbolic attractors of smooth cascades. Uspekhi Mat. Nauk. **39**(6) (1984) 75–113; English translation: Russian Math. Surveys **39**(6) (1984) 85–131.

Plykin, R. V., On the problem of topological classification of strange attractors of dynamical systems. Russ. Math. Surv. **576** (2002) 1163–1205.

Plykin, R. V. and Zhirov, A. Y., Some problems of attractors of dynamical systems. Topology. Appl. **54** (1993) 19–46.

Plykin, R. V., Sataev, E. A., and Shlyachkov, S. V., Strange attractors, Itogi Nauki i Tekhniki Sovremennye Problemy Mat. Fundamental'nye

Napravleniya **66** (VINITI, Moscow) (1991) 100–147; English translation: *Dynamical Systems IX*. Encyclopedia Math. Sci. **66** Springer-Verlag, Berlin (1995) 93–139.

Poincaré, H., Sur le problème des trois corps et les équations de la dynamique. Acta. Math. **13** (1890) 1–270.

Pollicott, M., *Lectures on Ergodic Theory and Pesin Theory on Compact Manifolds*. London Mathematical Society Lecture Note Series (1993).

Pomeau, Y. and Manneville, P., Intermittent transition to turbulence in dissipative dynamical systems. Comm. Math. Phys. **74** (1980) 189–197.

Pospisil, J. and Brzobohaty, J., Elementary canonical state models of Chua's circuit family. IEEE Trans. Circ. Syst. Fund. Theor. Appl. **43**(8) (1996) 702–705.

Pospisil, J., Brzobohaty, J., and Kolka, Z., Elementary canonical state models of the third-order autonomous piecewise-linear dynamical systems. *ECCTD '95 Proceedings of the 12th European Conference on Circuit Theory and Design. Istanbul Tech. University Part.*,**1** (1995) 463–466.

Pospisil, J., Brzobohaty, J., and Kolka, Z., Generalized canonical state models of third-order piecewise-linear dynamical systems and their applications. Radioengineering **8**(1) (1999a) 10–13.

Pospisil, J., Brzobohaty, J., Kolka, Z., and Horska, J., Decomposed canonical state models of the third-order piecewise-linear dynamical systems. In *Proceedings of the European Conference on Circuit Theory and Design. ECCTD'99. Politecnico di Torino. Part* **1** (1999b) 181–184.

Pospisil, J., Brzobohaty, J., Kolka, Z., and Horska-Kreuzigerova, J., Cascade state models of the higher-order piecewise-linear dynamical systems. In *9th Inter. Czech - Slovak Scientific Conference. Radioelektronika 99. Conference Proceedings Brno University Technol.* (1999c) 29–30.

Pospisil, J., Brzobohaty, J., Kolka, Z., and Horska-Kreuzigerova, J., New canonical state models of Chua's circuit family. Radioengineering **8**(3) (1999d) 1–5.

Pospisil, J., Brzobohaty, J., Kolka, Z., and Kreuzigerova, J., New canonical state models of the third-order piecewise-linear dynamical systems. In *Proceedings Eighteenth IASTED Inter. Conference Modelling, Identification and Control.* ACTA Press (1999e) 384–387.

Pospisil, J., Hanus, S., Michalek, V., and Dostal, T., Relation of canonical state models of linear and piecewise-linear dynamical systems. In *Proceedings Eighteenth IASTED Inter. Conference Modelling, Identification and Control.* ACTA Press (1999f) 388–389.

Pospisil, J., Kolka, Z., Horska, J., and Brzobohaty, J., Simplest ODE equivalents of Chua's equations. Int. J. Bifurcation and Chaos **10**(1) (2000) 1–23.

Potapov, A. and Ali, M. K., Robust chaos in neural networks. Phy. Lett. A **277**(6) (2000) 310–322.

Press, W. H., Teukolsky, S. A., Vettering, W. T., and Flannery, B. P., *Numerical Recipes in C*. Cambridge University Press (1992).

Priel, A. and Kanter, I., Robust chaos generation by a perceptron. Europhys. Lett. **51**(2) (2000) 230–236.

Prokopenko, V. G., Expanding the Chua's-circuit family. *Radiotekhnika i Elektronika* **45**(10) (2000) 1241–1244.

Przytycki, F., Construction of invariant sets for Anosov diffeomorphisms and hyperbolic attractors. Studia Math. **68** (1980) 199–213.

Pugh, C. and Shub, M., Stable ergodicity and Juliene quasi-conformality. JEMS **2** (2000) 1–52.

Pujals, E. R., Tangent bundles dynamics and its consequences. ICM **III** (2002) 1–3.

Pujals, E. R., On the density of hyperbolicity and homoclinic bifurcations for 3-D diffeomorphisms in attracting regions. Discrete and Continuous Dynamical Systems **16**(1) (2006) 179–226.

Pujals. E. R., Density of hyperbolicity and homoclinic bifurcations for topologically hyperbolic sets. Discrete and Continuous Dynamical Systems **20**(2) (2008) 337–408.

Pujals, E. R. and Sambarino, M., Homoclinic tangencies and hyperbolicity for surface diffeomorphisms. Annals Math. **151** (2000a) 961–1023.

Pujals, E. R. and Sambarino, M., On homoclinic tangencies, hyperbolicity, creation of homoclinic orbits and variation of entropy. Nonlinearity **13** (2000b) 921–926.

Pujals, E. and Sambarino, M., Integrability on codimension one dominated splitting. Bull. Braz. Math. Soc. **38** (2007) 1–19.

Pujals, E. R., Robert, L., and Shub, M., Expanding maps of the circle rerevisited – Positive Lyapunov exponents in a rich family. Ergod. Theor. Dyn. Syst. **26** (2006) 1931–1937.

Rabinovich, M. and Trubetskov, D., *The Introduction to the Theory of Oscillations and Waves*. Nauka, Moscow (1984).

Ramdani, S., Chua, L. O., Lozi, R., and Rossetto, B., A qualitative study comparing Chua and Lorenz systems. In *Proceedings of the 7th Inter. Specialist Workshop on Nonlinear Dynamics of Electronic Systems. Tech. University of Denmark* (1999) 205–208.

Rand, D., The topological classification of Lorenz attractors. Proc. Cambridge Phil. Soc. **83** (1978) 451–460; Russian translation: Strange Attractors (Mir, Moscow) (1981) 239–251.

Robbin, J., A structural stability theorem. Annals Math. **94** (1971) 447–493.

Robert, B. and Robert, C., Border collision bifurcations in a one-dimensional piecewise smooth map for a PWM current-programmed H-bridge inverter. Int. J. Control **75**(16–17) (2002) 1356–1367.

Robinson, C., Structural stability of vector fields. Annals Math. **99** (1974) 154–175; Errata in Robinson, C., Annals Math. **101** (1975) 368.

Robinson, C., Bifurcation to infinitely many sinks. Comm. Math. Phys. **90** (1983) 433–459.

Robinson, C., Transitivity and invariant measures for the geometric model of the Lorenz attractor. Ergod. Theor. Dyn. Syst. **4** (1984) 605–611.

Robinson, C., Homoclinic bifurcation to a transitive attractor of Lorenz type. Nonlinearity **2** (1989) 495–518.

Robinson, C., Nonsymmetric Lorenz attractor from a homoclinic bifuraction. SIAM J. Math. Analysis **32**(1) (2000) 119–141.

Robinson, C., *Dynamical Systems: Stability, Symbolic Dynamics, and Chaos.* CRC Press (2004).

Robinson, C. and Verjovsky, A., Stability of Anosov diffeomorphisms, "Seminario de Sistemas Dinamicos" (J. Palis, ed.) Monografias de Matematica **4** (1971). IMPA Rio de Janeiro. Brazil, Chapter 9.

Rodriguez Hertz, F., Stable ergodicity of certain linear automorphisms of the torus. Annals Math. **162**(1) (2005) 65–107.

Rolfsen, D., *Knots and Links.* DEPublish or Perish Press, Wilmington (1976) 287–288.

Rosenblatt, J. M. and Weirdl, M., Pointwise ergodic theorems via harmonic analysis. In *Ergodic Theory and its Connections with Harmonic Analysis, Proceedings of the 1993 Alexandria Conference* (K. E. Petersen and I. A. Salama, eds.) Cambridge University Press (1995).

Rössler, O. E., An equation for continuous chaos. Phys. Lett. A **57** (1976) 397–398.

Rôssler, O. E., Continuous chaos – Four prototype equations. Ann. New York Acad. Sci. **31** (1979) 376–392.

Rovella, A., The dynamics of perturbations of the contarcting Lorenz attractor. Bol. Soc. Bras. Mat. **24**(2) (1993) 233–259.

Ruelle D., Statistical mechanics of a one-dimensional lattice gas. Comm. Math. Phys. **9** (1968) 267–278.

Ruelle, D., A measure associated with axiom A attractors. Am. J. Math. **98** (1976) 619–654.

Ruelle, D., An inequality for the entropy of differentiable maps. Bol. Soc. Bras. Math. **9** (1978) 83–87.

Ruelle, D. and Takens, F., On the nature of turbulence. Comm. Math. Phys. **20** (1971) 167–192.

Russell, D. A., Hanson, J. D., and Ott, E., Dimension of strange attractors. Phys. Rev. Lett. **45** (1980) 1175–1178.

Russo, L., Altimari, P., Mancusi, E., Maffettone, P. L., and Crescitelli, S., Complex dynamics and spatio-temporal patterns in a network of three distributed chemical reactors with periodical feed switching. Chaos, Solitons & Fractals **28** (2006) 682–706.

Ruxton, G. D. and Rohani, P., Population floors and the persistence of chaos in ecological models. Theoretical Population Biology **53**(3) (1998) 175–183.

Rychlik, M., Bounded variation and invariant measures. Studia Math. **LXXVI** (1983a) 69–80.

Rychlik, M., Invariant measures and the variation principle for Lozi mappings. Ph.D. dissertation. University of California, Berkeley (1983b).

Rychlik,. M., Lorenz attractors through a Sil'nikov-type bifurcation, Part 1. Ergod. Theor. Dyn. Syst. **10** (1989) 793–821.

Sanchez-Salas, F. J., Sinai Ruclle–Bowen measures for piecewise hyperbolic transformations. Divulgaciones Matematicas **9**(1) (2001) 35–54.

Sannami, A., The stability theorems for discrete dynamical systems on two-dimensional manifolds. Nagoya Math. J. **90** (1983) 1–55.
Sannami, A., A topological classification of the periodic orbits of the Hénon Family. Japan J. Appl. Math. **6** (1989) 291–300.
Sannami, A., On the structure of the parameter space of the Hénon map. In *Towards the Harnessing of Chaos*. Elsevier, Amsterdam (1994) 289–303.
Sataev, E. A., Invariant measures for hyperbolic maps with singularities. Russian Math. Surveys **47**(1) (1992) 192–251.
Sataev, E. A., Non-existence of stable trajectories in non-autonomous perturbations of systems of Lorenz type. Math. Sbornik **196**(4) (2005) 561–594.
Sauer, T., Yorke, J. A., and Casdagli, M., Embedology. J. Stat. Phys. **65**(3–4) (1991) 579–616.
Schuster, H., *Deterministic Chaos*. Physik-Verlag GmbH, Weinheim, FRG (1984).
Sebesta, V., Predictability of chaotic signals. J. Elec. Engineering **50**(9–10) (1999) 302–304.
Sedgewick, R., *Algorithms*. Addison-Wesley, Advanced Book Program, Reading, MA, (1983).
Shanmugam, S. and Leung, H., A robust chaotic spread spectrum inter-vehicle communication scheme for ITS. IEEE International Conference on Intelligent Transportation Systems, Shanghai, China (2003) 1540–1545.
Shastry, M. C., Nagaraj, N., and Vaidya, P. G., The β-exponential map – A generalization of the logistic map, and its applications in generating pseudorandom numbers. Preprint arXiv:cs/0607069v2 (2006).
Shilnikov, L. P., A case of the existence of a countable number of periodic motions. Sov. Math. Docklady **6** (1965) 163–166 (translated by S. Puckette).
Shilnikov, L. P., A contribution of the problem of the structure of an extended neighborhood of rough equilibrium state of saddle-focus type. Math. USSR Shornik **10** (1970) 91–102 (translated by F. A. Cezus).
Shilnikov, L. P., The bifurcation theory and quasi-hyperbolic attractors. Uspehi Mat. Nauk. **36** (1981) 240–241.
Shilnikov, L. P., Bifurcations and chaos in the Shimizu–Marioka system. (in Russian) In *Methods and Qualitative Theory of Differential Equations*, Gorky State University (1986) 180–193; English translation: Selecta Mathematica Sovietica **10** (1991) 105–117.
Shilnikov, L. P., Strange attractors and dynamical models. J. Circuits Syst. Comput. **3**(1) (1993a) 1-10.
Shilnikov, L. P., Chua's circuit: Rigorous results and future problems. IEEE Trans. Circ. Syst. Fund. Theor. Appl. **40**(10) (1993b) 784–786.
Shilnikov, L. P., Bifurcations and strange attractors, ICM III (2002) 1–3.
Shilnikov, L. P. and Turaev, D. V., Simple bifurcations leading to hyperbolic attractors. Comput. Math. Appl. **34**(2–4) (1997) 173–193.
Shilnikov, A. L., Shilnikov, L. P., and Turaev, D. V., Normal forms and Lorenz attractors. Int. J. Bifurcation and Chaos **3**(5) (1993) 1123–1139.
Shimada, I. and Nagashima, T., A numerical approach to ergodic problem of dissipative dynamical systems. Prog. Theor. Phys. **61** (1979) 1605–1616.

Shishikura, M., The Hausdorff dimension of the boundary of the Mandelbrot set and Julia sets. Annals Math. **147** (1998) 225–267.

Shub, M., *Topological Transitive Diffeomorphism on* $\mathbb{T}^4$. Lecture Notes in Math. **206** Springer-Verlag, New York (1971).

Shub, M. and Sullivan, D., Expanding endomorphisms of the circle revisted. Ergod. Theor. Dyn. Syst. **5** (1985) 285–289.

Shub, M. and Wilkinson, A., Pathological foliations and removable zero exponents. Inv. Math. **139** (2000) 495–508.

Siegman, A. E., *Lasers*. University Science Books, Mill Valley, CA (1986).

Silva, C. P., Analytical study of the double-hook attractor. *Proceedings 34th Midwest Symposium on Circuits and Syst.* **2** (1991) 764–771.

Silva, C. P., Shilnikov theorem – A tutorial. IEEE Trans. Circ. Syst. Fund. Theor. Appl. **40** (2003) 675–682.

Simitses, G. J. and Hodges, D. H., *Fundamentals of Structural Stability*. Elsevier, Amsterdam (2006).

Simon, R., A 3-dimensional Abraham–Smale example. Proceedings Amer. Math. Soc. **34** (1972) 629–630.

Sinai, Y., Markov partitions and C-diffeomorphisms. Func. Anal. and its Appl. **2**(1) (1968a) 64–89.

Sinai, Y., Construction of Markov partitions. Func. Anal. and its Appl. **2**(2) (1968b) 70–80.

Sinai, Y., Dynamical systems with elastic collisions. Russian Math. Surveys **25** (1970) 141–192.

Sinai, Y., Gibbs measures in ergodic theory. Uspehi. Mat. Nauk. **27**(4) (1972) 21–64; English translation: Russian. Math. Surveys **27**(4) (1972) 21–69.

Sinai, Y., Stochasticity of dynamical systems. In *Nonlinear Waves* (A. V. Gaponov-Grekhov ed.) Nauka, Moscow (1979) 192 (in Russian).

Singer, D., Stable orbits and bifurcation of maps of the interval. SIAM J. Applied Mathematics **35**(2) (1978) 260–267.

Slodkowski, Z., Holomorphic motions and polynomial hulls. Proceedings AMS **111** (1991) 347–355.

Smale, S., Stable manifolds for differential equations and diffeomorphisms. Ann. Scuola Norm. Sup. Pisa **17** (1963) 97–116.

Smale, S., Diffeomorphisms with many periodic points. In *Differential and Combinatorial Topology*. A Symp. In Honor of Marston Morse (S. S. Cairns, ed.) Princeton University Press (1965) 63–80.

Smale, S., Differentiable dynamical systems. Bull. Amer. Math. Soc. **73** (1967) 747–817.

Smale, S., The Ω-stability Theorem. Proceedings Sympos. Pure. Math., Vol. **XIV**, Berkeley, CA, Amer. Math. Soc., Providence, RI (1968) 289–298.

Smale, S., Mathematical problems for the next century. Math. Intelligencer **20**(2) (1998) 7–15.

Smale, S., Mathematical problems for the next century. Mathematics, Frontiers and Perspectives (2000) (V. Arnold, M. Atiyah, P. Lax, and B. Mazur, eds.) Amer. Math. Soc., Providence, RI (2000).

Smillie, J., Complex dynamics in several variables. In *Flavors of Geometry* **31** of Math. Sci. Res. Inst. Pub. Cambridge University Press (1997) 117–150 (with notes by Gregery T. Buzzard).

Sompolinsky, H., Cristanti, A., and Sommers, H. J., Chaos in random neural networks. Phys. Rev. Lett. **61** (1988) 259–262.

Spany, V. and Pivka, L., Boundary surfaces in sequential circuits. Int. J. on Ckt. Th. and Appl. **18**(4) (1990) 349–360.

Spany, V. and Pivka, L., Invariant manifolds of sequential circuits. Elektrotechnicky Casopis **42**(6) (1991) 281–293.

Spany, V. and Pivka, L., Two-segment bistability and basin structure in three-segment PWL circuits. IEE Proceedings on Circuits Devices & Systems **140**(1) (1993) 61–67.

Sparrow, C., *The Lorenz Equations: Bifurcations, Chaos, and Strange Attractors.* Springer-Verlag, New York (1982).

Spivak, M., *A comprehensive Introduction to Differential Geometry.* Publish or Perish, Inc. Houston, TX (1999).

Sprott, J. C., Automatic generation of strange attractors. Comput. & Graphics **17**(3) (1993a) 325–332.

Sprott, J. C., *Strange Attractors: Creating Patterns in Chaos.* M&T Books, New York (1993b).

Sprott, J. C., How common is chaos? Phys. Lett. A **173** (1993c) 21–24.

Sprott, J. C., Some simple chaotic flows. Phys. Rev. E **50**(2) (1994a) R647–650.

Sprott, J. C., Predicting the dimension of strange attractors. Phys. Lett. A **192** (1994b) 355–360.

Sprott, J. C., *Chaos and Time-series Analysis.* Oxford University Press (2003).

Sprott, J. C., High-dimensional dynamics in the delayed Hénon map. Electronic Journal of Theoretical Physics **312** (2006) 19–35.

Sprott, J. C., Chaotic dynamics on large networks. Chaos **18** (2008) 023135.

Steriade, M., McCormick, D. A., and Sejnowski, T. J., Thalamocortical oscillations in the sleeping and aroused brain. Science **262** (1993) 679–685.

Sterling, D., Dullin, H. R., and Meiss, J. D., Homoclinic bifurcations for the Hénon map. Physica D **134** (1999) 2153–2184.

Stewart, I., The Lorenz attractor exists. Nature **406** (2000) 948–949.

Stoffer, D. and Palmer, K. J., Rigorous verification of chaotic behaviour of maps using validated shadowing. Nonlinearity **12**(6) (1999) 1683–1698.

Strelkova, G. and Anishchenko, V., Structure and properties of quasihyperbolic attractors. In Proceedings of Int. Conf. of COC'97 (St. Petersburg, Russia, August 27-29, 1997) **2** (1997) 345–346.

Sullivan, D., The universalities of Milnor, Feigenbaum and Bers. In *Topological Methods in Modern Mathematics.* SUNY at Stony Brook Proceedings Symp. held in honor of John Milnor's 60th birthday (1991) 14–21.

Sullivan, D., Bounds, quadratic differentials, and renormalization conjecture. AMS Centennial Publ. **2** (1992) 417–466.

Sullivan, M. C., The topology and dynamics of flows. In *Open Problems in Topology II* (Elliott Pearl, ed.) Elsevier, Amsterdam (2007).

Sun, J., Amellal, F., Glass, L., and Billette, J., Alternans and period-doubling bifurcations in atrioventricular nodal conduction. Journal Theoretical Biology **173** (1995) 79–91.

Svitanovic, P., *Universality in Chaos*, Adam Hilger, Bristol (1984).

Szasz, D., Hard ball systems and the Lorentz gas. Encycl. Math. Sciences **101** (2000).

Szpilrajn, E., La dimension et la mesure. Fundamenta Mathematica **28** (1937) 81–89.

Szymczak, A., A combinatorial procedure for finding isolating neighbourhoods and index pairs. Proceedings Roy. Soc. Edinburgh Sect. A **127** (1997) 1075–1088.

Tahzibi, A., Stably ergodic systems which are not partially hyperbolic. Isr. Journal of Math. **142** (2004) 315–344.

Taixiang, S. and Hongjian, X. On the basin of attraction of the two cycle of the difference equation. J. Difference Equations and Applications **13**(10) (2007) 945–952.

Takens, F., Multiplications in solenoids as hyperbolic attractors. Topology and its Applications **152** (2005) 219–225.

Tang, W. K. S., Zhong, G. Q., Chen, G., and Man, K. F., Generation of n-scroll attractors via sine function. IEEE Trans. Circ. Syst. Fund. Theor. Appl. **48**(11) (2001) 1369–1372.

Tang, K. S., Man, K. F., Zhong, G. Q., and Chen, G., Some new circuit design for chaos generation. In *Chaos in Circuits and Systems*. World Scientific, Singapore (2002) 171–189.

Tang, K. S., Man, K. F., Zhong, G. Q., and Chen, G. R., Modified Chua's circuit with $x|x|$, Control Theory & Applications **20**(2) (2003) 223–227.

Tapan, M. and Gerhard, S., On the existence of chaotic policy functions in dynamic optimization. Japanese Economic Review **50**(4) (1999) 470–484.

Tedeschini-Lalli, L. and Yorke, J. A., How often do simple dynamical processes have infinitely many coexisting sinks? Comm. Math. Phys. **106** (1985) 635–657.

Theiler, J., Eubank, S., Longtin, A., Galdrikian, B., and Farmer, J. D., Testing for nonlinearity in time series – The method of surrogate data. Physica D **58** (1992) 77–94.

Thomas, G. B. Jr. and Finney, R. L., Maxima, minima, and saddle points §12.8. In *Calculus and Analytic Geometry* (8th edn.) Addison-Wesley, Reading, MA (1992) 881–891.

Tibor, C., Barnabás, G., and Balázs, B., A verified optimization technique to locate chaotic regions of Hénon systems. J. Global Optimization **35**(1) (2006) 145–160.

Tirozzi, B. and Tsokysk, M., Chaos in highly diluted neural networks. Europhys. Lett. **14** (1991) 727–732.

Tovbis, A., Tsuchiya, M., and Jaffé, C., Exponential asymptotic expansions and approximations of the unstable and stable manifolds of singularly perturbed systems with the Hénon map as an example. Chaos **8** (1998) 665–681.

Tresser, C., Une theorème de Shilnikov en $C^{1,1}$. C. R. Acad. Paris **296**(1) (1983) 545–548.

Tse, C. K. and Chan, W. C. Y., Experimental verification of bifurcations in current-programmed dc/dc boost converters. in *Proceedings European Conf. Circuit Theory Design*, Budapest, Hungary (1997) 1274–1279.

Tsuji, R. and Ido. S., Computation of Poincaré map of chaotic torus magnetic field line using parallel computation of data table and its interpolation. In *Proceedings of Inter. Conference on Parallel Computing in Electrical Engineering* (2002) 386–390.

Tucker, W., Rigorous models for the Lorenz equations. UUDM Report 26, ISSN (1996a) 1101–3591.

Tucker, W., Transitivity of Lorenz-like maps and the tired baker's map. Preprint (1996b).

Tucker, W., The Lorenz attractor exists. C. R. Acad. Sci. Ser. I. Math. **328**(12) (1999) 1197–1202.

Tucker, W., A rigorous ODE solver and Smale's 14th problem. Found. Comput. Math. **2** (2002a) 53–117.

Tucker, W., Computing accurate Poincaré maps. Physica D **171**(3) (2002b) 127–137.

Tufillaro, N., Abbott, T., and Reilly, J., *An Experimental Approach to Nonlinear Dynamics and Chaos*. Addison-Wesley, Redwood City, CA (1992).

Turaev, D. V. and Shilnikov, L. P., On a blue sky catastrophe. Soviet. Math. Dokl. **342**(5) (1995) 596–599.

Turaev, D. V. and Shilnikov, L. P., An example of a wild strange attractor. Math. Sbornik **189**(2) (1996) 291–314.

Turaev, D. V. and Shilnikov, L. P., Pseudohyperbolicity and the problem on periodic perturbations of Lorenz-type attractors. Doklady Mathematics **77**(1) (2008) 17–21.

Umberger, D. K. and Farmer, J. D., Fat fractals on the energy surface. Phys. Rev. Lett. **55**(7) (1985) 661–664.

Ustinov, Y., Algebraic invariants of the topological conjugacy classes of solenoids. Mat. Zametki **421** (1987) 132–144; English translation: Math. Notes **42** (1987) 583–590.

Van Dantzig, D., Über topologisch homogene Kontinua. Fund. Math. **15** (1930) 102–125.

Vano, J. A., Wildenberg, J. C., Anderson, M. B., Noel, J. K., and Sprott, J. C., Chaos in low-dimensional Lotka–Volterra models of competition. Nonlinearity **19**(10) (2006) 2391–2404.

Van Strien, S., In *Dynamical Systems and Turbulence* (D. A. Rand and L. S. Young, eds.) LNM 898 Springer-Verlag, New York (1981).

Van Vleck, E. S., Numerical shadowing near hyperbolic trajectories. SIAM J. Sci. Comp. **16**(5) (1995) 1177–1189.

Verjovsky, A., Codimension one Anosov flows. Bol. Soc. Mat. Mexicana **19**(2) (1974) 49–77.

Viader, P., Paradıs, J., and Bibiloni, L., A new light on Minkowski $s?(x)$ function. J. Number Theory **73**(2) (1998) 212–227.

Viana, M., Strange attractors in higher dimensions. Bull. Braz. Math. Soc. **24** (1993) 13–62.
Viana, M., Dynamics: A probabilistic and geometric perspective. In *Proceedings of the International Congress of Mathematicians*, Vol. **I**. Berlin (1998) 557–578.
Viana, M., *Dynamics Beyond Uniform Hyperbolicity*. Lecture at Collége de France (2002).
Viana, R. L., Barbosa, J. R. R., and Grebogi, C., Unstable dimension variability and codimension-one bifurcations of two-dimensional maps. Phys. Lett. A **321**(4) (2004) 244–251.
Vietoris, L., Über den hoheren Zusammenhang kompakter Raume und eine Klasse von zusammenhangstreuen Abbildungen. Annals Math. **97** (1927) 454–472.
Vijayaraghavan, V. and Leung, H., A robust chaos radar for collision detection and vehicular ranging in intelligent transportation systems. ITSC (2004) 548–552.
Vikas, U. and Kumar, R., Chaotic population dynamics and biology of the top-predator. Chaos, Solitons & Fractals **21**(5) (2004) 1195–1204.
von Neumann, J., Proof of the Quasi-ergodic hypothesis. Proceedings Natl. Acad. Sci. USA **18** (1932a) 70–82.
von Neumann, J., Physical applications of the ergodic hypothesis. Proceedings Natl. Acad. Sci. USA **18** (1932b) 263–266.
Voronov, S. S., Kolpalrova, I. V., and Kuznetsov, V. A., Measurement methods using the properties of nonlinear dynamic systems. Izmeritel'naya Tekhnika **39**(12) (1996) 16–18.
Voronov, S. S., Kolpakova, L. V., and Kuznetsov, V. A., Chaotic oscillator method – Approaches to diagnosing the parameters of nonlinear chaotic systems. Izmeritel'naya Tekhnika **43**(4) (2000) 19–21.
Wagner, A., *Robustness and Evolvability in Living Systems*. Princeton Studies in Complexity. Princeton University Press (2005).
Walczak, P. G., Langevin, R., Hurder, S., and Tsuboi, T., Foliations. In *Proceedings of the International Conference in Lodz, Poland* (2005) 13–24.
Walters, P., *An Introduction to Ergodic Theory*. Springer, New York (1982).
Wang, X. M. and Fang, Z. J., The properties of borderlines in discontinuous conservative systems. The European Physical Journal D – Atomic, Molecular, Optical and Plasma Physics **37**(2) (2006) 247–253.
Wang, Z. X. and Guo, D. R, *Special Functions*. World Scientific, Singapore (1989).
Wang, Q. and Young, L. S., Strange attractors with one direction of instability. Comm. Math. Phys. **218**(1) (2001) 1–97.
Watkins, W. T., Homeomorphic classification of certain inverse limit spaces with open bonding maps. Pacific J. Math. **103** (1982) 589–601.
Weisstein, E., Homoclinic tangle. In *Smale Horseshoe. Homoclinic Point*. Eric Weisstein's World of Mathematics (2002).
Wen, L., Generic diffeomorphisms away from homoclinic tangencies and heterodimensional cycles. Bull. Braz. Math. Soc. New Series **35** (2004) 419–452.
Wen, L., The selecting lemma of Liao. Discrete Contin. Dyn. Syst. **20** (2008) 159–175.

Widrow, B. and Lehr, M. A., 30 years of adaptive neural networks – Perceptron. madaline, and backpropagation. Proceedings IEEE **78**(9) (1990) 1415–1442.

Williams, R. F., A note on unstable homeomorphisms. Proceedings Amer. Math. Soc. **6** (1955) 308–309.

Williams, R. F., One-dimensional nonwandering sets. Topology **6** (1967) 473–487.

Williams, R. F., Classification of one-dimensional attractors. Proceedings Symp. in Pure Math. (1970) 361–393.

Williams, R. F., Expanding attractors. Publ. Math. IHES **43** (1974) 169–203.

Williams, R. F., The Structure of Lorenz attractors. Publ. Math. IHES **50** (1979) 321–347.

Wolf, C., Generalized physical and SRB measures for hyperbolic diffeomorphisms. J. Stat. Phys. **122**(6) (2006) 1111–1138.

Wolf, D. M., Varghese, M., and Sanders, S. R., Bifurcation of power electronic circuits. J. Franklin Inst. **331B**(6) (1994) 957–999.

Wu, C. W. and Rul'kov, N. F., Studying chaos via 1-D maps – A tutorial. IEEE Trans. Circ. Syst. Fund. Theor. Appl. **40**(10) (1993) 707–721.

Xu, M., Chen, G., and Tian, Y. T., Identifying chaotic systems using Wiener and Hammerstein cascade models. Math. & Computer Modelling **33**(4) (2001) 483–493.

Yakubu, A. A., Multiple attractors in Juvenile-adult single species models. J. Difference Equations and Applications **9**(12) (2003) 1083–1098.

Yalcin, M. E., Ozoguz, S., Suykens, J. A. K., and Vandewalle, J., n-scroll chaos generators – A simple circuit model. Electronics Letters **37**(3) (2001) 147–148.

Yalcin, M. E., Suykens, J. A. K., Vandewalle, J., and Ozoguz, S., Families of scroll grid attractors. Int. J. Bifurcation and Chaos **12**(1) (2002) 23–41.

Yamanaka, T. A., Characterization of dense vector fields in $G^1(M)$ on 3-manifolds. Hokkaido Math. J. **31** (2002) 97–105.

Yang, X. S., A proof for a theorem on intertwining property of attraction basin boundaries in planar dynamical systems. Chaos, Solitons & Fractals **15**(4) (2003a) 655–657.

Yang, X. S., Structure of basin boundaries of attractors of ODE's on $\mathbb{S}^2$. Chaos, Solitons & Fractals **16**(1) (2003b) 147–150.

Yang, J., Newhouse phenomenon and homoclinic classes. Preprint arXiv:0712.0513 (2007).

Yang, X. S. and Li, Q., On entropy of Chua's circuit. Int. J. Bifurcation and Chaos **15**(5) (2005) 1823–1828.

Yang, X. S. and Tang, Y., Horseshoes in piecewise continuous maps. Chaos, Solitons & Fractals **19** (2004) 841–845.

Yau, H. T. and Yan J. J., Design of sliding mode controller for Lorenz chaotic system with nonlinear input. Chaos, Solitons & Fractals **19**(4) (2004) 891–898.

Yesufu, T. K., Geocommercial circuit analysis – A paradigm for order. In *Proceedings of 1995 International Symposium on Nonlinear Theory and its Applications* **1** (1995) 595–598.

Ying-Cheng, L., Grebogi, C., Yorke, J. A., and Kan, I., How often are chaotic saddles nonhyperbolic? Nonlinearity **6** (1993) 779–797.

Yoccoz, J. C., On the local connectivity of the Mandelbrot set. Preprint (1990).

Yoccoz, J. C., Polynômes quadratiques et attracteur de Hénon. Séminaire Bourbaki **33** Exposé N° 734 (1990)–(1991).

Yoccoz, J. C., Introduction to hyperbolic dynamics – Real and complex dynamical systems. In *Proceedings of the NATO Advanced Study Institute* held in Hillerod, June 20-July 2, 1993 (B. Branner and P. Hjorth, eds.) NATO Advanced Science Institutes Series C: Mathematical and Physical Sciences **464**, Kluwer Academic Publishers, Dordrecht (1995) 265–291.

Yorke, J. A., Grebogi, C., Ott, E., and Tedeschini-Lalli, L., Scaling behavior of windows in dissipative dynamical systems. Phys. Rev. Lett. **54**(11) (1984) 1095–1098.

Yoshitake, Y. and Sueoka, A., Forced self-excited vibration with dry friction. In *Applied Nonlinear Dynamics and Chaos of Mechanical Systems with Discontinuities*. World Scientific, Singapore (2000) 237–259.

You, Z. P., Kostelich, E., and Yorke, J. A., Calculating stable and unstable manifolds. Int. J. Bifurcation and Chaos **1** (1991) 605–623.

Young, L. S., A Bowen–Ruelle measure for certain piecewise hyperbolic maps. Trans. Amer. Math. Soc. **287** (1985) 41–48.

Young, L. S., Decay of correlations for certain quadratic maps. Comm. Math. Phys. **146** (1992) 123–138.

Young, L. S., Developments in chaotic dynamics. Notices Amer. Math. Soc. **45**(10) (1998a) 1318–1328.

Young, L. S., Statistical properties of dynamical systems with some hyperbolicity. Annals Math. **147**(3) (1998b) 585–650.

Young, L. S., Recurrence times and rates of mixing. Israel J. Math. **110** (1999) 153–188.

Yu, W. W., Cao, J., Wong, K. W., and Lü, J., New communication schemes based on adaptive synchronization. Chaos **17**(3) (2007) 33–114.

Yuan, G. H., *Shipboard Crane Control, Simulated Data Generation and Border-collision Bifurcations*. Ph.D. dissertation, Univ of Maryland, College Park (1997).

Yuan, G. H., Banerjee, S., Ott, E., and Yorke, J. A., Border collision bifurcations in the buck converter. IEEE Trans. Circ. Syst. Fund. Theor. Appl. **45**(7) (1998) 707–716.

Yulmetyev, R., Emelyanova, N., Demin, S., Gafarov, F., Hänggi, P. and Yulmetyeva, D., Fluctuations and noise in stochastic spread of respiratory infection epidemics in social networks – Unsolved problems of noise and fluctuations UPoN 3rd International Conference (2002) 408–421.

Zeraoulia. E., A new chaotic attractor from 2-D discrete mapping via border-collision period doubling scenario. Discrete Dynamics in Nature and Society (2005) 235–238.

Zeraoulia, E., Analysis of a new chaotic system with three quadratic nonlinearities. Dynamics of Continuous, Discrete and Impulsive Systems **14**(b) (2007) 603–613.

Zeraoulia, E. and Hamri, N. E., A generalized model of some Lorenz-type and quasi-attractors type strange attractors in three-dimensional dynamical systems. International Journal of Pure & Applied Mathematical Sciences **2**(1) (2005) 67–76.

Zeraoulia, E. and Sprott, J. C., A two-dimensional discrete mapping with C^∞-multifold chaotic attractors. Electronic Journal of Theoretical Physics **5**(17) (2008a) 111–124.

Zeraoulia, E. and Sprott, J. C., A minimal 2-D quadratic map with quasi-periodic route to chaos. Int. J. Bifurcation and Chaos **18**(5) (2008b) 1567–1577.

Zeraoulia, E. and Sprott, J. C., On the robustness of chaos in dynamical systems: Theories and applications. Front. Phys. China **3** (2008c) 195–204.

Zeraoulia, E. and Sprott, J. C., The effect of modulating a parameter in the logistic map. Chaos **18** (2008d) 023119-1–023119-7.

Zeraoulia E. and Sprott, J. C., *2-D Quadratic Maps and 3-D ODE Systems: A Rigorous Approach*. World Scientific Series on Nonlinear Science Series A **73** (2010).

Zeraoulia, E. and Sprott, J. C., A unified chaotic mapping that contains the Hénon and the Lozi systems. Electronic Journal of Theoretical Physics (2011).

Zgliczynski, P., Fixed point index for iterations, topological horseshoe and chaos. Topological Methods in Nonlinear Analysis **8**(1) (1996) 169–177.

Zgliczynski, P., A computer assisted proof of chaos in the Rôssler equations and in the Hénon map. Nonlinearity **10**(1) (1997a) 243–252.

Zgliczynski, P., Computer assisted proof of the horseshoe dynamics in the Hénon map. Random & Computational Dynamics **5** (1997b) 1–17.

Zhirov, A. Y., Hyperbolic attractors of diffeomorphisms of oriented surfaces I – Coding, classification, and coverings. Mat. Sbornik **1856** (1994a) 3–50; English translation: Russian Acad. Sci. Sbornik Math. **82** (1995a) 135–174.

Zhirov, A. Y., Hyperbolic attractors of diffeomorphisms of oriented surfaces III – A classification algorithm. Mat. Sbornik **1862** (1995) 69–82; English translation: Sbornik Math. **186** (1995b) 221–244.

Zhirov, A. Y., Hyperbolic attractors of diffeomorphisms of oriented surfaces II – Enumeration and application to pseudo-Anosov diffeomorphisms. Mat. Sbornik **1859** (1994b) 29–80; English translation: Russian Acad. Sci. Sbornik Math. **83** (1996) 23–65.

Zhirov, A. Y., Complete combinatorial invariants of conjugacy of hyperbolic attractors of diffeomorphisms of surfaces. J. Dynam. Control Systems **6** (2000) 397–430.

Zhong, G. Q. and Ayrom, F., Periodicity and chaos in Chua's circuits. IEEE Trans. Circ. Syst. **CAS-32** (1985) 501–503.

Zuo, C., and Wang, X., Attractors and quasi-attractors of a flow. Journal of Applied Mathematics and Computing **23**(1–2) (2007) 411–417.

Zuppa, C., Regularisation C^∞ des champs vectoriels qui préservent l'élément de volume. Bulletin of the Brazilian Mathematical Society **10**(2) (1979) 51–56.

Index

1-D singular map, 349
2-D Lorenz attractor, 260
2-D Lorenz-type attractor, 267
2-D piecewise-smooth map, 359

admissible multivalued map, 217
Anishchenko–Astakhov oscillator, 286
Anosov automorphism, 102
Anosov diffeomorphism, 97
approximate trajectory, 85
Arnold cat map, 111
attractor, 28
attractors of a Smale system, 106
autocorrelation function (ACF), 51

B-exponential map, 335
baker's map, 224
Barreto–Hunt–Grebogi–Yorke conjecture, 317
Bernoulli diffeomorphism, 109
Bernoulli map, 108
Blaschke product, 107
border-collision bifurcation, 342
border-collision pair bifurcation, 346
border-collision pair bifurcation (2-D), 368
border-crossing bifurcation, 347
boundary crisis, 145
boundary point, 306
bubbling transition, 140

C^1-robust transitive sets with singularities, 259
Cantor set, 137
Cantor set of the solenoid type, 310
cardiac conduction model, 369
central limit theorem, 53
chain-recurrent, 248
chain-transitive, 248
chaos, 27
chaos in the double scroll, 294
chaotic in the sense of Smale, 14
chaotic saddle, 139
chaoticity, 223
Chua's circuit, 288
coexistence phenomenon, 28
Collet–Eckmann map, 312
combinatorially equivalent, 308
complex behavior, 27
computer-assisted computation, 214
computer-assisted proof, 4
Conley index theory, 215
connecting orbit shadowing theorem, 88
contracting Lorenz attractor, 239
correlation, 53
correlation coefficient, 54
correlation decay, 53
critical element, 220
critical point, 150
crossing bifurcation, 346

dangerous tangency, 283

density and robustness of chaos, 34
density of hyperbolicity, 138, 309
domain of attraction, 32
dominating splitting, 110

elementary transversality, 81
enumeration of attractors, 35
epsilon-stable, 248
Epstein class, 308
ergodic theory, 50
evaluation of the ACF, 52
existence of the Lorenz attractor, 211
expanding Lorenz attractor, 240
expanding map, 105
exponential instability, 263
Fibonacci, 111 (map)
floating point number, 155
forced damped pendulum, 153
forward invariant compact set, 310
fraction of nonhyperbolic parameter values, 149
Frobenius–Perron equation, 332

generalized hyperbolic attractor, 126
generalized Poincaré map, 3
generating hyperbolic attractors, 134
generic property, 62, 82
geodesic flow on compact smooth manifold, 115
geometric distribution, 223
geometric model, 298
geometric models of the Lorenz equation, 220
geometrical Lorenz attractor, 229
Gibbs u-measure, 131

Hénon map, 275
Hénon-like diffeomorphism, 281
Hölder continuous, 132
Hölder exponent, 132
Hausdorff dimension, 42
heteroclinic orbit, 92
heteroclinic Shilnikov theorem, 92
homoclinic bifurcation, 91
homoclinic case, 223
homoclinic orbit, 92

homoclinic orbit shadowing theorem, 89
homoclinic point, 15
homoclinic tangency, 15
horseshoe-type mapping, 23
hyperbolic, 96
hyperbolic chaos, 116
hyperbolic saddle focus, 92
hyperbolic set, 101
hyperbolicity of the logistic map, 121
hyperbolicity test, 139

induced holomorphic polynomial-like map, 308
intersect transversely, 81
interval arithmetic, 7
interval Newton's method, 5
invariant decomposition, 244
invariant foliation, 82
invariant manifold, 15
isolating block, 216
isolating neighborhood, 23, 215

Kupka–Smale diffeomorphism, 100

LCE spectrum, 264
Lebesgue measure, 39
limit image, 225
Liouville measure, 116
Lipschitz constant, 219
local product structure, 85
locally maximal hyperbolic set, 102
Lorenz manuscript, 230
Lorenz system, 210
Lorenz-like attractor, 209
Lorenz-like families with criticality, 228
Lorenz-type attractor, 208
Lozi map attractor, 253
Lyapunov exponent, 47

measure theoretic persistence, 228
method of fixed point index, 19
Milnor's problem, 310
minimum angle, 147
modeling hyperbolic attractors, 167

modeling Plykin's attractor, 202
modeling the Arnold cat map, 187
modeling the Bernoulli map, 194
modeling the Smale–Williams attractor, 168
Morse–Smale diffeomorphism, 100
multivalued map, 215

N-piece invariant set, 395
new kind of attractor, 259
Newhouse interval, 140
nilpotent point, 320
nonbifurcation, 343
nondegenerate, 305
nonhyperbolic, 227
noninvertible piecewise-linear map, 378
nonorientable, 234
nonrobust chaos, 362
nonwandering set, 29
normal form, 342
nowhere dense set, 310

onset of chaos, 382
orbitally stable, 248

parameter-shifted shadowing property, 87
Peixoto's theorem, 63
period doubling bifurcation, 287
period-n-tupling scenario, 332
periodic perturbation, 252
periodic point, 304
periodic points of the TS-map, 22
periodic window, 30
persistence, 35
perturbation, 266
plateau, 280
Plykin attractor, 120
Poincaré map, 1
Poincaré section, 2
principal nest of parapuzzle pieces, 313
proof of Anosov's theorem, 64
proof of robust chaos, 371
pseudo-hyperbolic, 243

pseudo-periodic orbit, 85

quadratic tangency, 282
quasi-attractor, 273
quasi-conformal deformation, 309
quasi-hyperbolic parameter, 158
quasi-hyperbolicity, 157
quotient-semi-flow, 247

regular map, 305
repeller, 106
representable multivalued map, 219
rescaling technique, 240
residual set, 62
restrictive, 310
riddled basin, 33
Riemannian manifold, 84
rigorous analysis, 211
rigorous proof, 4
robust chaos, 29
robust hyperbolicity, 115
robust statistical description, 313
robust strange attractor, 262

S-unimodal, 305
saddle basis set, 106
saddle separatrix loop, 235
semi-conjugacy, 23
semi-orientable, 232
sensitive dependence on initial conditions, 28
separatrix loop, 209
set of cardinals, 229
shadowing lemma, 84, 89
shadowing theorem, 87
shift automorphism, 16
Shilnikov criterion, 90
short periodic orbit, 236
Sinai–Ruelle–Bowen measure, 40
singular horseshoe, 227
singular-hyperbolic attractor, 259
Smale horseshoe, 10
Smale's conjecture, 258
Smale's horseshoe-type, 273
Smale–Moser theorem, 91
Smale–Williams attractor, 119

solenoid attractor, 117
spectral decomposition, 77
stability conjecture, 64
stability triangle, 319
statistical properties of unimodal
    map, 314
stochastic birth–death process, 150
stochastic Boolean network, 324
strange attractor, 28
Strelkova–Anishchenko map, 286
strong law of large numbers (SLLN),
    53
strong PSSP for the Lorenz flow, 88
strong stable subspace, 245
structural stability, 61
structurally nontransversal
    homoclinic trajectory, 249
structurally unstable, 210
structure of Anosov diffeomorphism,
    103
structure of the Lorenz attractor, 230
subcritical border-collision period
    doubling, 368
subdivision algorithm, 157
supercritical border-collision period
    doubling, 369
suspension, 260
symbolic dynamics, 17
symbolic prehistory, 229

tangency point, 145
ten-times-improvement test, 149
testing hyperbolicity, 176
three-piece invariant set, 390

topological classification, 229
topological entropy, 43
topological invariant, 213
topological transitivity, 226
transition function, 83
transition Lorenz
    attractor-quasi-attractor, 264
transitive cycle, 310
transitivity of the Lorenz attractor,
    224
transversal homoclinic loop, 249
transversal homoclinic point, 16, 91
transversality, 80
trapping region, 254, 262
turning orbit, 254
two-piece invariant set, 384

unicity of orbit, 370
uniform hyperbolicity, 115, 276
uniform hyperbolicity test, 154
unimodal map, 303
unit tangent bundle, 116
unstable horseshoe, 226

vertex, 156

weak attractor, 239
weak attractor in the Milnor sense, 34
weak inverse image, 215
weak PSSP for the Lorenz flow, 88
wild strange attractor, 241
windows extended, 319
windows limited, 319